SMART TEXTILES AND SMART POLYMERS

Kruger Brentt
Publishers

SMART TEXTILES AND SMART POLYMERS

Ning Lu

Kruger Brentt Publishers

2024

Kruger Brentt Publishers UK. LTD.
Company Number 9728962

Regd. Office: 68 St Margarets Road, Edgware, Middlesex HA8 9UU

ISBN: 978-1-78715-272-4

Disclaimer:

Every possible effort has been made to ensure that the information contained in this book is accurate at the time of going to press, and the publisher and author cannot accept responsibility for any errors or omissions, however caused. No responsibility for loss or damage occasioned to any person acting, or refraining from action, as a result of the material in this publication can be accepted by the editor, the publisher or the author. The Publisher is not associated with any product or vendor mentioned in the book. The contents of this work are intended to further general scientific research, understanding and discussion only. Readers should consult with a specialist where appropriate.

Every effort has been made to trace the owners of copyright material used in this book, if any. The author and the publisher will be grateful for any omission brought to their notice for acknowledgement in the future editions of the book.

For information on all our publications visit our website at http://krugerbrentt.com/

PREFACE

In the ever-evolving landscape of materials science and technology, the intersection of smart textiles and polymer materials has ushered in a new era of innovation. The amalgamation of advanced materials with wearable technologies has given rise to multifunctional fabrics capable of responding to diverse environmental stimuli, thereby revolutionizing industries ranging from fashion to energy harvesting. This compilation delves into the forefront of research and development within this dynamic field, presenting a comprehensive overview of recent advancements in smart textiles and polymer materials. The collection of papers explores a wide spectrum of topics, from the ingenious integration of triboelectric nanogenerators into human clothing for energy conversion to the intricacies of modifying electrospun polyacrylonitrile nanofibrous membranes to achieve both waterproof and breathable properties without resorting to fluorine-based materials.

The realms of fire-retardant scenarios and high-temperature air filtration applications with binary polyamide-imide fibrous superelastic aerogels and they conduct numerical analyses to unravel the complexities of fiber/air-coupling fields in annular jet configurations and showcase the exceptional mechanical properties of carboxylated carbon nanotube/polyimide films with low thermal expansion coefficients. From the synthesis of electrospun polyvinylidene fluoride-based nanofibers and their piezoelectric properties to the development of supercapacitors featuring a fiber-shaped structure composed of graphene, the chapters in this compilation span a vast array of groundbreaking research. The meticulous exploration of topics such as proton exchange membrane fabrication, vibrational emissions analysis in polymer composites, and the photocatalytic degradation characteristics of catalysts based on carbon nanofibers further enriches our understanding of the cutting-edge developments in this field.

As we navigate through the intricacies of optical algorithms for multilayer transparent mixed films, SEM-based analyses of helical-fiber nonwovens, and the review of calcium alginate fiber, it becomes evident that this compilation serves as a comprehensive resource for researchers, scientists, and enthusiasts alike. The exploration continues with the development of ultra-sensitive and flexible strain sensors, the modeling of dynamic equivalent resistance in knitted strain sensors, and the utilization of the finite element method for designing composite motor case connectors in solid rocket propulsion systems. The culmination of these research endeavors in the investigation into the tensile characteristics of carbon fiber composite laminates and the prediction of size and electrical performance in knitted strain sensors exemplifies the diverse and impactful nature of research within the realms of smart textiles and polymer materials. This book encompasses a diverse range of professionals, researchers, students, and enthusiasts in fields related to materials science, textile engineering, nanotechnology, and polymer chemistry.

We are grateful to all those persons as well as various books, manuals, periodicals, newsletters, magazines, journals etc. that helped in the preparation of this book. In spite of the best efforts, it is possible that some errors may have occurred into the compilation and editing of the book. Further queries, constructive suggestions and criticisms for the further improvement of the book are always welcome and shall be thankfully acknowledged.

Ning Lu

CONTENTS

CHAPTER-1

RECENT DEVELOPMENT OF WEARABLE TRIBOELECTRIC NANOGENERATORS: AT HUMAN CLOTHING FOR ENERGY CONVERSION

1. INTRODUCTION: AN OVERVIEW

In recent years, the quest for sustainable and innovative energy sources has propelled the exploration of novel technologies, and among them, wearable energy harvesting devices have emerged as promising contenders. One groundbreaking avenue in this domain is the development of wearable triboelectric nanogenerators (TENGs) seamlessly integrated into human clothing for efficient energy conversion. The convergence of nanotechnology, materials science, and energy harvesting has led to the creation of wearable TENGs that capitalize on the triboelectric effect, converting mechanical energy from human motion into electrical energy. This transformative technology not only harnesses the ubiquitous and abundant movements associated with daily activities but also offers a unique solution for powering portable electronic devices, thus paving the way for a sustainable and self-sufficient energy future.

Energy issues are attracting more and more attention due to the continuous growth of energy needs both in industry and in the daily life of people. Beyond the traditional technique of producing energy through fossil fuels, new renewable-energy sources such as wind [1], solar [2], acoustic [3], raindrop energy [4], and ocean energy [5] are being explored as alternatives. Different energy-harvesting and conversion strategies are suited to specific applications. For example, humans now rely heavily on portable electronic devices such as cell phones, tablets, and Bluetooth headsets, which need to be charged frequently. The necessity of frequent charging and the climate restrictions of these flexible energy-consuming devices greatly hinder their practicality, sustainability, and broad-range applications, even for wearable electronics [**6**]. As a result, continuous power supply has become a hot topic in energy-conversion research. It is estimated that if the motion or working time of a human body is 6 h per day, the peak energy generated by human motion could be up to 3.4 W h. This is enough to fully charge a battery with 3.4 V operation voltage and 1000 mA h capacity, close to the capacity of a smartphone battery [**7**]. Thus, collecting and converting the biomechanical energy from the human body into energy is a potential way to improve the convenience of sourcing power and offers a promising way to break the bottleneck of inconvenience created by the requirement of frequent charging.

The question is how to convert the mechanical energy of human motion into electricity as efficiently as possible. The wearable triboelectric generator is a new type of power generation device that converts kinetic energy into electric power [**8,9,10,11**]. Several attempts have been carried out to convert human kinetic energy into electrical energy, and the output of some generators has shown the potential for application in electronic devices [**12,13,14**]. Given the power output and portability of the devices, integrating such devices into clothes, shoes, hats, bags, or the fabrication of wearable generators (as is our focus here) is an attractive research direction for future device design. For wearable generators, not only is power output important, but wearability is also vital. The characteristics of comfort, breathability, skin sensitivity, washability, and moisture permeability of the fabric are the basic factors used to evaluate the standard of wearability for a given material [**15,16,17**]. Therefore, even though many two-dimensional generators that can be bent but cannot practically be worn on the human body have been designed, it's hard to define them as wearable devices because they lack fabric features. In addition, some studies have proposed the concept of a generator that can be utilized in clothes, but actual tests of such a generator on the human body as a garment are absent. As for devices that are parts of garments or integrated into garments, the output performance of their use in real-life scenarios is usually much lower than the output generated in a laboratory [**18**]. Many factors, including the motion strength, motion frequency, materials, and device structure, all influence the real output of the generators. Focusing on the fabric features, this paper summarizes the works of wearable triboelectric nanogenerators that have been truly tested in human clothing, based on the method of fabric manufacture and fabric structures of the devices.

To meet the needs of human activities, smart wearable devices used for human clothing must be light [19], soft [20], washable [21,22,23], breathable [24], foldable [25], tailorable [26], and stretchable [27] so that they can be directly manufactured or integrated into cloth. Many attempts at manufacturing stretchable fiber-based/

fabric-based generators that are able to be attached or jointed to cloth as wearable prototypes have been carried out. At present, due to the defects in the material and/or haptic experience [28], reports on such devices truly being used in human clothing are still limited, and some of the reported textile-based/-shaped devices were far from ideal in collecting human-motion energy [29].

In this review, we summarize the braided structure and performance of triboelectric nanogenerators (TENGs) over the last decade that were integrated into human clothing to collect human-motion energy in real-world circumstances and deeply analyze them from the perspective of the textile structure. Different fabric structure makes the fabrics suitable for different clothing applications. Woven fabric that is tight, windproof, durable, good for draping, and wrinkle resistant is usually used in shirts, jackets, suits, trousers, etc. [**30,31**]. Knitted fabric that is relatively loose, soft, and elastic; has good breathability [**32**]; is easily dispersed; and has an edge that is easy to roll is usually used to make underwear, socks, T-shirts, sweaters, leggings, and so on [**33,34**]. Textile-based generators or woven/knitted structural generators are ideal for wearable devices intended for practical applications in garments. In addition, different properties of yarns, such as the pliability, thickness, length, durability, etc., are suitable for different fabric structures. In light of this and the findings of our research, designing corresponding textile energy devices by using available materials or designing new materials can hopefully be accelerated.

2. WEARABLE GENERATORS, INCORPORATING INTELLIGENT TECHNOLOGY, SEAMLESSLY INTEGRATED INTO SMART GARMENTS

Thus far, integrating the devices into garments is the most popular method for fabricating wearable generators for energy capture and conversion. Since the electric materials are usually rigid and are difficult to fabricate into cloth directly, the most common strategy is modifying them into flexible fibers or yarns before manufacturing textile devices. In recent years, many research experiments have focused only on the flexibility of the devices, which is the basic requirement for wearable devices, whereas occasional works have paid attention to comfort assessment, which is strongly linked to the composition of the fabric or textile and is essential to the real use of such wearable generators in context. Many aspects contribute to the haptic experience of textiles, such as wettability, stiffness, smoothness, and so on, and the fabric structure is one of the most influential factors in the comfort assessment [**28**].

The TENG is usually based on two friction materials of opposite polarity, which have opposite charges when subjected to an external force to form an electrical potential. By connecting these two materials with conductive wires, the electrical energy generated by the charge movement is obtained. Some TENGs work with human skin or clothing, acting as one of the friction electrodes, so the TENG is designed to be composed of a single electrode. The fundamental working modes of TENGs can be divided into four categories: vertical contact-separation mode, lateral sliding mode, single-electrode mode, and free-standing mode [**35,36,37,38**]. Inspired by these, researchers have invented many kinds of fabric-based devices [**39,40,41,42,43,44,45,46**]. Some of these devices are based on woven structures [**47,48,49,50,51**] and others on knitted structures [**52,53**]. We summarize the fabric-based devices into these classifications, as shown in **Figure 1**: woven-structure TENG based on two types of one-dimensional (1D) electrodes (**Figure 1**b,c) and 1D devices (**Figure 1**d,e), knitted-structure TENG based on two types of one-dimensional (1D) electrodes (**Figure 1**f) and 1D devices (**Figure 1**g,h), and TENG based on coated fabric (**Figure 1**i–k).

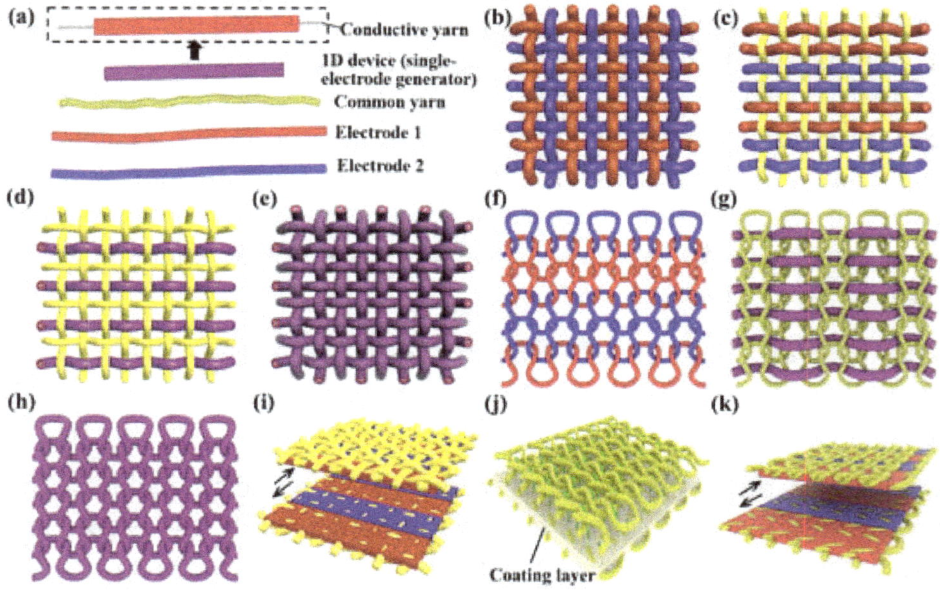

Figure 1. Schematic of the textile-based generators used in human clothing. (a) The basic elements of a textile-based generator. (b) Woven-structure generator based on two kinds of 1D electrodes. (c) Woven-structure generator based on two kinds of 1D electrodes and common yarns. (d) Woven-structure generator based on two kinds of 1D devices and common yarns. (e) Woven-structure generator based on 1D devices. (f) Knitted-structure generator based on two kinds of 1D electrodes. (g) 1D device sewn into common knitted fabric. (h) Knitted-structure generator based on 1D devices. (i) Woven-structure generator based on fabric coated with functional layers. (j) 3D spacer fabric-structure generator and (k) knitted-structure generator based on fabric coated with functional layers.

The yarns in woven fabrics are interwoven and extend horizontally and vertically, so the woven fabric is highly compact and has good wind resistance. The yarns in knitted fabrics exist in the form of nested loops. Each loop can be stretched and extended in all directions, so the knitted fabric has good elasticity and breathability. The differences in fabric properties due to different structures are enormous. Here, we propose a simplified classification of TENG, which contains only two categories. In this paper, we analyze in detail the design and performance of TENGs that have truly been used as part of garments to harvest human kinetic energy over the last decade and discuss the challenges faced by TENGs.

2.1. TENGs Based on Woven Textiles

Woven fabric is a textile formed by weaving. Most woven fabrics were produced on a loom and are made of many threads woven on a warp and a weft. TENGs with a structure woven from different weaving units will be discussed in this section.

2.1.1. Woven-Structure Generators Based on Fibers/Yarns

The yarns in woven fabric are nearly immobile, leading to an almost unextendible fabric sheet with limited deformations in the yarn structure. Thus, normal woven fabric does not need a great deal of flexibility or bending in the warp and weft threads. Therefore, a rigid, inflexible, but conductive metal electrode can be woven into fabric-based devices as warp or weft, even when they are not that satisfactory for weaving. Many devices have modified the electrode instead of using metal wires directly to increase softness and comfort. In 2016, Wen et al. [**54**] fabricated a TENG by using a Cu-coated ethylene vinyl acetate (EVA) electrode and a polydimethylsiloxane (PDMS)-covered Cu-coated EVA tubing electrode as warp and weft, respectively. The Cu-coated EVA tubing in this work acts not only as the electrode for the TENG but also as the holder for fabricating a fiber-shaped dye-sensitized solar cell. Therefore, it can also be fabricated into a textile-based supercapacitor to store the electric energy collected by the TENG after rectification. Although the single EVA tubing was flexible, after copper deposition and PDMS coating, the diameter of the single TENG unit was about 3 mm (**Figure 2**a). The metal yarn-based fabric structure made of such thick yarns makes the device inflexible and bulky. As a result, the device can only be integrated in limited positions on garments, such as by attaching it onto a T-shirt.

Dong et al. [**55**] overcame the inflexibility of thick yarn-based fabric. In their study, a 3-ply-twisted stainless steel/polyester fiber-blended yarn was used as the warp thread and a PDMS-coated energy-harvesting yarn was used as the weft thread. Although the diameter of the warp and weft yarns was even larger than that of the previously mentioned electrode, the flexibility of the textile was maintained, allowing it to be stretched, folded, or crimped (**Figure 2**b). This effect relies on two advanced improvements in the materials and structure design. One is that the warp and weft yarns were all extremely flexible and could be stretched, twisted, bent, and knotted. The other improvement is that they interwove Z-directional cotton yarns with the weft yarn along the warp direction to make the fabric soft and skin-friendly. This compound structure can be categorized as the fourth structure in **Figure 1**d. The cotton yarns not only increase flexibility, but also play a role in absorbing the sweat of human skin or moisture in the environment to balance the humidity between the skin and the outer environment. In tests conducted with the fabric worn on the forearm, the average voltage amplitude reached up to 125 V.

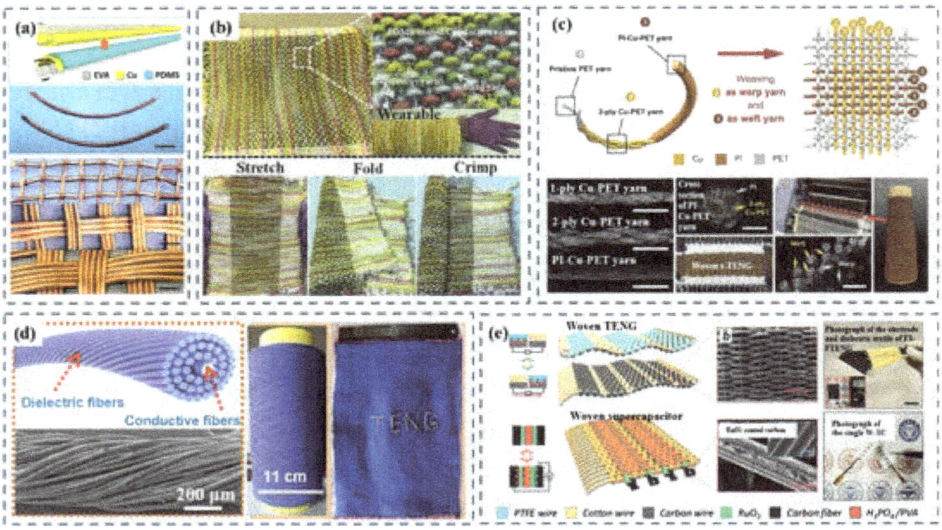

Figure 2. (a) TENG fabricated from a Cu-coated ethylene vinyl acetate (EVA) electrode and a polydimethylsiloxane (PDMS) covered Cu-coated EVA tubing electrode. Reproduced under the terms of the CC-BY Creative Commons Attribution 4.0 International License (https://creativecommons.org/licenses/by/4.0) (accessed on 28 September 2022) [54]. Copyright 2016, American Association for the Advancement of Science. (b) Digital photographs of a large-area wearable textile TENG (top view). Bottom views are photographs of the TENG under various mechanical deformations, including stretching, folding, and crimping. Reproduced with permission [55]. Copyright 2017, John Wiley and Sons. (c) TENG fabricated from Cu-coated PET warp yarns and 2-ply PI-coated Cu-PET weft yarns. Reproduced with permission [56]. Copyright 2016, John Wiley and Sons. (d) The core–shell yarn manufactured by 200 elastic spandex fibers tightly twined around two parallel stainless-steel fibers and digital photographs of the TENG. Reproduced with permission [57]. Copyright 2017, American Chemical Society. (e) Schematic illustration of the free-standing-mode fabric TENG. Reproduced with permission [58]. Copyright 2018, Elsevier.

Similarly, Zhao et al. [**56**] also used twisted yarn to fabricate a functional textile. They designed a TENG that was fabricated by directly weaving together Cu-coated polyethylene terephthalate (PET) warp yarns (300 μm in diameter) and 2-ply polyimide (PI)-coated Cu-PET weft yarns (350 μm in diameter) (**Figure 2**c) on a weaving loom. They stitched the prepared TENG into a white, nonelastic cotton chest strap and used analogue-to-digital conversion and a filter for monitoring the respiration of the tester. A real-time respiratory pattern could be recorded with four different breathing states, including deep, shallow, rapid, and slow using the prepared textile triboelectric nanogenerators (t-TENGs).

Yu et al. [57] further improved the method of twisting the blended yarns by adopting the stainless-steel fibers as the core and twisting the dielectric fibers to form the sheath, creating core–shell yarns through a commercial machine. The manufactured TENG was obtained by weaving the yarn composed using this method. To fabricate core–shell yarn, 200 elastic spandex fibers were tightly twined around two parallel stainless-steel fibers (Figure 2d). The high ratio of elastic dielectric fibers can reduce the rigidity of the yarn and improve its softness. The resulting TENG textiles are skin-friendly and flexible, and their fabrication processes is compatible with industrial manufacturing technology for large-scale textile production. In addition to the woven-structure textile, they also made a knitted-structure textile whose loops were very loose to collect the kinetic friction. By using this knitted TENG textile, they were able to realize more complicated and fashionable garment designs. A greater number of positions on the human body could be used for attaching the TENG textiles onto clothes. In their test, two pieces of TENG textiles woven from spandex-fiber-based core–shell yarns were sewn under the arm of a sweater and under the thenar of a sock. The output performance of the TENG textile under the thenar was significantly higher than that of the TENG textile under the arm. It achieved outputs as high as ~125 V open-circuit voltage and ~4 mA m−2 short-circuit current density when running.

Unlike the methods discussed above, in which metal coatings or metal wires were used, Chen et al. **[58]** used commercial polytetrafluoroethylene (PTFE), carbon, and cotton wires to fabricate a TENG, and the process was carried out on a traditional shuttle-flying weaving loom. In this work, they designed the TENG with two different work modes: a vertical contact-separation mode and a lateral sliding mode. For the contact-separation-mode TENG, the electrode textile was prepared by using nonconductive cotton threads as the warp and conductive carbon wires as the weft. A dielectric textile was woven from the carbon warp threads and PTFE weft wires. By assembling a woven supercapacitor into the TENG with a full-wave rectifier, a device was made for harvesting arm-swaying energy that could continually charge an electric watch as a self-powered system (**Figure** 2e). The long-term stability of the self-powered device was tested in this study, and after 15,000 cycles, ~80% of the original output voltage remained. The mosaic-pattern knitted textile allowed for the flexibility and aesthetic appearance of the fabric device. Additionally, though the underarm is one of the few places on the human body where both contact-separation and sliding modes of generating electricity can be applied, it is also the place where the body is most likely to produce sweat.

The use of cotton greatly inhibits the performance of the TENG by exposing it to high humidity and liquid contact [**59,60,61**]. If the cotton becomes conductive after absorbing moisture when the wearer is in motion or in a humid environment, no power energy will be generated. Therefore, a waterproofing treatment should be considered to improve the device's performance. Gong et al. [**62**] invented an amphibious triboelectric textile, which was woven from super stretchable and flexible triboelectric yarns consisting of intrinsically elastic silicone rubber tubes and built-in helical-structure stainless-steel yarns. In this study the single-electrode triboelectric yarn was capable of lighting up a liquid-crystal display device underwater. The triboelectric textile woven from the yarns can be worn on the tester's elbow and harvest biomechanical energy from bending the elbow. The elbow-bending test confirmed the flexibility and durability of the yarns. Additionally, the ultralong yarns makes it possible for the device to be fabricated on a knitting machine.

The weaving method requires the lowest performance of functional yarns, because even short, coarse, or poorly flexible yarns can be woven into fabrics by the weaving method. As long as the functional yarns are used as the weft yarns and the flexible yarns are used as the warp, a woven fabric can be made. Modifying the fibers/yarns endowed the material with a special function while simultaneously enhancing the stability and life cycle of the fabric device. However, because the modified electronic fibers/yarns were not as soft as traditional yarns, the textile devices became bulky and inflexible. Further effort is needed to fabricate soft, flexible, and functional fibers/yarns for the construction of generator devices.

2.1.2. Woven-Structure Generators Based on Textile Strips

Beyond the fiber-based/yarn-based woven-structure generators, some studies used textile strips as warp and weft to fabricate textile-based generators. The use of woven fabric strips as units to create new fabrics is a method of fabric production that is unique to weaving. Woven fabric strips usually are soft, have a flat surface, and can be produced in large quantities. The length and width are controllable and can be easily remanufactured into garment fabrics. Using woven fabric strips as the basic unit and modifying them to obtain the desired properties and making functional two-dimensional fabrics from the treated strips is a relatively simple and efficient method. The resulting fabric is as soft and flexible as normal clothes and can easily be integrated into garments. The same method cannot be applied to knitted fabrics because their properties of common dispersion, hemming [**34**], etc. prevent them from being cut into strips. It is also possible to manufacture knitted strips specifically, but this method is tedious and the resulting individual strips have uneven surfaces and tend to curl, making it difficult to manufacture garments.

In 2015, Pu et al. [**7**] fabricated a woven TENG by using 10 Ni-coated polyester strips as longitude lines and 10 parylene-Ni-coated strips as latitude lines (**Figure 3**a). The TENG cloth was worn in different positions on the human body—namely, under the foot, under the arm, and at the elbow joint. When doing activities, the TENG cloth worn under the foot and arm generated enough power to light up 37 and 17 LEDs, respectively. However, this device has some limitations. First, when using this device, it was difficult to integrate the storage of a lithium-ion battery (LIB) into clothing. The safety and resource consumption of the device were also problematic. In 2016 [**63**], Pu et. al. proposed a new self-powered system that used the same materials and same fabric structure as their previous TENG, but a textile-structure supercapacitor was used instead of the LIB to store the generated power. The supercapacitor shared the same warp system as the TENG, forming a seamless fabric so that the TENG and the supercapacitor became a single unit without seams or imperfections (**Figure 3**b).

Zhou et al. [**64**] and Tian et al. [**65**] focused on the different power outputs generated by different human motions. In Zhou's work, the source materials for the TENG were a commercial nylon-fabric strip, a polyester-fabric strip, and a homemade conductive silver-fabric strip. The silver strip was pasted in the center of two nylon strips or two polyester strips. The resulting nylon–silver–nylon strips and polyester–silver–polyester strips were used as warp threads and weft threads, respectively, that were woven into a fabric (**Figure 3**c). The homemade conductive silver-fabric strips made of silver fibers and cotton fibers were shown to greatly improve the durability and life of the device. To demonstrate its potential applications, the textile-structure TENG was integrated into shoes, coats, and trousers to harvest different kinds of mechanical energy from human motions. The currents generated from footsteps, the shaking of clothes, bending leg joints, and bending arm joints were 0.3 μA, 0.75 μA, 0.9 μA, and 0.75 μA, respectively (the effective area of the TENG was 20 cm^2). These test results further confirmed the TENG's practical application potential as a wearable device. Tian et al. [**65**] fabricated a TENG by using Ni-coated polyester conductive textile strips and silicone rubber-Ni-coated polyester strips using a traditional "plain weave" method (**Figure 3**d). The currents of the TENG fixed under the arm, at the elbow joint, under the foot, and at the knee joint reached 30 μA, 4 μA, 40 μA, and 15 μA, respectively. The thickness of the strip was 750 μm, which caused the device to be bulky and inflexible. Additionally, the device's thick coating layer eventually fell off following subsequent use of the device, reducing its output and life.

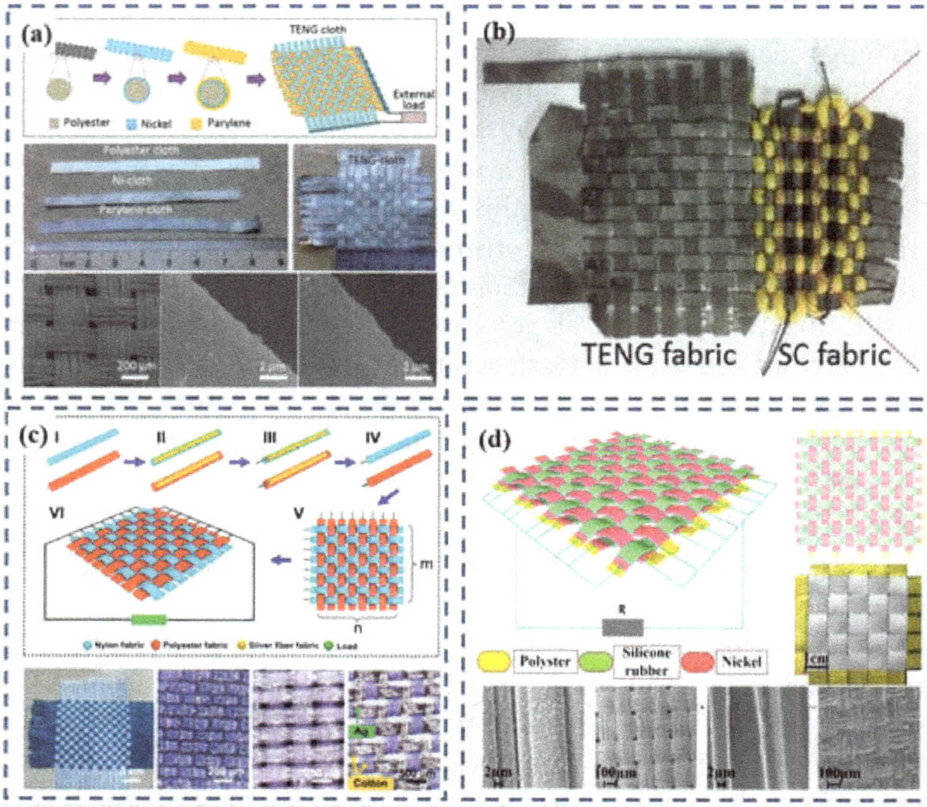

Figure 3. (a) TENG formed by using 10 Ni-coated polyester strips as longitude lines and 10 parylene-Ni-coated strips as latitude lines. Reproduced with permission [7]. Copyright 2015, John Wiley and Sons. (b) Self-power system that used a fabric-structure TENG and textile-structure supercapacitor. Reproduced with permission [63]. Copyright 2015, John Wiley and Sons. (c) TENG made of a commercial nylon-fabric strip, polyester-fabric strip, and homemade conductive silver-fabric strip. Reproduced with permission [64]. Copyright 2014, American Chemical Society. (d) TENG manufactured using Ni-coated polyester conductive-textile strips and silicone rubber-Ni-coated polyester strips by the traditional "plain grain" method. Reproduced with permission [65]. Copyright 2017, Elsevier.

The strip-based woven-structure TENG has a performance similar to that of the fiber-based/yarn-based woven-structure TENG. In addition, it is flexible and durable, making it possible to integrate such fabric devices into human clothing. However, this structure requires two weaving processes, which makes the device-fabrication process more complex and time-consuming than other structural TENGs. For strip-based woven textile devices, the best and only use is to integrate them into clothing as part of the garment for wearable applications. Therefore, it is important to find more ways to make fabric devices that have a simple fabrication process and can be manufactured on a large scale.

2.1.3. Generators Based on Woven Coated Fabric

In 2019, Qiu et al. [26] utilized the electro-spinning and electrospray methods to obtain PET fabric coated with PTFE nanoparticles and PVDF nanofibers (Figure 4a). The resulting fabric was then firmly attached to a conductive fabric with double-sided tape. By integrating the power-generating fabric with a daily-wear garment, they found that mechanical energy could easily be harvested from human movement. After biomechanical excitation, this all-fabric textile produced a power density of 80 mW/m2 at a load of 50 MΩ. The flexibility, light weight, air permeability, and durability of the power-generating fabric was tested, and the results showed that the power-generating fabric could be applied to clothing. A series of common fabrics such as nylon, silk, cotton, T/C (terylene and cotton mixed to a specific ratio), and PP (polypropylene), along with PET, were chosen as raw materials and served as triboelectric layers to demonstrate the applicability of diverse fabrics in power-generating fabric construction. The results demonstrated that this coating strategy is universally applicable to most commonly used fabrics, and the surface-coating technique successfully modified the composite fabrics. It is worth noting that the power-generating fabrics maintained their basic function even after being cut into pieces. What is more, research found that the device can work well even when tailored, halved, or spliced. This characteristic has exciting potential, opening up the prospect of designing tailor-made, power-generating clothes in different styles and sizes.

From this point of view, electro-spinning nanofibers into fabrics is an ideal method for manufacturing wearable devices. Whereas a traditional coat layer reduces the breathability of the fabric [**24**], the electro-spinning coating has the advantage of both air permeability and flexibility, and furthermore allows the surface of the fabric to be modified so it can achieve the desired function [**68,69,70,71,72**].

Guo et al. [**66**] also used electro-spinning coatings in the manufacture of TENG. They designed a hybrid triboelectric and piezoelectric generator to maximize the collection of human kinetic energy. The hybrid generator consisted of two fabric electrodes: conductive fabric covering silk electro-spinning nanofibers and conductive fabric covering PVDF electro-spinning nanofibers (**Figure 4**b). The study demonstrated the applicability of the hybrid generator to monitor human activity and personal medical care, and the cycle stability of the device was also verified. Compared to cold, rigid medical monitors, soft, fabric-based detectors can provide a more comfortable experience for patients.

In addition to the electro-spinning method, electroless deposition (ELD) can also be used to fabricate functional coatings for TENG. In fabricating TENG, Pu et al. [**67**] proposed a laser-scribing mask and the ELD nickel-plating method for synthesizing conductive circuits/patterns on fabrics. They fabricated TENG fabrics with interlaced grating structures (**Figure** 4c) and integrated them with SC to realize a fabric-based energy-harvesting system. The thin coating of Ni does not significantly increase the weight of the fabric, maintaining the light weight and softness of the pristine textile. As shown in **Figure** 1d, TENG fabric can be easily bent, wrapped, and immersed in water without damage, which shows that it has good flexibility and that household clothes can be used and washed. In addition, the TENG fabric has been proven to have good air permeability and shape retention.

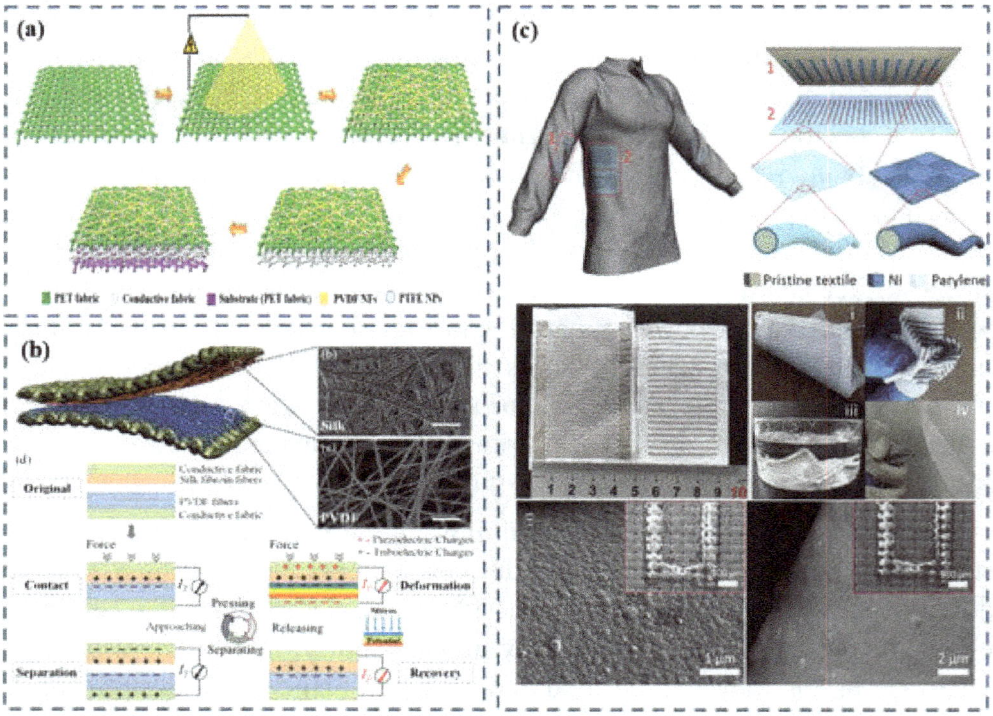

Figure 4. (a) Schematic illustration of modified fabrication process of power-generating fabrics. Reproduced with permission [26]. Copyright 2019, Elsevier. (b) Schematic diagram of an all-fiber hybrid triboelectric nanogenerator, which consists of two electrodes (conductive fabric) and electrospun silk nanofibers and PVDF nanofibers serving as a triboelectric pair. Schematic view of the operating principle of the hybrid nanogenerator. Reproduced with permission [66]. Copyright 2018, Elsevier. (c) Schematic of a power-textile with a pair of TENG fabrics consisting of a slider fabric (1) in the sleeve and a stator fabric (2) underneath the arm. Reproduced with permission [67]. Copyright 2016, John Wiley and Sons.

The thickness of the coating affects the breathability of the fabric, and the adhesion condition of the coating to the fabric substrate determines the washability of the fabric. Attention needs to be paid to the breathability of the fabric and the adhesion of the coating when designing the wearable TENG. The coating method of the grating structure retains the breathability of original fabric. The use of nanoscale particles and fibers as coatings enhances the adhesion between the coating and the fabric substrate. All these designs are well considered for the performance of the generator based on woven coated fabric. In addition, the lifetime of the coating has to be considered, because once the generator is made/integrated into a garment, the fabric will be affected by high-frequency external forces during wearing or washing.

2.2. Generators utilizing knitted textiles as their foundation

Knitted fabric created by interlacing yarn in a series of connected loops using straight eyeless needles or by machine has high flexibility and breathability [**73,74**]. The knitting loops are arranged by suspension in a horizontal (course) or vertical (wale) direction. These meandering and suspended loops can be easily stretched in different directions, meaning they have more elasticity than other types of textiles [**75,76**]. However, harsh deformations during the fabrication of a knitted fabric can possibly damage the fiber/yarn. During the process of manufacturing knitted fabrics, each loop has to withstand manipulation by mechanical external forces, so the yarn must be durable.

2.2.1. Knitted-Structure Generators Based on Fibers/Yarns

Dong et al. [**77**] developed a knitted-fabric TENG by using a single energy-harvesting yarn that is fabricated by coating silicone rubber over the surface of three-ply twisted stainless-steel/polyester-fiber blended yarn (**Figure 5**a). By taking advantage of the weft-knitting technique, the resulting fabric TENG possessed high elasticity, flexibility, and stretchability so it could be elongated, widened, or distorted by external or internal forces in any direction. As a prototype, the knitted power-generating fabric could be worn on the body directly, for example as an insole inside a shoe or as a bracelet worn on the wrist. After 50,000 cycles of repeated contact–separation motion at a contact frequency of 4 Hz, the open-circuit voltage and short-circuit current showed no obvious degradation. The electrical outputs of the knitted TENG fabric experienced no decrease after multiple washing cycles. The study demonstrated that the electrical output performance was positively correlated with yarn diameter. However, the problem is that given the current industry trend of pursuing lightness, thinness, and transparency, a fabric TENG made from yarns that are large in diameter results in thick and bulky clothing. The compatibility of the thick fabric with common fabrics, which tend to be thin, is also an issue. Therefore, the application of wearable devices in real-life situations should be further explored to create a balance between diminished thickness and high output.

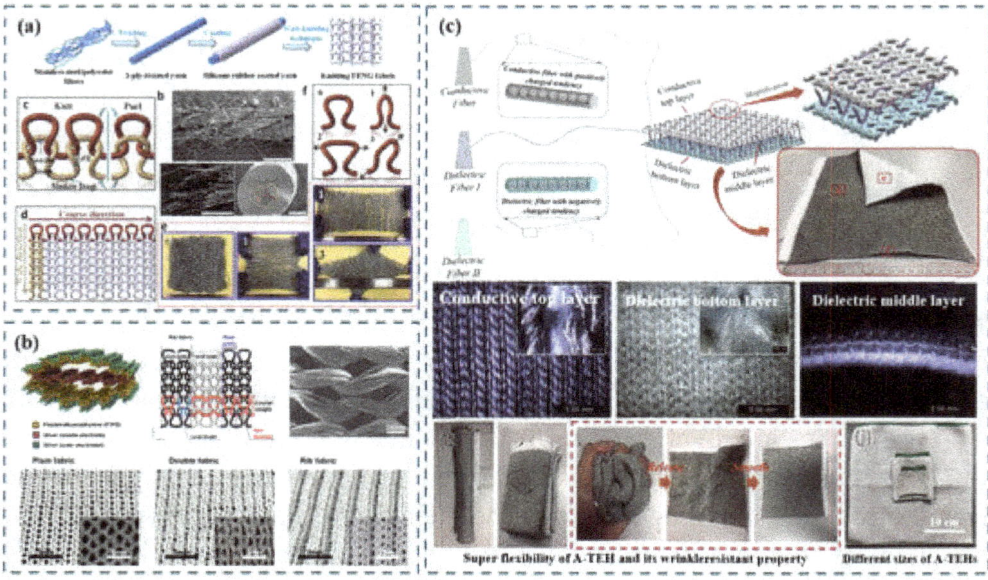

Figure 5. (a) Knitted-structure generator based on 1D devices. Reproduced with permission [77]. Copyright 2017, American Chemical Society. (b) Knitted-structure generator based on two kinds of 1D electrodes. Reproduced with permission [78]. Copyright 2017, American Chemical Society. (c) Schematic illustration of the realization of the energy-harvesting mode as 3D full-fabric structural integrity using all fiber materials by a computerized knitting programming and loop structures. Reproduced with permission [41]. Copyright 2019, Elsevier.

Using a different approach than the single-electrode work mode discussed above, Kwak et al. [**78**] invented a knitted-fabric TENG that harvests motion energy using a contact-separation mode. The TENG was made from five knitted-fabric layers (**Figure 5**b), which consisted of knitted PTFE fabrics for the top and bottom triboelectric layers and knitted Ag fabrics for the electrode in the middle and on the back of the top and bottom triboelectric layers. They investigated plain-, double-, and rib-fabric structures (**Figure 5**b) and analyzed their potential for textile-based energy harvesting. Compared with the fabrics' original states, increasing the amount of surface contact during the stretching is crucial. Although it is possible to extend the plain fabric in both the wale and course directions, its capacity to extend is lower than that of the double and rib structures because the contact between each loop is maintained even during stretching. Double-knitted fabric and rib-knitted fabric showed significantly superior stretchability and output. The latter showed significantly improved stretchability, which dramatically enhanced the triboelectric power-generation's performance due to the increased contact surface. The durability of the rib-knitted TENG was also superb; the voltage was maintained even after 1800 stretching cycles under 30% strain. After integrating the TENG into a sportscoat and running, the TENG was able to light 15 green LEDs.

Gong et al. [**41**] adopted a new kind of all-textile energy harvester with three sets of fibers by using computer programming. In this device, the conductive silver-plated nylon fibers with a positively charged tendency were knitted as the top layer, the dielectric polyacrylonitrile (PAN) fibers with a negatively charged tendency formed the bottom layer, and the dielectric PAN fibers without post-treatment were used to directly knit the top layer and bottom layer together (**Figure 5**c). This compound-structure fabric overcomes the shortcomings of traditional knitted fabric, which is prone to dislodgement and edge curl, and it can be tailored into any desired shape. In their research, a shoe-insole-shaped harvester was tailored to collect human kinetic energy. When the fabric becomes dirty, leading to performance degradation, the electric power generation can be recovered again after simply rinsing the fabric with tap water like common clothes and air-drying it. The textile energy harvester's wearable performance, including its comfort, breathability, washability, and unique advantage of tailorability, lends it to versatile product design.

The capacity of the knitted loops to stretch and deform greatly improves the elasticity of the fabric device and makes up for the inflexibility of the functional fibers/yarns. Moreover, the modification of the fibers/yarns does not affect the air permeability of the knitted fabric. Compared with a woven device, a knitted device can be softer and more compatible with the design requirements of garments.

2.2.2. Generators Based on Knitted, Coated Fabric

The high elasticity and porosity of knitted fabrics contribute to the breathability and softness of the garment. If a coating is applied to the top of a knitted fabric, when the knitted fabric is being stretched, the breathability of the fabric may be reduced and the adhesion of the coating may be compromised. Much work has been carried out to investigate ways to enhance knitted/coated-based generators' garment properties, such as breathability, stretchability, and washability.

Huang et al. [**6**] studied the influence of different knitted structures on the output performance of TENGs. The TENG they used was fabricated by using a heat-welding adhesive net to coat the expanded PTFE membrane on top of knitted cloth via a thermal-calendaring process (**Figure 6**a). They designed five types of knitted cloth with different stitch densities on each side. PTFE was coated on to all 10 sides of the five knitted cloths. The result showed that the TENG textiles with higher stitch densities showed a voltage of 900 V and current of 19 μA, which are approximately twice as high as those of alternate sides with lower stitch density, and higher than other types of TENG textiles with lower stich density. Clearly, the superiority of these TENG textiles comes from the larger contact area. The TENG fabric developed during this experiment could be folded like normal clothes. Its good durability was demonstrated from the voltage–current curves of the TENG-textile electrode under deformed states. It retained good conductivity after being twisted or stretched for 500 cycles. There were no obvious changes to its voltage–current curves after it was washed in up to 10 cycles, which shows good washability.

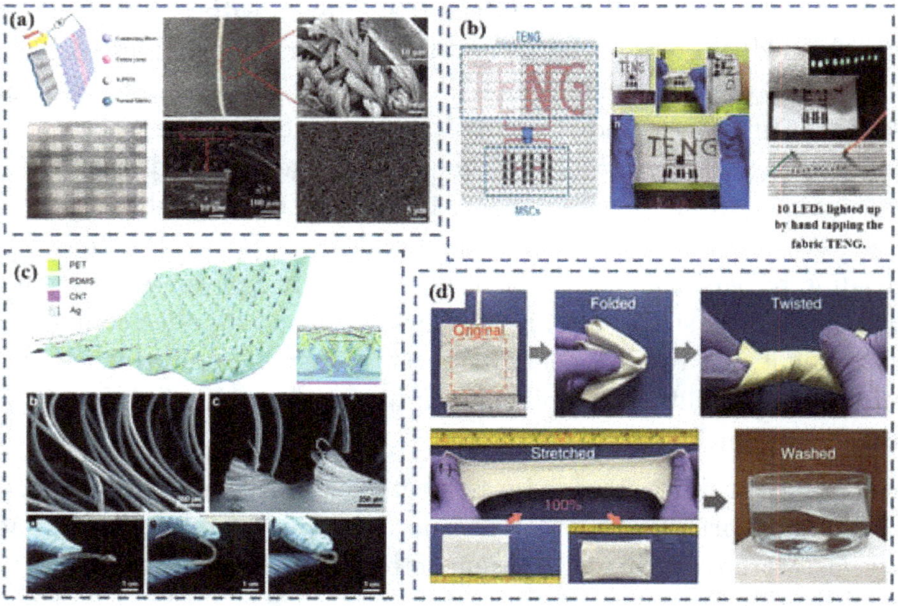

Figure 6. (a) The free-standing-mode TENG textile, which is made up of a laminated composite fabric consisting of an expanded polytetrafluoroethylene (E-TFE) film and a common fabric. Reproduced with permission [6]. Copyright 2017, Elsevier. (b) Stretchable coplanar self-charging power knitted textiles with a triboelectric nanogenerator (TENG) and microsupercapacitors (MSC). Reproduced with permission [75]. Copyright 2017, American Chemical Society. (c) TENG made of 3D-space fabric on the market as the base material by coating PDMS, conductive silver paste, and CNT. Reproduced with permission [79]. Copyright 2017, Royal Society of Chemistry. (d) Photographs demonstrating that the textile triboelectric nanogenerator (textile-TENG) possesses excellent endurance for successively experiencing deformations of folding, twisting, and stretching, as well as severe washing. Reproduced under the terms of the CC-BY Creative Commons Attribution 4.0 International License (https://creativecommons.org/licenses/by/4.0) (accessed on 21 September 2022). [80].

Cong et al. [75] developed a stretchable knitted TENG using the resist dyeing-analogous method. In this method, Ni-coated textile with two in-plane electrodes and an elastomeric PDMS thin layer is coated onto one of the Ni-coated textiles. The TENG generates power when a polyester textile does repeated touching–separating motions. Using the same method, they also invented a microsupercapacitor that was fabricated into the same textile (**Figure 6**b) with the TENG. The impact of different strains and humidities on the output was studied. After being stretched to 100% strain along the course direction, there was no significant decrease in the short-circuit current, whereas the open-circuit voltage exhibited a slight decrease. The output increased in a dry environment (20% relative humidity) and decreased in a humid environment (50% relative humidity). The adhesion stability of the coating was analyzed through the scanning electron microscopy (SEM) image of the Ni layer combined with the resistance-strain trend. They found that the coating was quite stable under a tensile strain of less than 50%, whereas damage appeared in the coating under a tensile strain of 100%. Based on this, the cycling performance of the TENG was tested under 50% strain. The result was promising, demonstrating that the current of the TENG textile did not decrease after 4000 cycles at 50% stretched strain. It is important to note that when the hydrophobic material was replaced by a hydrophilic material during the experiments, the device was damaged. Therefore, two factors limit the application of the TENG: strain and the material's performance when wet.

The development of multilayer knitted fabrics is another attractive solution for manufacturing wearable flexible generators. Liu et al. **[79]** utilized a commercially available 3D-space fabric with a three-dimensionally penetrated structure that can offer spontaneous elastic space for pressing and releasing to fabricate a TENG. Thin PDMS film was carefully coated onto one surface of the 3D PET fabric directly. The CNT sheets were closely stacked onto the outer surface of the treated layer to serve as one electrode, whereas the other untreated surface was coated with a conductive silver paste to serve as the other electrode (**Figure 6**c). However, because the two layers were not friendly to human skin and the coating reduced the breathability, comfort, washability, stretchability, etc. of the fabric, the potential application of the device in daily wear is limited. During the experiment, the device was only made into insoles to collect people's kinetic energy while walking.

Xiong et al. **[80]** proposed a synergetic triboelectric trapping layer of black phosphorus (BP) protected by a hydrophobic coating of cellulose oleoyl ester nanoparticles (HCOENPs) to alleviate degradation. By using these chemicals, they developed a durable skin-touch-triggered textile-TENG with a sandwiched structure. It was constructed using three fabric layers: a triboelectric fabric, a fabric electrode, and a waterproof fabric. By dip-coating or spray-coating, BP and HCOENPs were coated onto the PET fabric layer by layer in order to make the resulting triboelectric fabric, known as HCOENPs/BP/PET fabric (HBP-fabric). The silver flake mixed with PDMS was used as a as conducting medium and was coated onto the PET fabric via dip-coating to attain the fabric electrode. The waterproof fabric was created by dip-coating the HCOENPs onto a PET fabric. All three fabrics were stuck together with the aid of double-side tape, with the HBP-fabric in the middle. The all-fabric-based configuration delivered a conformable textile-TENG with extreme deformability, which worked well under 100% stretch conditions and had high durability. The output was maintained even after suffering 500 cycles of extreme deformations and 72 h of severe washing (**Figure 6**d). When mounted directly on skin or cloth, the TENG could fully fit different body regions, and it produced stable output voltages and current densities at different body locations.

The cover of the coating layer may make knitted devices lose their breathability and elasticity easily, and could even lead to skin-safety problems. The coating that Xiong et. al. developed solves the problems listed above, and even increases the aesthetic property of the knitted fabric device, making it possible to make patterned designs.

2.2.3. 1D Devices Sewn into a Single Knitted Fabric

In addition to the above-mentioned wearable devices based on different fabric structures or textile technologies, there are some special generators that are single threads and were integrated into the fabric in the form of a stitch [**81,82**], cross stitch [**83**], or embroidery pattern [**81**]. As long as the mechanical properties of the yarn satisfy the sewing requirements, it can be fabricated into sophisticated textile structures and patterns.

In 2014, Zhong et al. [**84**] fabricated a metal-free fiber-based generator (FBG) using a cost-effective method that involved commercial cotton thread, a PTFE aqueous suspension, and carbon nanotubes as source materials (**Figure 7**a). This paper establishes the first proof of the idea that FBG can be sewn into knitted textiles (**Figure 7**a). FBG can extract energy from biomechanical motion to power a mobile medical system, making self-powered smart clothing possible. It can also convert biomechanical motion/vibration energy into electrical energy by using electrostatic effects, with an average output-power density of approximately 0.1 μW/cm^2.

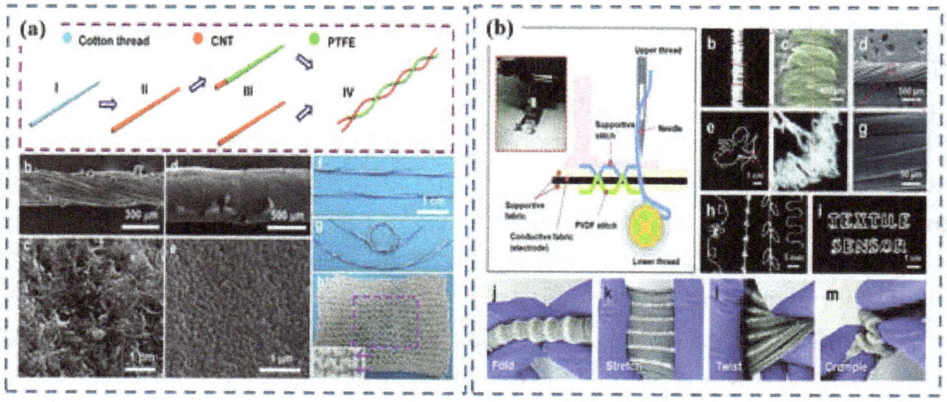

Figure 7. (a) Schematic diagram illustrating the fabricating process of an FBG that is made from CNT-coated cotton thread and PTFE thread. Reproduced with permission [84]. Copyright 2014, American Chemical Society. (b) PVDF stitch-based triboelectric textile sensors. Reproduced with permission [81]. Copyright 2018, The Royal Society of Chemistry.

Shin et al. [**81**] reported using a sewing machine to stitch PVDF into programmable textile patterns for wearable self-powered triboelectric sensors. During their research, the PVDF thread was fabricated by dry-jet wet spinning and spin–draw. Then the PVDF thread was used as the lower thread and a commercial PET thread was used as the upper thread to stitch a design into knitted conductive fabrics, such as the simple stripe, embroidered pattern, embroidered lines, and even letters, indicating that PVDF threads are mechanically strong enough to be sewn into arbitrary patterns (**Figure** 7b). The stitch pattern is composed of twisted 5-ply PVDF threads. When stitched on stretchable fabric, the PVDF stitch patterns showed a good mechanical stability for folding, stretching, twisting, and crumpling deformations. The conductive fabric was sandwiched between supportive cotton fabrics, which were used as insulating layers. In order to demonstrate their applications in wearable devices, smart gloves and joint pads were manufactured based on PVDF-stitch triboelectric sensors. These wearable sensors can detect and distinguish different gestures and body actions by generating inherent signal patterns that represent specific gestures and actions. The device performance of the stitch sensor was retained without significant decrease even after repeated washing cycles, demonstrating that the proposed PVDF stitch sensor can be utilized for practical applications. In addition, compared to patterns that were coated on the fabric, the pattern in this wearable generator has better stability because it is stitched into the fabric. Moreover, combining this stitching technology with other wearable fabrics that can be tailored [**26**], manufactured on a large scale, and produced in large quantities [**57**] makes it possible to design garments that generate electricity.

The interlocking-loop structure of knitted fabric allows the coating material to cover the textile firmly and deeply. The existence of air gaps in knitted fabric can block the unlimited spread of a liquid coating material, meaning that it can be painted into various patterns as desired to meet the aesthetic requirements of various types of clothing. The addition of functional materials can directionally improve the relevant characteristics of the fabric and give new functions to the device. The patterned coating method instead of full coverage can not only retain the clothing characteristics of the original fabric, but also increase the aesthetic design. The knitted/coated-based generator shows a diversified development prospect.

2.3. Other Generators

Different fabric structures allow for different functions, so in some special and complex situations it is necessary to apply multiple fabrics at the same time [**85**]. Thus, hybrid textiles that are composed of a variety of structural fabrics have emerged. Each fabric in a hybrid textile performs its own duties to jointly create a resistance to the influences of a harsh environment. Inspired by this, Kim et al. [**86**] made a hybrid TENG that is based on woven fabric and knitted fabric (**Figure 8**a). The TENG is intended to be used as sportswear, so it has to make full use of the tightness and durability of woven fabrics and the soft and breathable properties of knitted fabrics. The basic structure of the waterproof and breathable textile-based TENG described in this study is composed of three distinctive layers. Using fabric woven with nylon as an outer layer grants both wind and water protection. A 50 μm-thick PTFE membrane (Gore-Tex) is the center layer, and the knitted polyurethane-based bottom lining with a breathable structure allows moisture to escape through the center-layer PTFE membrane. The flexible components of these TENGs were demonstrated easily when attached to clothes, effectively harvesting the motion energy of the human body.

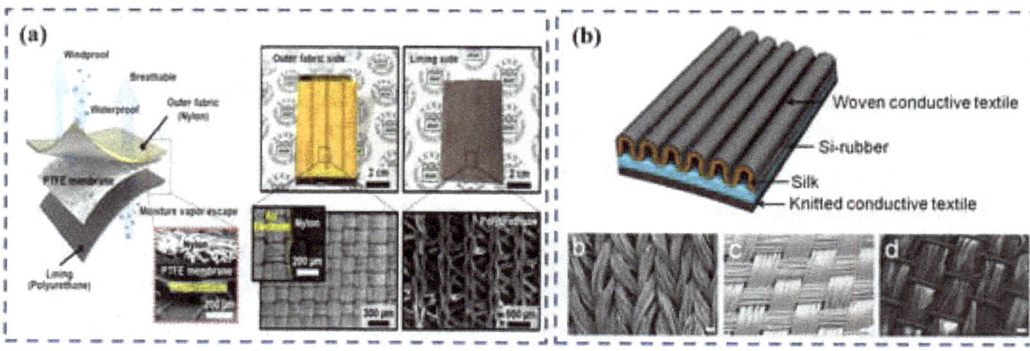

Figure 8. (a) Schematic of textile-TENG components (i.e., Gore-Tex); a woven-fabric composite consists of a nylon outer fabric, PTFE membrane (center layer), and polyurethane lining. Reproduced with permission [86]. Copyright 2014, Elsevier. (b) Corrugated textile-based triboelectric-structure generator consisting of a woven conductive textile, knitted conductive textile, and silk. Reproduced under the terms of the CC-BY Creative Commons [87].

Another study that analyzed the potential of a hybrid textile was that of Choi et al. [87], who proposed a corrugated textile-based triboelectric generator that can generate energy by stretching. The generator consists of woven conductive textile, silk, Si-rubber, and knitted conductive textile. The stretching process of the generator brings the silk and Si-rubber into full contact. The more friction and deformation that the silk and Si-rubber experience under high external mechanical force, the higher the quantity of triboelectric charges. This design cleverly puts the elastic characteristics of different fabric structures to use. Woven fabric is generally very poor in elasticity, so the woven components are made into a corrugated shape in this device to avoid breakage or fatigue and to increase the friction, resulting in a higher output. Knitted fabrics and silicone rubber are elastic and can be stretched from the original state to the fully stretched state, then spontaneously recover. This movement not only increases the friction, but also provides the driving force behind the device's deformation (Figure 8b). Overall, this study provides a unique example of a hybrid woven-and-knitted generator that takes advantage of the elasticity of knitted fabric and bypasses the disadvantage of the inelasticity of woven fabrics. Moreover, the device can generate considerable energy from various deformations not only through pressing and rubbing but also by stretching. Additionally, the experiments demonstrated the generation of sufficient energy from various activities of a human body to power about 54 LEDs. These results demonstrate the potential application of the textile-based generator for self-powered systems. Table 1 provides a comparison of the electrical outputs of TENGs that were integrated into human clothing to collect human-motion energy in real-world circumstances.

Table 1. List of the reported textile-based TENGs with different performance.

Ref	Location of Energy Collection	Current	Voltage	Charge Accumulation	Materials	Device Substrate	Combination with Garment
[6]		19 μA	900 V	203 mWm−2	PTFE	Polyester fabric	Knitted
[7]	Under foot, under arm, elbow joint	4 μA	50 V	3.7 μC min−1	Ni-cloth, parlyene	Polyester fabric	Attached
[26]	Underneath the arm	1.4 μA	113.21 V	80 mW/m2	PTFE, PVDF	Nylon, silk, cotton, T/C, PET, PP fabric	
[54]	Underneath the arm	0.91 mA	12.6 V	11.92 mA cm−2	EVA, PDMS	EVA tubes	Attached, woven
[58]	Underneath the arm	1.5 μA	~118 V	48 nC	Conductive carbon wires	Carbon and PTFE textile	
[63]	Underneath the arm	40 μA			Ni and parlyene	Polyester yarns	Woven
[66]	Elbow	12 μA	500 V	310 μW/cm2	Silk fibroin, PVDF	Conductive fabrics	
[67]	Sleeve, underneath the arm	55 μA	100 V		Ni and parlyene	Polyester fabric	Attached
[75]		2.9 μA	150 V	85 mW·m−2	Stainless steel, polyester	Silicone, rubber	Knitting
[77]	Chest	1.8 μA	49 V	50.6 mF cm−2, 94.5 mW m−2	Ni, rGO-Ni	Stretchable polyester fabric	Sewn
[88]	Hand	0.25 mA	80 V		PTFE and copper, ZnO and copper	Polymer textile	
[89]	Between forearm, human body	0.2 mA	2 kV	69 μC/s	Nylon, Dacron	Cotton	

Ref	Location of Energy Collection	Current	Voltage	Charge Accumulation	Materials	Device Substrate	Combination with Garment
[90]	Under arm	0.4 μA	40 V	0.18 μW/cm2, 85.2 mF/cm2	PI, PU, Al, PDMS, CNT/RuO2 PVA/H3PO4	Carbon fabric	
[91]	Pocket, sleeve	65 μA	120 V		ZnO, PDMS	Ag-coated knitted textile	Attached
[92]	Elbow, knee		4.16 V		Metal, conductive fiber, AgNWs	Nylon fiber, silicone, rubber tube	
[93]	Chest, hand, wrist				CNT	Polyester, nylon textile	Sewn
[86]		4 μA	120 V	68 μW m−2	Au nanodots, polyurethane/PTFE	Nylon woven-fabric	
[94]		399.42 mA	17 V		Copper	Polyester filament	Sewn
[81]	Wrist, elbow, ankle, knee	190 nA	1.8 V		PVDF, Al	Nylon fabric	Sewn, stitched
[95]	Butt, underneath the arm, arm, knee		540 V	2 Wm−2	PEDOT: PSS, PTFE or silicon rubber	Cotton textile	Attached
[96]	Sleeve	3 μA	60 V	≈78.1 μWh cm−2, 14 mW cm−2	Ni/Cu, rGO/CNT, NiCo BOH	Polyester yarn	Woven, knitted

The tightness, durability, and elasticity of knitted-structure fabrics differ greatly compared to those of woven-structure fabrics, which results in huge differences in water resistance, wind resistance, breathability, and stretch ability of the two fabrics. The composite use of the two structures of fabrics can complement each other's strengths and achieve a perfect synergy effect.

3. CONCLUSION

Research in the field of wearable triboelectric generators is increasing, and pioneering research into real applications of this technology is a growing need in both scientific and industry research. In addition to the two key characteristics of wearable triboelectric generators of flexibility and generating friction, features such as softness, breathability, washability, and wear resistance have also attracted a lot of attention from the research community. This paper reviews wearable triboelectric generators that are used in human clothing for energy conversion. The study focuses on analyzing fabric structure and examining the integration method of flexible generators and common fibers/yarns/textiles. Compared to the knitting method, the woven method has fewer restrictions on the flexibility and thickness of the yarn. Remaining challenges and perspectives are also investigated to suggest how to bring fully generated clothing to practical applications in the near future.

4. CONCISE SUMMARY

Functional yarns with different properties are suitable for fabricating devices with different structures and designs. A woven structure has low requirements for the flexibility and length of the yarn. Woven fabrics have high durability and wear resistance, keep their shape well, and are often made into shirts, suits, denims, jackets, etc. Knitted structures have high requirements for the strength of the yarn because the yarn frequently suffers from the application of external force during the weaving process. Knitted fabrics have high elasticity and good air permeability, and are usually made into T-shirts, underwear, sportswear, socks, etc. Some special high-strength and high-flexibility functional yarns can not only be made into woven or knitted fabric devices but also into many more complex structures such as embroidery patterns, which greatly improves the ornamental potential and practicality of wearable devices.

For coated-fabric devices, coating on different fabric structures produces different fabric characteristics. The warp and weft yarns of the woven structure are usually arranged tightly, which gives the fabric a smooth surface that can easily be coated with functional materials. Unfortunately, the coverage of the coating generally results in a loss of air permeability for woven fabrics. As for knitted fabrics, the air gap between the loops provides enough space for the coating, allowing the coating to be applied to the yarn more deeply. The air permeability of coated/knitted fabrics will also decrease, but not as severely as for woven fabrics. Furthermore, most coatings are accompanied by drawbacks such as shedding, lack of breathability, and being unfriendly to the skin. The resulting issues of shortened life cycle, safety, and comfort need to be optimized by solving these problems, allowing yarn-based fabric generators to exhibit better mechanical strength and processability.

No matter what kind of structure the wearable device is, it must ultimately be integrated into clothing. Hence, the softness, breathability, comfort, washability, abrasion resistance, etc. of the device all need to be considered. Up to this point, the majority of research in this field has focused exclusively on the flexibility of the device, overlooking other practical properties. In recent years, more and more research projects have begun to test water washability, breathability, etc. with fabric pieces. So far, no one has been able to produce a garment where power-generating fabric forms the whole or main part of the clothing. The size, sweat safety, abrasion resistance, etc. are all obstacles to make power-generating garments. On one hand, increasing the output of the device is the main purpose of the structure design and material choice; on the other hand, the safety of the device is an issue that must be considered. For the human body, a safe voltage is below 36 V, a safe alternating current is 10 mA, and a safe direct current is 30 mA. If the voltage and current exceed the limit values at the same time, the device will be unsafe for human use. Although the power output produced by most devices up to this point has not been harmful to the human body, as further development pursues high-output devices, management and design used to control the limitation must be included in the device system.

Some efforts have been made in the field of device safety and practicality, such as the use of a waterproof coating layer [**62,80,97,98**], the application of a buck converter, and the integration of wearable supercapacitors [**96,99**], all of which are attempts to increase the safety and comfort of TENGs. In 2014, Zhao [**100**] set up a buck converter in the collection device to balance the gap between high voltage and low current, providing a possible reference for developing safety features in future electronic devices. The issues of humidity resistance [**27**], life cycle, and energy-conversion efficiency are also a concern. The research and development discussed above have increased TENGs' resilience to wear, shear, water washing, extrusion, etc., but there is still area for improvement and increased testing.

Most of the research discussed in this paper focuses on integrating the generator into the upper garment; however, Proto's [**18**] work showed that the kinetic energy produced by the lower body of humans is much higher than that of the upper body. Therefore, integrating generators into trousers to collect electricity is a possible direction for future development. Moreover, the energy collected from the human body is quite different from the energy collected in the laboratory, and increasing the comfort of a wearable generator is generally accompanied by a reduction in the amount of energy harvested [**100**]. A good balance between the comfort and actual output needs to be further explored. The current wearable devices under review have only been integrated into clothing in the form of small-area sheet fabrics, and the possibility of developing a fabric device that can be directly made into clothing should be investigated further. Smart power-generating clothing gives a new meaning to our clothing and provides a green and renewable way to collect electricity.

REFERENCES

1. Zhang, C.; Liu, Y.; Zhang, B.; Yang, O.; Yuan, W.; He, L.; Wei, X.; Wang, J.; Wang, Z.L. Harvesting Wind Energy by a Triboelectric Nanogenerator for an Intelligent High-Speed Train System. *ACS Energy Lett.* **2021**, *6*, 1490–1499. **[Google Scholar] [CrossRef]**
2. Kim, J.Y.; Lee, J.W.; Jung, H.S.; Shin, H.; Park, N.G. High-Efficiency Perovskite Solar Cells. *Chem. Rev.* **2020**, *120*, 7867–7918. **[Google Scholar] [CrossRef] [PubMed]**
3. Yuan, M.; Li, C.; Liu, H.; Xu, Q.; Xie, Y. A 3D-Printed Acoustic Triboelectric Nanogenerator for Quarter-Wavelength Acoustic Energy Harvesting and Self-Powered Edge Sensing. *Nano Energy* **2021**, *85*, 105962. **[Google Scholar] [CrossRef]**
4. Wang, L.; Wang, Y.; Wang, H.; Xu, G.; Doring, A.; Daoud, W.A.; Xu, J.; Rogach, A.L.; Xi, Y.; Zi, Y. Carbon Dot-Based Composite Films for Simultaneously Harvesting Raindrop Energy and Boosting Solar Energy Conversion Efficiency in Hybrid Cells. *ACS Nano* **2020**, *14*, 10359–10369. **[Google Scholar] [CrossRef] [PubMed]**
5. Wang, Z. Catch Wave Power in Floating Nets. *Nature* **2017**, *542*, 159–160. **[Google Scholar] [CrossRef]**
6. Huang, T.; Zhang, J.; Yu, B.; Yu, H.; Long, H.; Wang, H.; Zhang, Q.; Zhu, M. Fabric Texture Design for Boosting the Performance of a Knitted Washable Textile Triboelectric Nanogenerator as Wearable Power. *Nano Energy* **2019**, *58*, 375–383. **[Google Scholar] [CrossRef]**
7. Pu, X.; Li, L.; Song, H.; Du, C.; Zhao, Z.; Jiang, C.; Cao, G.; Hu, W.; Wang, Z.L. A Self-Charging Power Unit by Integration of a Textile Triboelectric Nanogenerator and a Flexible Lithium-Ion Battery for Wearable Electronics. *Adv. Mater.* **2015**, *27*, 2472–2478. **[Google Scholar] [CrossRef]**

8. Zhang, D.; Yang, W.; Gong, W.; Ma, W.; Hou, C.; Li, Y.; Zhang, Q.; Wang, H. Abrasion Resistant/Waterproof Stretchable Triboelectric Yarns Based on Fermat Spirals. *Adv. Mater.* **2021**, *33*, 2100782. **[Google Scholar] [CrossRef]**

9. Qu, X.; Liu, Z.; Tan, P.; Wang, C.; Liu, Y.; Feng, H.; Luo, D.; Li, Z.; Wang, Z. Artificial Tactile Perception Smart Finger for Material Identification Based on Triboelectric Sensing. *Sci. Adv.* **2022**, *8*, 2521. **[Google Scholar] [CrossRef]**

10. Meng, J.; Guo, Z.; Pan, C.; Wang, L.; Chang, C.; Li, L.; Pu, X.; Wang, Z. Flexible Textile Direct-Current Generator Based on the Tribovoltaic Effect at Dynamic Metal-Semiconducting Polymer Interfaces. *ACS Energy Lett.* **2021**, *6*, 2442–2450. **[Google Scholar] [CrossRef]**

11. Walden, R.; Aazem, I.; Babu, A.; Pillai, S.C. Textile-Triboelectric Nanogenerators (T-TENGs) for Wearable Energy Harvesting Devices. *Chem. Eng. J.* **2023**, *451*, 138741. **[Google Scholar] [CrossRef]**

12. Jeong, S.Y.; Hwang, W.S.; Cho, J.Y.; Jeong, J.C.; Ahn, J.H.; Kim, K.B.; Hong, S.D.; Song, G.J.; Jeon, D.H.; Sung, T.H. Piezoelectric Device Operating as Sensor and Harvester to Drive Switching Circuit in Led Shoes. *Energy* **2019**, *177*, 87–93. **[Google Scholar] [CrossRef]**

13. Chang, C.C.; Shih, J.F.; Chiou, Y.C.; Lee, R.T.; Tseng, S.F.; Yang, C.R. Development of Textile-Based Triboelectric Nanogenerators Integrated with Plastic Metal Electrodes for Wearable Devices. *Int. J. Adv. Manuf. Technol.* **2019**, *104*, 2633–2644. **[Google Scholar] [CrossRef]**

14. Yin, Z.; Gao, S.; Jin, L.; Guo, S.; Wu, Q.; Li, Z. A Shoe-Mounted Frequency Up-Converted Piezoelectric Energy Harvester. *Sens. Actuators A Phys.* **2021**, *318*, 112530. **[Google Scholar] [CrossRef]**

15. Li, Z.; Zhu, M.; Qiu, Q.; Yu, J.; Ding, B. Multilayered Fiber-Based Triboelectric Nanogenerator with High Performance for Biomechanical Energy Harvesting. *Nano Energy* **2018**, *53*, 726–733. **[Google Scholar] [CrossRef]**

16. Wang, C.; Shim, E.; Chang, H.K.; Lee, N.; Kim, H.R.; Park, J. Sustainable and High-Power Wearable Glucose Biofuel Cell Using Long-Term and High-Speed Flow in Sportswear Fabrics. *Biosens. Bioelectron.* **2020**, *169*, 112652. **[Google Scholar] [CrossRef]**

17. Peng, X.; Dong, K.; Ye, C.; Jiang, Y.; Zhai, S.; Cheng, R.; Liu, D.; Gao, X.; Wang, J.; Wang, Z. A Breathable, Biodegradable, Antibacterial, and Self-Powered Electronic Skin Based on All-Nanofiber Triboelectric Nanogenerators. *Sci. Adv.* **2020**, *6*, eaba9624. **[Google Scholar] [CrossRef]**

18. Proto, A.; Penhaker, M.; Bibbo, D.; Vala, D.; Conforto, S.; Schmid, M. Measurements of Generated Energy/Electrical Quantities from Locomotion Activities Using Piezoelectric Wearable Sensors for Body Motion Energy

Harvesting. *Sensors* **2016**, *16*, 524. **[Google Scholar]** **[CrossRef]****[Green Version]**

19. Bishop, D.P. Fabrics: Sensory and Mechanical Properties. *Text. Prog.* **1996**, *26*, 1–62. **[Google Scholar]** **[CrossRef]**
20. Fuzek, J.F. Some Factors Affecting the Comfort Assessment of Knit T-Shirts. *Ind. Eng. Chem. Prod. Res. Dev.* **1981**, *20*, 254–259. **[Google Scholar]** **[CrossRef]**
21. Guan, X.; Xu, B.; Wu, M.; Jing, T.; Yang, Y.; Gao, Y. Breathable, Washable and Wearable Woven-Structured Triboelectric Nanogenerators Utilizing Electrospun Nanofibers for Biomechanical Energy Harvesting and Self-Powered Sensing. *Nano Energy* **2021**, *80*, 105549. **[Google Scholar]** **[CrossRef]**
22. Busolo, T.; Szewczyk, P.K.; Nair, M.; Stachewicz, U.; Kar-Narayan, S. Triboelectric Yarns with Electrospun Functional Polymer Coatings for Highly Durable and Washable Smart Textile Applications. *ACS Appl. Mater. Interfaces* **2021**, *13*, 16876–16886. **[Google Scholar]** **[CrossRef]** **[PubMed]**
23. Lou, M.; Abdalla, I.; Zhu, M.; Wei, X.; Yu, J.; Li, Z.; Ding, B. Highly Wearable, Breathable, and Washable Sensing Textile for Human Motion and Pulse Monitoring. *ACS Appl. Mater. Interfaces* **2020**, *12*, 19965–19973. **[Google Scholar]** **[CrossRef]** **[PubMed]**
24. Jiang, F.; Zhou, X.; Lv, J.; Chen, J.; Chen, J.; Kongcharoen, H.; Zhang, Y.; Lee, P.S. Stretchable, Breathable, and Stable Lead-Free Perovskite/Polymer Nanofiber Composite for Hybrid Triboelectric and Piezoelectric Energy Harvesting. *Adv. Mater.* **2022**, *34*, e2200042. **[Google Scholar]** **[CrossRef]** **[PubMed]**
25. Jeong, S.Y.; Shim, H.R.; Na, Y.; Kang, K.S.; Jeon, Y.; Choi, S.; Jeong, E.G.; Park, Y.C.; Cho, H.-E.; Lee, J. Foldable and Washable Textile-Based Oleds with a Multi-Functional near-Room-Temperature Encapsulation Layer for Smart E-Textiles. *npj Flex. Electron.* **2021**, *5*, 15. **[Google Scholar]** **[CrossRef]**
26. Qiu, Q.; Zhu, M.; Li, Z.; Qiu, K.; Liu, X.; Yu, J.; Ding, B. Highly Flexible, Breathable, Tailorable and Washable Power Generation Fabrics for Wearable Electronics. *Nano Energy* **2019**, *58*, 750–758. **[Google Scholar]** **[CrossRef]**
27. Wang, J.; He, J.; Ma, L.; Yao, Y.; Zhu, X.; Peng, L.; Liu, X.; Li, K.; Qu, M. A Humidity-Resistant, Stretchable and Wearable Textile-Based Triboelectric Nanogenerator for Mechanical Energy Harvesting and Multifunctional Self-Powered Haptic Sensing. *Chem. Eng. J.* **2021**, *423*, 130200. **[Google Scholar]** **[CrossRef]**
28. Wiskott, S.; Weber, M.O.; Heimlich, F.; Kyosev, Y. Effect of Pattern Elements of Weft Knitting on Haptic Preferences Regarding Winter Garments. *Text. Res. J.* **2017**, *88*, 1689–1709. **[Google Scholar]** **[CrossRef]**

29. Jiang, C.; Wu, C.; Li, X.; Yao, Y.; Lan, L.; Zhao, F.; Ye, Z.; Ying, Y.; Ping, J. All-Electrospun Flexible Triboelectric Nanogenerator Based on Metallic Mxene Nanosheets. *Nano Energy* **2019**, *59*, 268–276. **[Google Scholar] [CrossRef]**

30. Chen, S.F.; Hu, J.L.; Teng, J.G. A Finite-Volume Method for Contact Drape Simulation of Woven Fabrics and Garments. *Finite Elem. Anal. Des.* **2001**, *37*, 513–531. **[Google Scholar] [CrossRef]**

31. Naveed, T.; Zhong, Y.; Yu, Z.; Naeem, M.A.; Kai, L.; Xie, H.; Farooq, A.; Abro, Z.A. Influence of Woven Fabric Width and Human Body Types on the Fabric Efficiencies in the Apparel Manufacturing. *Autex Res. J.* **2020**, *20*, 484–496. **[Google Scholar] [CrossRef]**

32. Xu, F.; Dong, S.; Liu, G.; Pan, C.; Guo, Z.H.; Guo, W.; Li, L.; Liu, Y.; Zhang, C.; Pu, X.; et al. Scalable Fabrication of Stretchable and Washable Textile Triboelectric Nanogenerators as Constant Power Sources for Wearable Electronics. *Nano Energy* **2021**, *88*, 106247. **[Google Scholar] [CrossRef]**

33. Jost, K.; Stenger, D.; Perez, C.R.; McDonough, J.K.; Lian, K.; Gogotsi, Y.; Dion, G. Knitted and Screen Printed Carbon-Fiber Supercapacitors for Applications in Wearable Electronics. *Energy Environ. Sci.* **2013**, *6*, 2698. **[Google Scholar] [CrossRef]**

34. Kaldor, J.M.; James, D.L.; Marschner, S. Simulating Knitted Cloth at the Yarn Level. *ACM Trans. Graph.* **2008**, *27*, 1–9. **[Google Scholar] [CrossRef]**

35. Singh, E.; Singh, P.; Kim, K.S.; Yeom, G.Y.; Nalwa, H.S. Flexible Molybdenum Disulfide (MoS_2) Atomic Layers for Wearable Electronics and Optoelectronics. *ACS Appl. Mater. Interfaces* **2019**, *11*, 11061–11105. **[Google Scholar] [CrossRef]**

36. Dong, K.; Peng, X.; Wang, Z.L. Fiber/Fabric-Based Piezoelectric and Triboelectric Nanogenerators for Flexible/Stretchable and Wearable Electronics and Artificial Intelligence. *Adv. Mater.* **2020**, *32*, e1902549. **[Google Scholar] [CrossRef]**

37. Kim, W.G.; Kim, D.W.; Tcho, I.W.; Kim, J.K.; Kim, M.S.; Choi, Y.K. Triboelectric Nanogenerator: Structure, Mechanism, and Applications. *ACS Nano* **2021**, *15*, 258–287. **[Google Scholar] [CrossRef]**

38. Yang, B.; Xiong, Y.; Ma, K.; Liu, S.; Tao, X. Recent Advances in Wearable Textile-Based Triboelectric Generator Systems for Energy Harvesting from Human Motion. *EcoMat* **2020**, *2*, e12054. **[Google Scholar] [CrossRef]**

39. Zhao, Z.; Pu, X.; Du, C.; Li, L.; Jiang, C.; Hu, W.; Wang, Z.L. Freestanding Flag-Type Triboelectric Nanogenerator for Harvesting High-Altitude Wind Energy from Arbitrary Directions. *ACS Nano* **2016**, *10*, 1780–1787. **[Google Scholar] [CrossRef]**

40. Siddiqui, S.; Lee, H.B.; Kim, D.; Duy, L.; Hanif, A.; Lee, N. An Omnidirectionally Stretchable Piezoelectric Nanogenerator Based on Hybrid Nanofibers and Carbon Electrodes for Multimodal Straining and Human Kinematics Energy Harvesting. *Adv. Energy Mater.* **2018**, *8*, 1701520. **[Google Scholar] [CrossRef]**

41. Gong, J.; Xu, B.; Guan, X.; Chen, Y.; Li, S.; Feng, J. Towards Truly Wearable Energy Harvesters with Full Structural Integrity of Fiber Materials. *Nano Energy* **2019**, *58*, 365–374. **[Google Scholar] [CrossRef]**

42. Wu, C.; Kim, T.W.; Li, F.; Guo, T. Wearable Electricity Generators Fabricated Utilizing Transparent Electronic Textiles Based on Polyester/Ag Nanowires/ Graphene Core-Shell Nanocomposites. *ACS Nano* **2016**, *10*, 6449–6457. **[Google Scholar] [CrossRef] [PubMed]**

43. Zhang, Z.; Chen, Y.; Guo, J. Zno Nanorods Patterned-Textile Using a Novel Hydrothermal Method for Sandwich Structured-Piezoelectric Nanogenerator for Human Energy Harvesting. *Phys. E Low-Dimens. Syst. Nanostructures* **2019**, *105*, 212–218. **[Google Scholar] [CrossRef]**

44. Li, L.; Chen, Y.; Hsiao, Y.; Lai, Y. Mycena Chlorophos-Inspired Autoluminescent Triboelectric Fiber for Wearable Energy Harvesting, Self-Powered Sensing, and as Human–Device Interfaces. *Nano Energy* **2022**, *94*, 106944. **[Google Scholar] [CrossRef]**

45. Chen, W.; Fan, W.; Wang, Q.; Yu, X.; Luo, Y.; Wang, W.; Lei, R.; Li, Y. A Nano-Micro Structure Engendered Abrasion Resistant, Superhydrophobic, Wearable Triboelectric Yarn for Self-Powered Sensing. *Nano Energy* **2022**, *103*, 107769. **[Google Scholar] [CrossRef]**

46. Mao, Y.; Li, Y.; Xie, J.; Liu, H.; Guo, C.; Hu, W. Triboelectric Nanogenerator/ Supercapacitor in-One Self-Powered Textile Based on Ptfe Yarn Wrapped PDMS/MnO_2NW Hybrid Elastomer. *Nano Energy* **2021**, *84*, 105918. **[Google Scholar] [CrossRef]**

47. Liu, S.; Xuan, W.; Jin, H.; Zhang, L.; Xu, L.; Zhang, Z.; Dong, S.; Luo, J. Self-Powered Multi-Parameter Sensing System without Decoupling Algorithm Needed Based on Flexible Triboelectric Nanogenerator. *Nano Energy* **2022**, *104*, 107889. **[Google Scholar] [CrossRef]**

48. Bai, S.; Zhang, L.; Xu, Q.; Zheng, Y.; Qin, Y.; Wang, Z.L. Two Dimensional Woven Nanogenerator. *Nano Energy* **2013**, *2*, 749–753. **[Google Scholar] [CrossRef]**

49. Fu, Y.; Wu, H.; Ye, S.; Cai, X.; Yu, X.; Hou, S.; Kafafy, H.; Zou, D. Integrated Power Fiber for Energy Conversion and Storage. *Energy Environ. Sci.* **2013**, *6*, 805. **[Google Scholar] [CrossRef]**

50. Zhang, M.; Gao, T.; Wang, J.; Liao, J.; Qiu, Y.; Yang, Q.; Xue, H.; Shi, Z.; Zhao, Y.; Xiong, Z.; et al. A Hybrid Fibers Based Wearable Fabric Piezoelectric

Nanogenerator for Energy Harvesting Application. *Nano Energy* **2015**, *13*, 298–305. **[Google Scholar] [CrossRef]**

51. Lai, Y.; Deng, J.; Zhang, S.L.; Niu, S.; Guo, H.; Wang, Z.L. Single-Thread-Based Wearable and Highly Stretchable Triboelectric Nanogenerators and Their Applications in Cloth-Based Self-Powered Human-Interactive and Biomedical Sensing. *Adv. Funct. Mater.* **2017**, *27*, 1604462. **[Google Scholar] [CrossRef]**
52. Li, M.; Xu, B.; Li, Z.; Gao, Y.; Yang, Y.; Huang, X. Toward 3D Double-Electrode Textile Triboelectric Nanogenerators for Wearable Biomechanical Energy Harvesting and Sensing. *Chem. Eng. J.* **2022**, *450*, 137491. **[Google Scholar] [CrossRef]**
53. Dong, S.; Xu, F.; Sheng, Y.; Guo, Z.; Pu, X.; Liu, Y. Seamlessly Knitted Stretchable Comfortable Textile Triboelectric Nanogenerators for E-Textile Power Sources. *Nano Energy* **2020**, *78*, 105327. **[Google Scholar] [CrossRef]**
54. Wen, Z.; Yeh, M.-H.; Guo, H.; Wang, J.; Zi, Y.; Xu, W.; Deng, J.; Zhu, L.; Wang, X.; Hu, C.; et al. Self-Powered Textile for Wearable Electronics by Hybridizing Fiber-Shaped Nanogenerators, Solar Cells, and Supercapacitors. *Sci. Adv.* **2016**, *2*, e1600097. **[Google Scholar] [CrossRef][Green Version]**
55. Dong, K.; Deng, J.; Zi, Y.; Wang, Y.-C.; Xu, C.; Zou, H.; Ding, W.; Dai, Y.; Gu, B.; Sun, B.; et al. 3D Orthogonal Woven Triboelectric Nanogenerator for Effective Biomechanical Energy Harvesting and as Self-Powered Active Motion Sensors. *Adv. Mater.* **2017**, *29*, 1702648. **[Google Scholar] [CrossRef]**
56. Zhao, Z.; Yan, C.; Liu, Z.; Fu, X.; Peng, L.M.; Hu, Y.; Zheng, Z. Machine-Washable Textile Triboelectric Nanogenerators for Effective Human Respiratory Monitoring through Loom Weaving of Metallic Yarns. *Adv. Mater.* **2016**, *28*, 10267–10274. **[Google Scholar] [CrossRef]**
57. Yu, A.; Pu, X.; Wen, R.; Liu, M.; Zhou, T.; Zhang, K.; Zhang, Y.; Zhai, J.; Hu, W.; Wang, Z.L. Core-Shell-Yarn-Based Triboelectric Nanogenerator Textiles as Power Cloths. *ACS Nano* **2017**, *11*, 12764–12771. **[Google Scholar] [CrossRef]**
58. Chen, J.; Guo, H.; Pu, X.; Wang, X.; Xi, Y.; Hu, C. Traditional Weaving Craft for One-Piece Self-Charging Power Textile for Wearable Electronics. *Nano Energy* **2018**, *50*, 536–543. **[Google Scholar] [CrossRef]**
59. Zhang, H.; Yang, Y.; Su, Y.; Chen, J.; Hu, C.; Wu, Z.; Liu, Y.; Wong, C.P.; Bando, Y.; Wang, Z.L. Triboelectric Nanogenerator as Self-Powered Active Sensors for Detecting Liquid/Gaseous Water/Ethanol. *Nano Energy* **2013**, *2*, 693–701. **[Google Scholar] [CrossRef]**
60. Guo, H.; Chen, J.; Tian, L.; Leng, Q.; Xi, Y.; Hu, C. Airflow-Induced Triboelectric Nanogenerator as a Self-Powered Sensor for Detecting Humidity and Airflow Rate. *ACS Appl. Mater. Interfaces* **2014**, *6*, 17184–17189. **[Google Scholar] [CrossRef]**

61. Zhang, C.; Zhang, W.; Du, G.; Fu, Q.; Mo, J.; Nie, S. Superhydrophobic Cellulosic Triboelectric Materials for Distributed Energy Harvesting. *Chem. Eng. J.* **2023**, *452*, 139259. **[Google Scholar] [CrossRef]**

62. Gong, W.; Hou, C.; Zhou, J.; Guo, Y.; Zhang, W.; Li, Y.; Zhang, Q.; Wang, H. Continuous and Scalable Manufacture of Amphibious Energy Yarns and Textiles. *Nat. Commun.* **2019**, *10*, 868. **[Google Scholar] [CrossRef] [PubMed] [Green Version]**

63. Pu, X.; Li, L.; Liu, M.; Jiang, C.; Du, C.; Zhao, Z.; Hu, W.; Wang, Z.L. Wearable Self-Charging Power Textile Based on Flexible Yarn Supercapacitors and Fabric Nanogenerators. *Adv. Mater.* **2016**, *28*, 98–105. **[Google Scholar] [CrossRef] [PubMed]**

64. Zhou, T.; Zhang, C.; Han, C.B.; Fan, F.R.; Tang, W.; Wang, Z.L. Woven Structured Triboelectric Nanogenerator for Wearable Devices. *ACS Appl. Mater Interfaces* **2014**, *6*, 14695–14701. **[Google Scholar] [CrossRef] [PubMed] [Green Version]**

65. Tian, Z.; He, J.; Chen, X.; Zhang, Z.; Wen, T.; Zhai, C.; Han, J.; Mu, J.; Hou, X.; Chou, X.; et al. Performance-Boosted Triboelectric Textile for Harvesting Human Motion Energy. *Nano Energy* **2017**, *39*, 562–570. **[Google Scholar] [CrossRef]**

66. Guo, Y.; Zhang, X.; Wang, Y.; Gong, W.; Zhang, Q.; Wang, H.; Brugger, J. All-Fiber Hybrid Piezoelectric-Enhanced Triboelectric Nanogenerator for Wearable Gesture Monitoring. *Nano Energy* **2018**, *48*, 152–160. **[Google Scholar] [CrossRef]**

67. Pu, X.; Song, W.; Liu, M.; Sun, C.; Du, C.; Jiang, C.; Huang, X.; Zou, D.; Hu, W.; Wang, Z.L. Wearable power-textiles by integrating fabric triboelectric nanogenerators and fiber-shaped dye-sensitized solar cells. *Adv. Energy Mater.* **2016**, *6*, 1601048. **[Google Scholar] [CrossRef]**

68. Liu, L.; Xu, W.; Ding, Y.; Agarwal, S.; Greiner, A.; Duan, G. A Review of Smart Electrospun Fibers toward Textiles. *Compos. Commun.* **2020**, *22*, 100506. **[Google Scholar] [CrossRef]**

69. Guo, H.; Chen, Y.; Li, Y.; Zhou, W.; Xu, W.; Pang, L.; Fan, X.; Jiang, S. Electrospun Fibrous Materials and Their Applications for Electromagnetic Interference Shielding: A Review. *Compos. Part A Appl. Sci. Manuf.* **2021**, *143*, 106309. **[Google Scholar] [CrossRef]**

70. Lu, T.; Cui, J.; Qu, Q.; Wang, Y.; Zhang, J.; Xiong, R.; Ma, W.; Huang, C. Multistructured Electrospun Nanofibers for Air Filtration: A Review. *ACS Appl. Mater. Interfaces* **2021**, *13*, 23293–23313. **[Google Scholar] [CrossRef]**

71. Xie, Y.; Ma, Q.; Yue, B.; Chen, X.; Jin, Y.; Qi, H.; Hu, Y.; Yu, W.; Dong, X.; Jiang, H. Triboelectric Nanogenerator Based on Flexible Janus Nanofiber

Membrane with Simultaneous High Charge Generation and Charge Capturing Abilities. *Chem. Eng. J.* **2023**, *452*, 139393. **[Google Scholar]** **[CrossRef]**

72. Babu, A.; Aazem, I.; Walden, R.; Bairagi, S.; Mulvihill, D.M.; Pillai, S.C. Electrospun Nanofiber Based Tengs for Wearable Electronics and Self-Powered Sensing. *Chem. Eng. J.* **2023**, *452*, 139060. **[Google Scholar]** **[CrossRef]**
73. Mikučionienė, D.; Čiukas, R.; Mickevičienė, A. The Influence of Knitting Structure on Mechanical Properties of Weft Knitted Fabrics. *Mater. Sci.* **2010**, *16*, 221–225. **[Google Scholar]**
74. Sala de Medeiros, M.; Chanci, D.; Moreno, C.; Goswami, D.; Martinez, R.V. Waterproof, Breathable, and Antibacterial Self-Powered E-Textiles Based on Omniphobic Triboelectric Nanogenerators. *Adv. Funct. Mater.* **2019**, *29*, 1904350. **[Google Scholar]** **[CrossRef]**
75. Cong, Z.; Guo, W.; Guo, Z.; Chen, Y.; Liu, M.; Hou, T.; Pu, X.; Hu, W.; Wang, Z.L. Stretchable Coplanar Self-Charging Power Textile with Resist-Dyeing Triboelectric Nanogenerators and Microsupercapacitors. *ACS Nano* **2020**, *14*, 5590–5599. **[Google Scholar]** **[CrossRef]**
76. Rezaei, J.; Nikfarjam, A. Rib Stitch Knitted Extremely Stretchable and Washable Textile Triboelectric Nanogenerator. *Adv. Mater. Technol.* **2021**, *6*, 2000983. **[Google Scholar]** **[CrossRef]**
77. Dong, K.; Wang, Y.C.; Deng, J.; Dai, Y.; Zhang, S.L.; Zou, H.; Gu, B.; Sun, B.; Wang, Z.L. A Highly Stretchable and Washable All-Yarn-Based Self-Charging Knitting Power Textile Composed of Fiber Triboelectric Nanogenerators and Supercapacitors. *ACS Nano* **2017**, *11*, 9490–9499. **[Google Scholar]** **[CrossRef]**
78. Kwak, S.S.; Kim, H.; Seung, W.; Kim, J.; Hinchet, R.; Kim, S.W. Fully Stretchable Textile Triboelectric Nanogenerator with Knitted Fabric Structures. *ACS Nano* **2017**, *11*, 10733–10741. **[Google Scholar]** **[CrossRef]**
79. Liu, L.; Pan, J.; Chen, P.; Zhang, J.; Yu, X.; Ding, X.; Wang, B.; Sun, X.; Peng, H. A Triboelectric Textile Templated by a Three-Dimensionally Penetrated Fabric. *J. Mater. Chem. A* **2016**, *4*, 6077–6083. **[Google Scholar]** **[CrossRef]**
80. Xiong, J.; Cui, P.; Chen, X.; Wang, J.; Parida, K.; Lin, M.F.; Lee, P.S. Skin-Touch-Actuated Textile-Based Triboelectric Nanogenerator with Black Phosphorus for Durable Biomechanical Energy Harvesting. *Nat. Commun.* **2018**, *9*, 4280. **[Google Scholar]** **[CrossRef][Green Version]**
81. Shin, Y.; Lee, J.; Park, Y.; Hwang, S.; Chae, H.G.; Ko, H. Sewing Machine Stitching of Polyvinylidene Fluoride Fibers: Programmable Textile Patterns for Wearable Triboelectric Sensors. *J. Mater. Chem. A* **2018**, *6*, 22879–22888. **[Google Scholar]** **[CrossRef]**

82. Zhao, Z.; Huang, Q.; Yan, C.; Liu, Y.; Zeng, X.; Wei, X.; Hu, Y.; Zheng, Z. Machine-Washable and Breathable Pressure Sensors Based on Triboelectric Nanogenerators Enabled by Textile Technologies. *Nano Energy* **2020**, *70*, 104528. **[Google Scholar]** **[CrossRef]**
83. Lee, J.; Shin, Y.; Lee, G.; Kim, J.; Ko, H.; Chae, H.G. Polyvinylidene Fluoride (PVDF)/Cellulose Nanocrystal (CNC) Nanocomposite Fiber and Triboelectric Textile Sensors. *Compos. Part B Eng.* **2021**, *223*, 109098. **[Google Scholar]** **[CrossRef]**
84. Zhong, J.; Zhang, Y.; Zhong, Q.; Hu, Q.; Hu, B.; Wang, Z.L.; Zhou, J. Fiber-based generator for wearable electronics and mobile medication. *ACS Nano* **2014**, *8*, 6273–6280. **[Google Scholar]** **[CrossRef]** **[PubMed]**
85. Cao, Y.; Shao, H.; Wang, H.; Li, X.; Zhu, M.; Fang, J.; Cheng, T.; Lin, T. A Full-Textile Triboelectric Nanogenerator with Multisource Energy Harvesting Capability. *Energy Convers. Manag.* **2022**, *267*, 115910. **[Google Scholar]** **[CrossRef]**
86. Kim, T.; Jeon, S.; Lone, S.; Doh, S.J.; Shin, D.; Kim, H.K.; Hwang, Y.; Hong, S.W. Versatile Nanodot-Patterned Gore-Tex Fabric for Multiple Energy Harvesting in Wearable and Aerodynamic Nanogenerators. *Nano Energy* **2018**, *54*, 209–217. **[Google Scholar]** **[CrossRef]**
87. Choi, A.Y.; Lee, C.J.; Park, J.; Kim, D.; Kim, Y.T. Corrugated Textile Based Triboelectric Generator for Wearable Energy Harvesting. *Sci. Rep.* **2017**, *7*, 45583. **[Google Scholar]** **[CrossRef][Green Version]**
88. Chen, J.; Huang, Y.; Zhang, N.; Zou, H.; Liu, R.; Tao, C.; Fan, X.; Wang, Z.L. Micro-Cable Structured Textile for Simultaneously Harvesting Solar and Mechanical Energy. *Nat. Energy* **2016**, *1*, 16138. **[Google Scholar]** **[CrossRef]**
89. Cui, N.; Liu, J.; Gu, L.; Bai, S.; Chen, X.; Qin, Y. Wearable triboelectric generator for powering the portable electronic devices. *ACS Appl. Mater. Interfaces* **2015**, *7*, 18225–18230. **[Google Scholar]** **[CrossRef]**
90. Jung, S.; Lee, J.; Hyeon, T.; Lee, M.; Kim, D.H. Fabric-based integrated energy devices for wearable activity monitors. *Adv. Mater.* **2014**, *26*, 6329–6334. **[Google Scholar]** **[CrossRef]**
91. Seung, W.; Gupta, M.K.; Lee, K.Y.; Shin, K.S.; Lee, J.H.; Kim, T.Y.; Kim, S.; Lin, J.; Kim, J.H.; Kim, S.W. Nanopatterned textile-based wearable triboelectric nanogenerator. *ACS Nano* **2015**, *9*, 3501–3509. **[Google Scholar]** **[CrossRef]**
92. Gong, W.; Hou, C.; Guo, Y.; Zhou, J.; Mu, J.; Li, Y.; Zhang, Q.; Wang, H. A wearable, fibroid, self-powered active kinematic sensor based on stretchable sheath-core structural triboelectric fibers. *Nano Energy* **2017**, *39*, 673–683. **[Google Scholar]** **[CrossRef]**

93. Liu, M.; Pu, X.; Jiang, C.; Liu, T.; Huang, X.; Chen, L.; Du, C.; Sun, J.; Hu, W.; Wang, Z.L. Large-area all-textile pressure sensors for monitoring human motion and physiological signals. *Adv. Mater.* **2017**, *29*, 1703700. **[Google Scholar]** **[CrossRef]**

94. Lee, H.; Roh, J.S. Wearable electromagnetic energy-harvesting textiles based on human walking. *Text. Res. J.* **2018**, *89*, 2532–2541. **[Google Scholar]** **[CrossRef]**

95. He, T.; Shi, Q.; Wang, H.; Wen, F.; Chen, T.; Ouyang, J.; Lee, C. Beyond energy harvesting multi-functional triboelectric nanosensors on a textile. *Nano Energy* **2019**, *57*, 338–352. **[Google Scholar]** **[CrossRef]**

96. Liu, M.; Cong, Z.; Pu, X.; Guo, W.; Liu, T.; Li, M.; Zhang, Y.; Hu, W.; Wang, Z.L. High-Energy Asymmetric Supercapacitor Yarns for Self-Charging Power Textiles. *Adv. Funct. Mater.* **2019**, *29*, 1806298. **[Google Scholar]** **[CrossRef]**

97. Yang, Y.; Sun, N.; Wen, Z.; Cheng, P.; Zheng, H.; Shao, H.; Xia, Y.; Chen, C.; Lan, H.; Xie, X.; et al. Liquid-Metal-Based Super-Stretchable and Structure-Designable Triboelectric Nanogenerator for Wearable Electronics. *ACS Nano* **2018**, *12*, 2027–2034. **[Google Scholar]** **[CrossRef]**

98. Xiong, J.; Lin, M.; Wang, J.; Gaw, S.L.; Parida, K.; Lee, P.S. Wearable All-Fabric-Based Triboelectric Generator for Water Energy Harvesting. *Adv. Energy Mater.* **2017**, 7, 1701243. **[Google Scholar]** **[CrossRef]**

99. Chen, C.; Guo, H.; Chen, L.; Wang, Y.C.; Pu, X.; Yu, W.; Wang, F.; Du, Z.; Wang, Z.L. Direct Current Fabric Triboelectric Nanogenerator for Biomotion Energy Harvesting. *ACS Nano* **2020**, *14*, 4585–4594. **[Google Scholar]** **[CrossRef]**

100. Zhao, J.; You, Z. A Shoe-Embedded Piezoelectric Energy Harvester for Wearable Sensors. *Sensors* **2014**, *14*, 12497–12510. **[Google Scholar]** **[CrossRef]**

CHAPTER-2

INTRODUCTION TO PERSONAL COOLING GARMENTS

1. INTRODUCTION: AN OVERVIEW

In the relentless pursuit of comfort and adaptability in our daily lives, personal cooling garments have emerged as a revolutionary solution, blending fashion with functionality. As global temperatures continue to rise and traditional cooling methods face challenges, these innovative garments offer a promising avenue to enhance personal well-being. Designed with advanced materials and cutting-edge technologies, these garments go beyond mere aesthetics, incorporating cooling elements that provide a refreshing escape from the heat. This convergence of fashion and thermoregulation not only addresses the immediate need for temperature control but also aligns with the broader goals of sustainable and personalized approaches to comfort.

Thermal comfort is vital to human beings, not only because we are homothermic and need a consistent body temperature to survive, but also because thermal discomfort can lead to psychological problems, such as heat stress, which can result in tedium and exhaustion in mentality and physicality, causing personal health issues and the curtailment of work productivity [**1**,**2**]. According to a statement from the American Society of Heating, Refrigeration, and Air-Conditioning Engineers (ASHRAE), thermal comfort is a mind condition that conveys satisfaction with the ambient environment temperature [**3**]. Generally, the narrow range of comfortable temperature for human beings is between 20 °C to 27 °C [**4**]; outside of this range, we would need additional means, such as garments or air-conditioning, to maintain thermal comfort.

Garments are a paramount part of personal thermoregulation, which can directly influence the heat-exchanging process between the human body and the environment [**5**], although the human body has a certain ability to automatically regulate its temperature through various activities, such as altering the rates of metabolism and blood flow, sweating, pore-shrinking, etc. [**4**] When the human body is in hot weather, the sweating and blood flow rates can be increased to enhance thermal diffusion. However, during extreme weather or harsh conditions, which exceed the ability to self-regulate, it is necessary to utilize functional garments to achieve thermal comfort.

The personal thermal garment is an auspicious technology that can avoid thermally-induced health issues and provide a sustainable solution to reduce energy consumption [**6**]. With the global warming effect, there has been a rising demand for personal thermal comfort in high temperature weather. Energy consumption has been increasing continuously and dramatically, and is expected to triple by 2050 [**7**]. Therefore, it is urgent to meet the demand for the development of advanced thermoregulation garments. In consideration of these reasons, over the past few decades, a considerable amount of research of advanced thermal devices has been conducted to design and fabricate intelligent thermoregulation garments. At present, personal heating garments have enjoyed rapid development, from heating materials to heating techniques, strategies, and heating effect evaluation, and they have captured a dominant share of the thermal regulating textile market. There are various heating garments using different heating materials, such as carbon nanotube (CNT) [**8,9**], carbon fiber (CF) [**10,11**], graphene [**12,13**], metallic nanowire meshes [**14,15**], etc. Thermal regulating garments with cooling functions are also strongly demanded in numerous application conditions, including exercising or working in hot weather, personal portable cooling equipment in the medical area, or in special protective garments fields, such as astronaut garments, medical protective garments, and firefighter garments. However, compared with the large-scale commercial production of personal heating garments, the progress of personal cooling garments (PCGs) has been lagging behind.

The concept of personal cooling garments was first put forward in 1958 [**16**]. However, it was not until 1962 that the first personal refrigeration garment was produced for the aerospace industry. Since then, many different kinds of cooling garments, such as air-cooled, liquid-cooled, radiative-cooled, and thermoelectric (TE)-cooled garments, have continuously been fabricated and tested. Due to the technical difficulty in achieving a highly efficient cooling effect on garments, the PCG market has developed at a much slower pace.

In recent years, with the rapid evolution of smart and functional textiles, as well as the global demand of energy-saving solutions, more efforts have been devoted to the development of PCGs, with many new materials, cooling techniques, and device structure optimizations being examined. Therefore, a systematic review of recent advances in personal cooling garments is in timely demand, which needs to cover the classification and the forefront development of personal cooling materials and strategies in detail. In this review, we have focused on the state-of-the-art developments of personal cooling garments. We divide cooling garments into two categories based on whether electric power is required for the cooling functional: non-electric cooling and electric cooling. As shown in **Figure 1**, non-electric cooling contains ice cooling, phase change materials (PCMs) cooling, and radiative cooling, while electric cooling techniques include TE cooling, liquid cooling, and air-cooling. Furthermore, we summarized the major challenges of the development of PCGs and pointed out the future perspectives, trying to facilitate the progress of personal cooling garments.

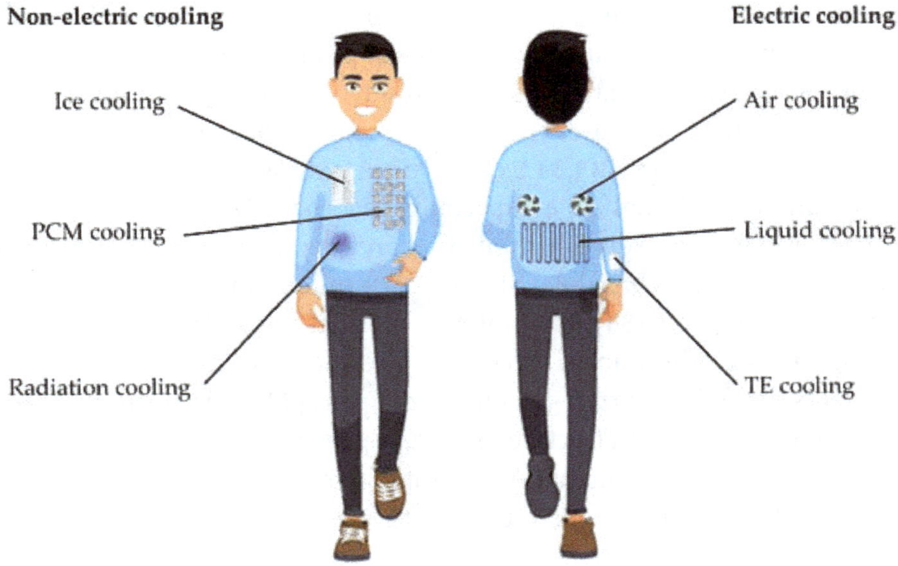

Figure 1. Cooling strategies of personal cooling garments.

2. NON-ELECTRIC COOLING

A stable core body temperature is essential for maintaining optimal functions of the human body. To reduce the health risks caused by hazardous heat environments, microclimate cooling technologies have been developed to enhance the heat exchange between the human body and the environment. Non-electric cooling strategies, which do not require a power supply system [**17,18,19**], can be divided into three main types: ice cooling [**20**], PCMs cooling [**21**], and radiation cooling [**22**].

2.1. Ice Cooling

In the early development of PCGs, an ice cooling garments [**23**] was one of the most common types due to the simplicity of preparation and low cost. It was designed with specific pockets to hold the ice, as shown in **Figure 2**. Through the ice melting process, the heat generated from the skin surface is absorbed, resulting in a temperature drop and thermal comfort in hot weather. Based on the high latent heat, availability, and low cost, the ice cooling garment was widely applied in protective garment industries [**24,25**], athletic wear [**26,27,28**], and military uniforms [**29,30**]. Juhani et al. confirmed that an ice vest effectively cooled the skin temperature, which was beneficial to both physiology and subjectivity [**25**]. Furthermore, the thermal comfort provided by the ice cooling vest remarkably increased work efficiency by 10%. The same conclusion was drawn by Cooter et al. [**31**]. From their study, the effectiveness of the ice vest in improving endurance performance during sustained heavy work was closely examined. However, contact with ice for a long time can lead to tissue irritation [**32**]. Additionally, the heavy weight, large volume, and cooling interruption have strongly restricted the application of ice cooling garments in daily life. Therefore, the investigation of flexible cooling sources represents a potential approach to solve these problems. Gel ice cooling technology [**33,34,35**] has been extensively explored lately as a promising candidate. Dehghan et al. compared the effectiveness of an ice gel cooling vest under two exercise intensities and demonstrated a noticeable skin temperature drop, and lower heat strain score index during light activities as a result of the cooling vest [**33**]. It is noteworthy that under heavy activities, there were no significant differences. Moreover, Chesterton et al. compared the skin-cooling effects of flexible frozen gel and frozen peas [**34**]. The result indicates that flexible frozen gel is not as effective as frozen peas in skin freezing and calming. Therefore, ice cooling garments with satisfactory cooling performance and flexibility is still a challenge, which requires further studies on advanced materials and technologies.

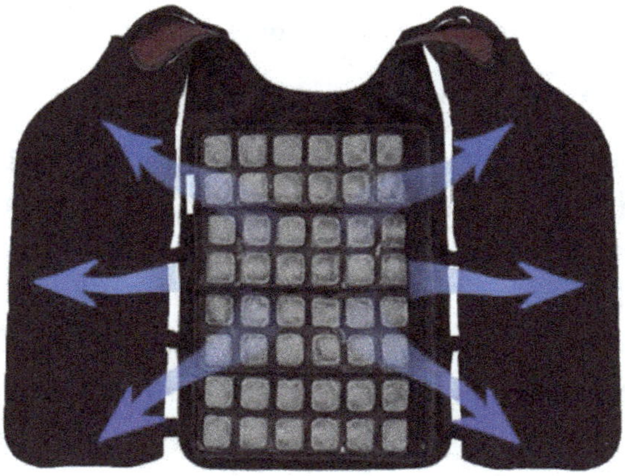

Figure 2. Illustration of an ice cooling vest.

2.2. PCMs Cooling

PCMs can directly use the latent heat from the body or the environment to lower the temperature of the microclimate between the garment and the body, without any extra energy consumption [36]. With the exploration of materials, the inherent defects of the above-mentioned frozen materials (as a typical phase change material) have been revealed gradually, such as an uncontrollable cooling effect and limited cooling duration. Given these circumstances, the demand for seeking more comfort and effective PCM for wearable cooling application has been ceaselessly rising [37,38,39]. Li et al. developed a novel PCM cooling garment with a designed placement of PCM through multi-scenario experiments, which is light and convenient for daily use [40]. With the calculation method for the minimum charge of PCM, the final cooling garment is only 1.39 kg with obvious cooling performance, which exhibited a maximum 1.78 °C temperature drop during indoor walking. Besides packaging a PCM module in the garment, it is feasible for PCM to compound with yarns in various ways, such as impregnation and filling with hollow fibers [41], melt spinning [42], coating [43], and microcapsule encapsulation [44]. In the last decade, using these properties of PCMs to fabricate adaptive textiles or fibers has been conspicuous and noticeable [45,46,47]. V. Skurkyte-Papieviene et al. explored a type of PCM microcapsules MPCM32D, adding multiwall carbon nanotubes (MWCNTs) and poly (3,4-ethylenedioxyoxythiophene) poly (styrene sulphonate) (PEDOT: PSS) as thermally conductive additives to improve the outer shell thermal conductivity, which relatively enhanced the heat storage and release capability of PCMs microcapsules [44]. Yang et al. developed a 3D-printed flexible phase-change nonwoven fabric with excellent stability and durability [46]. Owing to the poly (ethylene glycol) (PEG)-grafted TPU prepolymer and embedded single-walled carbon nanotubes, the nonwoven fabric exhibited adequate thermal regulation and radiation resistance performance, even at cycles up to 2000. Moreover, a dramatic breakthrough in the commercial application of wearable PCM has been made by Outlast® technologies [48]. The Outlast® fiber exhibits the desired thermo-regulating performance through incorporating PCM by microencapsulation, remarkably reducing sweat formation by 48% in a hot temperature environment. In summary, the extraordinary breathability and flexibility ensure the PCM textile has a large-scale implementation in PCM, which is the critical development direction of PCM cooling garments [47].

2.3. Radiation Cooling

Based on the highly transparent atmospheric window in the infrared (IR) wavelength range between 8 to13 μm, which coincides with the thermal radiation of humans, radiation cooling (RC) can dump human heat directly into outer space without external energy [**49**]. As a promising alternative technology to locally controlling skin temperature, RC has aroused wide interest in personal cooling garments by taking advantage of an atmospheric window [**50,51,52**]. Hsu et al. developed an RC textile with good transparency to mid-infrared human body radiation based on nanoporous polyethylene (nanoPE) substrate, which can be used as an infrared radiation heat dissipation fabric for an individual cooling strategy [**51**]. Due to nanoPE having high transparency and transmittance, it can be appropriately used as a heat dissipation material. Their textile promoted effective radiative cooling and lower temperature around 2.7 °C, which provides a promising textile for personal cooling garments. After development of their textile, they utilized nanoPE embedded with a bilayer emitter to fabricate a dual-mode textile, which has both cooling and heating modes just by flipping the textile, as shown in **Figure 3** [**53**]. The bilayer emitter with the carbon side (high emissivity) and the copper side (low emissivity) noticeably provided a 6.5 °C temperature difference in artificial skin by flipping the same textile, showing a significant technological advancement in wearable RC application.

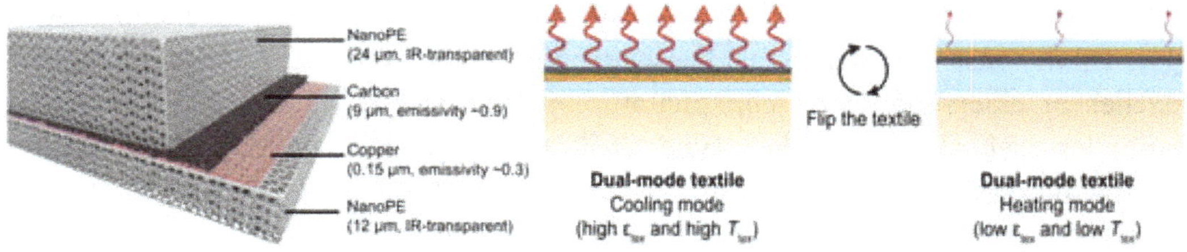

Figure 3. Layered structure of the textile and schematic diagram of the two modes, reproduced from [53].

However, when there are other intense thermal radiation sources, such as direct sunlight in hot weather, it is critical to investigate RC textiles with high reflection to the solar radiation band [**22,54,55**]. Because the solar spectrum is mainly between 0.3 to 4.0 μm [**56**], it is greatly prospective and feasible to both enhance solar reflection and dissipate human thermal radiation. Cai et al. proposed the first spectrally-selective RC textile for outdoor cooling with more than 90% reflection of solar irradiance and satisfactory transmission of the human body thermal radiation, which enabled simulated skin to avoid overheating by 5–13 °C compared with cotton fabric [**57**]. Irfan et al. developed a nanofabric by nanoparticle-doped polymer (zinc oxide and polyethylene) materials and electrospinning technology, which offered 91% solar reflectivity and 81% mid-infrared transmissivity, resulting in a 9 °C cooling performance compared with cotton textile [**58**]. Through numerous studies, it has been solidly confirmed that RC garments have a great potential for mass adoption of effective and energy-saving cooling technology in daily life.

3. ELECTRIC COOLING

With the vigorous developments of wearable electronic devices, the utilization of electronic devices with cooling materials and strategies is the main development direction of intelligent cooling garments. Under these circumstances, the personal cooling garment under electric power supplied has enjoyed a sustained evolution. At present, the main types of this area are air cooling garments (ACGs) [**59**], liquid cooling garments (LCGs) [**60**], and TE cooling garments [**61**].

3.1. Air Cooling

As a traditional cooling strategy, an ACG provides thermal comfort by forcing air to flow through the microclimate between the clothing and the human body [**62**]. The major advantages of ACGs are low cost, light weight, and portability [**63**]. At the early stage of development, ACGs were applied in aerospace and military fields, such as protective garments for pilots and soldiers to reduce heat stress [**64,65,66**]. Hadid et al. reported that during the same intensity of activity, an ACG led to a lower body temperature and reduction of perspiration by 20% [**67**]. However, more complex living and working conditions require higher cooling performance with better comfort and security of the ACG. Additionally, ACG configurations are normally heavy and large, limiting their widespread applications in our daily life. Therefore, optimization of wearing comfort and thermal comfort is in great demand [**68,69**]. Yang et al. investigated the influence of clothing size and the air ventilation rate on the cooling performance of ACGs, which demonstrated that air ventilation greatly reduced the predicted core temperatures in two garment sizes; however, there was almost no impact of garment size on the predicted thermo-physiological responses in high ventilation [**70**]. Similarly, Zhao et al. enhanced the cooling performance through clothing eyelet designs, which provided an alternative method to optimize ACGs in practical use [**71**].

The cooling effect provided by a single air-cooling strategy can be further improved to satisfy the ever-growing cooling needs. Consequently, a lot of efforts have been made in studying the possibility of combining an air-cooling strategy with other cooling methods to explore novel cooling approaches. Ni et al. developed a novel hybrid personal cooling vest (PCV), as shown in **Figure 4**a [**72**]. Their novel PCV was incorporated with PCMs and ventilation fans, which indicates the applicability and reliability of this hybrid cooling garment. Through experimental studies, the cooling efficacy in a hot, humid climate chamber was examined. Lou et al. investigated the relationship between the cooling effect with different body positions based on an air tubing network and TE cooling plates, which is helpful to improve the combination of the cooling system and the garment in an effective and comfortable way in daily life [**73**]. Based on large numbers of studies, the ACG is verifiably suitable to meet the practical wearing demand based on the notable portability and simplicity of operation and usage.

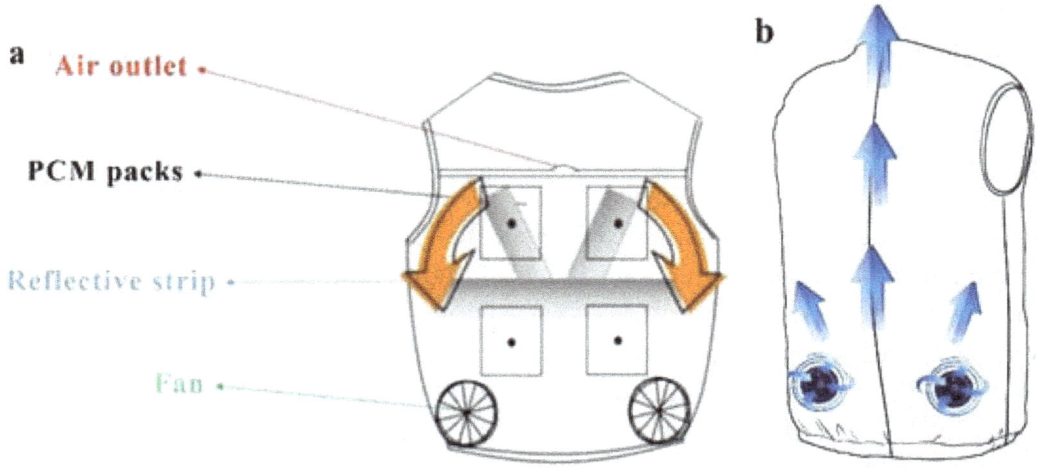

Figure 4. (a) Illustration of air cooling with a PCM cooling vest (PCV), adapted from [72], (b) Schematic diagrams of an air cooling garment.

3.2. Liquid Cooling

Liquid cooling garments are generally embedded with circulating water tubes filled with a cold liquid resource and a micro water pump device at the inner layer to drive the liquid flowing in the tube to reduce the temperature [65,74], as shown in Figure 5a. The first LCG was supplied to an astronaut as a protective garment to lower the body temperature [75]. After numerous explorations, the LCG has been proven to be one of the most promising technologies in the wearable cooling arena and is used in many fields, such as military [76], mining [77], and sporting [78]. Guo et al. proposed a heat transfer model of LCGs to analyze the effects of different factors on the LCG performance and optimize the design of LCGs [74]. With the optimization they made, the max cooling rate reached 243.2 W/m2 with a maximum work duration time of 3.36 h. However, due to the embedded heavy device and the risk of skin burn caused by the stream formation [79], the LCG needs more efforts to modify the cooling systems for safety and comfort. Grazyna et al. developed a novel LCG with a sensor to adjust the microclimate temperature, and modular knitted fabric, which can be directly worn on the human body [80], resulting in great safety and comfort for the user. The notable tube system they proposed fitted to the skin nicely, which enhanced the thermal conductivity and cooling efficiency. Shu et al. proposed an intelligent temperature control system in LCGs [81], which verified that the new smart system can regulate temperature accurately and extend the duration of more than 30% of the cooling devices.

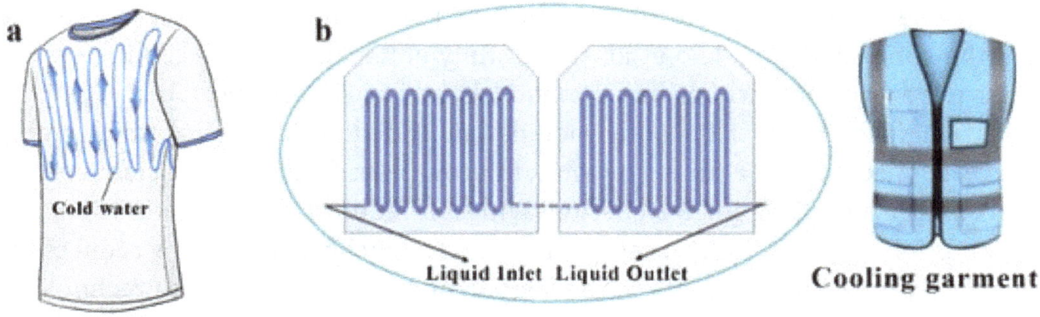

Figure 5. (a) Schematic diagrams of a liquid cooling garment. (b) Liquid cooling garment with a TE device, reproduced or adapted from [82], with permission from Elsevier, 2022.

To achieve significant amelioration of the LCG, combining other cooling technologies with LCGs exhibits remarkable potential in practical usage [**55**]. Zhang et al. reported a novel LCG with TE materials, which considerably alleviates the thermal stress hazard, as shown in **Figure** 5b [**82**]. They explored the effect of ambient temperature and heat dissipation on cooling performance, which also proposed an accurate method of assessing LCGs.

3.3. TE Cooling

For individual wearable cooling devices, air cooling and liquid cooling garments are inconvenient and heavy with bulky air or fluidic channels, and the cooling performance of them is not stable and reliable [**83**]. Given these intrinsic disadvantages, homeostatic solid-state cooling strategies, such as electrocaloric cooling [**84,85**], magnetocaloric [**86**], and TE cooling [**87**], have garnered significant attention. Particularly, TE cooling exhibits great potential for practical industrial cooling application owing to its reliable cooling performance and small device dimensions [**88**]. Therefore, wearable TE cooling provides a desired alternative method for personal thermal comfort [**89**].

TE materials consisting of different types (N- and P-type) of conductors or semiconductors can be used to directly convert heat energy to electricity and vice versa [**90**]. Based on the Peltier effect [**91,92**], when a direct current passes through different TE materials, heat is absorbed or dissipated at the junctions, resulting in a hot side and a cold side [**93**]. TE devices are lightweight and have no moving parts or noise, which can be considered as the cooling strategy with the most potential in the future [**94**].

For small solid-state cooling, the TE cooling plate (TECP) is a good solution for industrial TE application due to their large-scale commercial production, low cost, and light weight [95]. Therefore, cooling devices based on TECP were conceived to explore practical applications of TEC, such as the Embr wave bracelet and Sony

Reno pocket that can be worn on the arm or put in the pocket of a garment to cool the connected skin. Luo et al. successfully embedded a TECP cooling system in an undergarment, which has dramatic light weight and portability [96]. As shown in Figure 6, the TE cooling module embedded with heatsink as a cooling source, connected with a tubing network to provide uniform and sufficient cooling performance, resulting in 15% energy savings of indoor heating, ventilation, and air conditioning (HAVC). However, the rigid and bulky heatsink greatly reduces the wearability of TE and obstructs the development of flexible TE cooling technology. Therefore, a flexible heatsink has been predicated as a hopeful approach to ameliorate the flexibility of TE cooling. Jaeyoo et al. proposed a flexible cooling device with a designed heatsink that was composed of silicone elastomer, phase change material, and graphite powders. Their device could lower temperatures by around 5 °C and maintained temperatures for more than 5 h under ambient air temperature with commercial TECPs. Recently, an innovative mask integrated with thermoelectric devices and a 3D printed framework has been reported [97]. Through the test, a notable reduction in the temperature was around 3.5 °C with a low voltage application.

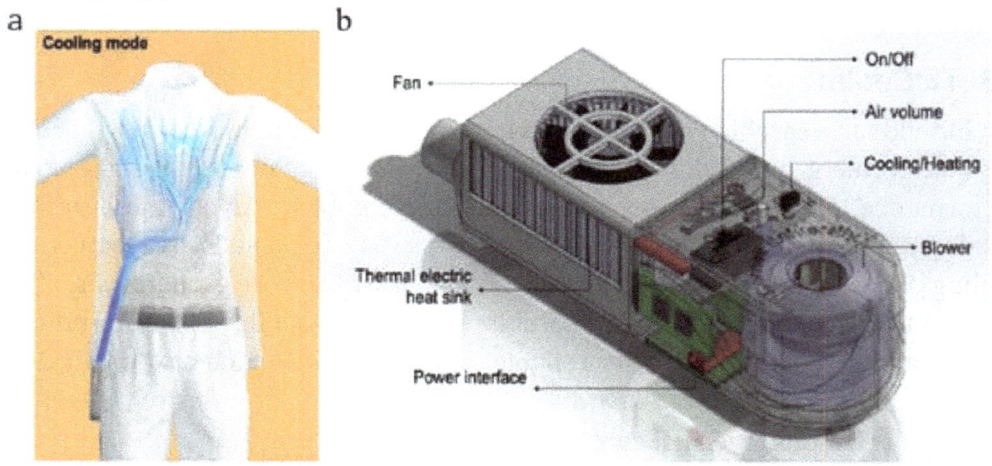

Figure 6. Schematic image of (a) a TE cooling undergarment and (b) illustration of a TE cooling module, reproduced or adapted from [96], with permission from Elsevier, 2020.

Nevertheless, more efforts need to be made to seek efficient and wearable thermoelectric devices with prospective flexibility. Therefore, fabricating and softening the TEC structure with a flexible heatsink is clamant to be explored. Hong et al. developed a flexible and portable thermoelectric cooler that has a scalable application, as seen in **Figure 7 [98]**. Inorganic semiconductor thermoelectric pillars were placed on two stretchable flexible Ecoflex films, which were filled with thermally conductive filler aluminum nitride to improve thermal conductivity.

The device could work continuously for 8 h without any heatsink equipment and delivers more than 7.6 °C cooling effects. Zhang et al. proposed a wearable TEC based on a two-layer flexible heatsink [**99**], which was composed of hydrogel and nickel foam as phase change material to absorb the heat, and thermally conductive material to conduct heat, respectively. Furthermore, the discrete heatsink they explored ensured the desired flexibility and achieved a large temperature drop of 10 °C under 0.3 A input current.

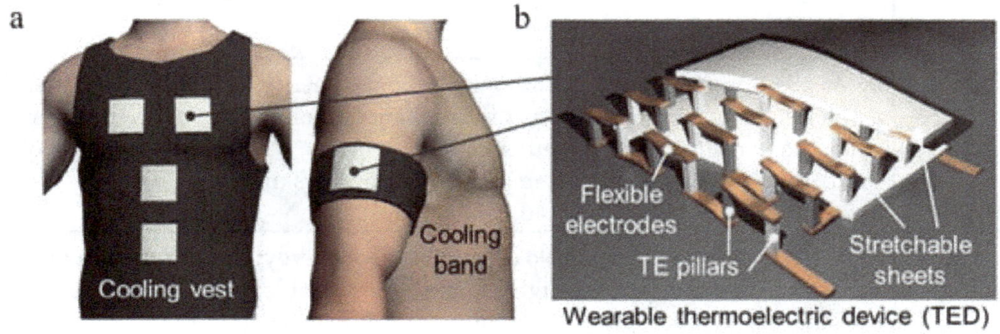

Figure 7. Schematic images of (a) cooling garments with wearable TE devices (TEDs) and (b) the structure of a TED, [98].

To further ameliorate flexibility and promote the flexible TEC evolution, TE fiber has been widely investigated to provide better wearing comfort. Zhang et al. prepared super long flexible TE micro/nanowires by hot drawing technology [**100**]. They fabricated both N- and P-type semiconducting materials into a fiber and covered them with borosilicate glass to protect and prevent the fiber from being oxidized. The fiber was flexible and had a high TE property, and could be easily woven into the textile and create 6.2 °C cooling performance. Zheng et al. reported a novel design of fiber-based thermoelectric textiles (TETs) [**101**]. Their TE fiber was fabricated by inorganic TE materials and liquid metal, which were encapsulated by polydimethylsiloxane. The TET provided a stable cooling performance of 3.1 °C with notable stretchability and flexibility.

4. CONCLUSIONS

This work has summarized the latest development of PCGs and reviewed the existing research on cooling strategies and materials. Depending on the requirement of electric power, PCGs can be divided into non-electric and electric cooling garments. Each type of PCGs has its intrinsic advantages and drawbacks. As for non-electric cooling garments, the primary advantage is that they are energy-saving technologies, but the cooling performance is generally lower than electric cooling. However, with a power source, the portability tends to decrease, and there is a more complex requirement on system integration. The detailed comparison of different PCGs is shown in **Table 1**.

Table 1. The summary of different cooling strategies.

Cooling Strategy	Mechanism	Advantages	Disadvantage(s)
Ice cooling	Ice absorbs the body heat during melting to cool it down.	High cooling performance; no consumption of energy; reach -20 °C low temperature [19]	Limited cooling duration; heavy and no flexibility; penetration of condensate; risk of tissue damage
PCMs cooling	PCMs absorb heat through a phase change process when the temperature rises to a certain range.	No consumption of energy; easy maintenance; cooling efficiency: 1.78–3 °C temperature drop of skin [40,102]	High cost of materials and manufacturing; limited cooling duration
Radiative cooling	High solar reflection and infrared emissivity to prevent heat absorption and enhance self-heat release	No consumption of energy; light and breathable; cooling efficiency: temperatures 2–13 °C lower than normal textile [51,57]	High cost of materials and manufacturing
Air cooling	Blowing air into a garment to enhance sweat evaporation	Long-term cooling; easy manufacturing; cooling efficiency: 0.3–1 °C temperature drop of skin [59,103]	The impermeable fabric aggravates thermal discomfort
Liquid cooling	Water tubes in the garment circulate cooled water and lower the temperature	Long-term cooling; easy manufacturing; cooling efficiency: 1.2–2.5 °C temperature drop of skin [104]	Large and heavy devices; poor comfort; risk of security
TE cooling	When direct current passes through circuits composed of different semiconductors, heat is absorbed or dissipated at the junctions, resulting in a hot side and a cold side	Stable, reliable, and regulated cooling performance; small and light; cooling efficiency: 38 °C temperature drop of skin [98,101]	High cost of materials and manufacturing; no commercial flexible devices

The use of PCGs can significantly improve human comfort and health during various warm conditions. It also contributes to global efforts in saving energy and reducing environmental pollution; however, with the increasing complexity of environmental change and people's demands, more and more factors need to be considered when designing and evaluating cooling garments. These factors need to be comprehensively explored from the three dimensions of environment, clothing, and the human body. In terms of environment, different use scenarios should be considered, such as changing ambient temperature and humidity. From the clothing itself, the selection of fabrics, the weight, size, location, and structure of the cooling device, the combination of fabrics, and the overall design of clothing need to be studied more. For the human, the thermal balance ability of the human body itself, the different cold and heat perception abilities of different parts of the human body, and even different postures will affect the cooling effect of the refrigerated clothing. Therefore, numerous variables affect the design and performance of personal cooling garments.

Nevertheless, there are still restrictions existing in the developments so far. Thus, firstly, it is paramount to improve the cooling performance of the materials and explore the new advanced strategies, which are supposed to be more light, low-cost, environment-friendly, and easy to prepare. Furthermore, optimizing and modifying the personal thermal comfort evaluation system is required to consider more parameters to assess the cooling efficiency of different PCGs. Simultaneously, broadening the application fields is another crucial element requiring further development, so that these garments can not only be used in special protective garment fields, such as aerospace, but also in personal daily life, including refrigeration sportswear, personal portable medical cooling devices, etc. Apart from this, some of the PCGs only exist in laboratories, and large-scale commercial production should be developed in the near future. Finally, studying and analyzing the novel structures of PCGs with the consideration of enhanced wearability may be strongly needed for future development of personal cooling garments.

5. CONCISE SUMMARY

Thermal comfort is of critical importance to people during hot weather or harsh working conditions to reduce heat stress. Therefore, personal cooling garments (PCGs) is a promising technology that provides a sustainable solution to provide direct thermal regulation on the human body, while at the same time, effectively reduces energy consumption on whole-building cooling. This paper summarizes the current status of PCGs, and depending on the requirement of electric power supply, we divide the PCGs into two categories with systematic instruction on the cooling materials, working principles, and state-of-the-art research progress. Additionally, the application fields of different cooling strategies are presented. Current problems hindering the improvement of PCGs, and further development recommendations are highlighted, in the hope of fostering and widening the prospect of PCGs.

REFERENCES

1. Sajjad, U.; Hamid, K.; Tauseef ur, R.; Sultan, M.; Abbas, N.; Ali, H.M.; Imran, M.; Muneeshwaran, M.; Chang, J.-Y.; Wang, C.-C. Personal thermal management—A review on strategies, progress, and prospects. *Int. Commun. Heat Mass Transf.* **2022**, *130*, 105739. **[Google Scholar] [CrossRef]**
2. Ebi, K.L.; Capon, A.; Berry, P.; Broderick, C.; de Dear, R.; Havenith, G.; Honda, Y.; Kovats, R.S.; Ma, W.; Malik, A.; et al. Hot weather and heat extremes: Health risks. *Lancet* **2021**, *398*, 698–708. **[Google Scholar] [CrossRef] [PubMed]**
3. Rozmi, S.A.; Zakaria, M.A. Thermal Comfort Assessment in Office under High Air Conditioner Setting Temperature: A Review. *Recent Trends Civ. Eng. Built Environ.* **2021**, *2*, 226–235. **[Google Scholar]**
4. Zhang, X.; Chao, X.; Lou, L.; Fan, J.; Chen, Q.; Li, B.; Ye, L.; Shou, D. Personal thermal management by thermally conductive composites: A review. *Compos. Commun.* **2021**, *23*, 100595. **[Google Scholar] [CrossRef]**
5. Zhu, F.; Feng, Q. Recent advances in textile materials for personal radiative thermal management in indoor and outdoor environments. *Int. J. Therm. Sci.* **2021**, *165*, 106899. **[Google Scholar] [CrossRef]**
6. Farooq, A.S.; Zhang, P. Fundamentals, materials and strategies for personal thermal management by next-generation textiles. *Compos. Part A Appl. Sci. Manuf.* **2021**, *142*, 106249. **[Google Scholar] [CrossRef]**
7. Shi, X.-L.; Zou, J.; Chen, Z.-G. Advanced thermoelectric design: From materials and structures to devices. *Chem. Rev.* **2020**, *120*, 7399–7515. **[Google Scholar] [CrossRef]**
8. Choi, H.N.; Jee, S.H.; Ko, J.; Kim, D.J.; Kim, S.H. Properties of Surface Heating Textile for Functional Warm Clothing Based on a Composite Heating Element with a Positive Temperature Coefficient. *Nanomaterials* **2021**, *11*, 904. **[Google Scholar] [CrossRef]**
9. Lepak-Kuc, S.; Taborowska, P.; Tran, T.Q.; Duong, H.M.; Gizewski, T.; Jakubowska, M.; Patmore, J.; Lekawa-Raus, A. Washable, colored and textured, carbon nanotube textile yarns. *Carbon* **2021**, *172*, 334–344. **[Google Scholar] [CrossRef]**
10. Tian, T.; Wei, X.; Elhassan, A.; Yu, J.; Li, Z.; Ding, B. Highly flexible, efficient, and wearable infrared radiation heating carbon fabric. *Chem. Eng. J.* **2021**, *417*, 128114. **[Google Scholar] [CrossRef]**
11. Reese, J.; Vorhof, M.; Hoffmann, G.; Böhme, K.; Cherif, C. Joule heating of dry textiles made of recycled carbon fibers and PA6 for the series production of thermoplastic composites. *J. Eng. Fibers Fabr.* **2020**, *15*, 5828. **[Google Scholar] [CrossRef]**

12. Ruiz-Calleja, T.; Bonet-Aracil, M.; Gisbert-Payá, J.; Bou-Belda, E. Analysis of the influence of graphene and phase change microcapsules on thermal behavior of cellulosic fabrics. *Mater. Today Commun.* **2020**, *25*, 101557. **[Google Scholar] [CrossRef]**

13. Bramhecha, I.; Sheikh, J. Antibacterial and waterproof breathable waterborne polyurethane functionalised by graphene to develop UV and NIR-protective cotton fabric. *Carbon Trends* **2021**, *4*, 100067. **[Google Scholar] [CrossRef]**

14. Klochko, N.; Klepikova, K.; Zhadan, D.; Kopach, V.; Chernyavskaya, S.; Petrushenko, S.; Dukarov, S.; Lyubov, V.; Khrypunova, A. Thermoelectric textile with fibers coated by copper iodide thin films. *Thin Solid Film.* **2020**, *704*, 138026. **[Google Scholar] [CrossRef]**

15. Hsu, P.C.; Liu, X.G.; Liu, C.; Xie, X.; Lee, H.R.; Welch, A.J.; Zhao, T.; Cui, Y. Personal Thermal Management by Metallic Nanowire-Coated Textile. *Nano Lett.* **2015**, *15*, 365–371. **[Google Scholar] [CrossRef] [PubMed]**

16. Billingham, J. Heat exchange between man and his environment on the surface of the moon. *J. Br. Interplanet. Soc.* **1959**, *17*, 297–300. **[Google Scholar]**

17. Raad, R.; Itani, M.; Ghaddar, N.; Ghali, K. A novel M-cycle evaporative cooling vest for enhanced comfort of active human in hot environment. *Int. J. Therm. Sci.* **2019**, *142*, 1–13. **[Google Scholar] [CrossRef]**

18. Byun, S.H.; Yun, J.H.; Heo, S.Y.; Shi, C.; Lee, G.J.; Agno, K.C.; Jang, K.I.; Xiao, J.; Song, Y.M.; Jeong, J.W.J.A.S. Self-Cooling Gallium-Based Transformative Electronics with a Radiative Cooler for Reliable Stiffness Tuning in Outdoor Use. *Adv. Sci.* **2022**, *9*, 2202549. **[Google Scholar] [CrossRef]**

19. Luomala, M.J.; Oksa, J.; Salmi, J.A.; Linnamo, V.; Holmér, I.; Smolander, J.; Dugué, B. Adding a cooling vest during cycling improves performance in warm and humid conditions. *J. Therm. Biol.* **2012**, *37*, 47–55. **[Google Scholar] [CrossRef]**

20. Zare, M.; Dehghan, H.; Yazdanirad, S.; Khoshakhlagh, A.H. Comparison of the Impact of an Optimized Ice Cooling Vest and a Paraffin Cooling Vest on Physiological and Perceptual Strain. *Saf. Health Work* **2019**, *10*, 219–223. **[Google Scholar] [CrossRef]**

21. Yousefi, S.; Jamekhorshid, A.; Tahmasebi, S.; Sadrameli, S.M. Experimental and numerical performance evaluation of a cooling vest subtending phase change material under the extremely hot and humid environment. *Therm. Sci. Eng. Prog.* **2021**, *26*, 101103. **[Google Scholar] [CrossRef]**

22. Song, Y.-N.; Ma, R.-J.; Xu, L.; Huang, H.-D.; Yan, D.-X.; Xu, J.-Z.; Zhong, G.-J.; Lei, J.; Li, Z.-M. Wearable Polyethylene/Polyamide Composite Fabric for Passive Human Body Cooling. *ACS Appl. Mater. Interfaces* **2018**, *10*, 41637–41644. **[Google Scholar] [CrossRef] [PubMed]**

23. Kamon, E.; Kenney, W.; Deno, N.; Soto, K.; Carpenter, A. Readdressing personal cooling with ice. *Am. Ind. Hyg. Assoc. J.* **1986**, *47*, 293–298. **[Google Scholar] [CrossRef] [PubMed]**

24. Muir, I.H.; Bishop, P.A.; Ray, P. Effects of a novel ice-cooling technique on work in protective clothing at 28 °C, 23 °C, and 18 °C WBGTs. *Am. Ind. Hyg. Assoc. J.* **1999**, *60*, 96–104. **[Google Scholar] [CrossRef]**

25. Smolander, J.; Kuklane, K.; Gavhed, D.; Nilsson, H.; Holmér, I. Effectiveness of a light-weight ice-vest for body cooling while wearing fire fighter's protective clothing in the heat. *Int. J. Occup. Saf. Ergon.* **2004**, *10*, 111–117. **[Google Scholar] [CrossRef]**

26. Duffield, R.; Dawson, B.; Bishop, D.; Fitzsimons, M.; Lawrence, S. Effect of wearing an ice cooling jacket on repeat sprint performance in warm/humid conditions. *Br. J. Sport. Med.* **2003**, *37*, 164–169. **[Google Scholar] [CrossRef] [Green Version]**

27. Kenny, G.P.; Schissler, A.R.; Stapleton, J.; Piamonte, M.; Binder, K.; Lynn, A.; Lan, C.Q.; Hardcastle, S.G. Ice cooling vest on tolerance for exercise under uncompensable heat stress. *J. Occup. Environ. Hyg.* **2011**, *8*, 484–491. **[Google Scholar] [CrossRef]**

28. Taylor, L.; Stevens, C.J.; Thornton, H.R.; Poulos, N.; Chrismas, B.C. Limiting the rise in core temperature during a rugby sevens warm-up with an ice vest. *Int. J. Sport. Physiol. Perform.* **2019**, *14*, 1212–1218. **[Google Scholar] [CrossRef] [Green Version]**

29. Lee, J.K.; Kenefick, R.W.; Cheuvront, S.N. Novel cooling strategies for military training and operations. *J. Strength Cond. Res.* **2015**, *29*, S77–S81. **[Google Scholar] [CrossRef]**

30. Countryman, J.D.; Dow, D.E. Historical development of heat stroke prevention device in the military. In Proceedings of the 2013 35th Annual International Conference of the IEEE Engineering in Medicine and Biology Society (EMBC), Osaka, Japan, 3–7 July 2013; pp. 2527–2530. **[Google Scholar]**

31. Cotter, J.D.; Sleivert, G.G.; Roberts, W.S.; Febbraio, M.A. Effect of pre-cooling, with and without thigh cooling, on strain and endurance exercise performance in the heat. *Comp. Biochem. Physiol. Part A Mol. Integr. Physiol.* **2001**, *128*, 667–677. **[Google Scholar] [CrossRef]**

32. Konz, S. Personal cooling garments: A review. *ASHRAE Trans.* **1984**, *90*, 499–518. **[Google Scholar]**

33. Dehghan, H.; Gharebaei, S. Effectiveness of Ice Gel Cooling Vest on Physiological Indices in Hot and Dry Conditions in a Climate Chamber. *Health Syst. Res.* **2017**, *13*, 10–13. **[Google Scholar]**

34. Chesterton, L.S.; Foster, N.E.; Ross, L. Skin temperature response to cryotherapy. *Arch. Phys. Med. Rehabil.* **2002**, *83*, 543–549. **[Google Scholar]** **[CrossRef]** **[PubMed]**

35. Teunissen, L.; Janssen, E.; Schootstra, J.; Plaude, L.; Jansen, K. Evaluation of Phase Change Materials for Personal Cooling Applications. *Cloth. Text. Res. J.* **2021**, 887302X211053007. **[Google Scholar]** **[CrossRef]**

36. Erkan, G. Enhancing the thermal properties of textiles with phase change materials. *Res. J. Text. Appar.* **2004**, *8*, B008. **[Google Scholar]** **[CrossRef]**

37. Zhang, C.; Lin, W.; Zhang, Q.; Zhang, Z.; Fang, X.; Zhang, X. Exploration of a thermal therapy respirator by introducing a composite phase change block into a commercial mask. *Int. J. Therm. Sci.* **2019**, *142*, 156–162. **[Google Scholar]** **[CrossRef]**

38. Jing, Y.; Zhao, Z.; Zhang, N.; Cao, X.; Sun, Q.; Yuan, Y.; Li, T. Ultraflexible, Cost-Effective and Scalable Polymer-Based Phase Change Composites for Wearable Thermal Management. *Soc. Sci. Res. Netw.* **2022**, 4150536. **[Google Scholar]** **[CrossRef]**

39. Itani, M.; Ghaddar, N.; Ouahrani, D.; Ghali, K.; Khater, B. An optimal two-bout strategy with phase change material cooling vests to improve comfort in hot environment. *J. Therm. Biol.* **2018**, *72*, 10–25. **[Google Scholar]** **[CrossRef]**

40. Li, W.; Liang, Y.; Liu, C.; Ji, Y.; Cheng, L. Study of ultra-light modular phase change cooling clothing based on dynamic human thermal comfort modeling. *Build. Environ.* **2022**, *222*, 109390. **[Google Scholar]** **[CrossRef]**

41. Yan, Y.; Li, W.; Zhu, R.; Lin, C.; Hufenus, R. Flexible Phase Change Material Fiber: A Simple Route to Thermal Energy Control Textiles. *Materials* **2021**, *14*, 401. **[Google Scholar]** **[CrossRef]**

42. Tomaszewski, W.; Twarowska-Schmidt, K.; Moraczewski, A.; Kudra, M.; Szadkowski, M.; Pałys, B. Nonwoven with Thermal Storage Properties Based on Paraffin–Modified Polypropylene Fibres. *Fibres Text. East. Eur.* **2012**, *6B*, 64–69. **[Google Scholar]**

43. Lei, J.; Kumarasamy, K.; Zingre, K.T.; Yang, J.; Wan, M.P.; Yang, E.-H. Cool colored coating and phase change materials as complementary cooling strategies for building cooling load reduction in tropics. *Appl. Energy* **2017**, *190*, 57–63. **[Google Scholar]** **[CrossRef]**

44. Skurkyte-Papieviene, V.; Abraitiene, A.; Sankauskaite, A.; Rubeziene, V.; Baltusnikaite-Guzaitiene, J. Enhancement of the Thermal Performance of the Paraffin-Based Microcapsules Intended for Textile Applications. *Polymers* **2021**, *13*, 1120. **[Google Scholar]** **[CrossRef]** **[PubMed]**

45. Shi, J.; Aftab, W.; Liang, Z.; Yuan, K.; Maqbool, M.; Jiang, H.; Xiong, F.; Qin, M.; Gao, S.; Zou, R. Tuning the flexibility and thermal storage capacity of solid–solid phase change materials towards wearable applications. *J. Mater. Chem. A* **2020**, *8*, 20133–20140. **[Google Scholar] [CrossRef]**

46. Yang, Z.; Ma, Y.; Jia, S.; Zhang, C.; Li, P.; Zhang, Y.; Li, Q. 3D-printed flexible phase-change nonwoven fabrics toward multifunctional clothing. *ACS Appl. Mater. Interfaces* **2022**, *14*, 7283–7291. **[Google Scholar] [CrossRef]**

47. Niu, Z.; Yuan, W. Smart nanocomposite nonwoven wearable fabrics embedding phase change materials for highly efficient energy conversion–storage and use as a stretchable conductor. *ACS Appl. Mater. Interfaces* **2021**, *13*, 4508–4518. **[Google Scholar] [CrossRef]**

48. Zhang, Y.F.; Sun, B.K. Development of Outlast Fiber and Study about Its Character of Thermoregulation. *Adv. Mater. Res.* **2012**, *557–559*, 979–982. **[Google Scholar]**

49. Liu, J.; Zhou, Z.; Zhang, J.; Feng, W.; Zuo, J. Advances and challenges in commercializing radiative cooling. *Mater. Today Phys.* **2019**, *11*, 100161. **[Google Scholar] [CrossRef]**

50. Tong, J.K.; Huang, X.; Boriskina, S.V.; Loomis, J.; Xu, Y.; Chen, G. Infrared-Transparent Visible-Opaque Fabrics for Wearable Personal Thermal Management. *ACS Photonics* **2015**, *2*, 769–778. **[Google Scholar] [CrossRef] [Green Version]**

51. Hsu, P.C.; Song, A.Y.; Catrysse, P.B.; Liu, C.; Peng, Y.C.; Xie, J.; Fan, S.H.; Cui, Y. Radiative human body cooling by nanoporous polyethylene textile. *Science* **2016**, *353*, 1019–1023. **[Google Scholar] [CrossRef][Green Version]**

52. Anderson, D.M.; Fessler, J.R.; Pooley, M.A.; Seidel, S.; Hamblin, M.R.; Beckham, H.W.; Brennan, J.F. Infrared radiative properties and thermal modeling of ceramic-embedded textile fabrics. *Biomed. Opt. Express* **2017**, *8*, 1698–1711. **[Google Scholar] [CrossRef] [PubMed][Green Version]**

53. Hsu, P.-C.; Liu, C.; Song, A.Y.; Zhang, Z.; Peng, Y.; Xie, J.; Liu, K.; Wu, C.-L.; Catrysse, P.B.; Cai, L. A dual-mode textile for human body radiative heating and cooling. *Sci. Adv.* **2017**, *3*, e1700895. **[Google Scholar] [CrossRef] [PubMed] [Green Version]**

54. Zhao, B.; Hu, M.; Ao, X.; Chen, N.; Pei, G. Radiative cooling: A review of fundamentals, materials, applications, and prospects. *Appl. Energy* **2019**, *236*, 489–513. **[Google Scholar] [CrossRef]**

55. Peng, Y.; Cui, Y. Advanced Textiles for Personal Thermal Management and Energy. *Joule* **2020**, *4*, 724–742. **[Google Scholar] [CrossRef]**

56. Gueymard, C.A. The sun's total and spectral irradiance for solar energy applications and solar radiation models. *Sol. Energy* **2004**, *76*, 423–453. **[Google Scholar] [CrossRef]**

57. Cai, L.; Song, A.Y.; Li, W.; Hsu, P.-C.; Lin, D.; Catrysse, P.B.; Liu, Y.; Peng, Y.; Chen, J.; Wang, H.; et al. Spectrally Selective Nanocomposite Textile for Outdoor Personal Cooling. *Adv. Mater.* **2018**, *30*, 1802152. **[Google Scholar] [CrossRef] [PubMed]**

58. Iqbal, M.I.; Lin, K.; Sun, F.; Chen, S.; Pan, A.; Lee, H.H.; Kan, C.-W.; Lin, C.S.K.; Tso, C.Y. Radiative Cooling Nanofabric for Personal Thermal Management. *ACS Appl. Mater. Interfaces* **2022**, *14*, 23577–23587. **[Google Scholar] [CrossRef] [PubMed]**

59. Al Sayed, C.; Vinches, L.; Dupuy, O.; Douzi, W.; Dugue, B.; Hallé, S. Technology. Air/CO2 cooling garment: Description and benefits of use for subjects exposed to a hot and humid climate during physical activities. Int. J. Min. Sci. Technol. 2019, 29, 899–903. [Google Scholar] [CrossRef]

60. Kotagama, P.; Phadnis, A.; Manning, K.C.; Rykaczewski, K. Rational Design of Soft, Thermally Conductive Composite Liquid-Cooled Tubes for Enhanced Personal, Robotics, and Wearable Electronics Cooling. *Adv. Mater. Technol.* **2019**, *4*, 1800690. **[Google Scholar] [CrossRef]**

61. Xu, S.; Li, M.; Dai, Y.; Hong, M.; Sun, Q.; Lyu, W.; Liu, T.; Wang, Y.; Zou, J.; Chen, Z.-G.; et al. Realizing a 10 °C Cooling Effect in a Flexible Thermoelectric Cooler using a Vortex Generator. *Adv. Mater.* **2022**, *34*, 2204508. **[Google Scholar] [CrossRef]**

62. Choudhary, B.; Wang, F.; Ke, Y.; Yang, J. Development and experimental validation of a 3D numerical model based on CFD of the human torso wearing air ventilation clothing. *Int. J. Heat Mass Transf.* **2020**, *147*, 118973. **[Google Scholar] [CrossRef]**

63. Heqing, L.; Liying, G.; Bo, Y.; Tianyu, L.; Congying, O. Experimental Study on the Effect of Air Cooling Garment on Skin Temperature and Microclimate. In Proceedings of the 11th International Mine Ventilation Congress, Xi'an, China, 14–20 September 2018; Springer: Singapore, 2019; pp. 742–752. **[Google Scholar]**

64. Harrison, M.; Gibson, T. *The History of the IAM: Protecting Against the Elements*; Report; Royal Air Force Institute of Aviation Medicine: Bedfordshire, UK, 1982. **[Google Scholar]**

65. Yazdi, M.M.; Sheikhzadeh, M. Personal cooling garments: A review. *J. Text. Inst.* **2014**, *105*, 1231–1250. **[Google Scholar] [CrossRef]**

66. Zhu, R.; Wang, H.; Liu, S.; Pan, G. Advances in the Soviet/Russian EVA Spacesuit Technology. *Manned Spacefl.* **2009**, *1*, 25–45. **[Google Scholar]**

67. Hadid, A.; Fuks, Y.; Erlich, T.; Yanovich, R.; Heled, Y.; Azriel, N.; Moran, D. Effect of a personal ambient ventilation system on physiological strain during heat stress wearing body armour. In Proceedings 13th International Conference on Environmental Ergonomics, Boston, MA, USA, 2–7 August 2009; pp. 252–254. **[Google Scholar]**

68. Wu, G.; Liu, H.; Wu, S.; Liu, Z.; Mi, L.; Gao, L. A study on the capacity of a ventilation cooling vest with pressurized air in hot and humid environments. *Int. J. Ind. Ergon.* **2021**, *83*, 103106. **[Google Scholar] [CrossRef]**

69. Choudhary, B.; Udayraj. A coupled CFD-thermoregulation model for air ventilation clothing. *Energy Build.* **2022**, *268*, 112206. **[Google Scholar] [CrossRef]**

70. Yang, J.; Wang, F.; Song, G.; Li, R.; Raj, U. Effects of clothing size and air ventilation rate on cooling performance of air ventilation clothing in a warm condition. *Int. J. Occup. Saf. Ergon.* **2022**, *28*, 354–363. **[Google Scholar] [CrossRef]**

71. Zhao, M.; Yang, J.; Wang, F.; Udayraj; Chan, W.C. The cooling performance of forced air ventilation garments in a warm environment: The effect of clothing eyelet designs. *J. Text. Inst.* **2022**, 1–10. **[Google Scholar] [CrossRef]**

72. Ni, X.; Yao, T.; Zhang, Y.; Zhao, Y.; Hu, Q.; Chan, A.P. Experimental study on the efficacy of a novel personal cooling vest incorporated with phase change materials and fans. *Materials* **2020**, *13*, 1801. **[Google Scholar] [CrossRef] [Green Version]**

73. Lou, L.; Wu, Y.S.; Zhou, Y.; Fan, J. Effects of body positions and garment design on the performance of a personal air cooling/heating system. *Indoor Air* **2022**, *32*, e12921. **[Google Scholar] [CrossRef]**

74. Guo, T.; Shang, B.; Duan, B.; Luo, X. Design and testing of a liquid cooled garment for hot environments. *J. Therm. Biol.* **2015**, *49*, 47–54. **[Google Scholar] [CrossRef]**

75. Chambers, A.B. Controlling Thermal Comfort in the EVA Space Suit. 1970. Available online: **https://ntrs.nasa.gov/citations/19700048296** (accessed on ٢ December ٢٠٢٢).

76. Revaiah, R.; Kotresh, T.; Kandasubramanian, B. Technical textiles for military applications. *J. Text. Inst.* **2019**, *111*, 273–308. **[Google Scholar] [CrossRef]**

77. Al Sayed, C.; Vinches, L.; Hallé, S. Towards optimizing a personal cooling garment for hot and humid deep mining conditions. *Open J. Optim.* **2016**, *5*, 35. **[Google Scholar] [CrossRef][Green Version]**

78. Cvetanović, S.G.; Rutić, S.Z.; Krstić, D.D.; Florus, S.; Otrisal, P. The influence of an active microclimate liquid-cooled vest on heat strain alleviation. *Therm. Sci.* **2021**, *25*, 3837–3846. **[Google Scholar] [CrossRef]**

79. Teunissen, L.P.J.; Wang, L.-C.; Chou, S.-N.; Huang, C.-h.; Jou, G.-T.; Daanen, H.A.M. Evaluation of two cooling systems under a firefighter coverall. *Appl. Ergon.* **2014**, *45*, 1433–1438. **[Google Scholar] [CrossRef] [PubMed][Green Version]**

80. Bartkowiak, G.; Dabrowska, A.; Marszalek, A. Assessment of an active liquid cooling garment intended for use in a hot environment. *Appl. Ergon.* **2017**, *58*, 182–189. **[Google Scholar] [CrossRef] [PubMed]**

81. Shu, W.; Zhang, X.; Yang, X.; Luo, X. A Smart Temperature-Regulating Garment for Portable, High-Efficiency and Comfortable Cooling. *J. Electron. Packag.* **2021**, *144*, 31010. **[Google Scholar] [CrossRef]**

82. Zhang, M.; Li, Z.; Wang, Q.; Xu, Y.; Hu, P.; Zhang, X. Performance investigation of a portable liquid cooling garment using thermoelectric cooling. *Appl. Therm. Eng.* **2022**, *214*, 118830. **[Google Scholar] [CrossRef]**

83. Yang, B.; Ding, X.; Wang, F.; Li, A. A review of intensified conditioning of personal micro-environments: Moving closer to the human body. *Energy Built Environ.* **2021**, *2*, 260–270. **[Google Scholar] [CrossRef]**

84. Valant, M. Electrocaloric materials for future solid-state refrigeration technologies. *Prog. Mater. Sci.* **2012**, *57*, 980–1009. **[Google Scholar] [CrossRef]**

85. Ma, R.; Zhang, Z.; Tong, K.; Huber, D.; Kornbluh, R.; Ju, Y.S.; Pei, Q. Highly efficient electrocaloric cooling with electrostatic actuation. *Science* **2017**, *357*, 1130–1134. **[Google Scholar] [CrossRef][Green Version]**

86. Gottschall, T.; Gracia-Condal, A.; Fries, M.; Taubel, A.; Pfeuffer, L.; Manosa, L.; Planes, A.; Skokov, K.P.; Gutfleisch, O. A multicaloric cooling cycle that exploits thermal hysteresis. *Nat. Mater.* **2018**, *17*, 929–934. **[Google Scholar] [CrossRef] [PubMed]**

87. Ding, J.; Zhao, W.; Jin, W.; Di, C.-a.; Zhu, D. Advanced Thermoelectric Materials for Flexible Cooling Application. *Adv. Funct. Mater.* **2021**, *31*, 2010695. **[Google Scholar] [CrossRef]**

88. Sun, W.; Liu, W.-D.; Liu, Q.; Chen, Z.-G. Advances in thermoelectric devices for localized cooling. *Chem. Eng. J.* **2022**, *450*, 138389. **[Google Scholar] [CrossRef]**

89. Chen, W.-Y.; Shi, X.-L.; Zou, J.; Chen, Z.-G. Thermoelectric Coolers: Progress, Challenges, and Opportunities. *Small Methods* **2022**, *6*, 2101235. **[Google Scholar] [CrossRef]**

90. He, W.; Zhang, G.; Zhang, X.; Ji, J.; Li, G.; Zhao, X. Recent development and application of thermoelectric generator and cooler. *Appl. Energy* **2015,** *143,* 1–25. **[Google Scholar] [CrossRef]**

91. Yang, Z.; Zhu, C.; Ke, Y.J.; He, X.; Luo, F.; Jianl, W.; Wang, J.F.; Sun, Z.G. Peltier effect: From linear to nonlinear. *Acta Phys. Sin.* **2021,** *70,* 108402. **[Google Scholar] [CrossRef]**

92. Zhao, D.; Tan, G. A review of thermoelectric cooling: Materials, modeling and applications. *Appl. Therm. Eng.* **2014,** *66,* 15–24. **[Google Scholar] [CrossRef]**

93. Alaoui, C. Peltier thermoelectric modules modeling and evaluation. *Int. J. Eng.* **2011,** *5,* 114. **[Google Scholar]**

94. Lv, S.; Qian, Z.; Hu, D.; Li, X.; He, W. A comprehensive review of strategies and approaches for enhancing the performance of thermoelectric module. *Energies* **2020,** *13,* 3142. **[Google Scholar] [CrossRef]**

95. Xu, Y.; Li, Z.; Wang, J.; Zhang, M.; Jia, M.; Wang, Q. Man-portable cooling garment with cold liquid circulation based on thermoelectric refrigeration. *Appl. Therm. Eng.* **2022,** *200,* 117730. **[Google Scholar] [CrossRef]**

96. Lou, L.; Shou, D.; Park, H.; Zhao, D.; Wu, Y.S.; Hui, X.; Yang, R.; Kan, E.C.; Fan, J. Thermoelectric air conditioning undergarment for personal thermal management and HVAC energy saving. *Energy Build.* **2020,** *226,* 110374. **[Google Scholar] [CrossRef]**

97. Suen, W.S.; Huang, G.; Kang, Z.; Gu, Y.; Fan, J.; Shou, D. Development of wearable air-conditioned mask for personal thermal management. *Build. Environ.* **2021,** *205,* 108236. **[Google Scholar] [CrossRef] [PubMed]**

98. Hong, S.; Gu, Y.; Seo, J.K.; Wang, J.; Liu, P.; Meng, Y.S.; Xu, S.; Chen, R.K. Wearable thermoelectrics for personalized thermoregulation. *Sci. Adv.* **2019,** *5,* 11. **[Google Scholar] [CrossRef] [PubMed][Green Version]**

99. Zhang, Y.; Gao, J.; Zhu, S.; Li, J.; Lai, H.; Peng, Y.; Miao, L. Wearable Thermoelectric Cooler Based on a Two-Layer Hydrogel/Nickel Foam Heatsink with Two-Axis Flexibility. *ACS Appl. Mater. Interfaces* **2022,** *14,* 15317–15323. **[Google Scholar] [CrossRef] [PubMed]**

100. Zhang, T.; Li, K.; Zhang, J.; Chen, M.; Wang, Z.; Ma, S.; Zhang, N.; Wei, L. High-performance, flexible, and ultralong crystalline thermoelectric fibers. *Nano Energy* **2017,** *41,* 35–42. **[Google Scholar] [CrossRef]**

101. Zheng, Y.; Han, X.; Yang, J.; Jing, Y.; Chen, X.; Li, Q.; Zhang, T.; Li, G.; Zhu, H.; Zhao, H.; et al. Durable, stretchable and washable inorganic-based woven thermoelectric textiles for power generation and solid-state cooling. *Energy Environ. Sci.* **2022,** *15,* 2374–2385. **[Google Scholar] [CrossRef]**

102. Gao, C.; Kuklane, K.; Wang, F.; Holmér, I. Personal cooling with phase change materials to improve thermal comfort from a heat wave perspective. *Indoor Air* **2012**, *22*, 523–530. **[Google Scholar]** **[CrossRef]**

103. Barwood, M.J.; Newton, P.S.; Tipton, M.J. Ventilated vest and tolerance for intermittent exercise in hot, dry conditions with military clothing. *Aviat. Space Environ. Med.* **2009**, *80*, 353–359. **[Google Scholar]** **[CrossRef]**

104. Yuan, M.; Wei, Y.; An, Q.; Yang, J. Effects of a liquid cooling vest on physiological and perceptual responses while wearing stab-resistant body armor in a hot environment. *Int. J. Occup. Saf. Ergon.* **2022**, *28*, 1025–1032. **[Google Scholar]** **[CrossRef]**

CHAPTER-3

MODIFY ELECTROSPUN POLYACRYLONITRILE NANOFIBROUS MEMBRANES FOR HYDROPHOBICITY WITHOUT USING FLUORINE, ACHIEVING BOTH WATERPROOF AND BREATHABLE PROPERTIES

1. INTRODUCTION: AN OVERVIEW

In the realm of advanced materials engineering, the pursuit of hydrophobic modifications in nanofibrous membranes stands as a crucial objective for enhancing their functionality. Electrospun polyacrylonitrile (PAN) nanofibrous membranes have garnered significant attention due to their unique properties, but the incorporation of hydrophobicity without resorting to fluorine-based compounds remains a challenge. This pursuit is particularly relevant for applications where achieving both waterproof and breathable characteristics is essential. In this context, innovative modifications to electrospun PAN nanofibrous membranes present a promising avenue to address this challenge, paving the way for materials with superior water resistance and breathability. This endeavor not only expands the applicability of such membranes in diverse fields but also aligns with the growing emphasis on sustainable and environmentally friendly material development.

With the proposal of healthy, green and environmentally friendly lifestyles, outdoor sports have been highly sought after by the general public in recent years, which stimulates the market demand for sports/outdoors textiles to grow sharply [**1**,**2**,**3**]. In order to adapt to the complex and changeable environment of outdoor sports, scientists in academia have done tremendous work in recent years in studying the functional properties, such as waterproofness and breathability, cold resistance, heat preservation or UV protection of the sports/outdoors textiles [**4**,**5**,**6**]. Possessing the functional membranes as the core layer, waterproof and breathable fabric is the representative of the outdoor sports textile products. The hydrophobic microporous membranes contain more holes and interconnected channels, such as polytetrafluoroethylene (PTFE) bidirectional stretch membranes, which make them become the mainstream of the products, but their application is limited by some default properties that make them hard to degrade and easy to enrich [**7**,**8**]. Therefore, the development of fluorine-free and environmentally friendly functional waterproof breathable membranes is an important development direction at present.

Electrospinning has prosperous prospect in the field of waterproof breathable membranes because of its unique advantages, such as simple spinning equipment, diversified raw material combinations, strong scalability and controllable spinning process [9,10,11]. Gu et al. [12] fabricated electrospun polyurethane (PU) microporous membranes through changing the volume ratios of the mixture solvents. Furthermore, the resultant membranes presented a modest water vapor transmission rate (WVTR) of 10.1 kg m−2 d−1, as well as robust mechanical properties with a tensile strength of 6.6 MPa, which is desirable in protective clothing. However, the hydrostatic pressure was only 2.1 kPa and could not meet the actual application requirements [13]. In consideration of improving the waterproofness, the hydrophobic modification of the membranes for waterproof and breathable application has brought about widespread attention [14,15,16]. In order to construct rough structures for acquiring hydrophobicity, Wang et al. [17] prepared PU/SiO2 nanofibrous membranes by enriching SiO2 particles on the surface. The final polyurethane/SiO2 nanofibrous membranes presented a water contact angle (WCA) of 154°, and the hydrostatic pressure reached 8.02 kPa.

To further increase the hydrophobicity as well as the waterproofness of the functional membranes, low surface energy substances including fluorine-containing compounds were selected for hydrophobic finishing [**18**,**19**]. Wang et al. [**20**] fabricated electrospun polyacrylonitrile (PAN) fibrous membrane which was modified with waterborne fluorinated polyurethane (WFPU). After the WFPU modification, the pristine PAN fibrous membranes possessed high waterproofness of up to 83.4 kPa. However, during the use of long-chain fluorocarbon-type finishing agents, some toxic substances are usually left behind. This is because the covalent bond energy of the structure is extremely strong due to the presence of fluorine atoms, and it is not easy for it to decompose in the natural environment [**21**,**22**,**23**]. Therefore, in the selection of finishing agent, the first consideration is to choose short-chain fluorocarbon or fluorine-free finishing agents. Zhang et al. [**24**] prepared environmentally friendly siliceous polyurethane/stearic acid nanofiber membranes for waterproof breathable application. By adjusting the addition amount of the stearic acid, the waterproofness of SIPU/SA nanofiber membranes was greatly improved with the hydrostatic pressure of 79 kPa. However, the tensile strength was 6.2 MPa and could not meet the needs of practical applications.

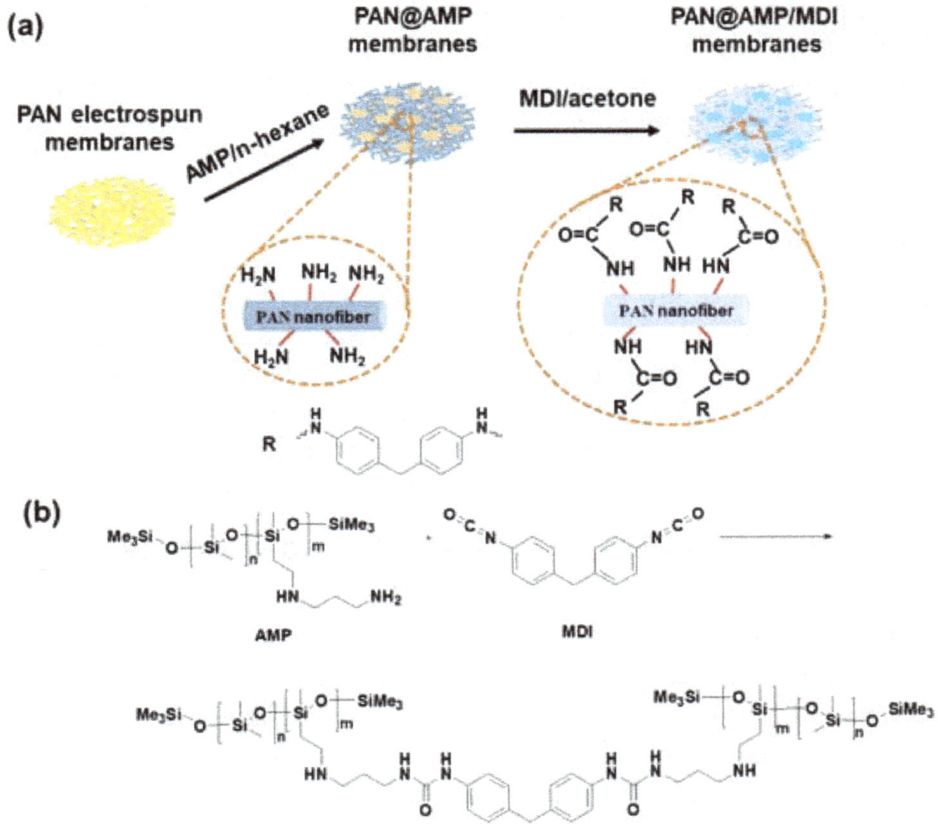

Figure 1. (a) Schematic illustration of the preparation procedure of fluoride-free PAN@AMP/MDI nanofibrous membranes. (b) The cross-linking reaction between AMP and MDI.

The key to our design was to fabricate composite nanofibrous membranes with robust waterproof and mechanical properties via fluorine-free hydrophobic modification and in situ cross-linking, as illustrated in **Figure 1.** Due to the small pore diameter, fine tortuous pore channels, high porosity and easy access to be modified, polyacrylonitrile (PAN) nanofibrous membranes were chosen as the backbone for fabricating the functional membrane. After the fluorine-free modification with amino functional modified poysiloxane (AMP), there was hydrophobic adhesion structure interspersed in the nanofibrous membranes, which would not only endow the membranes with hydrophobicity, but also enhance the mechanical performance of the membranes at the same time. In order to further improve the strength of the PAN@AMP membranes, 4, 4'-methyl diphenylene diisocyanate (MDI) were introduced subsequently for the in situ cross-linking reaction with AMP. By adjusting the concentration of AMP and MDI, the effects on the morphology structure, wetting behavior, mechanical property and waterproof breathable performance were investigated. This design may offer a new avenue for the development of a functional waterproof breathable nanofibrous membranes that meet the needs for wide application, such as protective clothing, a separation process, and self-cleaning materials.

2. MATERIALS AND METHODS

2.1. Materials

Polyacrylonitrile (PAN, Mw = 90,000 g/mol) was acquired from Kaneka Co., Ltd., Tokyo, Japan. Amino functional modified polysiloxane (AMP, surface energy is 20 J m−2) and 4, 4'-methyl diphenylene diisocyanate (MDI, Mw = 250.25) were supplied from Aladdin Chemical Reagent Co. N, N-dimethylformamide (DMF), acetone, and n-hexane were brought by Sinopharm Chemical Reagent Co. Ltd., Shanghai, China. All reagents were employed as supplied without any processing.

2.2. Fabrication of PAN@AMP/MDI Nanofibrous Membranes

The PAN powders were dissolved in DMF and stirred for 12 h to make a uniform and transparent electrospun solution with a concentration of 9 wt%. PAN nanofibrous membranes were then fabricated under a needle electrospinning environment (25 ± 2 °C and 45 ± 2%) through a DXES-3 electrospinning machine (SOF Nanotechnology Co., Ltd., Shanghai, China), while the solution perfusion rate was 1.0 mL/h, the receiving distance was 21 cm and the voltage was 25 kV.

Subsequently, the as-spun PAN nanofibrous membranes were immersed in AMP/n-hexane solutions, where the solution concentrations were 0.5, 1, 2 and 4 wt%, respectively. Thereafter, the modified nanofibrous membranes were dried in an oven at 100 °C for 30 min, so that the hydrophobic agent formed a hydrophobic and stable adhesion structure in the nanofibrous membranes. For the convenience of subsequent representation, the membranes modified with AMP solution with a concentration of x wt% were defined as PAN@AMP-x.

Finally, the PAN@AMP nanofibrous membranes were modified with different concentrations of MDI/acetone solutions through coating treatment, where the solution concentrations were 0.5, 1, 2 and 4 wt%, respectively. Then, the PAN@ AMP/MDI nanofibrous membranes were prepared by then being dried in an oven at 100 °C for 30 min for the cross-linking reaction. Relevantly, the PAN@AMP/ MDI nanofibrous membranes were defined as PAN@AMP-x/MDI-y, where y is the concentration of MDI/acetone.

2.3. Structure Characterization and Performance Measurement

A field emission scanning electron microscope (FE-SEM, Hitachi S-4800, Chiyoda City, Japan) was used to characterize the morphology and structure of the nanofibrous membranes. The pore size and distribution of the nanofibrous membranes were analyzed through the gas permeation pore size analyzer (CFP-1100AX, PMI Inc., Newtown Square, PA, USA). Porosity of the samples was performed using the equation below [25]:

$$Porosity = (1 - \frac{m}{t \times S \times \rho}) \times 100\% \quad (1)$$

where m, t, and S are the mass, thickness, and area of per unit measured membrane, respectively. Furthermore, ϱ is the density of the raw material. A Nicolet 8700 FT-IR spectrometer was used to verify the presence of the modification agents on the modified membranes, and the ATR total reflection mode of the Fourier transform infrared spectrometer was used to characterize the prepared membranes. X-ray photoelectron spectroscopy (XPS, Thermo Fisher Scientific, Escalab 250Xi, Waltham, MA, USA) was utilized to test the surface chemical compositions. The surface roughness of the prepared membranes was obtained by atomic force microscopy (AFM, Dimension Icon, Bruker, Billerica, MA, USA).

An XQ-1C tensile tester (Shanghai New Fiber Instrument Co., Ltd., Shanghai, China) was used to investigate the tensile strength of the as-prepared membranes. The WCA was usually used to characterize the surface wettability of the nanofibrous membranes using an optical contact angle measuring instrument (KRUSS DSA30). Taking the waterproof property into consideration, hydrostatic pressure was measured based on AATCC 127 test criterion with a water pressure increasing rate of 6 kPa min^{-1}. The breathable performance of the membranes was determined by the testing water vapor transmission rate (WVTR) and air permeability. WVTR testing was performed according to the ASTM E96 evaporation method standard with the temperature of 38 °C, relative humidity of 50% and a wind velocity of 1 m s^{-1}. Following the ASTM D 737 criterion, the air permeable performance was tested under a differential pressure of 100 Pa. For each sample, hydrostatic pressure, WVTR and air permeability were tested at least three times.

3. RESULTS AND DISCUSSION

3.1. Effect of AMP Concentration

3.1.1. Morphology and Structure

Due to its small surface tension, AMP can spread easily and form a film on the surface of the PAN nanofibrous membranes, which changes the morphology and surface properties with endowing the hydrophobic properties of the nanofibrous membranes. By comparing the SEM images of PAN@AMP nanofibrous membranes hydrophobically modified by different concentrations of AMP (**Figure 2**a–e), it can be seen that the untreated PAN nanofibrous membranes have three-dimensional irregular non-woven structures, and the nanofibers are horizontal and vertical with uniform fineness at 318 nm. With the increase of AMP concentration from 0.5 to 4 wt%, there was an increment from 348 to 411 nm of the average fiber diameter (**Figure S1a**). When the concentration of AMP was 0.5 wt%, some bonding points appear on the nanofibers and are not connected each other. With the increase of AMP concentration, it was obvious that the adhesion structures were gradually formed between nanofibers. When the concentration of AMP was increased continually to 4 wt%, a wide range of film-shaped adhesion structures were observed in the nanofibers, and the adhesion points were connected to each other.

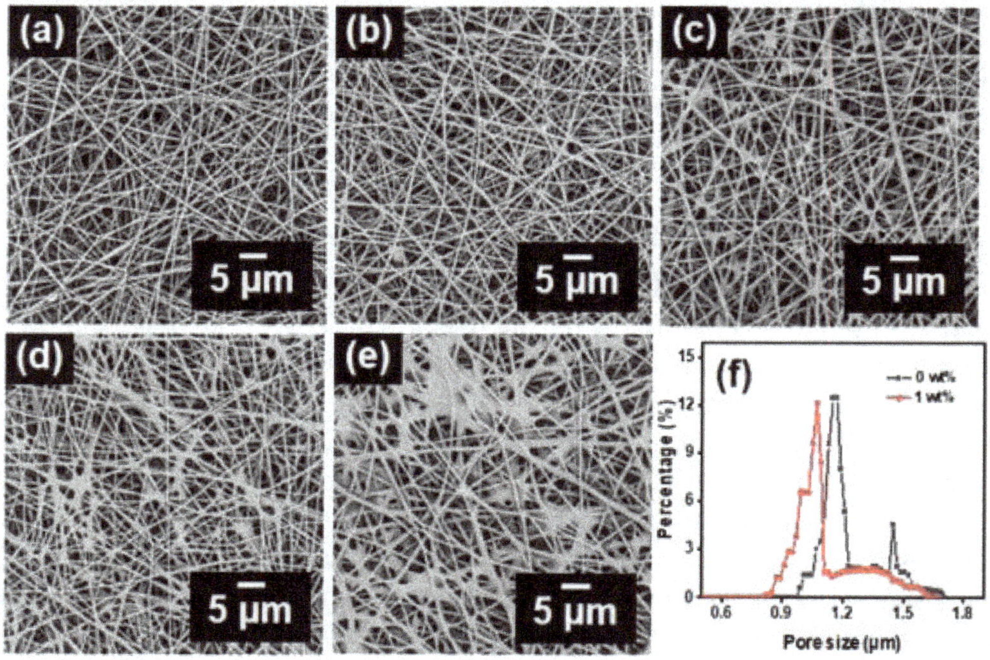

Figure 2. SEM images of hydrophobically-modified PAN@AMP membranes with different concentrations of AMP: (a) 0, (b) 0.5, (c) 1, (d) 2 and (e) 4 wt%. (f) Pore size distribution of PAN original membranes and PAN@AMP-1 membranes.

Considering the effect of the film-shaped adhesion structure on the porous structure of the membranes, the pore size distribution of PAN original membranes and PAN@AMP-1 membranes were investigated, as shown in **Figure 2**f. It was found that the average pore size of the untreated nanofibrous membranes was 1.18 μm. After hydrophobic modification with 1 wt% of AMP, the average pore size of the nanofibrous membranes was significantly reduced, and the pore size distribution was mainly concentrated in the vicinity of 1.07 μm, which was consistent with the phenomenon observed in the relevant SEM images.

3.1.2. Mechanical Properties

Considering the practical application of the PAN@AMP membranes, the mechanical performance of the modified membranes was measured. From the stress–strain curves in **Figure 3**, it can be seen that the strength of the untreated nanofibrous membranes were only 7.7 MPa and the strain was 73.4%. The modification of AMP brought about the obvious promotion of the strength of the PAN@AMP membranes. When the concentration of AMP increased from 0.5 to 2 wt%, the strength increased from 9.7 to 11.3 MPa, which increased by 46% compared with the original membranes. The significant promotion was due to the addition of AMP making the nanofibers stick together, endowing the nanofibers with a certain binding force, and increasing the strength of the nanofibrous membranes to a certain extent. However, the AMP concentration was further increased to 4 wt%, the strength of nanofibrous membranes decreased because of the large area film-shaped adhesion structure coving the PAN membranes, and the strength formed by AMP was poor [26]. Different from the law of strength, the strain always showed a decreasing trend from 73.4% to 40.7%, because the increase of the adhesion point made the nanofibers difficult to move and the strain decreased.

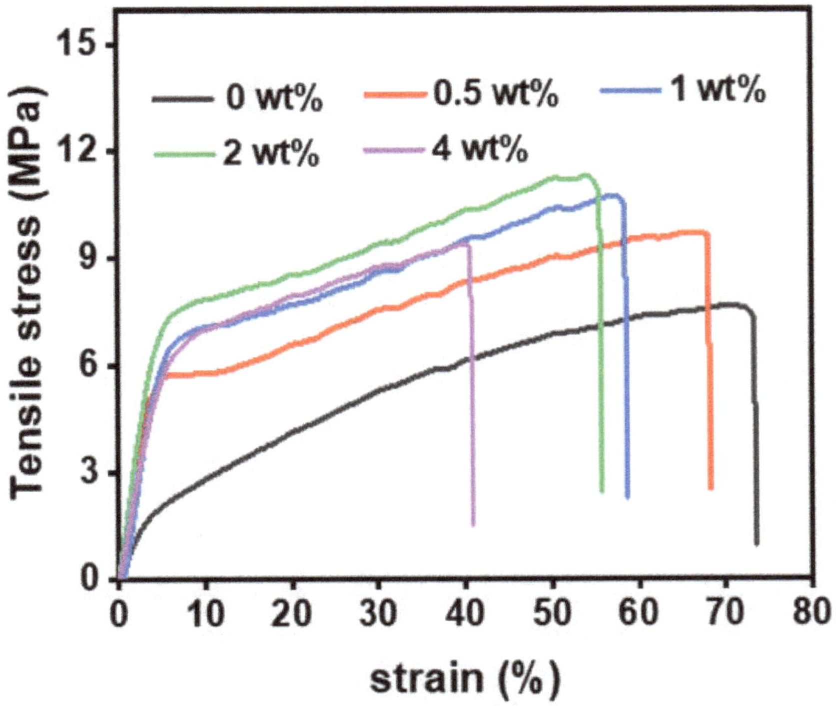

Figure 3. Stress–strain curves of hydrophobically-modified PAN@AMP nanofibrous membranes with different concentrations of AMP.

3.1.3. Waterproof Breathable Performance

In addition, in order to study the effect of AMP concentration on the wettability of the modified membranes surface, the static and dynamic wetting behaviors were tested. As presented in Figure 4a, the calculated Ra values of the PAN, PAN@AMP-1, PAN@AMP-2 and PAN@AMP-4 membranes were 518, 510, 470, 174 and 154 nm, respectively. The decreasing Ra values confirmed the decreased surface roughness of modified PAN@AMP membranes towards the AMP increasing. When tested with static water droplets, as a result of the existence of polarity cyanogroup, the untreated PAN nanofibrous membranes had the certain hydrophilicity with 69.5° of the contact angle. After the modification of AMP as low surface energy substances, the surface property was changed from hydrophilic to hydrophobic, and the WCA of PAN@AMP-0.5 membranes was 136.4°. When further increasing the concentration to 1 wt%, the WCA increased to 137.2°. However, when the concentration of AMP continued to increase to 2 wt%, the film-shaped adhesion structure resulted in the reduction of the surface roughness of the membranes, the WCA decreased to 129.3°, and remained unchanged at the concentration of 4 wt% (129.6°). This phenomenon was the synergistic effect between the low surface energy and rough surface.

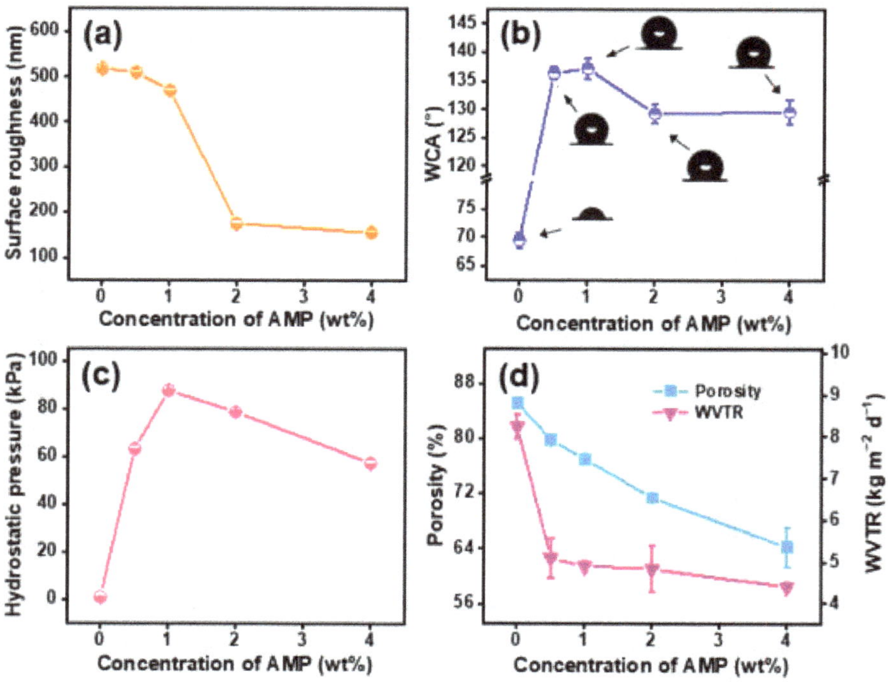

Figure 4. Surface wettability and waterproof breathable permeability of hydrophobically-modified PAN@AMP membranes with different concentrations of AMP: (a) Ra values, (b) contact angle, (c) hydrostatic pressure, (d) porosity and WVTR.

When the dynamic water method was used for the test, the nanofibrous membranes gradually became wet under a certain period of time, and were generally characterized by the hydrostatic pressure, as shown in **Figure 4**c. The PAN original nanofibrous membranes were wetting quickly due to their good hydrophilicity, and the hydrostatic pressure was only 1.1 kPa. While with the introduction of AMP, the hydrostatic pressure of the PAN@AMP nanofibrous membranes increased obviously. When the concentration was 1 wt%, the hydrostatic pressure increased to 87.6 kPa, which was caused by the increase of the contact angle and the decrease of the pore size of the nanofibrous membranes. When the concentration was further increased, the hydrostatic pressure decreased, and it dropped to 57.5 kPa when the concentration was 4 wt%. This was because the morphology and structure of the nanofibrous membranes were affected when the AMP content was too high, the surface roughness was reduced resulting in the decrease of the contact angle, and the hydrostatic pressure of the hydrophobically-modified nanofibrous membranes was reduced accordingly.

Meanwhile, the influence of different concentrations of AMP on the porosity and WVTR of the nanofibrous membranes was also explored, as shown in **Figure 4**d. The porosity of the original PAN membranes was 85.2% and decreased greatly with the increase of AMP concentration. When the concentration increased to 4 wt%, the porosity decreased to 64.3%. This decreasing trend was due to the introduction of AMP, which gradually produced a wide range of film-shaped adhesion structures on the surface of the nanofibrous membranes leading to the reduction of pores between nanofibers. Since the untreated nanofibrous membranes have good hydrophilicity and are easily wetted by water, the positive cup method was used to measure the WVTR of the original PAN and the PAN@AMP nanofibrous membranes. As shown in **Figure 4**d, it was found that the WVTR of the original membranes was 8.2 $kg{\cdot}m^{-2}{\cdot}d^{-1}$, and with the increase of AMP content, the WVTR showed a continuous downward trend. Finally, as the concentration increased to 4 wt%, the WVTR decreased to 4.4 $kg{\cdot}m^{-2}{\cdot}d^{-1}$, which was consistent with the change trend of porosity.

Comparing these testing data, the treatment with 1 wt% AMP could give the best comprehensive performance of the nanofibrous membranes with the strength of 10.7 MPa, the strain of 58.2%, the contact angle of 137.2°, the hydrostatic pressure of 87.6 kPa and a WVTR of 4.9 $kg{\cdot}m^{-2}{\cdot}d^{-1}$. Therefore, PAN@AMP-1 was selected as the substrate in subsequent experiments to study the effects of different concentrations of MDI on the structure and performance of nanofibrous membranes.

3.2. Effect of MDI Concentration

3.2.1. Morphology and Structure

It was proof-of-concept and reported in the previous literature that isocyanate (–NCO) in MDI may react with amino (–NH) in AMP to generate new substances under certain conditions [27]. To confirm this, PAN@AMP-1 was selected as the substrate for being modified through introducing MDI. As seen in **Figure 5**, after the introduction of MDI into the PAN@AMP-1 nanofibrous membranes, the adhesion structure slightly increased with the increase of MDI concentration. Moreover, it can be seen that most of the newly added MDI were captured on the film-shaped adhesion structure formed by AMP, facilitating the in situ cross-linking reaction between MDI and AMP.

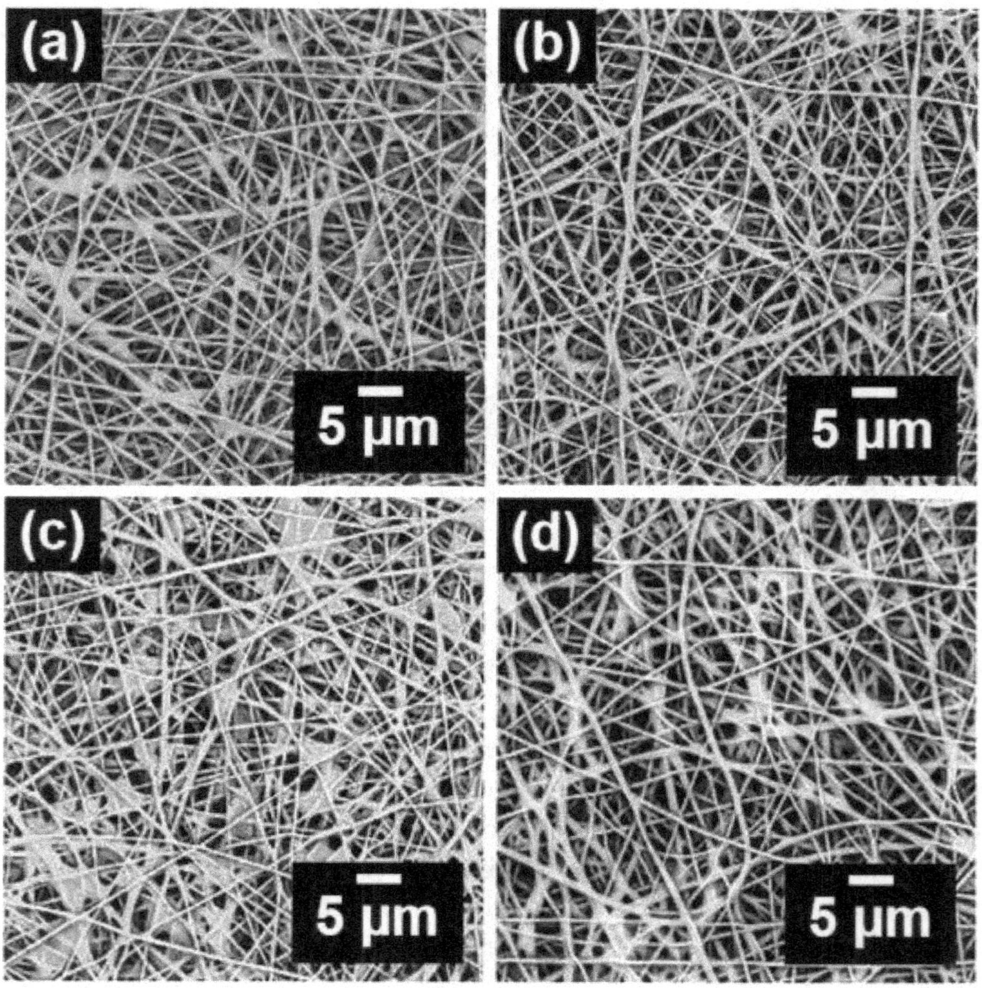

Figure 5. SEM images of PAN@AMP-1/MDI nanofibrous membranes modified by in situ cross-linking of MDI with different concentrations: (a) 0.5, (b) 1, (c) 2 and (d) 4 wt%.

3.2.2. FT-IR Spectral Characterization, XPS Analysis and Mechanical Properties

As shown in Figure 6a and Figure S2, only a typical stretching vibration absorption peak of cyanogroup (C≡N) was observed at 2243 cm^{-1} in regard to the PAN original nanofibrous membranes. The new peaks of 1259 cm^{-1}, 1089 cm^{-1} and 1014 cm^{-1} appeared in the PAN@AMP-0.5 nanofibrous membranes in the Figure S2. The bending vibration absorption peak of C-H in Si-CH^3 is 1259 cm^{-1}, and the double peaks at 1089 cm^{-1} and 1014 cm^{-1} are the characteristic peaks of the asymmetric stretching vibration of Si-O-Si group [28,29], which indicated that AMP was indeed attached to the nanofibrous membranes. Furthermore, the intensity of the three peaks was also enhanced with the increase of AMP concentration. Besides, the characteristic peak of the -NH group in AMP was at 1540 cm−1, which was not very obvious due to the relatively small content of amino in AMP.

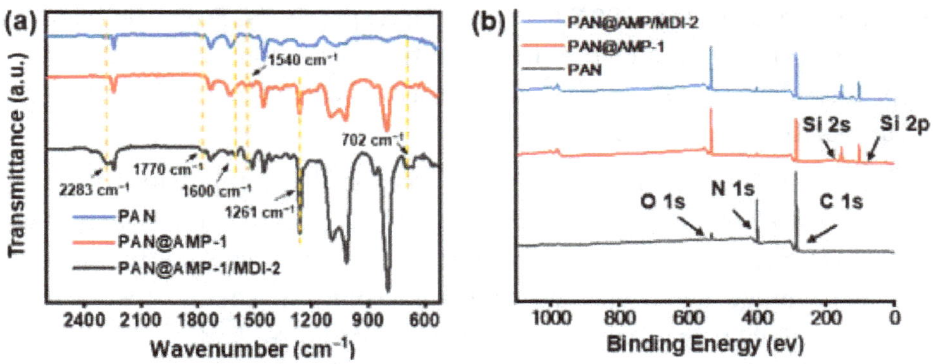

Figure 6. (a) FT-IR spectra and (b) XPS survey spectra of PAN original membranes, PAN@AMP-1 and PAN@AMP-1/MDI-2 nanofibrous membranes.

In order to verify the in situ cross-linking reaction between MDI and AMP, the infrared spectra of PAN original membranes, $PAN@AMP^{-1}$ nanofibrous membranes and PAN@AMP-1/MDI^{-2} membranes were analyzed in Figure 6a. The analysis showed that the characteristic peak of -NCO group in MDI was observed at 2283 cm^{-1} in the $PAN@AMP^{-1}$/MDI^{-2} nanofibrous membranes, indicating that MDI was indeed attached to the PAN@AMP-1/MDI-2 nanofibrous membranes. The characteristic peak of -NCO group at 2283 cm^{-1} suggested that MDI was not fully involved in the cross-linking reaction. Besides, the characteristic peak of the -NH group in AMP only existed in $PAN@AMP^{-1}$ nanofibrous membranes at 1540 cm^{-1}. After the introduction of 2 wt% MDI, the characteristic peak of the -NH group in AMP disappeared, and a new characteristic peak appeared at 1600 cm^{-1}, which was the stretching vibration absorption peak of the carbonyl group (-C=O) of the urea structure. The absorption peak of the amine group of the urea structure was 1317~1261 cm^{-1}, and 702 cm^{-1} was the bending vibration absorption peak of -NH. The alteration of the spectra indicated that the -NCO group in MDI and the -NH group in AMP indeed underwent in situ cross-linking reaction and generated polyurea. In addition, due to the strong activity of MDI, some monomers tended to self-polymerize into dimers [30]. The characteristic peak at 1770 cm^{-1} in Figure 6a was the stretching vibration absorption peak of the carbonyl group (–CO) of MDI dimers.

Figure 6b exhibited the typical XPS survey spectra of the PAN original membranes, PAN@AMP-1 and PAN@AMP-1/MDI-2 nanofibrous membranes. For the PAN original membranes, only carbon, nitrogen and oxygen were observed. The intensity of nitrogen in PAN@AMP-1 membranes was reduced due to the adhesion structures interspersing into the nanofibrous membranes. In order to verify the stability of the composition of the PAN@AMP-1 nanofibrous membranes depends on the interaction of PAN and AMP, XPS analysis of the samples before and after breathability tests using different pressures were investigated. Table S1 exhibited the atomic ratios of the samples before and after breathability tests using different pressures at 100 Pa and 200 Pa, respectively. From the results, there was no significant change of the atomic ratios of carbon, nitrogen, oxygen and silicon of the relevant membranes, which verified the stability of the composition of the composite nanofibrous materials.

Subsequently the modification of MDI increased the intensity of nitrogen in PAN@AMP-1/MDI^{-2} nanofibrous membranes, which were constant with the measured compositions for those samples, as summarized in Table 1. In order to further define the surface coverage, the details of the high-resolution N 1s peak were presented in Figure S3. The PAN original membranes exhibited a peak at 399.2 eV, which are assigned to –C≡N group [31,32]. After being modified by AMP with the adhesive structure on the surface of the PAN membranes, the peak area of –C≡N group at 399.2 eV was obviously smaller than that of the PAN original membranes. As shown in Figure S3b, PAN@AMP-1 nanofibrous membranes demonstrated the binding energies at about 399.9 eV, which is attributed to –C–N group; the result was in good agreement with that previously reported [33]. PAN@AMP-1/MDI-2 membranes presented the peaks at 399.9 and 400.9 eV, which were assigned to –C–N group and amide group (–CONH–), which was consistent with the FT-IR results (Figure 6a).

Table 1. Atomic ratios of carbon, nitrogen, oxygen and silicon on the surface of PAN, PAN@AMP^{-1} and PAN@AMP^{-1}/MDI^{-2} nanofibrous membranes. Data are calculated from XPS.

Samples	Atomic Percent (%)			
	C	O	N	Si
PAN nanofibrous membranes	75.31	3.55	21.13	-
PAN@AMP-1 nanofibrous membranes	52.38	23.48	2.16	21.98
PAN@AMP-1/MDI-2 nanofibrous membranes	54.27	22.34	3.51	19.89

Based on FT-IR and XPS analysis, polyurea was generated by an in situ cross-linking reaction between isocyanate components and amino compounds, which has good strength and wear resistance [**34**]. Therefore, the influence of different MDI contents on the mechanical properties of PAN@AMP/MDI membranes were investigated, as shown in **Figure 7.** From **Figure 3,** the tensile stress of PAN@AMP-1 nanofibrous membranes was 10.7 MPa, and the strain was 58.2%. Comparing with the mechanical property of PAN@AMP-1 nanofibrous membranes, the modification of MDI improved the tensile strength, while on the contrary the strain decreased. When the MDI concentration increased to 4 wt%, the strength increased to 12.7 MPa, and the strain decreased to 39.2%. Compared with the PAN@AMP-1 nanofibrous membranes, the strength increased by 19%, indicating that the introduction of MDI had a certain improvement in the tensile stress of nanofibrous membranes, which is mainly due to the production of polyurea which increased the strength of the modified nanofibrous membranes. The decreasing trend of the strain was because the adhesion structure limited the sliding of the nanofibers in the membranes.

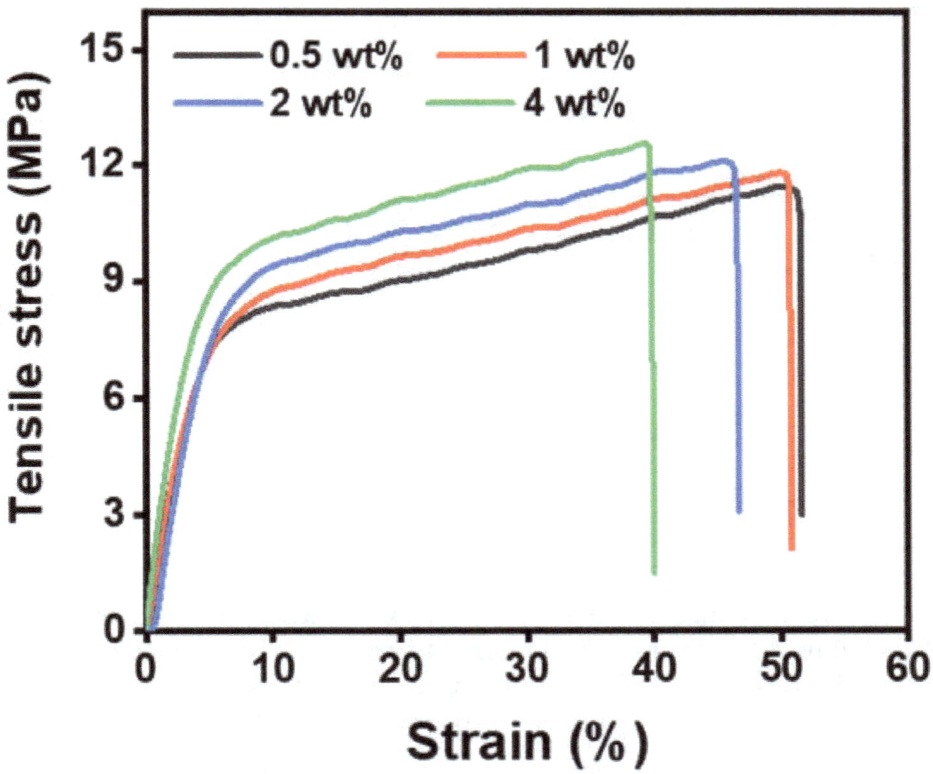

Figure 7. Stress–strain curves of PAN@AMP-1/MDI nanofibrous membranes modified with different concentrations of MDI.

3.2.3. Pore Structure, Surface Wettability and Waterproof Properties

After the apparent morphology of the PAN@AMP/MDI membranes was observed by SEM and their chemical structure was observed by infrared spectra, the microscopic pore structure of nanofibrous membranes was subsequently studied, as shown in **Figure 8**a and b. In the test of the previous system, it was found that the pore size distribution of PAN@AMP-1 was in the range of 0.81–1.66 μm, with an average pore size of 1.07μm. But after in situ cross-linking with MDI, the pore size of the nanofibrous membranes decreased obviously. Increasing the MDI concentration from 0.5 wt% to 4 wt%, the average pore size decreased from 0.98 μm to 0.70 μm, and the region of concentrated pore size distribution moved towards smaller pore size.

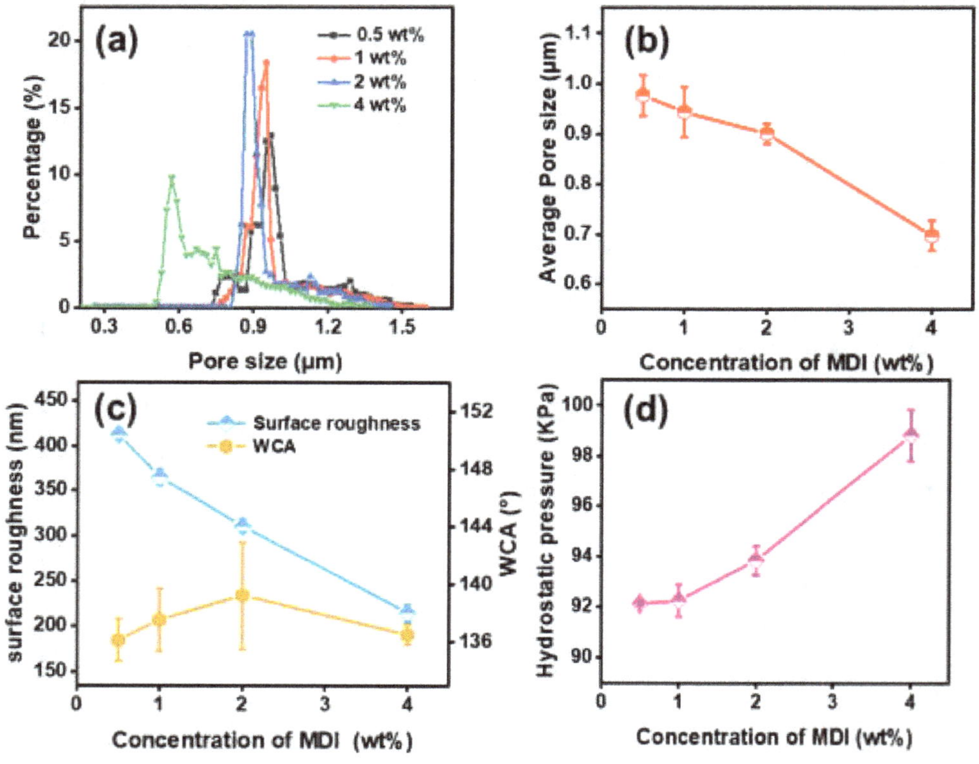

Figure 8. Pore structure, surface wettability and waterproof properties of PAN@ AMP-1/MDI membranes modified by in-situ cross-linking of MDI with different concentrations: (a) pore size distribution, (b) average pore size, (c) Ra values and contact angle, (d) hydrostatic pressure.

If the as-prepared membranes could be truly applied to the outdoor field, waterproofness is an essential property. In order to study the waterproof performance of the PAN@AMP-1/MDI nanofibrous membranes, the static contact angle and Ra values were measured, as shown in **Figure 8**c. In the case of increasing MDI content, the Ra values of the nanofibrous membranes surface were reduced from 413 to 215 due to the slightly increased adhesion structure, which was constant with the morphology observation (**Figure 5**). However, the value of the contact angle remained basically unchanged in a range of 136.1°–139.2°, which indicated that the low surface energy provided by the AMP with the concentration of 1 wt% conducted the main function. Thereafter, the hydrostatic pressure of the modified membranes was measured by the dynamic hydrostatic pressure method, as shown in **Figure 8**d. When the concentration of MDI increased, the hydrostatic pressure of the nanofibrous membranes also showed an increasing trend, increasing from 87.6 kPa of PAN@AMP-1 nanofibrous membranes to 98.8 kPa of PAN@AMP-1/MDI-4 nanofibrous membranes. This alteration was resulted from the combined effect of the pore size and surface wettability of the nanofibrous membranes. After the in situ cross-linking modification by MDI, the pore size of the nanofibrous membranes was reduced, so that the hydrostatic pressure was increased to a certain extent.

3.2.4. Moisture and Air Permeability

In addition, moisture and air permeability are also the necessary characteristics of outdoor sports clothing. Since the change of porosity can cause the change of the moisture and air permeability of the nanofibrous membranes, it is necessary to test the porosity of the modified membranes, as shown in **Figure 9**a. With the increase of MDI concentration from 0.5 wt% to 4 wt%, the porosity decreased slightly from 71.1% to 67.1%.

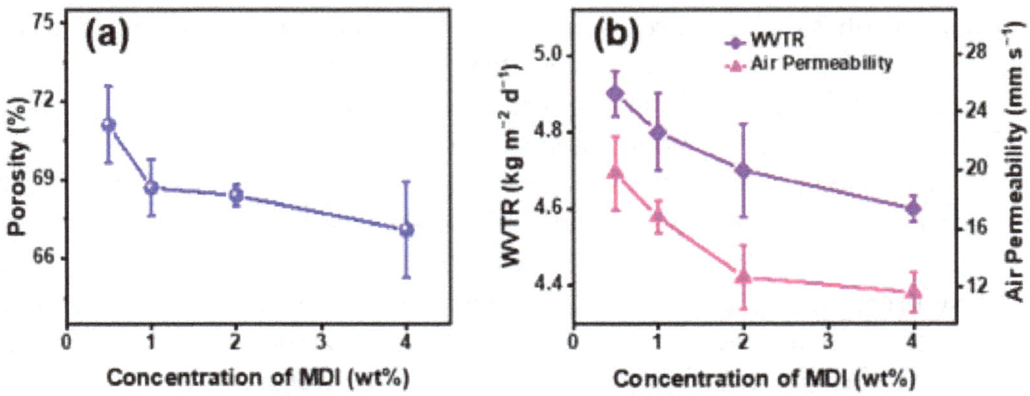

Figure 9. Moisture and air permeability of PAN@AMP-1/MDI nanofibrous membranes modified by in situ cross-linking of MDI with different concentrations: (a) porosity, (b) WVTR and air permeability.

Generally speaking, WVTR and air permeability are usually used to characterize the moisture and air permeability of nanofibrous membranes. As shown in Figure 9b, with the increase of MDI content, the air permeability presented a decreasing trend from 19.8 mm/s to 11.6 mm/s, which was consistent with the change of porosity. The decreasing trend of porosity also resulted in the WVTR reduction.

The WVTR of PAN@AMP-1 nanofibrous membranes was 4.9 kg·m^{-2}·d^{-1} and decreased to 4.6 kg·m^{-2}·d^{-1} after being modified with the concentration of MDI at 4 wt%.

The above experiments showed that the introduction of MDI content improved the hydrostatic pressure and stress to a certain extent, but the air permeability and WVTR decreased. Comprehensively, it was considered that the MDI concentration of 2 wt% had the best performance, with the tensile stress of 12.1 MPa, the contact angle of 139.2°, the hydrostatic pressure of 93.8 kPa, the WVTR of 4.7 kg·m−2·d−1 and the air permeability of 12.7 mm/s. Referring to the GB/T 40910-2021 standard "Evaluation of waterproof and breathable properties of textiles", PAN@AMP-1/MDI-2 nanofibrous membranes have excellent waterproof and breathable properties and were expected to be applied in the outdoor clothes field.

4. CONCLUSIONS

In this study, the electrospun PAN nanofibrous membranes were modified by AMP and MDI with fluorine-free hydrophobic modification and in situ cross-linking modification. Through performing the morphology structure and pore structure observation, mechanical properties, surface wettability and waterproof breathable performance of nanofibrous membranes, the effects of different concentrations of AMP and MDI were explored, so as to acquire waterproof breathable functional membranes with excellent performance. Firstly, after the hydrophobic modification of different concentrations of fluorine-free hydrophobic agents, a wide range of film-shaped adhesion structures were established among the nanofibers, and the adhesion points were connected to each other, so as to improve the waterproofness and strength of the membranes effectively. Considering the comprehensive performance, PAN@AMP-1 modified membranes were selected for the subsequent in situ cross-linking reaction via MDI modification. Subsequently, through the FT-IR spectrum analysis, it was found that the –NCO group in MDI reacted with the –NH group in AMP to form polyurea. When the MDI concentration was 2 wt%, the prepared PAN@AMP-1/MDI-2 membranes presented the relatively good property with the hydrostatic pressure of 93.8 kPa, the moisture permeability of 4.7 kg·m^{-2}·d^{-1}, the air permeability of 12.7 mm/s and the strength of 12.1 MPa, which indicated that the as-prepared membranes have extensive application prospects in the fields of waterproof breathable textiles.

5. CONCISE SUMMARY

Waterproof breathable functional membranes have broad application prospects in the field of outdoors textiles. The fluorine-containing microporous membranes of the mainstream functional products easily cause harm to the environment, and thus, the fluorine-free environmental nanofibrous membranes are an important development direction for functional membranes. In this subject, the electrospun polyacrylonitrile nanofibrous membranes were first hydrophobically modified by amino functional modified polysiloxane (AMP), followed by in situ cross-linking modified with 4, 4'-methyl diphenylene diisocyanate (MDI). The fluorine-free modification by AMP altered the surface of the membranes from hydrophilic to hydrophobic, and greatly improved the waterproof properties with the hydrostatic pressure reaching to 87.6 kPa. In addition, the formation of bonding points and the in situ preparation of polyuria through the reaction between the isocyanate in MDI and the amino group in AMP, could improve the mechanical properties effectively. When using AMP with the concentration of 1 wt% and MDI with the concentration of 2 wt%, the relatively good comprehensive performance was obtained with good water resistance (93.8 kPa), modest vapor permeability (4.7 kg m^{-2} d^{-1}) and air permeability (12.7 mm/s). Based on these testing data, the modified nanofibrous membranes had excellent waterproof and breathable properties, which has future potential in outdoor sports apparel.

REFERENCES

1. Ahn, H.W.; Park, C.H.; Chung, S.E. Waterproof and breathable properties of nanoweb applied clothing. *Text. Res. J.* **2011**, *81*, 1438–1447. **[Google Scholar]**
2. Meng, Q.B.; Lee, S.I.; Nah, C.; Lee, Y.S. Preparation of waterborne polyurethanes using an amphiphilic diol for breathable waterproof textile coatings. *Prog. Org. Coat.* **2009**, *66*, 382–386. **[Google Scholar] [CrossRef]**
3. Xing, L.L.; Zhou, Q.Q.; Chen, G.Q.; Sun, G.; Xing, T.L. Recent developments in preparation, properties, and applications of superhydrophobic textiles. *Text. Res. J.* **2022**, *92*, 00405175221097716. **[Google Scholar] [CrossRef]**
4. Zhou, W.; Gong, X.B.; Li, Y.; Si, Y.; Zhang, S.C.; Yu, J.Y.; Ding, B. Waterborne electrospinning of fluorine-free stretchable nanofiber membranes with waterproof and breathable capabilities for protective textiles. *J. Colloid Interface Sci.* **2021**, *602*, 105–114. **[Google Scholar] [CrossRef] [PubMed]**
5. Greszta, A.; Bartkowiak, G.; Dąbrowska, A.; Gliścińska, E.; Machnowski, W.; Kozikowski, P. Multilayer nonwoven inserts with aerogel/pcms for the improvement of thermophysiological comfort in protective clothing against the cold. *Materials* **2022**, *15*, 2307. **[Google Scholar] [CrossRef] [PubMed]**
6. Yue, Y.P.; Gong, X.B.; Jiao, W.L.; Li, Y.; Yin, X.; Si, Y.; Yu, J.Y.; Ding, B. In-situ electrospinning of thymol-loaded polyurethane fibrous membranes for waterproof, breathable, and antibacterial wound dressing application. *J. Colloid Interface Sci.* **2021**, *592*, 310–318. **[Google Scholar] [CrossRef]**
7. Mukhopadhyay, A.; Midha, V.K. A review on designing the waterproof breathable fabrics part I: Fundamental principles and designing aspects of breathable fabrics. *J. Ind. Text.* **2008**, *37*, 225–262. **[Google Scholar] [CrossRef]**

8. Saceviciene, V.; Juciene, M.; Bieliuniene, V.; Cepauskiene, V.; Urbelis, V. Investigation of the wettability of the hydrophobic textile after mechanical treatments. *Proc. Est. Acad. Sci.* **2015**, *64*, 118. **[Google Scholar] [CrossRef]**
9. Gorji, M.; Jeddi, A.; Gharehaghaji, A.A. Fabrication and characterization of polyurethane electrospun nanofiber membranes for protective clothing applications. *J. Appl. Polym. Sci.* **2012**, *125*, 4135–4141. **[Google Scholar] [CrossRef]**
10. Hong, S.K.; Lim, G.; Cho, S.J. Breathability enhancement of electrospun microfibrous polyurethane membranes through pore size control for outdoor sportswear fabric. *Sens. Mater.* **2015**, *27*, 77–85. **[Google Scholar]**
11. Li, Y.; Zhang, X.; Si, Y.; Yu, J.Y.; Ding, B. Super-elastic fluorinated polyurethane nanofibrous membranes with simultaneously waterproof and breathable performance. *ACS Appl. Polym. Mater.* **2022**, *4*, 5557–5565. **[Google Scholar] [CrossRef]**
12. Gu, X.; Li, N.; Luo, J.; Xia, X.; Gu, H.; Xiong, J. Electrospun polyurethane microporous membranes for waterproof and breathable application: The effects of solvent properties on membrane performance. *Polym. Bull.* **2018**, *75*, 3539–3553. **[Google Scholar] [CrossRef]**
13. Mukhopadhyay, A.; Midha, V.K. A review on designing the waterproof breathable fabrics Part II: Construction and suitability of breathable fabrics for different uses. *J. Ind. Text.* **2008**, *38*, 17–41. **[Google Scholar] [CrossRef]**
14. Cheng, X.Q.; Jiao, Y.; Sun, Z.K.; Yang, X.B.; Cheng, Z.J.; Bai, Q.; Zhang, Y.J.; Wang, K.; Shao, L. Constructing scalable superhydrophobic membranes for ultrafast water–oil separation. *ACS Nano* **2021**, *15*, 3500–3508. **[Google Scholar] [CrossRef] [PubMed]**
15. Feng, X.J.; Jiang, L. Design and creation of superwetting/antiwetting surfaces. *Adv. Mater.* **2006**, *18*, 3063–3078. **[Google Scholar] [CrossRef]**
16. Sheng, J.L.; Zhang, M.; Luo, W.J.; Yu, J.Y.; Ding, B. Thermally induced chemical cross-linking reinforced fluorinated polyurethane/polyacrylonitrile/ polyvinyl butyral nanofibers for waterproof-breathable application. *RSC Adv.* **2016**, *6*, 29629–29637. **[Google Scholar] [CrossRef]**
17. Wang, H.J.; Chen, L.; Yi, Y.Q.; Fu, Y.J.; Xiong, J.; Li, N. Durable polyurethane/ SiO_2 nanofibrous membranes by electrospinning for waterproof and breathable textiles. *ACS Appl. Nano Mater.* **2022**, *5*, 10686–10695. **[Google Scholar] [CrossRef]**
18. Sheng, J.L.; Li, Y.; Wang, X.F.; Si, Y.; Yu, J.Y.; Ding, B. Thermal inter-fiber adhesion of the polyacrylonitrile/fluorinated polyurethane nanofibrous membranes with enhanced waterproof-breathable performance. *Sep. Purif. Technol.* **2016**, *158*, 53–61. **[Google Scholar] [CrossRef]**

19. Wang, Y.; Guo, Q.; Li, Z.; Li, J.; He, R.; Xue, K.; Liu, S. Preparation and modification of PVDF membrane and study on its anti-fouling and anti-wetting properties. *Water* **2022,** *14,* 1704. **[Google Scholar] [CrossRef]**

20. Wang, J.Q.; Li, Y.; Tian, H.Y.; Sheng, J.L.; Yu, J.Y.; Ding, B. Waterproof and breathable membranes of waterborne fluorinated polyurethane modified electrospun polyacrylonitrile fibers. *RSC Adv.* **2014,** *4,* 61068–61076. **[Google Scholar] [CrossRef]**

21. Mariussen, E. Neurotoxic effects of perfluoroalkylated compounds: Mechanisms of action and environmental relevance. *Arch. Toxicol.* **2012,** *86,* 1349–1367. **[Google Scholar] [CrossRef] [PubMed]**

22. Bougourd, J.; McCann, J. Designing waterproof and water repellent clothing for wearer comfort—A paradigm shift. In *Waterproof and Water Repellent Textiles and Clothing*; Woodhead Publishing: Cambridge, UK, 2018; pp. 301–345. **[Google Scholar]**

23. Webster, G. Potential human health effects of perfluorinated chemicals (PFCs). *Blood* **2010,** *4,* 1–10. **[Google Scholar]**

24. Zhang, P.; Ren, G.H.; Tian, L.L.; Li, B.; Li, Z.Z.; Yu, H.Q.; Wang, R.W.; He, J.X. Environmentally friendly waterproof and breathable nanofiber membranes with thermal regulation performance by one-step electrospinning. *Fiber. Polym.* **2022,** *23,* 2139–2148. **[Google Scholar] [CrossRef]**

25. Wang, Z.; Zhao, C.; Pan, Z. Porous bead-on-string poly (lactic acid) fibrous membranes for air filtration. *J. Colloid Interface Sci.* **2015,** *441,* 121–129. **[Google Scholar] [CrossRef] [PubMed]**

26. He, L.; Li, W.; Chen, D.; Yuan, J.; Lu, G.; Zhou, D. Microscopic mechanism of amino silicone oil modification and modification effect with different amino group contents based on molecular dynamics simulation. *Appl. Surf. Sci.* **2018,** *440,* 331–340. **[Google Scholar] [CrossRef]**

27. Gao, C.; Jin, Y.Z.; Kong, H.; Whitby, R.; Acquah, S.; Chen, G.Y. Polyurea-functionalized multiwalled carbon nanotubes: Synthesis, morphology, and Raman spectroscopy. *J. Phys. Chem. B* **2005,** *109,* 11925–11932. **[Google Scholar] [CrossRef][Green Version]**

28. Pirzada, T.; Arvidson, S.A.; Saquing, C.D.; Shah, S.S.; Khan, S.A. Hybrid carbon silica nanofibers through sol-gel electrospinning. *Langmuir* **2014,** *30,* 15504–15513. **[Google Scholar] [CrossRef]**

29. Liu, Y.H.; Xu, J.B.; Zhang, J.T.; Hu, J.M. Electrodeposited silica film interlayer for active corrosion protection. *Corros. Sci.* **2017,** *120,* 61–74. **[Google Scholar] [CrossRef]**

30. Boros, R.Z.; Rágyanszki, A.; Csizmadia, I.G.; Fiser, B.; Guljas, A.; Farkas, L.; Viskolcz, B. Industrial application of molecular computations on the dimerization of methylene diphenyl diisocyanate. *React. Kinet. Mech. Cat.* **2018**, *124*, 1–14. **[Google Scholar] [CrossRef]**

31. Pels, J.R.; Kapteijn, F.L.; Moulijn, J.A.; Zhu, Q.; Thomas, K.M. Evolution of nitrogen functionalities in carbonaceous materials during pyrolysis. *Carbon* **1995**, *33*, 1641–1653. **[Google Scholar] [CrossRef]**

32. Takahagi, T.; Shimada, I.; Fukuhara, M.; Morita, K.; Ishitani, A. XPS studies on the chemical structure of the stabilized polyacrylonitrile fiber in the carbon fiber production process. *J. Polym. Sci. Pol. Chem.* **1986**, *24*, 3101–3107. **[Google Scholar] [CrossRef]**

33. Tang, J.X.; He, N.Y.; Tan, M.J.; He, Q.G.; Chen, H. A novel substrate for in situ synthesis of oligonucleotide: Plasma-treated polypropylene microporous membrane. *Colloid Surf. A* **2004**, *242*, 53–60. **[Google Scholar] [CrossRef]**

34. Iqbal, N.; Sharma, P.K.; Kumar, D.; Roy, P.K. Protective polyurea coatings for enhanced blast survivability of concrete. *Constr. Build. Mater.* **2018**, *175*, 682–690. **[Google Scholar] [CrossRef]**

CHAPTER-4

BINARY POLYAMIDE-IMIDE FIBROUS SUPERELASTIC AEROGELS ARE DESIGNED TO EXCEL IN FIRE-RETARDANT SCENARIOS AND EXCEL IN HIGH-TEMPERATURE AIR FILTRATION APPLICATIONS

1. INTRODUCTION: AN OVERVIEW

Introducing a groundbreaking advancement in materials science, binary polyamide-imide fibrous superelastic aerogels emerge as a cutting-edge solution for addressing both fire-retardant requirements and high-temperature air filtration needs. These innovative aerogels, characterized by their unique composition and superelastic properties, stand at the forefront of technological progress, offering enhanced performance in critical applications where fire resistance and efficient air filtration in elevated temperatures are paramount. This pioneering development not only signifies a significant leap in material engineering but also holds promising implications for industries seeking advanced solutions for safety and environmental control.

The 2019 coronavirus disease (COVID-19) still constitutes the forefront of public health concerns based on the fact that more than 600 million confirmed cases were reported and there are almost 40,000 new cases a day [1,2]. Particularly, environmental exposure to fine particles with a diameter of less than 2.5 μm (PM2.5) tends to increase the risk of COVID-19 attack because of the particles' ability to carry viruses and float on air [3,4]. To ensure the low level of PM2.5 in the air, it is important to remove particles from emission sources, involving power generation, coal combustion, industrial and agricultural emissions, and so on [5,6]. The common high temperature (50–250 °C) feature of emission sources lead to the high demand of air filtration materials with thermostable and fire-retardant properties [7,8].

Currently, the commercial high-temperature air filtration materials mainly include glass fibers, polyimide (PI) fibers, and aramid fibers, which all feature with micro-sized fiber diameters [**9**,**10**,**11**]. Resulting from the accumulation of fibers, the pores are relatively large, and lead to difficulties in achieving high filtration efficiency towards small but poisonous $PM_{2.5}$ [**12**,**13**]. As an alternative, electrospun nanofibers have gained the attention of researchers, owing to their small fiber diameters, which would contribute to small pores and large specific surface area, all being beneficial to high air-filtration performance [**14**,**15**]. However, it should not be ignored that electrospun fibrous assemblies tend to show a dense packing architecture, which would give rise to challenges in decreasing pressure drop and increasing the dust loading capacity. Therefore, fabricating thermostable electrospun nanofibrous assemblies with high porosity is highly recommended.

Fibrous aerogels have been acknowledged due to their ultrahigh porosity, which generally can easily achieve 99%. Furthermore, assembling electrospun nanofibers to create aerogels would enable the formation of interconnected channels, scalable pores, and large specific surface areas [16,17]. All of the above characteristics are beneficial for maintaining a relative balance between high filtration efficiency and low pressure drop, and also can contribute to a high dust-holding capacity [18,19]. Although PI fibrous filters are the leading product in the air filtration market, there are almost no PI nanofibrous aerogels reported, which might be because the intrinsic insoluble property of PI makes electrospun nanofibers hard to acquire [20,21]. Our previous work proved polyamide-imide (PAI) can be the alternative material of PI, owing to its similar molecular structure and thermostable property, and, more importantly, its good solubility. However, the already-prepared PAI nanofibrous aerogels have to contain stiff 15 wt% SiO2 fibers in order to meet the requirement of satisfying mechanical property. The tedious procedures and harsh conditions for preparing stiff SiO2 fibers limit their practical application in high-temperature air filtration [22].

Herein, we constructed dual-scale sized PAI/PBMI fibrous aerogels based on the fabrication of PAI/BMI filaments and fibers, which were obtained from wet spinning and electrospinning processes, respectively. PAI/BMI content in the wet spinning solution was regulated first to obtain PAI/BMI filaments with a large diameter and robust mechanical properties. PAI/BMI filaments were further combined with electrospun PAI/BMI fibers with small diameters and experienced dispersion, freeze-drying, and crosslinking procedures, resulting in dual-scale sized PAI/PBMI fibrous aerogels. PAI/BMI filaments, as the enhancing skeleton, were introduced into the electrospun PAI/BMI fiber networks with an optimized weight ratio. The resultant dual-scale sized PAI/BMI fibrous aerogels showed competitive comprehensive performances including high filtration efficiency, low air resistance, thermostability, and fire-retardance.

2. MATERIALS AND METHODS

2.1. Materials

Polyamide-imide (PAI) with a molecular weight of 20,000 was bought from Nantong Bolian Material Technology Co., Ltd. (Nantong, China), and another PAI with the molecular weight of 37,000 was purchased from Solvay S.A., USA; N,N-1,4-bismaleimide (BMI) and tertiary butanol were provided by Aladdin Chemistry Co., Ltd., Shanghai, China; N,N-dimethylformamide (DMF) was purchased from Tianjin Kailis Fine Chemical Co., Ltd., Tianjin, China. All chemicals were of analytical grade (AR) and used directly without further purification.

2.2. Preparation of Spinning Solution

The preparation of the spinning solution in this experiment involves two systems:

- Feasibility analysis of wet spinning. The specific operation steps are as follows: the PAI (Solvay, Mw = 37,000) and BMI powders with different mass ratios were added into DMF, and stirred for 0.5 h at room temperature to obtain a clear and transparent uniform gelatinous liquid.
- Electrospun solution preparation. The specific operation steps are as follows: weigh a certain quality of PAI and BMI powder and add them to DMF, stirring them for 10 h at room temperature to obtain a clear and transparent spinning precursor solution. The content of PAI (Bolian, Mw = 20,000) and BMI in the electrospinning solution were 34 wt% and 6.8 wt%, respectively.

Among them, BMI was used as a small molecule cross-linking agent in the later process to conduct thermal cross-linking with PAI to build PAI/BMI semi-interpenetrating polymer network (semi-IPN).

2.3. Preparation of Wet-Spun PAI/BMI Filaments

The spinning solution was pumped into a solidification tank with tap water at a rate of 2 mL h^{-1} using a 23 G needle, and then directly wound on a receiving roller operating at a constant speed of 5 rpm after solidification. After spinning, the filament was placed in a vacuum oven, heated to 60 °C, and kept at a constant temperature for 2 h to remove residual solvent. After natural cooling, the PAI/BMI filament was obtained.

2.4. Preparation of Electrospun PAI/BMI Fibers

The prepared spinning solution was loaded into three syringes capped with 20 G needles, and the pumping-out rate of solutions was fixed at 1 mL h^{-1}. A high and constant voltage of 20 kV was applied to trigger the transformation of solution to continuous jets. The charged jets experienced a flight distance of 20 cm, which is ensured by moving the needle tips far away from or near the collector. In addition, the sliding table carrying the rotating solution moved horizontally at a speed of 30 cm min^{-1}, and the speed of the collector was 50 rpm. PAI/BMI fibers were deposited on stainless steel ground-receiving rollers. During the electrospinning process, the ambient temperature was controlled at 25 ± 3 °C and the relative humidity was controlled at 30 ± 5%.

2.5. Preparation of Dual-Scale Filament/Fiber Aerogels

The fabrication of dual-scale sized PAI/BMI fibrous aerogel mainly involves three steps: homogeneous dispersion, freeze-drying, and bond cross-linking. The electrospun fiber membrane was cut into small pieces with side lengths of about 5 mm, and the wet-spun filaments were cut into short fibers of 5 mm, and these were immersed in 100 mL of water/tert-butanol mixture with a weight ratio of 1/4. The mixture was homogenized and dispersed at 2800 rpm for 0.3 min in a beater to obtain uniform fiber dispersion. The mixture was then transferred to a freezing mold in a liquid nitrogen environment at −196 °C to achieve rapid freezing of the dispersion. The uncross-linked fiber aerogels were obtained after the ice crystals were withdrawn from the frozen body by a vacuum dryer, which was further placed at 200 °C for 2 h to induce the in-suit polymerization of BMI. Thus, the dual-scale PAI/PBMI fibrous aerogels were obtained.

2.6. Characterization

The microstructure of wet-spun filaments, electrospun fiber, and dual-scale PAI/PBMI aerogels was observed by optical microscopy (K-ALPHA, Thermo Fisher Scientific, Waltham, MA, USA) and scanning electron microscopy (Phenom XL, 10 keV, Phenom-World, Eindhoven, The Netherlands). The thermal stability of the material was characterized by a comprehensive thermal analyzer (STA 449F5(TG-DSC). YG005E electronic single-yarn strength tester was used to test the tensile properties of wet-spun filament, test details are provided.

3. RESULT AND DISCUSSION

3.1. Fabrication of Wet-Spun PAI/BMI Filaments

In order to fabricate superelastic and durable aerogels, a new strategy is proposed to prepare PAI/BMI (Polyamide-imide/N,N-1,4-bismaleimide) fibrous aerogels with dual-scale fiber diameters. Firstly, high strength PAI filament was prepared by the wet spinning process as the reinforcement of fibrous aerogels. The fabrication process of wet-spun PAI filaments is shown in Figure 1a. During the wet spinning process, the bidirectional diffusion between the solvent (DMF) and coagulation bath (water) led to the phase separation between polymer and solvents, which resulted in the formation of filaments. At the same time, due to the slow diffusion between the DMF and water, the filaments' forming process was relatively mild, resulting in the formation of the circular cross-section. In addition, when the spinning solution met with the coagulation bath, the filaments cortex was formed quickly, and some defects or cracks appeared on the surface of the filaments. The cortex contraction rate was small, while the core contraction rate was large, producing a flaw and a cavity between the cortex and the core [23,24]. After the primary filaments were stretched by the subsequent process, the micropores were elongated in the shape of a shuttle. The obtained PAI filaments are shown in Figure 1b. The filaments presented with a yellow luster, smooth surface, circular cross-section, and porous structure inside.

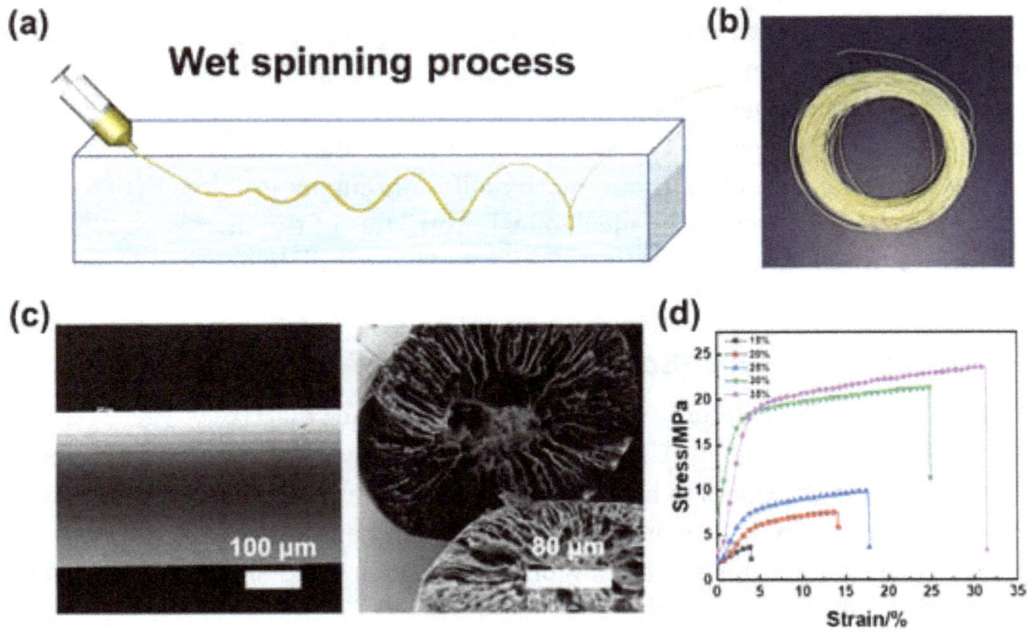

Figure 1. (a) The process of fabrication of PAI/BMI filaments. (b) PAI filaments. (c) The SEM images of PAI/BMI filament. (d) Stress-strain curves of filaments with different PAI content.

Figure 1c shows the electron microscopy (SEM) images of filaments and the SEM images of different PAI. The filaments' diameter showed negligible changes, however, the internal hole of the filaments decreased with the increase of PAI content. The reason for this was that when the concentration of the solution increased, the diffusion behavior of each molecule became harder, and then the diffusion coefficient of the solvent and coagulation bath decreased. Thus, as the PAI content increased, the higher the total solid content of the filament, the less likely it was to generate holes.

The stress-strain curves of filaments with different PAI contents are shown in **Figure 1**d. The tensile strength of the PAI filament first increased slightly when the PAI contents were less than 30 wt%. When the PAI content was more than 30 wt%, the viscosity of the spinning solution was too high, and the spinning could not be rapidly and stably produced. On the whole, when the PAI solute content was 30 wt%, the filaments had a breaking strength of 23 MPa, and the filaments preparation process still maintained good spinnability, which has both performance and cost advantages.

In addition, we further improved the mechanical properties of filaments by doping BMI into spinning solution. The tensile modulus of PAI filaments with different BMI contents before and after heat curing. The increased BMI content was beneficial to the improvement of filament strength, and the mechanical properties of filaments were greatly improved after heat curing. The reason for this was that, on the one hand, the in situ self-polymerization of the BMI monomer occurred in the thermal environment, generating a semi-interpenetrating polymer network (semi-IPN) that interweaves PAI and BMI on a molecular scale [**25**]. On the other hand, with the increase of BMI content, the solid content per unit length of filament increased, which improved the mechanical properties of the filaments. When the BMI content was 9 wt%, the tensile modulus had been greatly improved, and could reach 78.04 MPa.

3.2. Preparation of Dual-Scale Filament/Fiber Aerogels

PAI/BMI micro-sized fibers were prepared as the matrix of aerogels via electrospinning technology. The preparation process of the electrospun fibers is shown in **Figure 2**a. Under a high-voltage electric field, polymer solution pulled from the tip of the syringe would experience phase separation and an elongating process at the same time, depositing on the collectors and showing a disorderly arrangement morphology, as shown in **Figure 2**b.

Figure 2c depicts the synthetic route of the dual-scale PAI/BMI fibrous aerogels. The aerogels were prepared by blending wet-spun filaments and electrospun fibers. These two materials were first homogenized in water/tert-butanol mixture to form well-dispersed fiber dispersions, followed by freeze-drying to form an unjointed architecture pristine PAI/BMI aerogel. Driven by the moving solidification front, fibers in the dispersion were repelled and accumulated gradually in the unoccupied space of the growing cytosolic solvent crystals. This was controlled by complex and dynamic liquid fibers and fiber–fiber interactions [26]. In order to further promote the physical bonding between fibers, the uncross-linked PAI/BMI fibers after freeze-drying were heated at 220 °C for 2 h to form a cross-linked fiber network, resulting in the synthesized PAI/Polybismaleimide (PBMI) fibers with attractive compression recovery properties. The peak at 1608 cm^{-1} was caused by the tensile vibration of C=C, which was greatly reduced after in situ crosslinking, indicating that most of the C=C bonds had polymerized [27]. The intensity of the peaks at 748 cm^{-1} and 590 cm^{-1} was weakened, which can be attributed to out-of-plane C-H bending vibration and out-of-plane C-O bending vibration [28,29]. Notably, the wet-spun filaments in the PAI/PBMI fibrous aerogels acted as a rigid support, enhancing the structural stability of the aerogels.

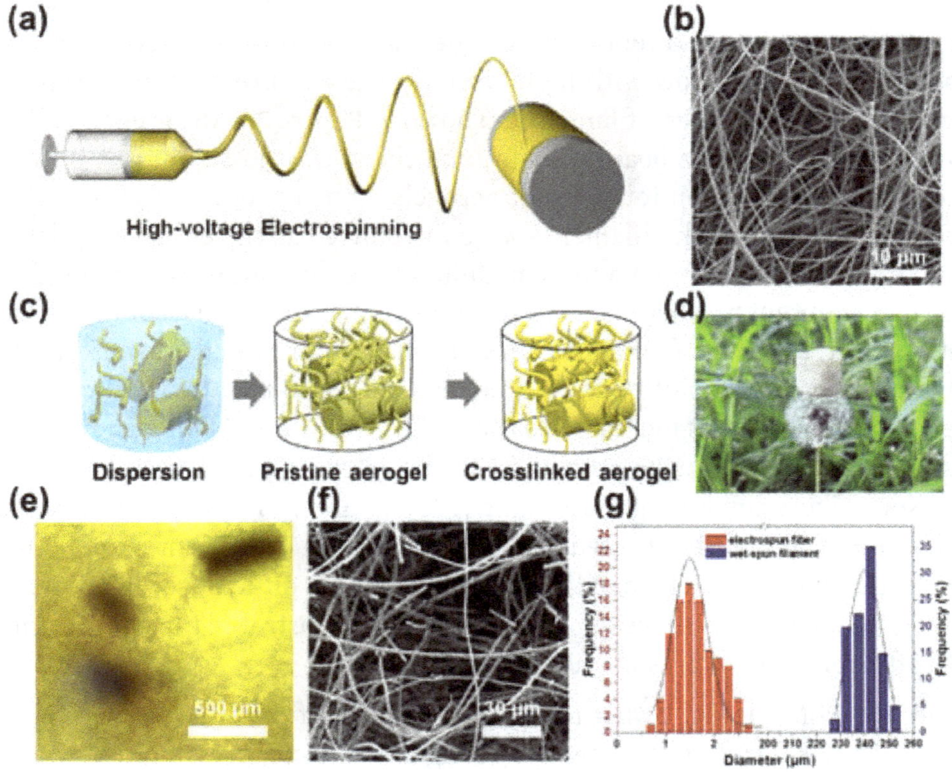

Figure 2. (a) The process of fabrication of PAI/BMI electrospun fiber. (b) The SEM images of PAI/BMI electrospun fibers. (c) The schematic diagram of PAI/PBMI aerogel preparation process. The PAI/PBMI aerogel (d) optical image and (e) SEM image. (f) Distribution of filaments and fibers within PAI/PBMI aerogel. (g) The diameter distribution of the PAI/PBMI aerogels.

The obtained aerogel showed a low bulk density of 0.011 g/cm3 and a high porosity of 99.2%, the white pom-poms of dandelion seeds showed negligible deformation, while supporting the ultralight aerogel (Figure 2d). Optical microscopy and SEM image of the microstructure of the obtained dual-scale PAI/

PBMI aerogels showed that filaments and fibers with totally different diameters existed in the aerogel. The micro-fibers inside the aerogel were firmly welded to the wet-spun filament scaffold to form a double network structure. The SEM image of the ultrafine fibers is shown in Figure 2f. The electrospun staple fibers are uniformly dispersed in the aerogel prepared after homogenous dispersion. The diameter distribution of the PAI/PBMI aerogels was evaluated, as shown in Figure 2g. The diameter of the electrospun fiber was mainly distributed from 2 to 4 μm, and the wet spinning filaments diameter was mainly distributed in 200~300 μm, illustrating that aerogels are composed of dual-scale sized fibers.

3.3. Compression Properties of Ultralight Composite Fiber Aerogel

The structural stability of filter media must be evaluated considering the continuous impaction from the airflow with high velocity. The compression property with a different mass ratio of fibers/filaments is shown in **Figure 3**a. All aerogels exhibited obvious nonlinear mechanical behavior, and there was no significant difference in the plastic deformation for the four aerogels after just one compression cycle. Interestingly, although the filaments were introduced as the enhancing skeleton, the maximum stress and the Young's modulus of the aerogels were not found when the filament content reached the largest proportion of 40%, which could be because the stiff, short, and thick filaments cannot form an intertwined network. With the change of the mass ratio of fibers/filaments from 6:4 to 8:2, the maximum stress and the Young's modulus increased to the highest value of 8.54 MPa and 17.36 MPa (**Figure 3**b), which might be attributed to the interaction network between the electrospun fibers combining more tightly. Afterwards, the compressive modulus of the aerogel was greatly reduced due to the reduction in the content of filaments used for the support structure in the aerogel, making it difficult to withstand large compressive stress. The above experimental results illustrated the contribution of "stiff-soft" fibrous networks on mechanical property.

The PAI/PBMI aerogel with mass ratio of fibers/filaments equal to 8:2 was subjected to 1000 loading and unloading fatigue cyclic compression tests with ε of 20% (which means the strain always maintains 20%) and a loading rate of 50 mm/min^{-1} (**Figure 3**c), which showed slight plastic deformation (3.34% for the 50th, 4.26% for the 1000th), and the structural robustness was outstanding. In contrast, typical polymer foams strained at 60% exhibited 20–30% plastic deformation, while other fibrous foams exhibited greater than 20% plastic deformation at similar strains [**30,31,32**]. Likewise, the stiffness or strength of PAI/PBMI aerogels did not decrease significantly after 1000 compression cycles, and retained more than 80% of the original Young's modulus and maximum stress (**Figure 3**d). The dual-scale composite aerogel material had good compression recovery and could be used for a long time.

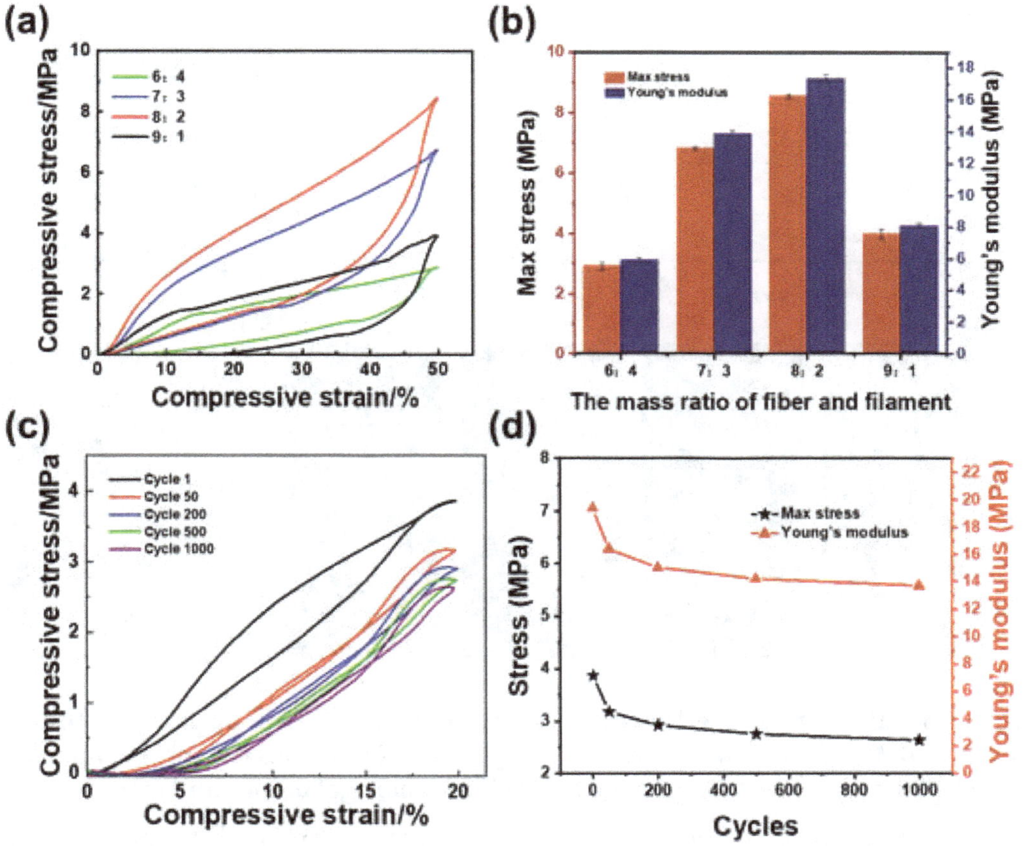

Figure 3. (a) The compression curves of different mass ratio of fibers/filaments aerogels. (b) The Young's modulus, maximum stress of PAI/PBMI aerogels. (c) The cyclic compression mechanism and (d) Young's modulus, maximum stress of PAI/PBMI aerogel.

3.4. High-Temperature Air Filtration Application of Aerogels

Considering the complex components of industrial dusts, we evaluated the air filtration of aerogels (the mass ratio of fibers/filaments equal to 8:2) towards particles with a wide range of sizes. Figure 4a shows the filtration efficiency of the PAI/PBMI aerogels for particles with particle sizes including 0.3, 0.5, 1, 3, 5, and 10 μm (PM0.3, PM0.5, PM1, PM3, PM5, and PM10). Obviously, higher filtration efficiencies could be reached while filtering larger particles, which illustrates the physical sieving mechanism of aerogels. In addition, for the PM3 accounting for the largest proportion within industrial dusts, aerogel materials exhibited a very high filtration efficiency of 90.78% (PM2.5) for fine particles while maintaining a low pressure drop of 20 Pa, and they showed a high quality-factor (QF) of 0.12 Pa-1, showing a cost-effective prospect in practical applications.

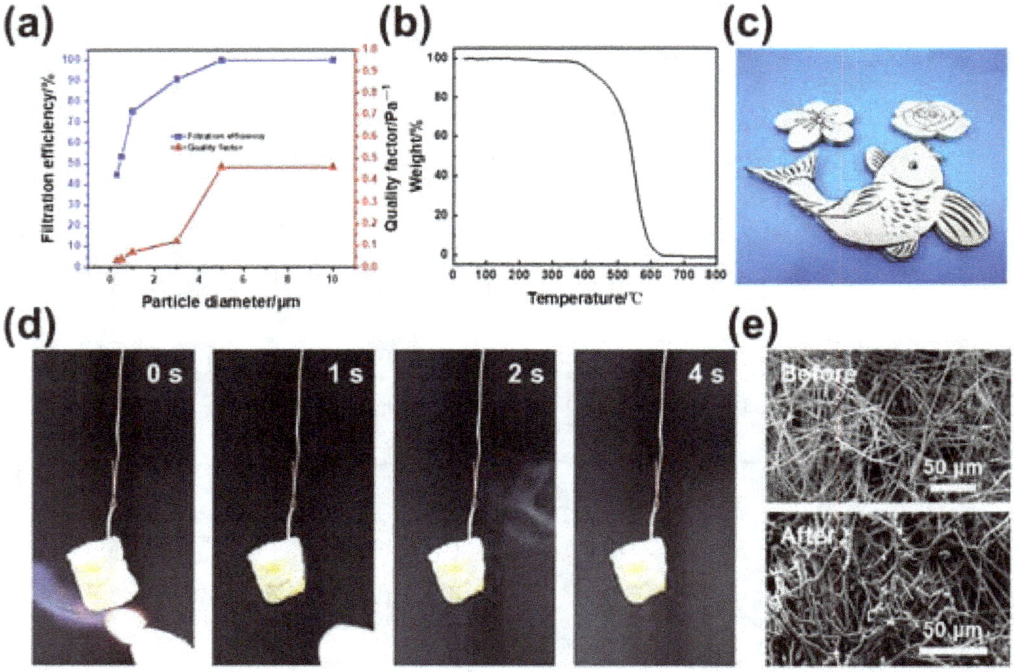

Figure 4. (a) Filtration efficiency. (b) TG curves. (c) An optical photograph of PAI/PBMI aerogel with diverse shapes. (d) Flame retardant performance demonstration. (e) The SEM image of PAI/PBMI aerogel before and after burning.

The thermostable property is very necessary for high-temperature air filters. **Figure 4**b shows the TG curve of the PAI/PBMI aerogel. The thermal decomposition of an aerogel is mainly divided into three stages. The first stage is 0~520 °C, during which the sample has only a slight weight loss, which is attributed to the loss of water molecules and other volatile small molecules in the sample. In the second stage (520~570 °C), significant weight loss occurs and the thermal degradation of aerogel components is serious. In the third stage, the residual carbon product continues to decompose at a high temperature to form the final residual carbon component. Because of the simplicity of the assembly process and the easy availability of electrospun fibers in our method, there is great versatility in controlling the shape of the PAI/PBMI aerogels. As shown in **Figure 4**c, PAI/PBMI integrated aerogels with desired shapes, such as plum blossom, peach blossom, and complex koi shapes, can be easily prepared. The demonstration process of the flame retardancy of aerogels is shown in **Figure 4**d. At the moment of contact with the high temperature flame, the aerogel burns slightly, but it goes out immediately after leaving the flame, and the fibers burn independently in the combustion process without dropping. According to the SEM images of aerogel materials before and after combustion (**Figure 4**e), most of the fibers in the aerogel materials remain independent as fibers after sintering at high temperature, while the conventional droplet polymer fibers show the form of molten film [**33**]. This is mainly because PAI polymer has excellent flame retardancy due to the existence of amide groups in the molecular chain.

4. CONCLUSIONS

In conclusion, we fabricated PAI/PBMI aerogels composed of dual-scale sized PAI/BMI filaments and fibers by exposing them to wet spinning and electrospinning process first. PAI/BMI filaments (with the average diameter of 239.83 μm) played the role of reinforcing skeletons, and PAI/BMI fibers (with an average diameter of 1.17 μm) served as the aerogel matrix. While the mass ratio of fibers/filaments was equal to 8:2, the maximum stress and Young's modulus exhibited the highest values of 8.54 MPa and 17.36 MPa, respectively. In addition, the composite aerogels could bear 1000 compressive cycles with only 4.26% plastic deformation, implying their good structural stability. Dual-scale sized PAI/PBMI filaments/fiber aerogels also showed a high filtration efficiency of 90.78%. The resultant dual-scale sized PAI/PBMI fibrous aerogels showed a compressive stress of 8.36 MPa, a high filtration efficiency of 90.78% (particle diameter of 2.5 μm); particles with a diameter over 5 μm had 99.99% ultra-high filtration efficiency, a low pressure drop of 20 Pa, high QF of 0.12 Pa^{-1}, and attractive thermostable and fire-retardant properties (thermal decomposition temperature up to 342.7 °C), showing great potential in the field of high temperature filtration.

5. CONCISE SUMMARY

Fibrous air filtration materials are highly desirable for particle removal from high-temperature emission sources. However, the existing commercial filter materials suffer from either low filtration efficiency or high pressure drop, due to the difficulty in achieving small fiber diameter and high porosity simultaneously. Herein, we report a facile strategy to fabricate mechanical robust fibrous aerogels by using dual-scale sized PAI/BMI filaments and fibers, which are derived from wet spinning and electrospinning technologies, respectively. The creativity of this design is that PAI/BMI filaments can serve as the enhancing skeleton and PAI/BMI fibers can assemble into high-porosity interconnected networks, enabling the improvement of both mechanical property and air filtration performance. The resultant dual-scale sized PAI/PBMI fibrous aerogels show a compressive stress of 8.36 MPa, a high filtration efficiency of 90.78% (particle diameter of 2.5 μm); for particle diameter over 5 μm, they have 99.99% ultra-high filtration efficiency, a low pressure drop of 20 Pa, and high QF of 0.12 Pa^{-1}, as well as thermostable and fire-retardant properties (thermal decomposition temperature up to 342.7 °C). The successive fabrication of this material is of great significance for the govern of industrial dust.

REFERENCES

1. Azouji, N.; Sami, A.; Taheri, M. EfficientMask-Net for face authentication in the era of COVID-19 pandemic. *Signal Image Video Process.* **2022,** *16*, 1991–1999. **[Google Scholar] [PubMed]**

2. Worby, C.J.; Chang, H. Face mask use in the general population and optimal resource allocation during the COVID-19 pandemic. *Nat. Commun.* **2020,** *11*, 4049. **[Google Scholar] [CrossRef]**

3. Marčenko, E.; Lampret, Ž.; Prek, M. Correlation between Air Pollution and the Spread and Development of COVID-19 Related Disease. *Stroj. Vestn.—J. Mech. Eng.* **2022,** *68*, 272–280. **[Google Scholar]**

4. Mousavi, E.S.; Kananizadeh, N.; Martinello, R.A.; Sherman, J.D. COVID-19 outbreak and hospital air quality: A systematic review of evidence on air filtration and recirculation. *Environ. Sci. Technol.* **2021,** *55*, 4134–4147. **[Google Scholar]**

5. Deng, Y.; Lu, T.; Cui, J.; Samal, S.K.; Xiong, R.; Huang, C. Bio-based electrospun nanofiber as building blocks for a novel eco-friendly air filtration membrane: A review. *Sep. Purif. Technol.* **2021,** *277*, 119623. **[Google Scholar] [CrossRef]**

6. Xiao, J.; Liang, J.; Zhang, C.; Tao, Y.; Ling, G.; Yang, Q. Advanced materials for capturing particulate matter: Progress and perspectives. *Small Methods* **2018,** *2*, 1800012. **[Google Scholar]**

7. Zhang, A.; Li, H.; Zhang, A.; Zhou, J.; Yan, Y. High-temperature bearable polysulfonamide/polyacrylonitrile composite nanofibers for high-efficiency $PM_{2.5}$ filtration. *Compos. Commun.* **2021,** *23*, 100582. **[Google Scholar]**

8. Wang, H.; Lin, S.; Yang, S.; Yang, X.; Song, J.; Wang, D.; Wang, H.; Liu, Z.; Li, B.; Fang, M.; et al. High-Temperature Particulate Matter Filtration with Resilient Yttria-Stabilized ZrO_2 Nanofiber Sponge. Small 2018, 14, 1800258. **[Google Scholar] [CrossRef]**

9. Xie, F.; Wang, Y.; Zhuo, L.; Jia, F.; Ning, D.; Lu, Z. Electrospun Wrinkled Porous Polyimide Nanofiber-Based Filter via Thermally Induced Phase Separation for Efficient High-Temperature PMs Capture. *ACS Appl. Mater. Interfaces* **2020,** *12*, 56499–56508. **[Google Scholar]**

10. Mendy, A.; Wu, X.; Keller, J.L.; Fassler, C.S.; Apewokin, S.; Mersha, T.B.; Xie, C.; Pinney, S.M. Air pollution and the pandemic: Long-term PM2.5 exposure and disease severity in COVID-19 patients. *Respirology* 2021, 26, 1181–1187. [Google Scholar] [CrossRef] [PubMed]

11. Zou, C.; Shi, Y.; Qian, X. Characterization of glass fiber felt and its performance as an air filtration media. *J. Ind. Text.* **2022,** *51* (Suppl. S1), 1186S–1206S. [Google Scholar]

12. Xu, K.; Deng, J.; Tian, G.; Zhan, L.; Ma, J.; Wang, L.; Ke, Q.; Huang, C. Downy feather-like para-aramid fibers and nonwovens with enhanced absorbency, air filtration and thermal insulation performances. *Nano Res.* **2022**, *15*, 5695–5704. [**Google Scholar**]

13. Xu, K.; Zhan, L.; Yan, R.; Ke, Q.; Yin, A.; Huang, C. Enhanced air filtration performances by coating aramid nanofibres on a melt-blown nonwoven. *Nanoscale* **2022**, *14*, 419–427. [**Google Scholar**] [**PubMed**]

14. Li, Y.; Cao, L.; Yin, X.; Si, Y.; Yu, J.; Ding, B. Ultrafine, self-crimp, and electret nano-wool for low-resistance and high-efficiency protective filter media against $PM_{0.3}$. *J. Colloid Interface Sci.* **2020**, *578*, 565–573. [**Google Scholar**]

15. Li, Y.; Yin, X.; Yu, J.; Ding, B. Electrospun nanofibers for high-performance air filtration. *Compos. Commun.* **2019**, *15*, 6–19. [**Google Scholar**]

16. Adhikary, S.K.; Ashish, D.K.; Rudžionis, Ž. Aerogel based thermal insulating cementitious composites: A review. *Energy Build.* **2021**, *245*, 111058. [**Google Scholar**]

17. Shah, S.N.; Mo, K.H.; Yap, S.P.; Radwan, M.K.H. Effect of micro-sized silica aerogel on the properties of lightweight cement composite. *Constr. Build. Mater.* **2021**, *290*, 123229. [**Google Scholar**]

18. Wang, T.; Long, M.; Zhao, H.; Liu, B.; Shi, H.; An, W.; Li, S.; Xu, S.; Wang, Y. An ultralow-temperature superelastic polymer aerogel with high strength as a great thermal insulator under extreme conditions. *J. Mater. Chem. A* **2020**, *8*, 18698–18706. [**Google Scholar**]

19. Qiao, S.; Kang, S.; Zhu, J.; Wang, Y.; Yu, J.; Hu, Z.J. Facile strategy to prepare polyimide nanofiber assembled aerogel for effective airborne particles filtration. *Hazard. Mater.* **2021**, *415*, 125739. [**Google Scholar**]

20. Jiang, S.; Hou, H.; Agarwal, S.; Greiner, A. Polyimide Nanofibers by "Green" Electrospinning via Aqueous Solution for Filtration Applications. *ACS Sustain. Chem. Eng.* **2016**, *4*, 4797–4804. [**Google Scholar**]

21. Qiao, S.; Kang, S.; Zhang, H.; Yu, J.; Wang, Y.; Hu, Z. Reduced shrinkage and mechanically strong dual-network polyimide aerogel films for effective filtration of particle matter. *Sep. Purif. Technol.* **2021**, *276*, 119393. [**Google Scholar**]

22. He, Y.; Itta, A.K.; Alwakwak, A.; Huang, M.; Rezaei, F.; Rownaghi, A.A. Aminosilane-Grafted SiO_2–ZrO_2 Polymer Hollow Fibers as Bifunctional Microfluidic Reactor for Tandem Reaction of Glucose and Fructose to 5-Hydroxymethylfurfural. ACS Sustain. Chem. Eng. 2018, 6, 17211–17219. [Google Scholar]

23. Weisser, P.; Barbier, G.; Richard, C.; Drean, J. Characterization of the coagulation process: Wet-spinning tool development and void fraction evaluation. *Text. Res. J.* **2016**, *86*, 1210–1219. **[Google Scholar] [CrossRef]**
24. Xu, W.; Jambhulkar, S.; Ravichandran, D.; Zhu, Y.; Lanke, S.; Bawareth, M.; Song, K. A mini-review of microstructural control during composite fiber spinning. *Polym. Int.* **2022**, *71*, 569–577. **[Google Scholar]**
25. Li, Y.; Cao, L.; Yin, X.; Si, Y.; Yu, J.; Ding, B. Interpenetrating Polymer Network Biomimetic Structure Enables Superelastic and Thermostable Nanofibrous Aerogels for Cascade Filtration of $PM_{2.5}$. *Adv. Funct. Mater.* **2020**, *30*, 1910426. **[Google Scholar] [CrossRef]**
26. Si, Y.; Yu, J.; Tang, X.; Ge, J.; Ding, B. Ultralight nanofibre-assembled cellular aerogels with superelasticity and multifunctionality. *Nat. Commun.* **2014**, *5*, 5802. **[Google Scholar]**
27. Fan, S.L.; Boey, F.Y.C.; Abadie, M.J.M. UV Curing of a Liquid Based Bismaleimide-Containing Polymer System. *Express Polym. Lett.* **2007**, *1*, 397–405. **[Google Scholar]**
28. Bhattacharyya, A.S.; Kumar, S.; Sharma, A.; Kumar, D.; Patel, S.B.; Paul, D.; Dutta, P.P.; Bhattacharjee, G. Metallization and APPJ treatment of bismaleimide. *High Perform. Polym.* **2017**, *29*, 816–826. **[Google Scholar] [CrossRef][Green Version]**
29. Lin, Q.; Li, J.; Yang, Y.; Xie, Z. Thermal behavior of coal-tar pitch modified with BMI resin. *J. Anal. Appl. Pyrolysis* **2010**, *87*, 29–33. **[Google Scholar]**
30. Yao, W.; Mao, R.; Gao, W.; Chen, W.; Xu, Z.; Gao, C. Piezoresistive effect of superelastic graphene aerogel spheres. *Carbon* **2020**, *158*, 418–425. **[Google Scholar]**
31. Xie, J.; Niu, L.; Qiao, Y.; Lei, Y.; Li, G.; Zhang, X.; Chen, P. The influence of the drying method on the microstructure and the compression behavior of graphene aerogel. *Diam. Relat. Mater.* **2022**, *121*, 108772. **[Google Scholar] [CrossRef]**
32. Mathur, R.B.; Dhami, T.L.; Bahl, O.P. Shrinkage behaviour of modified PAN precursors—Its influence on the properties of resulting carbon fibre. *Polym. Degrad. Stab.* **1986**, *14*, 179–187. **[Google Scholar]**
33. Chen, L.; Wang, Y. A review on flame retardant technology in China. Part I: Development of flame retardants. *Polym. Adv. Technol.* **2010**, *21*, 1–26. **[Google Scholar]**

CHAPTER-5

CONDUCTING A NUMERICAL ANALYSIS, WE EXPLORE THE FIBER/AIR-COUPLING FIELD IN ANNULAR JET CONFIGURATIONS

1. INTRODUCTION: AN OVERVIEW

In the realm of fluid dynamics and fiber-based systems, a comprehensive understanding of the intricate interactions within the fiber/air-coupling field is paramount. This study embarks on a numerical analysis focused on unraveling the complexities inherent in annular jet configurations. By delving into the dynamics of how fibers interface with the surrounding air in such geometric arrangements, we aim to glean insights that have implications for various industrial applications.

Melt-blowing technology has the advantages of short process flow and high production efficiency, and it is one of the most rapidly developing nonwoven technologies. Melt-blown fibers can be used for air and liquid filter materials, isolation materials, absorbent materials, mask materials, thermal insulation materials, oil-absorbing materials and wipes, etc. [**1**,**2**,**3**].

As shown in **Figure 1**, it is a common die head and is one of the core components of melt-blowing equipment. The high-speed and high-temperature jet, which is formed by the die head can rapidly draw the polymer melt extruded from the spinneret into micron-scale or nanofiber. The airflow field under the melt-blowing die not only affects the diameter of the fiber, but it also determines the strength of the fiber [**4**]. Therefore, a lot of research work was carried out on it. Uyttendaele et al. [**5**] measured the velocity field under a single-hole annular melt-blown die with a Reynolds number ranging from 3400 to 21,500 using a pitot tube. Shambaugh et al. [**6**,**7**] used a pitot tube to collect data on the airflow field under a common die. Through the statistical analysis of the experimental data, the empirical equations of the air velocity and air temperature distribution on the center line were obtained. Tate and Shambaugh [**8**] used the same tool to measure the flow field distribution of melt-blown dies with different geometric parameters. They found that the geometrical parameters of the slot die have a large effect on the temperature and speed on the spinning line. Chen Ting et al. [**9**] and Wang Xiaomei et al. [**10**] used a more advanced and accurate hot-wire anemometer to investigate the influence of geometric parameters on the flow field distribution under the die. They found that under the same initial conditions, decreasing the slot inclination and nose width and increasing the slot width resulted in higher airflow velocity and airflow temperature along the axis of the flow field. With the assistance of a hot-wire anemometer and a high-speed camera, Xie et al. [**11,12**] studied the relationship between the instantaneous velocity in the airflow field of the die and the whipping of the melt-blown fibers. Wang et al. [**13,14**] used a hot-wire anemometer to obtain the three-dimensional airflow field velocity distribution data of the common die and the new slot die online. Xie and his collaborators [**15**] used particle image velocimetry to obtain the instantaneous airflow velocity distribution of a slot die.

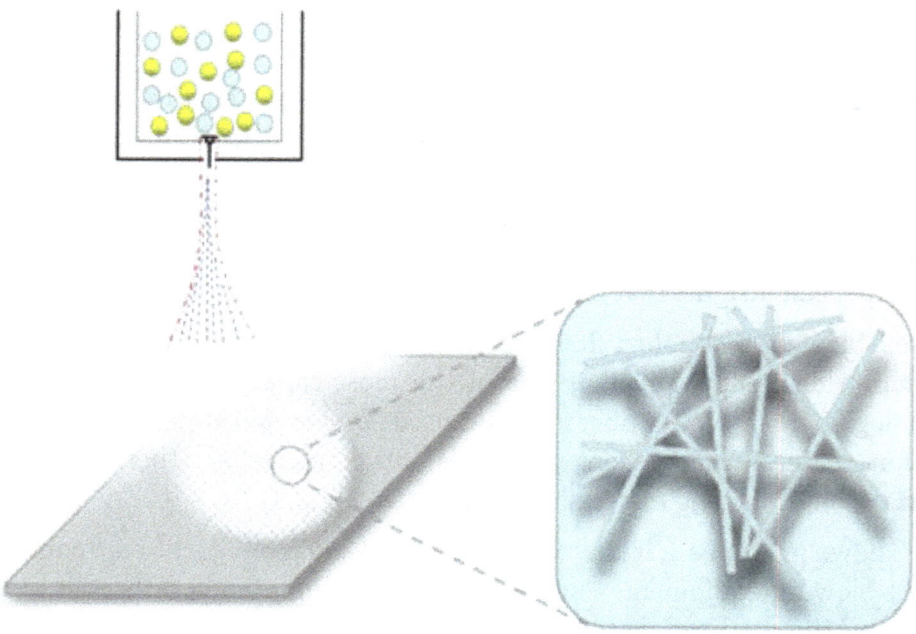

Figure 1. Schematic diagram of melt-blowing web formation.

Computational fluid dynamics (CFD) technology is a complement to experiments. Relying on CFD, the numerical solution of the above theoretical model can be obtained on the computer, and various details of the flow field can be obtained, such as the generation and propagation of vortices, the pressure distribution, and the magnitude of airflow drafting force. The cost of experimental measurement is relatively high, and it is quite convenient to use CFD technology to analyze the airflow field, which can be completely realized on the computer. Krutka et al. [16] used CFD technology for the first time to simulate the low-velocity airflow field of the slot die at room temperature. The maximum speed used in the numerical calculation process was only 34.6 m/s, which was much smaller than the air inlet velocity in actual industrial production. However, it was a milestone start for exploring the melt-blowing flow field. Moore et al. [17] used CFD technology to analyze the isothermal flow field of a two-dimensional annular melt-blowing die, and modified $C_{\varepsilon 1}$ and $C_{\varepsilon 2}$ in a Reynolds stress model. The Shambaugh team [18,19] used Fluent software to calculate the flow field under the melt-blown die. They found that changing the shape and size of the die had a certain influence on the air velocity on the spinning line in the flow field. Sun and Wang [20,21] optimized the airflow field of the slot dies with the help of Fluent software, and obtained the best structure of the common die. Wang et al. [22,23,24,25] designed a series of new slot dies and numerically analyzed the airflow field below them by means of CFD technology.

In the above numerical studies of the airflow field under the die, the presence of melt-blowing fibers was ignored. The main reasons are as follows. On one hand, the melt-blowing fibers move primarily near the axis of the spinneret orifice during the drawing process. On the other hand, melt-blown fibers occupy a small volume compared to air. Another condition that cannot be ignored is that the establishment and solution of the melt-blowing fiber/air-coupled field model is a huge challenge. Under the action of air force, the polymer melt is gradually thinned and transformed from a molten state to a solid state to form the final melt-blown microfiber. This process, although short, involves phase transition and deformation of the fiber-forming polymer, as well as complex turbulent flow and heat transfer.

Krutka and Shambaugh [26,27] were the first to investigate the effect of melt-blowing fibers on the flow field using CFD technology. In their study, the melt-blowing fibers were regarded as columnar solids, and three-dimensional numerical calculations were performed on the flow field under the melt-blowing die. They have made useful explorations on dealing with the two-phase flow in the melt-blowing flow field, and achieved important research results. However, there are some problems with their research. As shown in **Figure 2**, since the melt-blown fiber is a flexible substance, under the action of the surrounding unstable air force, the melt-blown fiber swings at a high frequency and has a certain amplitude during the drafting process. In the research of Krutka and Shambaugh, the columnar solid can only move along the spinning line direction in the melt-blown flow field without swinging, which is inconsistent with the actual situation. Therefore, in order to overcome the shortcomings of their research work, in this work the flexible melt-blown fiber/air-coupled field model was established, and Fluent software was used for the calculation. In addition, the volume of fluid (VOF) model would be tried for the first time for the numerical study the flow field in the presence of melt-blowing fibers in this paper.

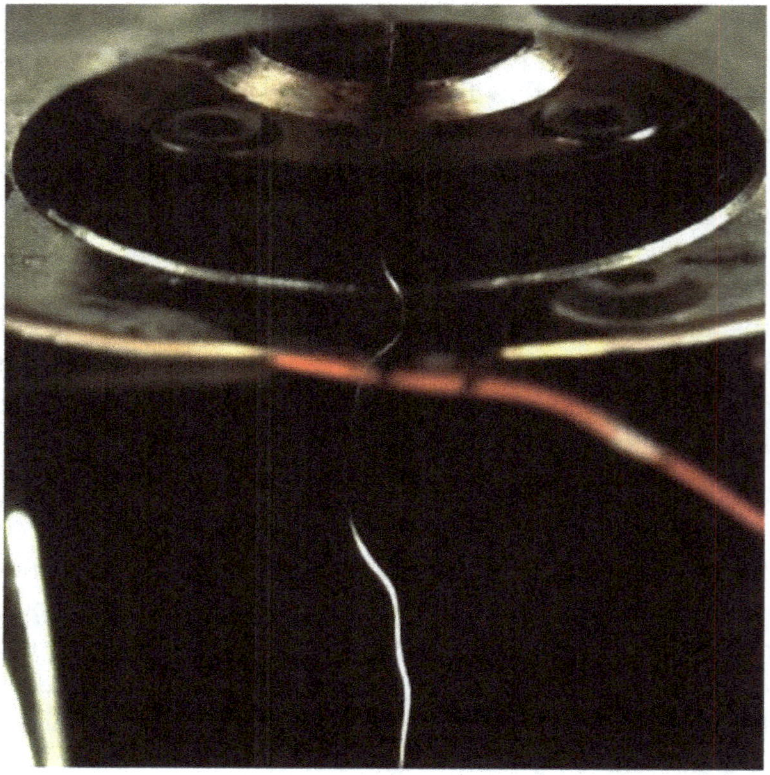

Figure 2. Whip trajectory of the melt-blown fiber under the following experimental conditions: air pressure—50,000 Pa, flow rate of polypropylene—7.8 cc/min, exposure frequency—5000 fps.

2. NUMERICAL SIMULATION

2.1. Structure and Dimensions of Annular Die

Compared to a slot die, the annular die has more orifices for polymer melt extrusion in the same size area and it has a wide range of applications in the factory. In this paper, the numerical calculation of the melt-blown fiber/air-coupling flow field under the ordinary annular die was carried out. Figure 3 shows a schematic structural diagram of an annular melt-blown die. The outer diameter (do) of the annular pore and the inner diameter (di) of the annular pore were 1.3 mm and 2.37 mm, respectively, which is exactly the same size as the die in Shambaugh's work [5,17]. The inner diameter (d) of the spinneret hole was 0.38 mm. The center point of the section at the exit of the spinneret was the origin of the coordinate system and the z-axis coincided with the axis of the spinneret hole.

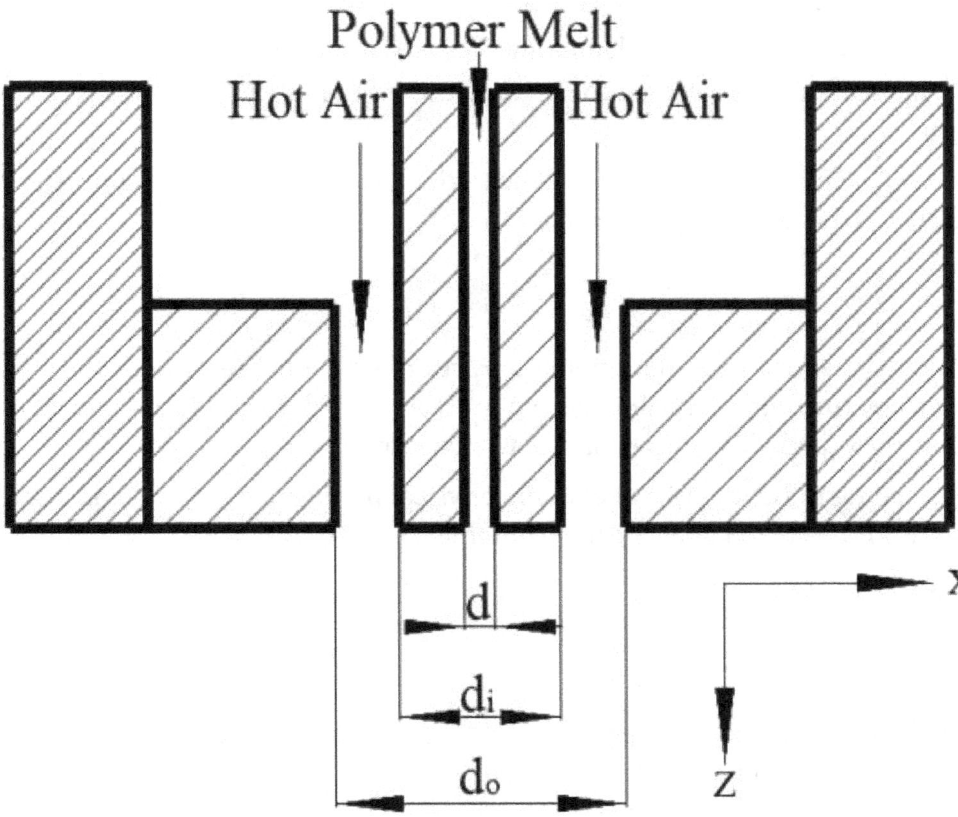

Figure 3. Annular die.

2.2. Conditional Assumption

Usually, because the air temperature value in the flow field near the melt-blown die is above 200 °C and very high, and the temperature of the polymer melt is also higher than 200 °C and the melting point of the polymer raw material, it will not solidify immediately and is in a liquid state. As the drawing progresses, the temperature in the flow field away from the melt-blown die decreases rapidly, causing a large amount of thermal energy of the polymer stream itself to be carried away by the gas flow. During this process, the temperature of the polymer gradually decreases, from a liquid state to a viscoelastic state, until it becomes a solid state, at which point the fiber drawing process ends. In this paper, only the area below the spinneret with z/do = 53.8 was studied. In this region, the melt-blown fiber was regarded as a continuous liquid flow, and the phase transition of the fiber was not considered. Compared to the columnar solid hypothesis of Krutka and Shambaugh [26,27], the melt-blown fiber was considered as a fluid, which was closer to the real drawing process.

2.3. Volume of Fluid (VOF) Model

The two phases, which are polymer and air, are not miscible. The volume of fluid (VOF) model [**28**] was selected to simulate the motion of the air and the polymer stream.

The VOF method uses the exponential function (F) to determine the free surface, and its governing equation is:

$$\frac{\partial F}{\partial t} + \nabla \cdot (u_0 F) = 0 \quad (1)$$

Among them, F is the fluid volume fraction, which is the ratio of the volume occupied by the fluid in the unit system to the mesh volume; t is the time; u0 is the fluid velocity; and ∇∇ is the Hamiltonian structure.

According to the definition of F, the physical parameters density and viscosity of the two fluid-mixed phases can be obtained:

$$\rho = F\rho_1 + (1 - F)\rho_g \quad (2)$$

$$u_0 = Fu_1 + (1 - F)u_g \quad (3)$$

In the formula, ρ is the density of the mixed phase of the two fluids, and the subscripts 1 and g denote the melt and the gas, respectively.

2.4. Computational Domain of Coupled Fields

Figure 4 reveals the three-dimensional computation domain of the coupled flow-field of the annular die. The coordinate systems in Figure 3 and Figure 4 were consistent. The z-axis, x-axis, and y-axis were perpendicular to each other. In the upper half of the computational domain, the heights of both the annular air holes and the spinneret holes were 5 mm. The lower half of the annular die computational domain was in the shape of a cone. The height of the cone-shaped flow-field region along the axis direction was 70 mm and its dimensionless ratio to do was 53.8. The diameter of the upper bottom surface was 10 mm and its dimensionless ratio to do was 7.7. The diameter of the lower bottom surface was 40 mm and its dimensionless ratio to do was 30.8.

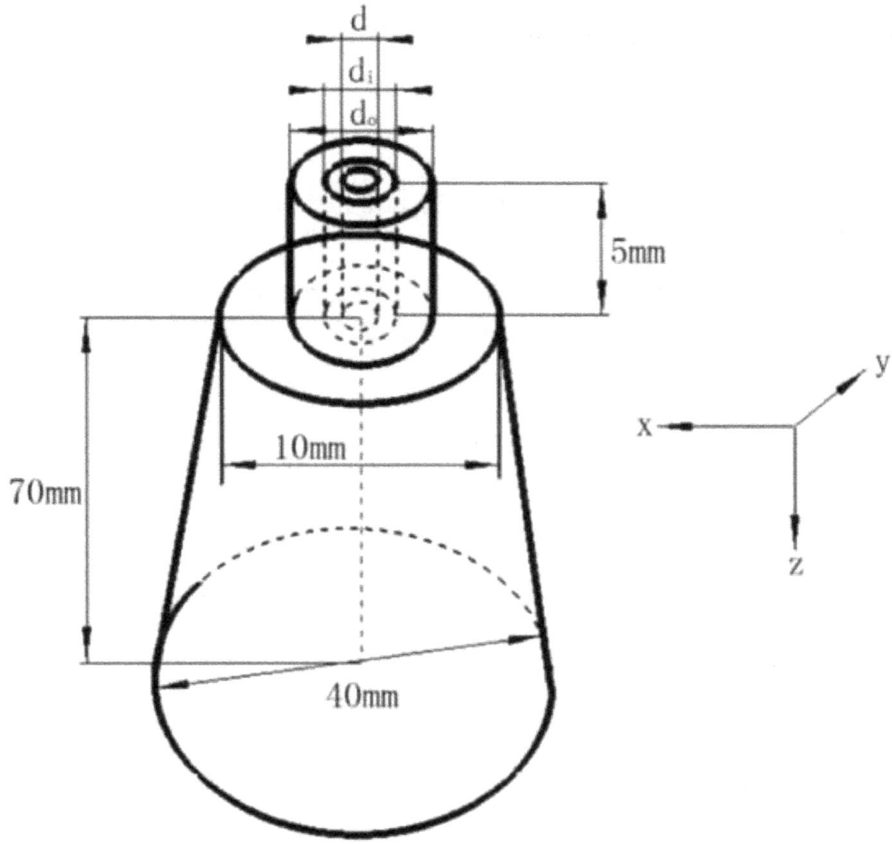

Figure 4. Computational domain.

2.5. Grid Division Method

According to its structural characteristics, the computational domain was divided into grids using Gambit. The annular jet-hole area and the spinning line area under the die were tetrahedral unstructured meshes; the rest areas were hexahedral structured meshes. The combination of unstructured grid and structured grid made the calculation result stable and not easy to diverge. After grid division, the minimum grid spacing of the computational domain was 0.05 mm; the maximum grid spacing was 0.2 mm; and the final total number of unit grids was 859,350.

2.6. Turbulence Model

In this paper, the standard $k\text{–}\varepsilon$ model [24] was chosen, which can reduce the computation time and cost. In addition, it was proved that the calculation results obtained by the standard $k\text{–}\varepsilon$ model were basically consistent with the data collected by the hot wire anemometer [29].

In order to solve the Reynolds stress term, the relationship between the Reynolds stress and the average velocity gradient was established, as follows:

$$-\rho\overline{u_i'u_j'} = \mu_t(\frac{\partial u_i}{\partial x_j} + \frac{\partial u_j}{\partial x_i}) - \frac{2}{3}(\rho k + \mu_t \frac{\partial u_i}{\partial x_i})\delta_{ij} \quad (4)$$

Among them, μt is turbulent viscosity; when I = j, δij = 1, when i~ = j, δij = 0; k is turbulent kinetic energy.

In the standard two-equation model, μt can be related to the turbulent kinetic energy k and the turbulent dissipation rate ε:

$$\mu_t = \rho C_\mu \frac{k^2}{\varepsilon} \quad (5)$$

$$\varepsilon = \frac{\mu}{\rho}\overline{(\frac{\partial u_i'}{\partial x_k})(\frac{\partial u_i'}{\partial x_k})} \quad (6)$$

The time-averaged form of the transport equation for *k* and ε is:

$$\frac{\partial(\rho k u_i)}{\partial x_i} = \frac{\partial}{\partial x_j}[(\mu + \frac{\mu_t}{\sigma_k})\frac{\partial k}{\partial x_j}] + G_k + G_b - \rho\varepsilon - Y_M + S_k \quad (7)$$

$$\frac{\partial(\rho \varepsilon u_i)}{\partial x_i} = \frac{\partial}{\partial x_j}[(\mu + \frac{\mu_t}{\sigma_\varepsilon})\frac{\partial \varepsilon}{\partial x_j}] + C_{1\varepsilon}\frac{\varepsilon}{k}(G_k + C_{3\varepsilon}G_b) - C_{2\varepsilon}\rho\frac{\varepsilon^2}{k} + S_\varepsilon \quad (8)$$

In these equations, C1ε, C2ε, and C3ε are empirical constants; σk is the Prandtl number corresponding to the turbulent kinetic energy k; σε is the Prandtl number corresponding to the turbulent dissipation rate ε; and Sk and Sε are the source terms.

The expressions corresponding to G_b and Y_M are:

$$G_b = \beta g_i \frac{\mu_t}{\sigma_t}\frac{\partial T}{\partial x_i} \quad (9)$$

$$Y_M = 2\rho\varepsilon M_t^2 \quad (10)$$

In these two equations, gi is the component of the gravitational acceleration of the gas in the i direction; β is the thermal expansion coefficient of the gas; Mt is the turbulent Mach number.

2.7. Boundary Condition Setting and Parameter Setting

The type of boundary set at the spinneret inlet was the velocity inlet boundary. The density of the polymer was 0.91 g/cm^3 and approximately equal to that of polypropylene, which represented the largest component of raw materials in the meltblowing process. The velocity of the melt was 5 m/s, and the melt entered the spinneret vertically. The hydraulic diameter at the entrance of the spinneret was equal to the diameter of the spinneret, and the turbulence intensity was set to 1%. The air inlet plane was set to "velocity inlet". The air entered the annular hole at a speed of 50 m/s, its direction was parallel to the axis of the annular hole, and the turbulence intensity was set to 3%. The hydraulic diameter of the annular air-hole inlet was the difference between the outer diameter and the inner diameter of the annular hole. The dynamic viscosity of air was 17.9 × 10−6 Pa·s and the dynamic viscosity of the polymer was set to 1.03 × 10−3 Pa·s. The Reynolds number of air at the inlet of annular pipe was 8174.6 and Reynolds number of polymer melt at the inlet of spinneret hole was 1678.64. The initial temperature of polymer melt and air was 300 K. The die head end, annular orifice, and spinneret conduits were designed to have no slip walls. The remaining planes or surfaces were designated as pressure outlets.

The gravitational acceleration in the z direction was set to 9.81 m/s^2. In solving, the PISO algorithm was chosen. For unsteady flow field, the choice of PISO algorithm had good convergence.

As a departure from the previous numerical simulation of the melt-blown flow field, in this paper, the unsteady flow field of the annular melt-blown die was calculated. In order to make the calculation result more accurate and stable, the time-step size was set to 0.1×10^{-9} s.

3. RESULTS

At t = 0 s, the melt-blown flow field was initialized, and the polymer melt and gas flow had not yet begun to move. Each calculation step (Δt) could solve the flow situation in the melt-blown flow field at the time $t = t + \Delta t$, so as to obtain the velocity and pressure distribution of the two fluids at the new time. At the same time, the position of the polymer melt at different times could also be obtained by calculation. With the accumulation of time, the motion trajectory of the polymer melt at any moment could be obtained.

3.1. Flow Field Distribution at Different Times

The geometry of the annular melt-blown die was different from that of the double-slot melt-blown die, and the steady-state flow field distribution on each symmetry plane was exactly the same [16]. However, due to the complex and changeable turbulent flow under the melt-blown die head, the speed and pressure of each point in the unsteady flow field were constantly changing. For an annular melt-blown die, the distribution on each of its symmetry planes may be completely different even at the same time. In this paper, the unsteady flow field distribution on two mutually perpendicular symmetry planes of the melt-blown die was investigated.

As shown in Figure 5 and Figure 6, the velocity distributions in the x–z plane and the y–z plane were constantly changing over time. In the early stage of the drafting process ($t = 1.000 \times 10^{-7}$ s to $t = 1.338 \times 10^{-4}$ s), the velocity distributions in the x–z plane and the y–z plane were basically the same. Moreover, the difference in air velocity in most of the area below the die was very small, except for the area near the die. Because the velocity of the air was much greater than that of the polymer melt, the high-velocity jet could rapidly move most of the air flow in the flow field, however, the air velocity near the polymer melt was lower. As the drafting progressed (t = 6.013 × 10-3 s to t = 1.543 × 10-2 s), the fluid velocity in the central region of the flow field was higher than that on both sides. From Figure 5 and Figure 6, it can be seen that the velocity distributions on the two symmetry planes were no longer the same at this time, and there were obvious differences.

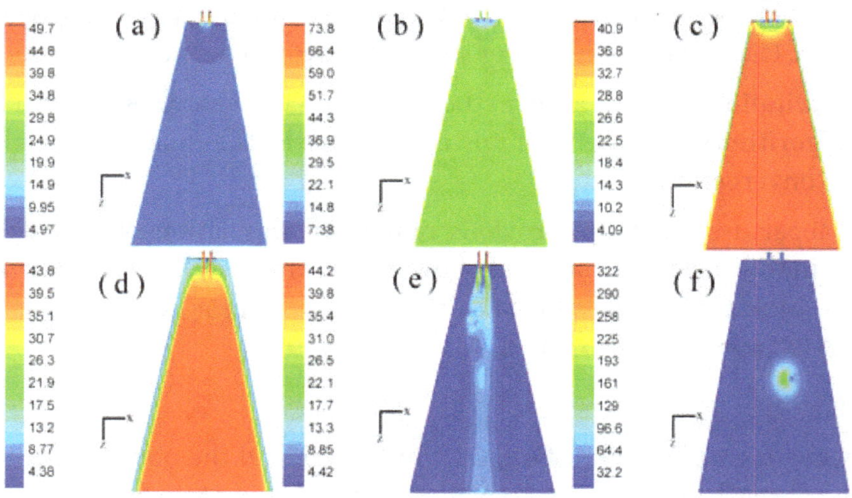

Figure 5. Velocity cloud distribution on the x–z plane (m/s): (a) velocity cloud distribution of flow field at time $t = 1.000 \times 10^{-7}$ s; (b) velocity cloud distribution of flow field at time $t = 1.600 \times 10^{-6}$ s; (c) velocity distribution of flow field at time $t = 2.979 \times 10^{-5}$ s; (d) velocity cloud distribution of flow field at time $t = 1.338 \times 10^{-4}$ s; (e) velocity cloud distribution of flow field at time $t = 6.013 \times 10^{-3}$ s; (f) velocity cloud distribution of flow field at time $t = 1.543 \times 10^{-2}$ s.

Of all the factors, the airflow velocity in the flow field had the greatest effect on the diameter of the melt-blown fibers. The pressure also had a certain effect on the fineness of the melt-blown fibers. The greater the air pressure the polymer melt is subjected to, the smaller the diameter of the melt-blown fibers, in theory. Figure 7 and Figure 8 show the static- pressure nephograms in the x–z and y–z planes. It can be seen from the figure that the pressure values at each point in the flow field below the annular melt-blown die were basically the same. Unlike velocity, at the same moment, the pressure distribution on the two symmetry planes was the same and did not change with the advancement of time.

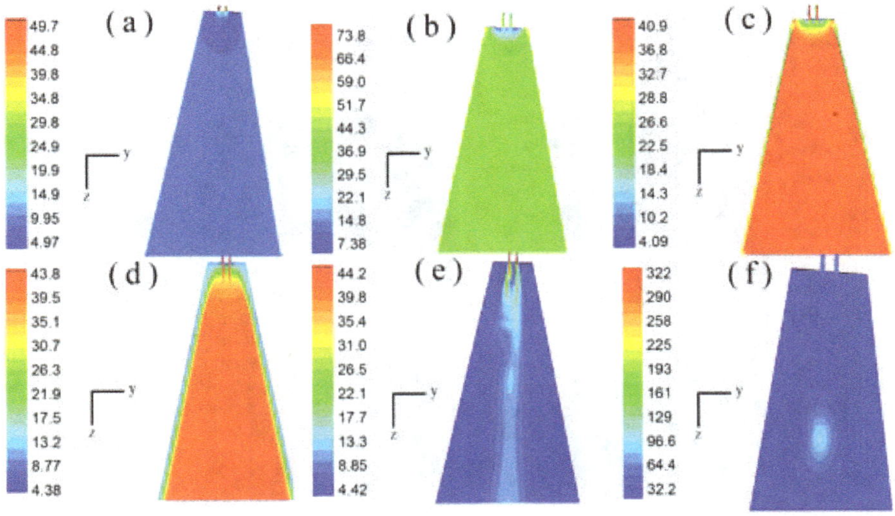

Figure 6. Velocity cloud distribution on the y–z plane (m/s): (a) velocity cloud distribution of flow field at time t = 1.000×10^{-7} s; (b) velocity cloud distribution of flow field at time t = 1.600×10^{-6} s; (c) velocity distribution of flow field at time t = 2.979×10^{-5} s; (d) velocity cloud distribution of flow field at time t = 1.338×10^{-4} s; (e) velocity cloud distribution of flow field at time t = 6.013×10^{-3} s; (f) velocity cloud distribution of flow field at time t = 1.543×10^{-2} s.

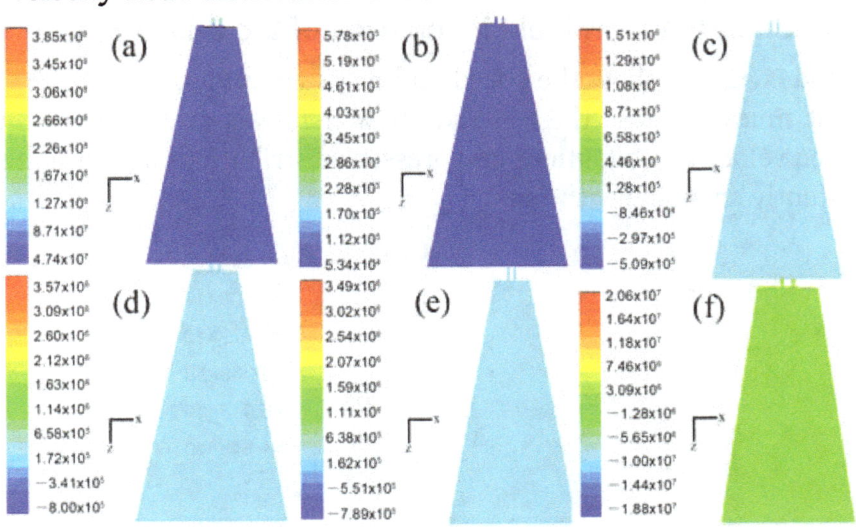

Figure 7. Pressure cloud distribution on the x–z plane (Pa): (a) velocity cloud distribution of flow field at time t = 1.000×10^{-7} s; (b) velocity cloud distribution of flow field at time t = 1.600×10^{-6} s; (c) velocity distribution of flow field at time t = 2.979×10^{-5} s; (d) velocity cloud distribution of flow field at time t = 1.338×10^{-4} s; (e) velocity cloud distribution of flow field at time t = 6.013×10^{-3} s; (f) velocity cloud distribution of flow field at time t = 1.543×10^{-2} s.

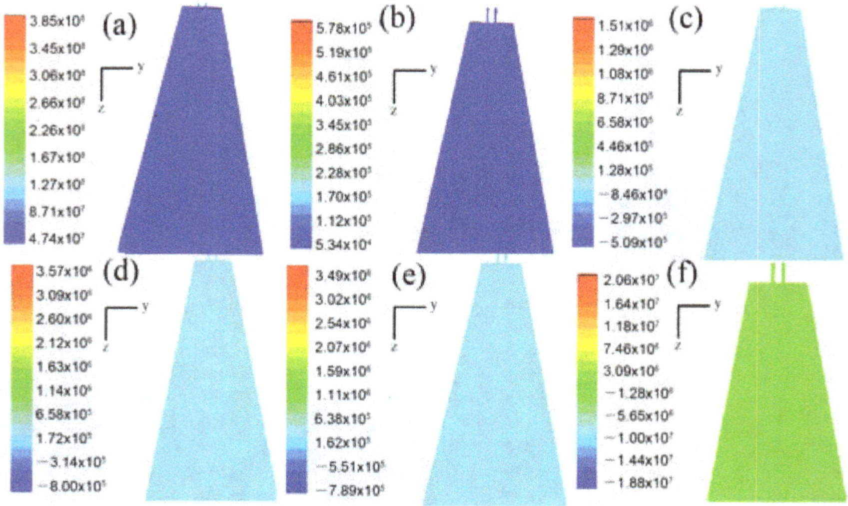

Figure 8. **Pressure cloud distribution on the y–z plane (Pa): (a) velocity cloud distribution of flow field at time t = 1.000 × 10^{-7} s; (b) velocity cloud distribution of flow field at time t = 1.600 × 10^{-6} s; (c) velocity distribution of flow field at time t = 2.979 × 10^{-5} s; (d) velocity cloud distribution of flow field at time t = 1.338 × 10^{-4} s; (e) velocity cloud distribution of flow field at time t = 6.013 × 10^{-3} s; (f) velocity cloud distribution of flow field at time t = 1.543 × 10^{-2} s.**

3.2. Variation of the Flow Field in the Area of the Spinning Line

Figure 9 and Figure 10 show the speed and pressure values on the spinning line at different times. Because the unsteady flow field was calculated in this work, the obtained velocity distributions and pressure distributions on the spinning line were constantly changing with time.

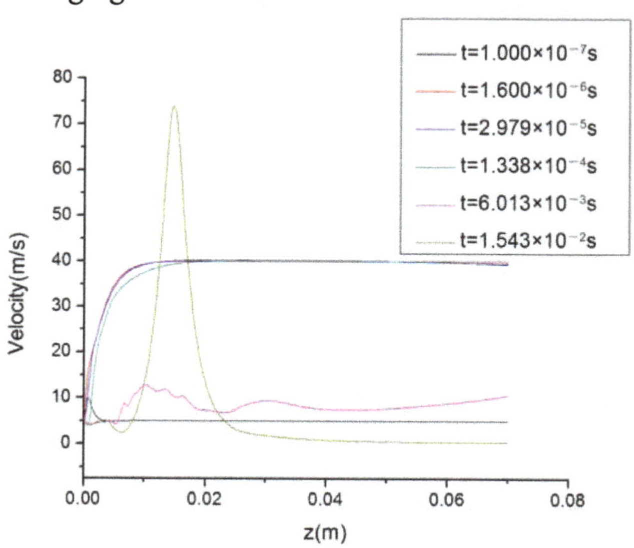

Figure 9. Variation of velocity on spinning line with time.

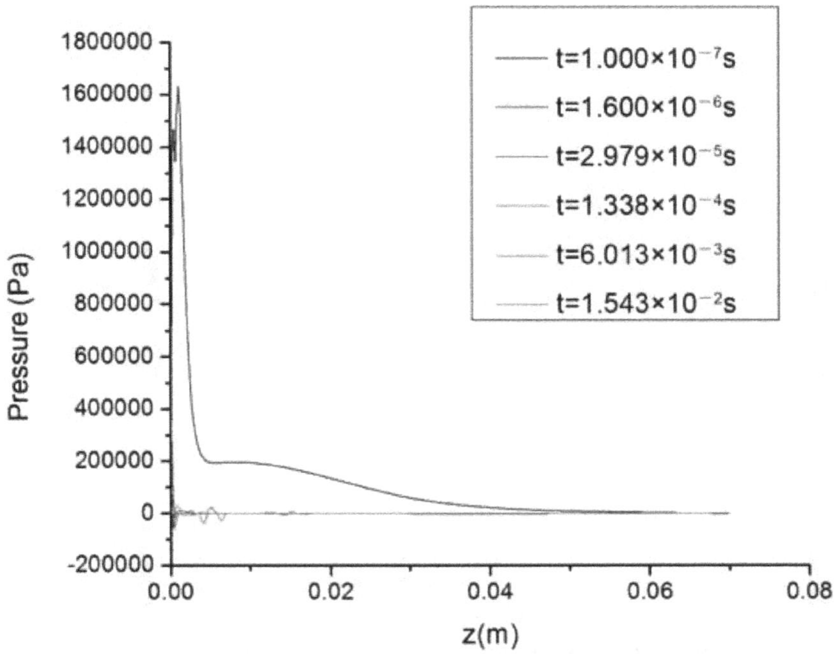

Figure 10. Variation of pressure on spinning line with time.

Compared with the previous pure air flow field [16,17,18], it can be seen from Figure 9 that in the coupled flow field containing the polymer melt, the air velocity curves on the spinning centerline were obviously different, and there was no regularity. For example, when $t = 6.013 \times 10^{-3}$ s, there was more than one speed peak on the spinning line and the difference between the peak and the speed in other regions was very small, while at $t = 1.600 \times 10^{-6}$ s, $t = 2.979 \times 10^{-5}$ s and $t = 1.338 \times 10^{-4}$ s, the air velocity on the spinning line rapidly increased to the maximum value and did not change.

In Figure 10, the pressure curve on the spinning line at t = 1.000 × 10−7 s was the largest. In the area close to the die, because the temperature of the polymer melt was usually above 200 °C and it was in a flowing state, the surrounding pressure was high at this time, which would affect the deformation of the cross section of the melt-blown fiber. In other areas, the static pressure values on the spinning line differed little and the distribution was almost uniform.

Figure 11a–f are the streamline diagrams of the spinning line at different times. In Figure 11a–f, the airflow around the spinning thread flows vertically downward. Although the movement directions of the airflows were the same at this time, it is clear from Figure 11e,f that there were differences in their velocity values. Consistent with Figure 9 and Figure 10, in Figure 11e,f, the air flow in the area of the melt-blown spinning line was chaotic and disordered, and its velocity magnitude and direction were inconsistent.

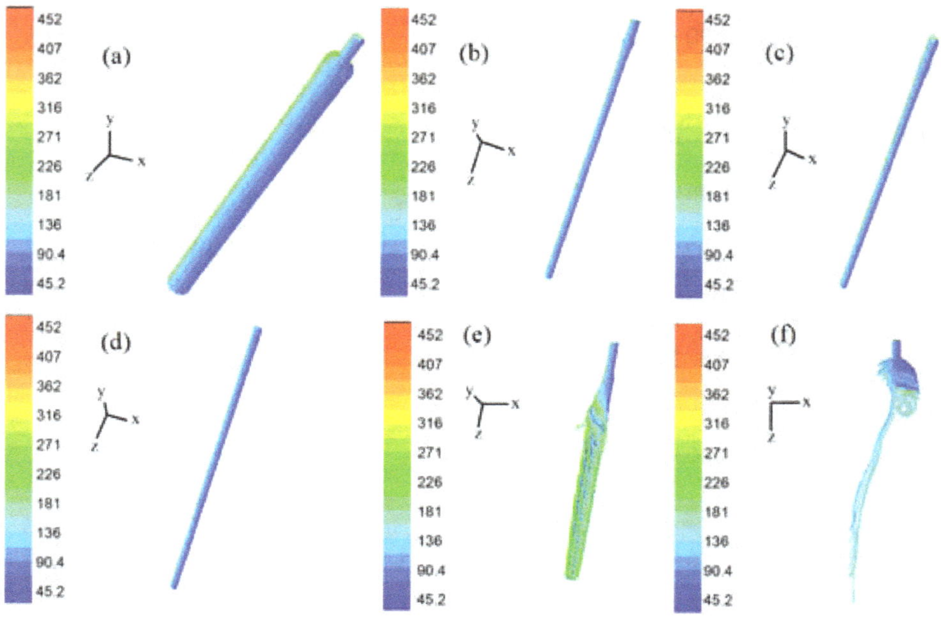

Figure 11. Pressure distribution on the y–z plane: (a) streamline diagram of the spinning line at t = 1.000 × 10^{-7} s; (b) streamline diagram of the spinning line at t = 1.600 × 10^{-6} s; (c) streamline diagram of the spinning line at t = 2.979 × 10^{-5} s; (d) streamline diagram of the spinning line at t = 1.338 × 10^{-4} s; (e) streamline diagram of the spinning line at t = 6.013 × 10^{-3} s; (f) streamline diagram of the spinning line at t = 1.543 × 10^{-2} s.

The velocity distribution of the airflow around the spinning line made the melt-blown fibers unevenly stressed during the drafting process, which not only affected the motion trajectory of the melt-blown fibers, but at the same time, also had a certain effect on the surface morphology of the melt-blown fibers. Therefore, it can be observed that the fiber surface was not smooth and had "reverse V-shaped" stripes under the fast-cooling condition [**30**].

Figure 12a,b show the instantaneous speeds at different points on the spinning line of the melt-blown die, which were measured by a hot-wire anemometer. As can be seen from the figure, the velocity of these two points on the center line of the die varied drastically. The instantaneous speed at each point fluctuated up and down around a certain speed average, which was similar to the experimental results of Yang and Zeng [**31**]. When z = 5 mm and z = 15 mm, basically all the speeds fluctuated up and down around 145 m/s and 140 m/s, respectively. The air flow under the melt-blown die was in a state of turbulence, so various physical parameters, such as the velocity, pressure, and temperature of the air in the flow field changed randomly with time and space. The experimental measurement data in **Figure 12a**,b demonstrate that the velocity in the centerline of the flow field below the flow field of the melt-blown die was constantly changing from moment to moment. Therefore, the turbulent flow in the melt-blown flow field was complex and changeable.

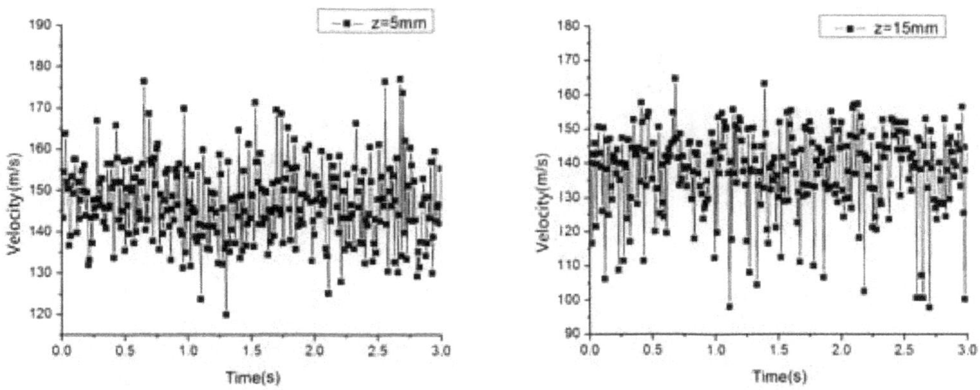

C (a) (b)

Figure 12. Instantaneous velocity fluctuations at two points on the spinning line: (a) z = 5 mm; (b) z = 15 mm. The pressure at the entrance was 50,000 Pa and the temperature was 300 K, ignoring the existence of melt-blown fibers.

In the melt-blown flow field, the drastic change of the instantaneous velocity of the air flow could easily cause the inconsistency of the stretching air velocity in the spinning line and its surrounding area, as shown in **Figure 11**a–f. Because the polymer melt was mainly active near the spinning centerline, the constant fluctuation of the airflow velocity on the centerline of the flow field with time affected the drafting stability of the melt-blown fiber and the whiplash of the fibers around the spinning line, resulting in fiber diameter inconsistencies and adhesions between adjacent fibers. It had a greater impact on the production of melt-blown nonwoven products.

3.3. The Trajectory of Polymer Melt

Figure 13a–f shows the trajectory of the polymer melt at different times. In **Figure 13**a, the extruded polymer melt is difficult to observe due to the short time. As can be seen from **Figure 13**b–f, the polymer melt was extruded from the spinneret holes of the annular die and gradually elongated. At this point the polymer melt occupied only a very small part of the spinning line and the surrounding area. Nonetheless, uneven forces on both sides of the polymer melt can be observed in **Figure 13**e,f due to the difference in air velocity around the spinning line.

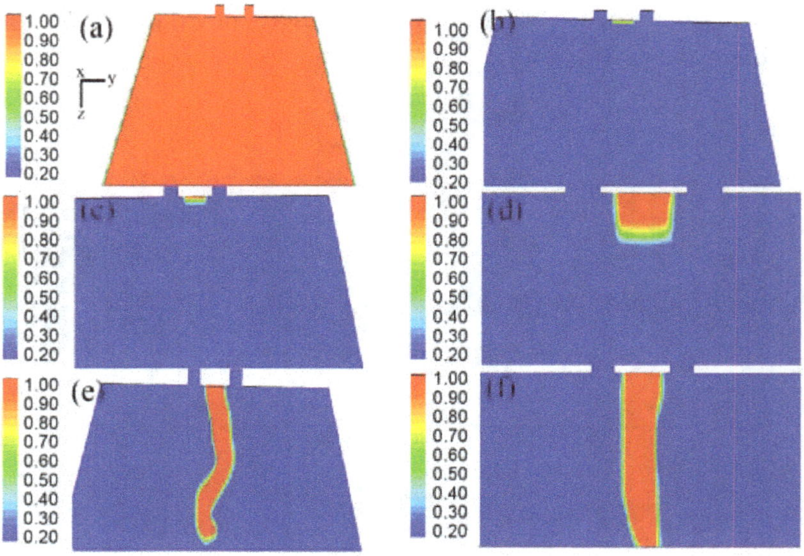

Figure 13. Movement trajectories of melt-blown fibers at different times, showing: (a) the y–z plane at time t = 1 × 10^{-7} s; (b) the y–z plane at time t = 1.6 × 10^{-6} s; (c) the y–z plane at time t = 2.9790 × 10^{-5} s; (d) the y–z plane at time t = 1.3379 × 10^{-4} s; (e) the y–z plane at time t = 6.0128 × 10^{-3} s; (f) the y–z plane at time t = 1.5430 × 10^{-2} s.

In Figure 13e,f, with the drafting of the fiber (t = 6.0128 × 10^{-3} s and t = 1.5430 × 10^{-2} s), the velocity in the melt-blown flow field was disordered, resulting in the polymer melt showing more obvious whiplash. In the study of Sun et al. [32], the amplitude of the melt-blown fiber was very small, with an order of magnitude of only 10^{-7} m; while it can be seen from Figure 13e that the amplitude of the polymer melt was slightly larger than the inner diameter (di) of the annular air hole, which was on the order of millimetres. The amplitude of the whipping of melt-blown fibers was often more than 0.5 mm. It was closer to the real motion of melt-blown fibers, which highlights the advantages of this research method. When the whipping amplitude of melt-blown fibers is more accurate, it can lay a foundation for predicting the impact point of melt-blown fibers and the direction of their alignment in the web [32]. In particular, the melt-blown fiber orientation information can be used to predict the web uniformity of melt blown nonwovens and guide the production of melt-blown fibers [33].

In **Figure 13**f, the diameter of the polymer melt decreased due to the tensile force of the surrounding air flow, which was consistent with the actual fiber drawing process.

In this work, the polymer melt was stretched under ideal conditions, free from accidental factors such as uneven raw material and abnormal production equipment.

Combined with the simulation results and experimental measurement results on the spinning line in **Figure 11** and **Figure 12**, it can be inferred that in the flow field below the die head, the whipping phenomenon of the melt-blown fibers was mainly due to drastic changes and inconsistencies in the airflow velocity near the spinning line. In addition, it is probably possible to draw velocity vectors of airflow in the vicinity of the oscillations of the melt core. Another interesting quantity is the vorticity in air that could also be plotted near the spinning, in order to detect vortices in air.

4. CONCLUSIONS

In this work, based on the VOF method, a three-dimensional geometric model of the annular die was established, and the unsteady air-coupled flow field of melt-blown fibers was numerically studied. This model provides a new idea for studying the air-coupled flow field of flexible melt-blown fibers, which can track the stretching and radial motion of the polymer melt.

The motion trajectory and flow field distribution of the polymer melt during the stretching process were obtained by numerical calculation. The study found that at the same time, the pressure distributions in different central symmetry planes of the annular melt-blown die were basically the same. However, the velocity distributions in different central symmetry planes may be different. The velocity and pressure on the centerline of the flow field change continuously with time, and the velocity changes more violently, while the pressure changes less. Different air velocities around the spinning line at the same time and their velocity fluctuations can cause fiber whipping and is the main reason for fiber whipping.

5. CONCISE SUMMARY

Melt-blowing technology is an important method for directly preparing micro-nanofiber materials by drawing polymer melts with high temperature and high velocity air flow. During the drawing process, the melt-blowing fiber not only undergoes a phase change, but also has an extremely complex coupling effect with the drawing airflow. Therefore, in the numerical calculation of the flow field, the existence of melt-blowing fibers is often ignored. In this paper, based on the volume of fluid method, a numerical study of the flexible fiber/air-coupling flow field of an annular melt-blowing die is carried out with the aid of computational fluid dynamics software. The results show that the pressure distribution in the different central symmetry planes of the ring die at the same time was basically the same. However, the velocity distribution may have been different; the velocity on the spinning line varied with time; the pressure changes on the spinning line were small; and velocity fluctuations around the spinning line could cause whiplash of the fibers.

REFERENCES

1. Wang, Z.F.; Macosko, C.W.; Bates, F.S. Fluorine-enriched melt-blown fibers from polymer blends of poly(butylene terephthalate) and a fluorinated multiblock copolyester. *Acs Appl. Mater. Inter.* **2016**, *8*, 3006–3012. **[Google Scholar] [CrossRef] [PubMed]**

2. Sun, F.; Li, T.-T.; Ren, H.; Jiang, Q.; Peng, H.-K.; Lin, Q.; Lou, C.-W.; Lin, J.-H. PP/TiO2 Melt-Blown Membranes for Oil/Water Separation and Photocatalysis: Manufacturing Techniques and Property Evaluations. *Polymers 2019*, 11, 775. **[Google Scholar] [CrossRef] [PubMed][Green Version]**

3. Xu, Y.Q.; Zhang, X.M.; Hao, X.B.; Teng, D.F.; Zhao, T.N.; Zeng, Y.C. Micro/nanofibrous Nonwovens with High Filtration Performance and Radiative Heat Dissipation Property for Personal Protective Face Mask. *Chem. Eng. J.* **2021**, *423*, 130175. **[Google Scholar] [CrossRef] [PubMed]**

4. Schmidt, J.; Usgaonkar, S.S.; Kumar, S.; Lozano, K.; Ellison, C.J. Advances in Melt Blowing Process Simulations. *Ind. Eng. Chem. Res.* **2019**, *61*, 65–85. **[Google Scholar] [CrossRef]**

5. Uyttendaele, M.A.J.; Shambaugh, R.L. The Flow Field of Annular Jets at Moderate Reynolds Numbers. *Ind. Eng. Chem. Res.* **1989**, *28*, 1735–1740. **[Google Scholar] [CrossRef]**

6. Harpham, A.S.; Shambaugh, R.L. Flow field of practical dual rectangular jets. *Ind. Eng. Chem. Res.* **1996**, *35*, 3776–3781. **[Google Scholar] [CrossRef]**

7. Harpham, A.S.; Shambaugh, R.L. Velocity and temperature fields of dual rectangular jets. *Ind. Eng. Chem. Res.* **1997**, *36*, 3937–3943. **[Google Scholar] [CrossRef]**

8. Tate, B.D.; Shambaugh, R.L. Temperature fields below melt-blowing dies of various geometries. *Ind. Eng. Chem. Res.* **2004**, *43*, 5405–5410. **[Google Scholar] [CrossRef]**

9. Chen, T.; Wang, X.H.; Huang, X.B. Modeling the air-jet flow field of a dual slot die in the melt-blowing nonwoven process. *Text. Res. J.* **2004**, *74*, 1018–1024. **[Google Scholar] [CrossRef]**

10. Wang, X.M.; Ke, Q.F. Empirical formulas for distributions of air velocity and temperature along the spinline of a dual slot die. *Polym. Eng. Sci.* **2005**, *45*, 1092–1097. **[Google Scholar] [CrossRef]**

11. Xie, S.; Zeng, Y.C. Turbulent air flow field and fiber whipping motion in the melt blowing process: Experimental study. *Ind. Eng. Chem. Res.* **2012**, *51*, 5346–5352. **[Google Scholar] [CrossRef]**

12. Xie, S.; Han, W.L.; Jiang, G.J.; Chen, C. Turbulent air flow field in slot-die melt blowing for manufacturing microfibrous nonwoven materials. *J. Mater. Sci.* **2018**, *53*, 6991–7003. **[Google Scholar] [CrossRef][Green Version]**
13. Wang, Y.D.; Ji, C.C.; Zhou, J.P. Experimental and numerical analysis of an improved melt-blowing slot-die. *e-Polymers* **2019**, *19*, 612–621. **[Google Scholar] [CrossRef]**
14. Ji, C.C.; Wang, Y.D. Experimental investigation on the three-dimensional flow field from a meltblowing slot die. *e-Polymers* **2020**, *20*, 724–732. **[Google Scholar] [CrossRef]**
15. Xie, S.; Jiang, G.J.; Ye, B.L.; Shentu, B.Q. Particle Image Velocimetry (PIV) Investigation of the Turbulent Airflow in Slot-Die Melt Blowing. *Polymers* **2020**, *12*, 279. **[Google Scholar] [CrossRef][Green Version]**
16. Krutka, H.M.; Shambaugh, R.L. Analysis of a melt-blowing die: Comparison of CFD and experiments. *Ind. Eng. Chem. Res.* **2002**, *41*, 5125–5138. **[Google Scholar] [CrossRef]**
17. Moore, E.M.; Shambaugh, R.L. Analysis of isothermal annular jets Comparison of Computational Fluid Dynamics and Experimental Data. *J. Appl. Poly. Sci.* **2004**, *94*, 909–922. **[Google Scholar] [CrossRef]**
18. Krutka, H.M.; Shambaugh, R.L.; Papavassiliou, D.V. Effects of die geometry on the flow field of the melt-blowing process. *Ind. Eng. Chem. Res.* **2003**, *42*, 5541–5553. **[Google Scholar] [CrossRef]**
19. Krutka, H.M.; Shambaugh, R.L.; Papavassiliou, D.V. Effects of temperature and geometry on the flow field of the melt blowing process. *Ind. Eng. Chem. Res.* **2004**, *43*, 4199–4210. **[Google Scholar] [CrossRef]**
20. Sun, Y.F.; Wang, X.H. Optimal geometry design of the melt-blowing slot die with high stagnation temperature via the orthogonal array method and numerical simulation. *J. Text. I* **2011**, *102*, 65–69. **[Google Scholar] [CrossRef]**
21. Sun, Y.F.; Wang, X.H. Optimization of air flow field of the melt blowing slot die via numerical simulation and genetic algorithm. *J. Appl. Polym. Sci.* **2010**, *115*, 1540–1545. **[Google Scholar] [CrossRef]**
22. Wang, Y.; Wang, X. Numerical analysis of new modified melt-blowing dies for dual rectangular jets. *Polym. Eng. Sci.* **2014**, *54*, 110–116. **[Google Scholar] [CrossRef]**
23. Wang, Y.; Zhou, J.; Gao, X. Numerical analysis of airflow fields from new melt-blowing dies for dual-slot jets. *ACS Omega* **2020**, *5*, 13409–13415. **[Google Scholar] [CrossRef] [PubMed]**

24. Ji, C.C.; Wang, Y.D.; Sun, Y.F. Numerical investigation on a melt-blowing die with internal stabilizers. *J. Ind. Text.* **2021**, *50*, 1409–1421. **[Google Scholar] [CrossRef]**
25. Wang, Y.D.; Qiu, Y.P.; Ji, C.C.; Wang, X.H.; Guan, F.W. The Effect of the Geometric Structure of the New Slot Die on the Flow Field Distribution. *Text. Res. J.* **2022**, *92*, 423–433. **[Google Scholar] [CrossRef]**
26. Krutka, H.M.; Shambaugh, R.L. Effects of Fiber on the Air Flow from an Annular Melt Blowing Die. *Ind. Eng. Chem. Res.* **2007**, *46*, 655–666. **[Google Scholar] [CrossRef]**
27. Krutka, H.M.; Shambaugh, R.L. Effects of the polymer fiber on the flow field from a slot melt blowing die. *Ind. Eng. Chem. Res.* **2008**, *47*, 935–945. **[Google Scholar] [CrossRef]**
28. Hirt, C.W.; Nichols, B.D. Volume of fluid (VOF) method for the dynamics of free boundaries. *J. Comput. Phys.* **1981**, *39*, 201–225. **[Google Scholar] [CrossRef]**
29. Sun, G.W.; Wang, Y.D.; Zhang, Y.J.; Han, W.L.; Shang, S.S. Formation Mechanism of Fibrous Web in the Solution Blowing Process. *ACS Omega* **2022**, *7*, 20584–20595. **[Google Scholar] [CrossRef]**
30. Xin, S.; Wang, X. Mechanism of Fiber Formation in Melt Blowing. *Ind. Eng. Chem. Res.* **2012**, *51*, 10621–10628. **[Google Scholar] [CrossRef]**
31. Yang, Y.; Zeng, Y. Measurement and Comparison of Melt-Blowing Airflow Fields: Nozzle Modifications to Reduce Turbulence and Fibre Whipping. *Polymers* **2021**, *13*, 719. **[Google Scholar] [CrossRef]**
32. Sun, Y.; Zeng, Y.; Wang, X. Three-Dimensional Model of Whipping Motion in the Processing of Microfibers. *Ind. Eng. Chem. Res.* **2011**, *50*, 1099–1109. **[Google Scholar] [CrossRef]**
33. Shkarin, R.; Shkarina, S.; Weinhardt, V.; Surmenev, R.A.; Surmeneva, M.A.; Shkarin, A.; Baumbach, T.; Mikut, R. GPU-accelerated Ray-casting for 3D fiber Orientation Analysis. *PLoS ONE* **2020**, *15*, e0236420. **[Google Scholar] [CrossRef]**

CHAPTER-6

CARBOXYLATED CARBON NANOTUBE/POLYIMIDE FILMS EXHIBIT EXCEPTIONAL MECHANICAL PROPERTIES AND A LOW THERMAL EXPANSION COEFFICIENT

1. INTRODUCTION: AN OVERVIEW

Introducing a noteworthy advancement in materials engineering, films comprising carboxylated carbon nanotubes and polyimide emerge as a remarkable solution with a unique combination of attributes. Characterized by both a low thermal expansion coefficient and outstanding mechanical properties, these films hold immense promise in various technological applications. This study delves into the synthesis and characterization of carboxylated carbon nanotube/polyimide films, aiming to unlock their potential for use in industries where superior mechanical resilience and thermal stability are crucial. The convergence of these properties positions these films as a compelling candidate for enhancing performance in diverse fields, from aerospace to advanced electronics.

With the fast development of material science, the fields of aerospace, electronic engineering, communication, and architecture have undergone rapid changes, which put forward higher requirements for applied materials to have a variety of superior properties, such as improved mechanical properties and thermal stability. Therefore, functional polymers with superlative comprehensive properties have attracted more and more attention [1]. The density of polymer materials is far lower than that of metals and ceramics, and with the development of material technology, their other properties have also been greatly improved.

Among these polymer materials, polyimide (PI) film materials have outstanding properties [**2,3,4,5,6,7,8**]: low density, high mechanical properties, great thermal stability, radiation resistance, corrosion resistance, excellent dielectric properties, and so on. Therefore, PI films can be used in microelectronics, aerospace, and other fields [**9,10,11,12**]. However, the coefficient of thermal expansion (CTE) of polyimide is much higher than that of inorganic materials such as metals and ceramics. Composites made of polyimide and inorganic materials will warp, crack or delaminate at high temperatures due to their mismatched CTE [**13,14,15,16**]. In addition, the mechanical properties of polyimide films are also insufficient in some applications requiring high strength [**17**]. Therefore, high-strength and low-CTE PI films are required to achieve applications in areas with higher performance requirements.

In recent years, carbon nanotubes (CNT) have begun to be applied to polymer composites. Since the advent of CNT [**18**], more and more scholars have studied them. CNT has a high aspect ratio, thermal conductivity, thermal stability, conductivity, and unique mechanical properties [**19,20**]. CNT can improve the thermal properties, mechanical properties, and other properties of polymer materials, with a significant enhancement even at very low concentrations (0.5 vol% [**21**], 0.79–5 wt.% [**22,23,24,25,26,27,28,29,30**]). Therefore, CNT has been well applied in aerospace and other fields. For example, aluminum was replaced by carbon nanotube fiber reinforced polymer (CNRP) on the Reusable Launch Vehicle to reduce the payload, leading to a weight reduction of 82% [**31**]. Adding CNT to the pure resin used by satellites can obtain better mechanical properties and thermal stability [**32**]. Poly (ionic liquid) s (PILs), as a critical material in the new generation of intelligent electromechanical equipment, has improved its electrical and mechanical properties after adding CNT [**33**]. Ju et al. [**34**] reported the fabrication of multiwalled CNTs/polyacrylonitrile fibers by electrospinning, which exhibited an electrical conductivity of 96 S/m at a nanotube loading of 3 wt.%. Jiang et al. [**35**] prepared multiwalled CNTs/PI composites, including 3,3′,4,4′-biphenyltetracarboxylic dianhydride (BPDA), p-phenylenediamine (p-PDA), which showed the effective absorption bandwidth of 2.72 GHz with the matching thickness of only 2.0 mm when the content of CNTs was 6 wt.%, and the tensile modulus and tensile strength of the composites all reached their maximum values of 4.9 GPa and 281.3 MPa when the content of CNTs was 0.5 wt.%, respectively.

In general, CNT can improve the thermal and mechanical properties of polymer materials. However, the main problems encountered in the preparation of CNT-reinforced composites are the aggregation and the weak interface interaction between CNT and the matrix with a high CNT content [**36,37,38**]. Self-organization of carbon nanotubes in the matrix can be realized through proper surface engineering, thus utilizing the aggregation of carbon nanotubes [**39,40**].

This controlled aggregation in a series of CNT systems can achieve splendid optoelectronic properties and promote the development of new nanotube sensors [**41,42**]. However, in the physical or chemical method commonly used to mix CNT and polymer, the phenomenon of agglomeration will lead to the deterioration of the dispersion of CNT, thus leading to the degradation of the properties of composites with high content of CNT. Therefore, the functionalization of CNT has been studied [**43,44,45,46**], and the surface chemical modification of CNT is used to reduce agglomeration. Zhang et al. [**47**] prepared acid-treated single-walled carbon nanotubes/PI films, which exhibited the best gas separation performance than the single-walled carbon nanotubes/PI composites without acid treatment at a filler content of 2 wt.%. Lee et al. [**48**] reported the unmodified, oxidized, and silanized CNTs/epoxy composites, the elastic modulus of the silanized CNTs/epoxy composite was 34% higher than that of the unmodified CNTs/epoxy composites, and 18% higher than that of the oxidized CNTs/epoxy composites. The surface functionalized CNT has improved dispersion in organic solvents and resin matrix, and the high dispersion of CNT can enhance the performance of CNT/PI films, thus giving CNT/PI films a broader application prospect.

Here, CNT was functionalized by acid treatment to improve its dispersion in polyamic acid (PAA) solution and PI film; thus, carboxylated CNT (C-CNT) was achieved. Next, CNT/PI and C-CNT/PI films were prepared with different contents of CNT and C-CNT, respectively. The dispersion of CNT and C-CNT, as well as the effects of filler content on the thermal and mechanical properties of the films, were systematically studied. Moreover, the thermal and mechanical performance differences between CNT/PI and C-CNT/PI films were compared under the same filler content.

2. MATERIALS AND METHOD

2.1. Materials

Polyamic acid (PAA, cas No. 25036-53-7, >99.6 wt.%, synthesized by 1,2,4,5-benzenetetracarboxylic anhydride (PMDA) and 4,4'-Oxydianiline (ODA)) was purchased from Xinsheng Plastic Company (Dongguan, China), which was dissolved in NMP at a ratio of 18 wt.%. N-methylpyrrolidone (NMP, cas No. 872-50-4, 99.5 wt.%) was obtained from Shanghai Adamas Reagents Company (Shanghai, China). Sulfuric acid (H_2SO_4, cas No. 7664-93-9, 95–98 wt.%) and nitric acid (HNO_3, cas No. 7697-37-2, 65–68 wt.%) were purchased from Sinopharm Chemical Reagent Company (Shanghai, China). Multiwalled carbon nanotube (cas No. 308068-56-6, 95 wt.%, length 10–30 μm, diameter 10–20 nm) was purchased from Jiangsu XFNANO Materials Technique Company (Nanjing, China).

2.2. Preparation of C-CNT

C-CNT was prepared by acid treatment; 5 g CNT was dispersed in 400 mL of a mixture of concentrated H_2SO_4/HNO_3 solution with a volume ratio of ٣:١ and stirred at ٥٠ °C for ٢٠ h [45]. The product was filtered and collected through 0.45 μm pore-sized PTFE film. Afterward, the product was washed with deionized water and dried at room temperature for 12 h, repeated three times. Finally, ground the obtained C-CNT into powder and screened with a 280-mesh sieve (0.055 mm).

2.3. Preparation of CNT/PI and C-CNT/PI Films

The preparation of C-CNT/PI films consisted of two steps, as shown in **Figure 1.** In the first step, the predetermined amount of C-CNT was added to the NMP solution and dispersed in the following steps: stirred for 30 min, sonicated for 30 min, stirred for 6 h, and sonicated for 30 min. Then the C-CNT suspension was added to the PAA solution and dispersed by the same stirring and ultrasonic steps as above. At this time, all PAA solutions were diluted to 10 wt.% by NMP. In the second step, the C-CNT/PAA suspensions were cast onto the glass plate and evacuated for 30 min for degassing (≤-0.1 MPa). Then the CNT/PI films were prepared by thermal imidization at the following temperatures: 80 °C, 120 °C, 160 °C, 200 °C, 250 °C, 300 °C, all times were 1 h for each temperature, and finally cooled to room temperature (25–30 °C). Through this process, a series of C-CNT/PI films with C-CNT concentrations of 0, 1, 3, 5, 7, and 9 wt.% were synthesized. The average thickness of the films was about 80 μm. The preparation process for CNT/PI films was the same as the above steps.

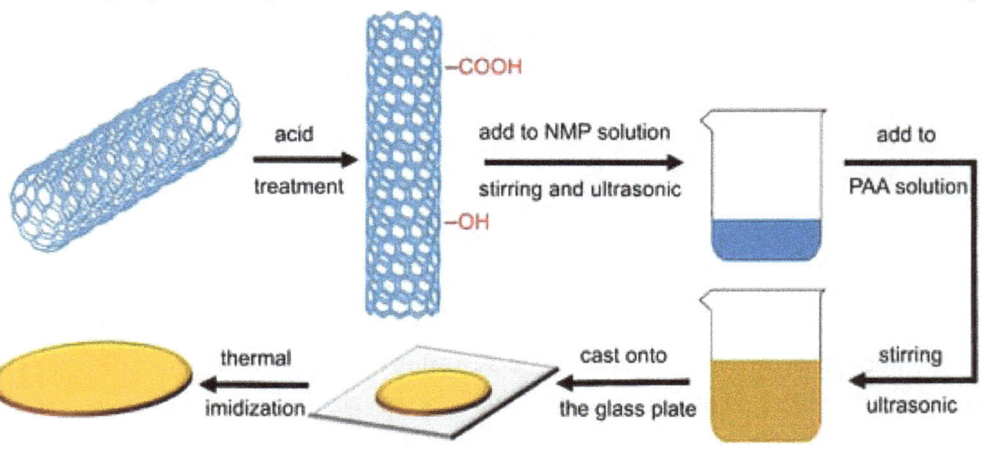

Figure 1. Preparation procedures of films.

2.4. Characterizations

Fourier transform infrared (FTIR) spectra of CNT and C-CNT were studied with a Fourier infrared spectrometer (FTIR, Nicolet6700, Benton, AR, USA) to detect any new absorption bands during the acid treatment. For sample preparation, the infrared KBr tableting method was applied. The spectrums were recorded with a resolution of 2 cm-1 in the range of 400–4000 cm-1. The surface chemistry of CNT and C-CNT was monitored by X-ray photoelectron spectroscopy (XPS, ESCALAB 250, Richardson, TX, USA). The pass energy was set at resolution 20 eV, and the X-ray source was Al-Ka (1486.6 eV). The water contact Angle of the film surface was measured by an optical surface analyzer (OSA200, Ningbo Scientific Instruments Company, Ningbo, China). The morphology of CNT and C-CNT in NMP solution was analyzed by a transmission electron microscope (TEM, JEM 2100F, Tokyo, Japan). The transmittance spectra of CNT/PI and C-CNT/PI films were measured by a UV-visible spectrophotometer (UV950, JASCO INTERNATIONAL Co., Ltd., Tokyo, Japan). The surface morphologies of films were observed by a scanning electron microscope (SEM, TM3000, Tokyo, Japan). Thermal gravimetric analysis (TGA, TGA 8000, New York, NY, USA) was used for thermal analysis of the CNT/PI films and C-CNT/PI films under a nitrogen atmosphere. The test samples were heated from 50 °C to 800 °C at the rate of 10 °C/min. CTEs of the films were performed on thermo-mechanical analysis (TMA, TMAQ400, Chicago, IL, USA) with a tension force of 0.05 N. The sample spacing was 16 mm, the temperature range was -10–150 °C, and the heating rate was 10 °C/min. Mechanical properties were tested by a microcomputer control electron universal testing machine (MTS, MTS-E42.503, Xiamen, China). The sample size was 100 mm × 10 mm, the gauge length was 50 mm, and the test was conducted at the crosshead speed of 5 mm/min. At least five samples were tested to analyze the mechanical properties of the films.

3. RESULTS AND DISCUSSION

3.1. Chemical Composition of CNT and C-CNT

Initially, the performance of CNT and C-CNT was compared and analyzed. The functional group difference between CNT and C-CNT was investigated by FTIR spectroscopy, as shown in Figure 2. There were features at wavenumbers 1000–1200 cm^{-1}, 1600–1680 cm^{-1}, and 2840–3000 cm^{-1} for CNT and C-CNT, corresponding to C–O, C=C, –CH stretching, respectively. The features at wavenumbers 1406 cm^{-1} for CNT and 1455 cm^{-1} for C-CNT were attributed to –CH bend vibration. Compared with CNT, C-CNT showed additional characteristic peaks at 1164 cm^{-1} and 1735 cm^{-1}, which can be attributed to the C–O (1000–1200 cm^{-1}) and C=O (1650–1800 cm^{-1}) stretching vibrations of the carboxylic group. Moreover, the peak observed for C-CNT at 3434 cm^{-1} originated from the stretching frequency of –OH (3200–3500 cm^{-1}). The results show that –COOH and more –OH functional groups appeared on the surface of C-CNT compared with CNT.

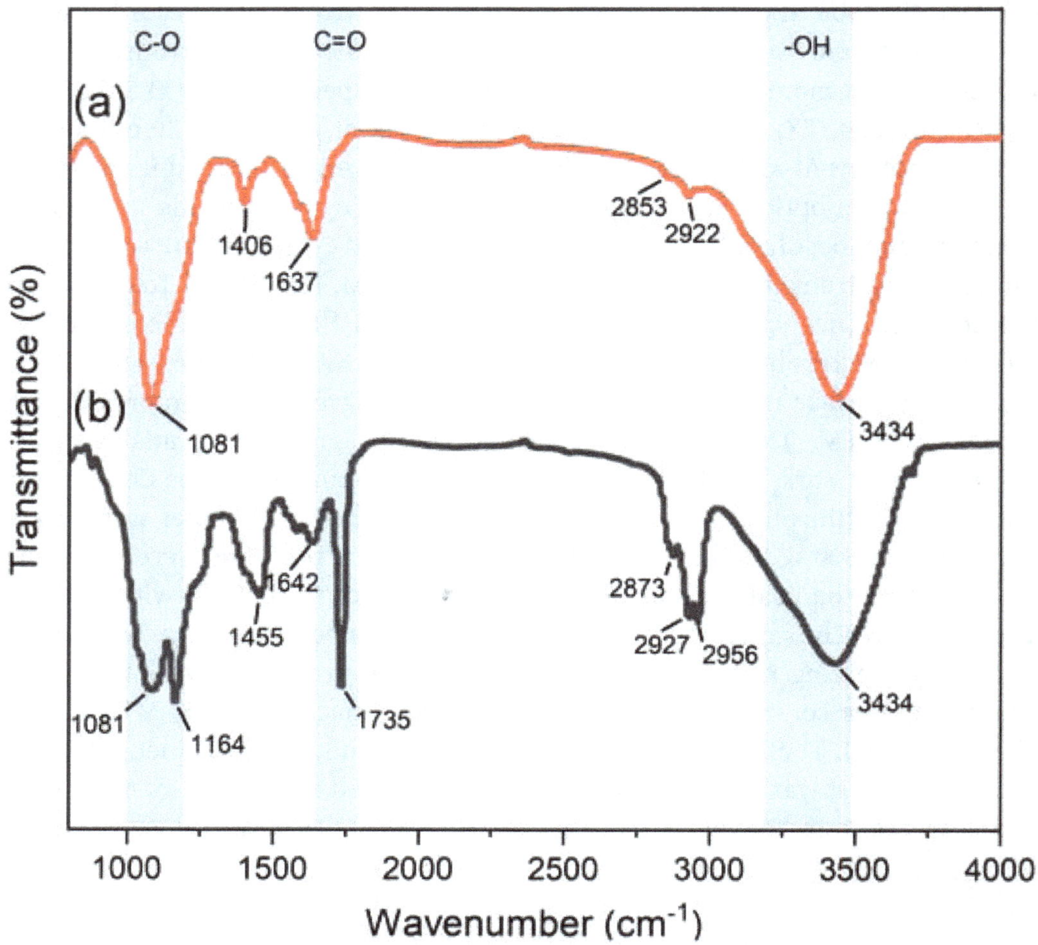

Figure 2. FTIR spectra of (a) CNT and (b) C-CNT.

XPS analysis was used to study the differences between elements and functional groups on the surface of CNT and C-CNT. Figure 3 presents C1s XPS spectra and XPS survey scans of CNT and C-CNT. The C1s peak was deconvoluted by XPS-PEAK software and divided according to four binding energy positions in order to analyze the content of different functional groups, as shown below: Sp^2 C (284.8 eV), Sp3 C (285.1 eV), –C–O– (286.2 eV) and –CO–O– (290.1 eV). The content of the functional groups and element content of CNT and C-CNT were summarized in Table 1 and Table 2, respectively. More O atom content was observed in C-CNT, with a twofold increase in the O/C ratio compared to CNT, implying that C-CNT had more oxygen-containing functional groups on the surface. Compared with CNT, the content of –C–O– of C-CNT increased from 7.04% to 11.74%, and the content of –CO–O– also increased, which indicated that C-CNT had more –COOH and –OH functional groups, which was also in agreement with the results of FTIR analysis.

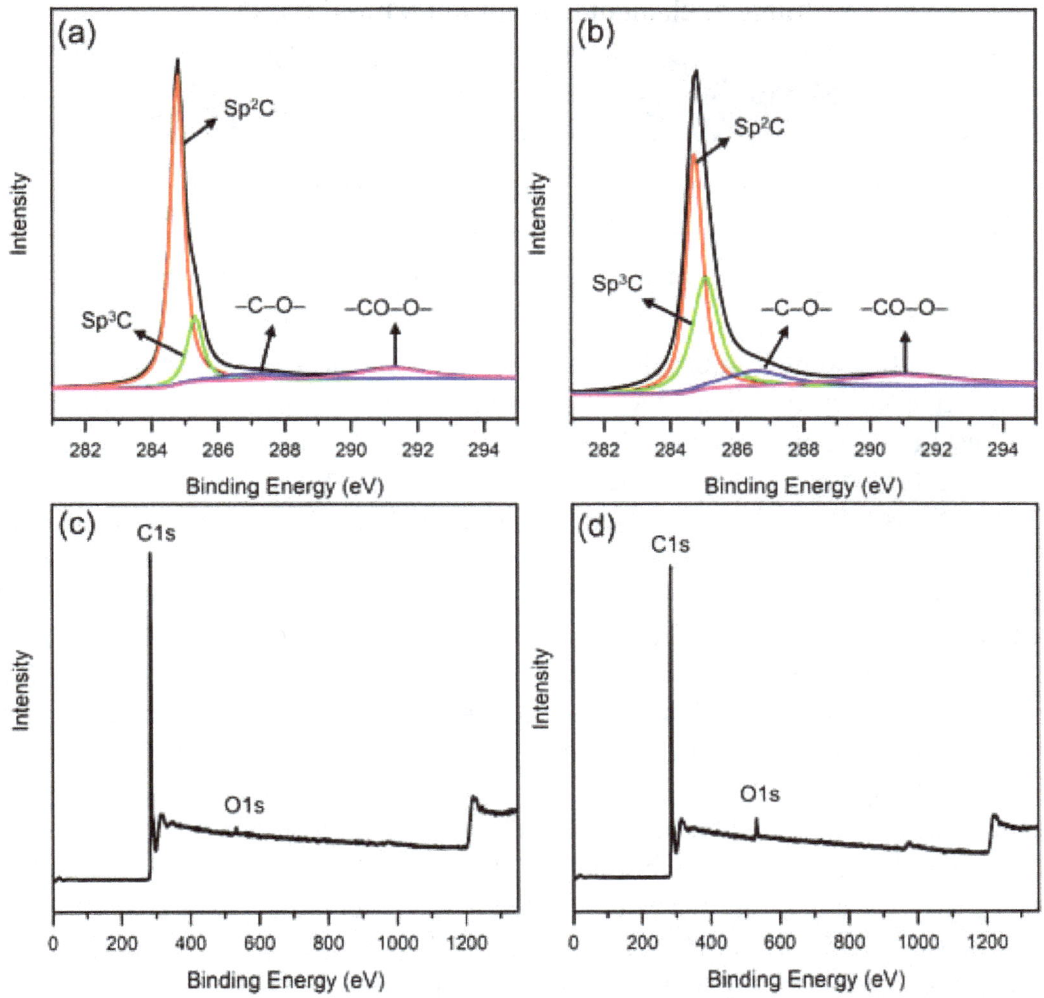

Figure 3. C1s XPS spectra of (a) CNT and (b) C-CNT, and XPS survey scans of (c) CNT and (d) C-CNT.

Table 1. Content of the functional groups of CNT and C-CNT.

Samples	Contents of the Functional Groups			
	Sp2 C(%)	Sp3 C(%)	-C-O-(%)	-CO-O- (%)
CNT	65.93	15.70	7.04	11.33
C-CNT	40.46	31.92	11.74	15.88

"CNT" is the abbreviation of carbon nanotube, "C-CNT" is the abbreviation of carboxylated CNT.

Table 2. Element content of CNT and C-CNT

Samples	Element Content (%)				Proportion
	C	O	N	S	O/C
CNT	97.30	1.27	1.28	0.15	0.01
C-CNT	95.20	3.79	0.85	0.15	0.04

"CNT" is the abbreviation of carbon nanotube, "C-CNT" is the abbreviation of carboxylated CNT.

The contact angle can provide information about surface chemistry because it is sensitive to surface chemistry. Therefore, the water contact angle of the films was tested and summarized in Figure 4 and Table 3. The contact angle of CNT/PI and C-CNT/PI films increased with the increase in CNT and C-CNT content, and CNT/PI films exhibited larger contact angles than CNT/PI films. The main chain of polyimide contains carbonyl groups, which can form hydrogen bonds with hydroxyl groups, so the hydrophilicity was good (55.74°). As a super hydrophobic material with a water contact angle of up to 158° [49], CNT can improve the hydrophobicity of PI films. When the CNT content was 9 wt.%, the contact angle of CNT-PI film was increased to 81.17°, which is 45.62% higher than that of pure PI film. However, C-CNT was not so effective in improving the contact angle of PI films because the carboxyl group in C-CNT was hydrophilic, so the hydrophobicity of C-CNT was not as good as that of CNT [50]. As a result, the contact angle of C-CNT/PI films only increased to 76.23 when the content of C-CNT was 9%, which was 36.76% higher than that of pure PI films. In a word, CNT and C-CNT can improve the contact angle and hydrophobicity of PI films, and the effect is best when 9 wt.% CNT is added. As a result, the self-cleaning, antifouling, and anti-corrosion properties of the films have been improved, which has a broader prospect in fields such as aerospace and medical care.

Table 3. The water contact angles of PI, CNT/PI and C-CNT/PI films.

Samples	Contact Angle (°)	Samples	Contact Angle (°)
PI	55.74 ± 2.31	-	-
1 wt.% CNT/PI	62.87 ± 1.53	1 wt.% C-CNT/PI	58.23 ± 1.97
3 wt.% CNT/PI	72.90 ± 0.21	3 wt.% C-CNT/PI	69.72 ± 0.74
5 wt.% CNT/PI	74.09 ± 0.32	5 wt.% C-CNT/PI	70.19 ± 1.52
7 wt.% CNT/PI	74.69 ± 0.62	7 wt.% C-CNT/PI	72.79 ± 0.81
9 wt.% CNT/PI	81.17 ± 0.91	9 wt.% C-CNT/PI	76.23 ± 1.65

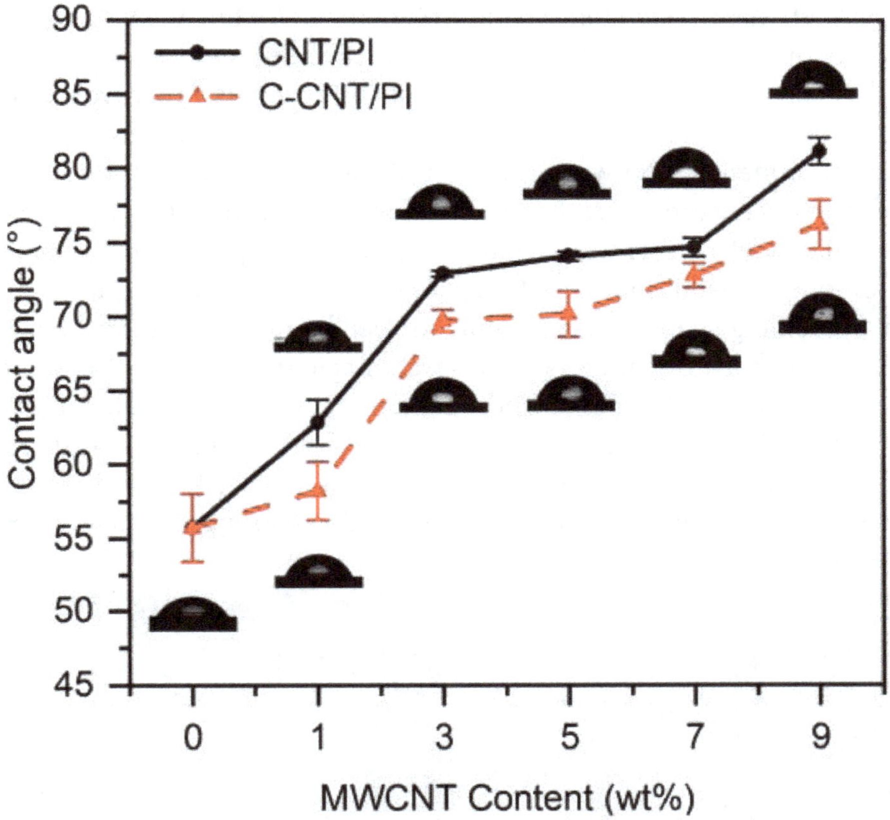

Figure 4. Summary of water contact angle measurements for PI, CNT/PI and C-CNT/PI films.

"PI" is the abbreviation of polyimide, "CNT" is the abbreviation of carbon nanotube, "C-CNT" is the abbreviation of carboxylated CNT; The "±" symbol is used to connect the mean and standard deviation of the measurements.

3.2. Morphology and Dispersion of CNT and C-CNT in Solution and Film

TEM examination of CNT and C-CNT was performed to investigate the morphology and dispersion. For the CNT and C-CNT showed in **Figure 5**, the average diameter was in the range of 10–20 nm and had similar profiles. However, C-CNT with a shorter length appeared (at the white arrow in **Figure** 5d). Notably, CNT was usually closed-ended (indicated by the circles in **Figure** 5c), whereas C-CNT was typically open-ended (indicated by the circles in **Figure** 5d) **[51]**. The reason is that the reaction usually starts at the defect sites such as the heptatomic rings, the $-CH_2$ and –CH groups [46], which often exist at both ends and surface of CNT, leading to the cutting and opening of carbon nanotubes after strong acid treatment. The dispersion of CNT and C-CNT in the NMP solution, as seen by TEM, was very different; CNT was clustered into bundles (**Figure 5**c), while C-CNT was randomly and loosely dispersed in the NMP solution (**Figure 5**d), neither of them had any particulate impurities. Due to the tiny size of CNT, there is a relatively strong van der Waals force, which makes it easy to become entangled or reunited into bundles. After acid treatment, C-CNT was cut short, reducing the agglomeration and bonding state of long fibers, and the carboxyl group on the surface of C-CNT increases the surface polarity of C-CNT, which makes C-CNT well dispersed in a polar solvent (NMP solution). In addition, C-CNT was observed to be locally agglomerated and present a buckling shape. This phenomenon can be related to them being mostly open-ended anchoring near CNTs because, after acid treatment, C-CNT was grafted with carboxyl groups (both ends and weak points on the surface), and hydrogen bonds were easily formed between carboxyl groups. Moreover, the structure of C-CNT was not damaged because both CNT and C-CNT showed a multiwall structure. In conclusion, C-CNT was grafted with new functional groups to open the fracture, and the tube wall structure was not damaged.

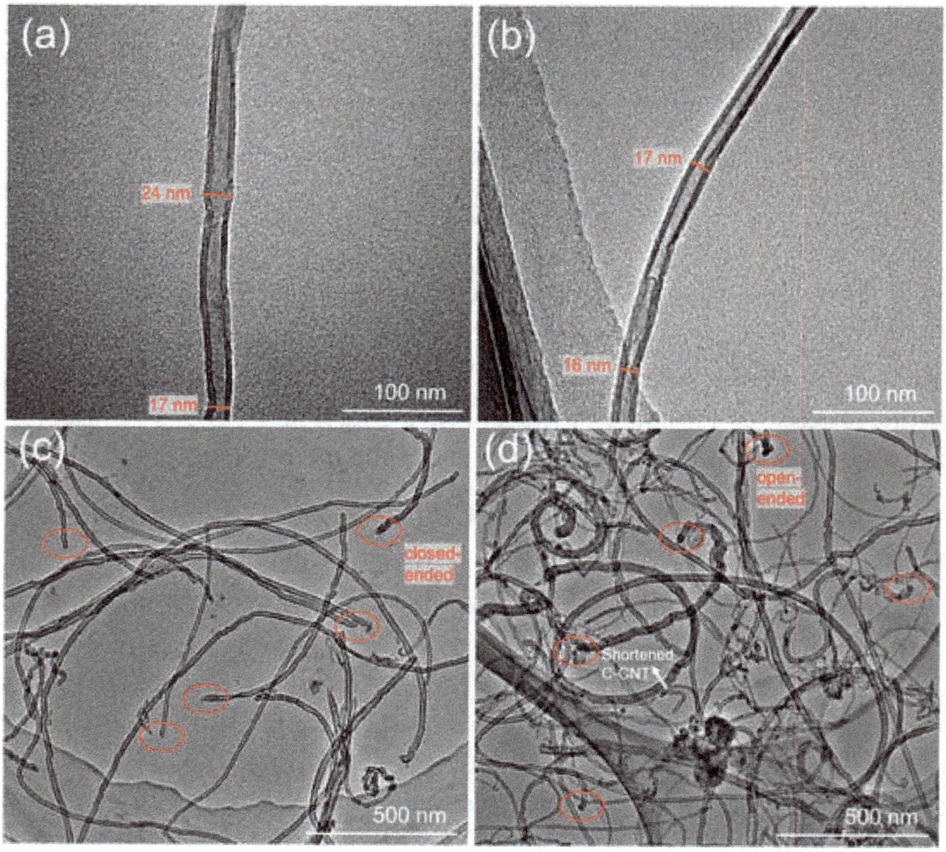

Figure 5. TEM image of (a) single CNT, (b) single C-CNT, (c) CNT with closed ends, (d) C-CNT with opened ends.

The dispersion of CNT and C-CNT in 10 wt.% PAA/NMP solutions is shown in **Figure 6.** It can be seen that the dispersion of C-CNT was much better than that of CNT. After standing for 36 h, the CNT suspensions containing 3 wt.% and 9 wt.% CNT began to settle, while C-CNT suspensions were still well dispersed. Even after standing for 72 h, 3 wt.% C-CNT suspensions had not entirely settled. This is due to the fact that –COOH groups on the surface of C-CNT belong to polar groups and are readily soluble in polar NMP solvents. In the process of curing the solution into a film, the great dispersion and stability of C-CNT in the solution make it maintain great dispersion in the film.

Figure 6. Comparison of photographs of CNT and C-CNT dispersed in solution for different hours.

In order to analyze the influence of filler content on the apparent morphology of films, photographs of PI films filled with CNT or C-CNT with different contents were taken. As shown in **Figure** 7a,b, with the increase in CNT or C-CNT content, the colour of the films gradually turned black. The surfaces of CNT/PI films and C-CNT/PI films were extremely smooth, within the range of 1–5 wt.% filler content. However, with the increase in filler content, at 9 wt.%, the surface of C-CNT/PI film was still smooth (**Figure** 7e), while the surface of CNT/PI film had visible particles (**Figure** 7d), which was due to the agglomeration of CNT. The behavior of CNT agglomeration can also be observed inside the film with high filler content [**39,48**]. Additionally, the film still exhibited good flexibility after adding filler, as shown in 9 wt.% C-CNT/PI film (**Figure** 7c).

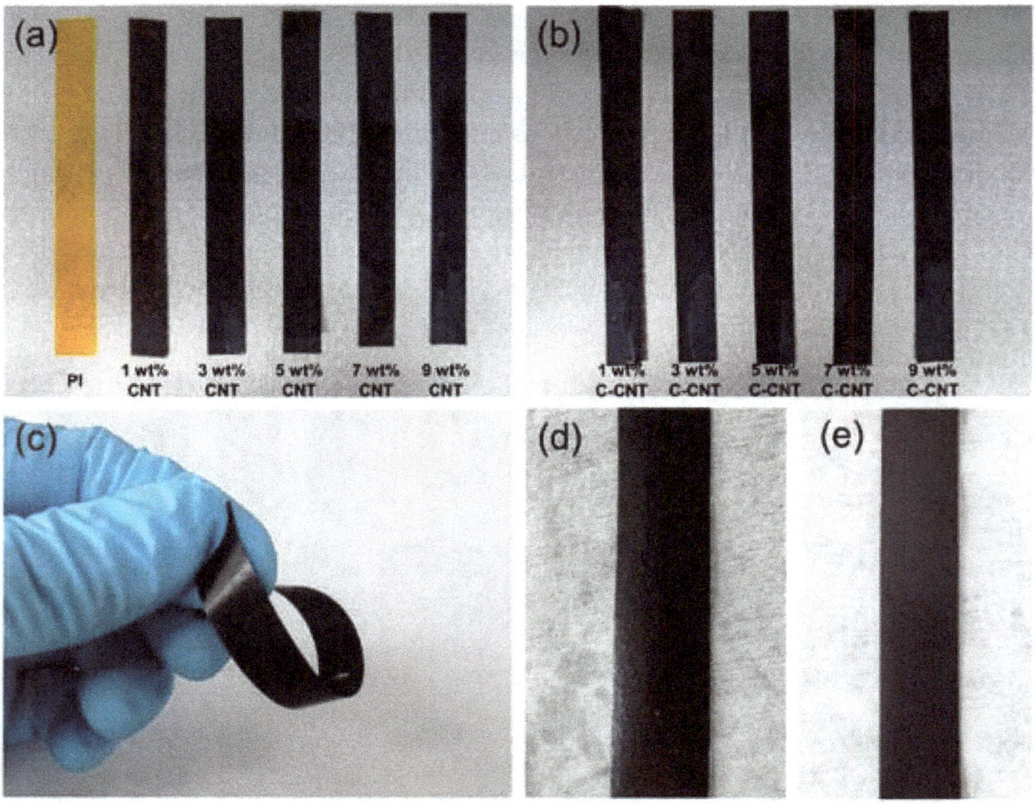

Figure 7. Photographs of PI films with different contents of (a) CNT and (b) C-CNT, (c) the flexible film, (d) 9 wt.% CNT/PI film, and (e) 9 wt.% C-CNT/PI film.

With UV-visible spectroscopy, it is possible to detect an absorption band caused by the electron transfer (Cm^1 to Vm^1 at 650 nm (1.9 eV)) within the van Hove singularities [52]. Here we looked at the high-energy transition at 650 nm to investigate the dispersion of CNT and C-CNT in PI films. The results are shown in Table 4 and Figure 8. The transmittance of CNT/PI and C-CNT/PI films decreased first and then increased with the increase in CNT and C-CNT content. When the filler content was 1%, CNT or C-CNT was not enough to cover the whole film (Figure 5a,b), so the film still maintained a certain transmittance. With the increase in filler content, the transmittance of the films decreased due to the outstanding light absorption of CNT and C-CNT. However, when the content of CNT and C-CNT reached 7 wt.%, the transmittance of the films increased because the high filler content led to a decrease in dispersion. In addition, it is worth noting that the transmittance of C-CNT/PI films was lower than that of CNT/PI films, especially when the filler content was 1 wt.% because the dispersion of C-CNT in the membrane was higher than that of CNT at the same concentration [53]. When CNT or C-CNT agglomerates, the PI film will have areas with few fillers, increasing the light passing through, thus improving the transmittance of the film [54]. The better dispersion of C-CNT than CNT results in lower transmittance of C-CNT/PI films than CNT/PI films and higher transmittance of the film under high filler concentration.

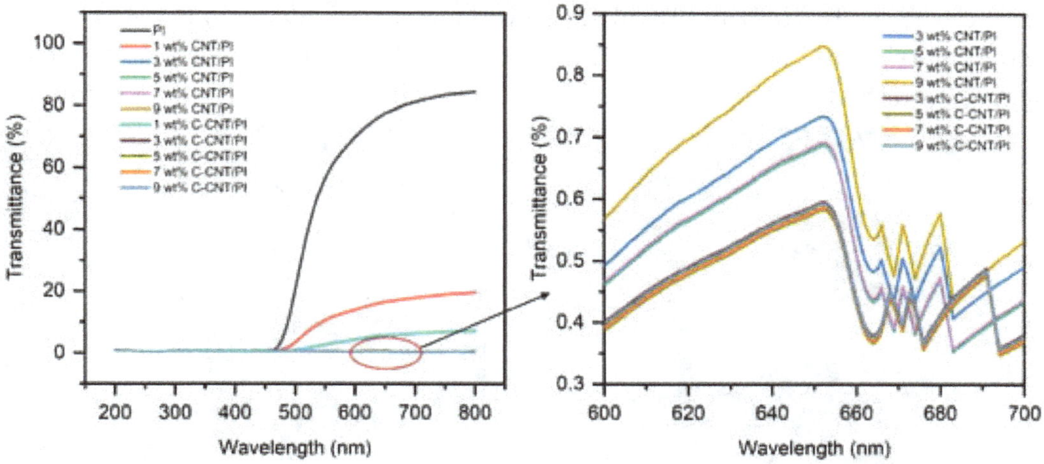

Figure 8. UV–visible spectra of PI, CNT/PI and C-CNT/PI films.

Table 4. Transmittance of PI, CNT/PI and C-CNT/PI films.

Samples	T650 (‰)	Samples	T650 (‰)
PI	774.18	-	-
1 wt.% CNT/PI	164.98	1 wt.% C-CNT/PI	59.45
3 wt.% CNT/PI	7.28	3 wt.% C-CNT/PI	5.90
5 wt.% CNT/PI	6.82	5 wt.% C-CNT/PI	5.81
7 wt.% CNT/PI	6.83	7 wt.% C-CNT/PI	5.81
9 wt.% CNT/PI	8.41	9 wt.% C-CNT/PI	5.82

"T_{650}" is the transmittance at 650 nm, "‰" means one over one thousand.

In order to study the dispersion and morphology of CNT and C-CNT in the films, the cross-sections of CNT/PI and C-CNT/PI films with filler contents of 0, 1, 3, 5, 7, and 9 wt.% were observed by SEM (**Figure S1**). The cross-sections were the fracture surfaces of the broken films after the mechanical properties test. As shown in **Figure 9**, CNT and C-CNT are marked with red circles. When the film is stretched, the strong adhesive interaction between carbon nanotubes and polymer matrix will produce stress, leading to local plastic deformation and rough fracture surface [47]. The C-CNT was mainly broken near the surface of the PI matrix, and compared with CNT/PI film, C-CNT/PI film had less filler exposed outside the matrix and a shorter filler length. This phenomenon means that the interfacial adhesion between C-CNT and PI matrix was better than CNT, which made C-CNT harder to be pulled out when the film broke [54]. Because of the hydrogen bonds formed by the H atom of the –COOH group on the surface of C-CNT and the C=O bond of the PI molecule, C-CNT was less agglomerated than CNT. As a result, C-CNT can be well wetted by the PI matrix, so the interfacial adhesion between C-CNT and PI matrix was better than CNT. In brief, C-CNT had better dispersion and less aggregation than CNT in the film, so it had better interaction with the PI matrix.

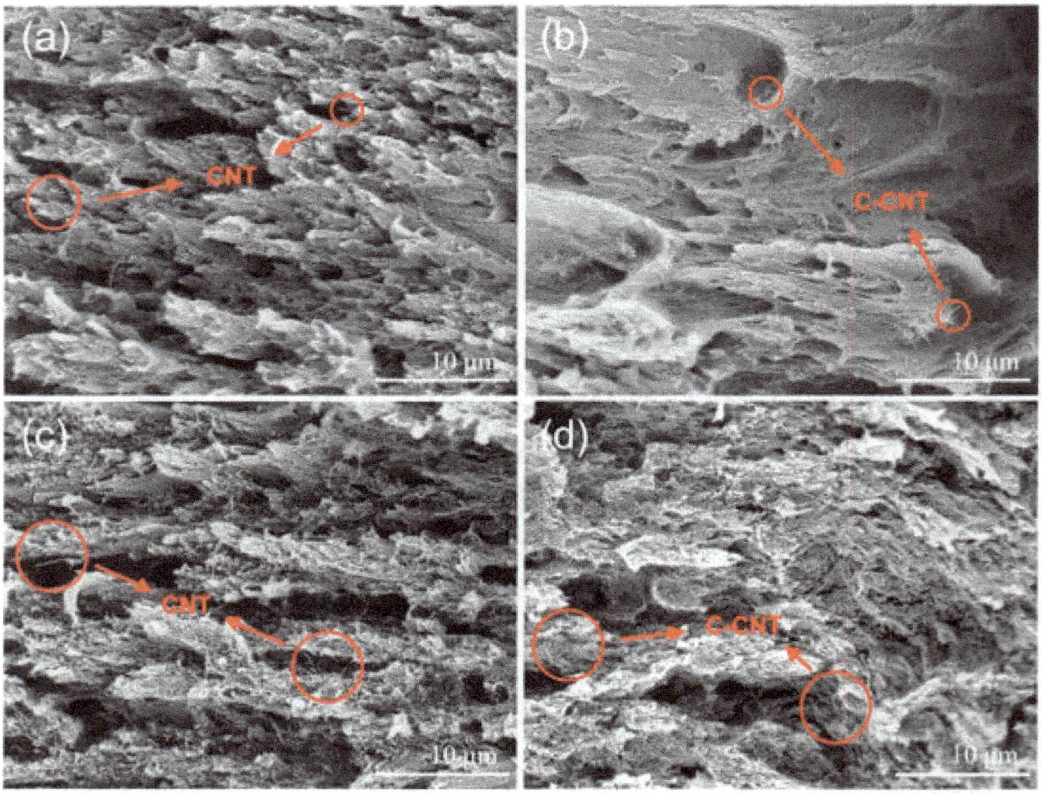

Figure 9. SEM cross-sectional images of (a) 1 wt.% CNT/PI film, (b) 1 wt.% C-CNT/PI film, (c) 9 wt.% CNT/PI film, (d) 9 wt.% C-CNT/PI film.

3.3. Thermal Properties of CNT/PI and C-CNT/PI Films

The thermal behaviors of PI films with various CNT and C-CNT contents were examined by TGA to evaluate the thermal stability, and the results are shown in Figure 10. The thermal decomposition of the PI film started at about 500 °C [55]. It can be seen from Table 5 that the 5 wt.% loss temperature of the pure PI film was only 551.63 °C. After mixing with CNT or C-CNT, the 5 wt.% loss temperature of the films was significantly increased and reached the highest value when the content was 7 wt.%. With the further increase in the content of CNT or C-CNT, the 5 wt.% loss temperature of the films decreased. This behavior may be due to the fact that excessive fillers tend to agglomerate and hinder heat transfer, generating high-temperature spots, thus affecting the continued improvement of thermal stability. At the same content, C-CNT showed a better effect of improving the thermal stability of the film than CNT, and the 5 wt.% loss temperature could reach up to 595.49 °C (7 wt.% C-CNT/PI film), which may be related to the high thermal stability of C-CNT and good filler-polymer affinity [56]. Moreover, the improved thermal resistance of PI films after mixing with CNT and C-CNT also led to a higher residual rate at 800 °C, from 49.93% (pure PI film) to 59.61% (9 wt.% C-CNT/PI film).

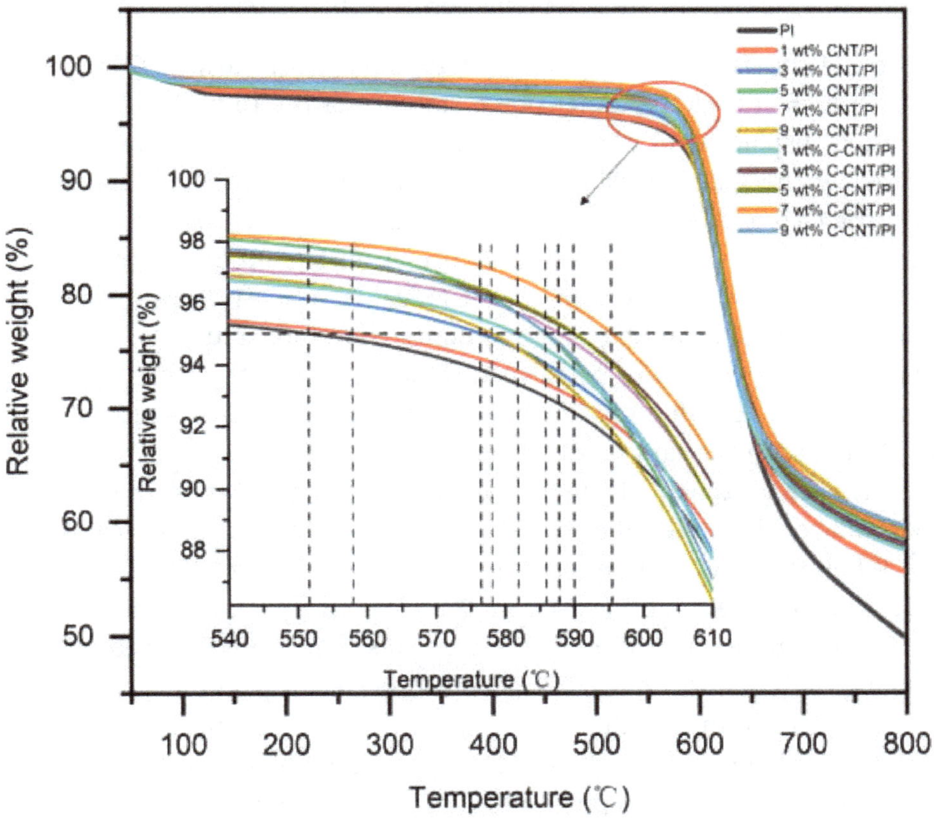

Figure 10. TGA results of PI films with different contents of CNT and C-CNT.

Table 5. The thermal properties of PI, CNT/PI and C-CNT/PI films.

Samples	5 wt.% Loss Temperature (°C)	800 °C Residual Rate (%)
PI	551.63	49.93
1 wt.% CNT/PI	557.97	55.65
3 wt.% CNT/PI	576.47	58.07
5 wt.% CNT/PI	585.93	58.61
7 wt.% CNT/PI	587.85	58.72
9 wt.% CNT/PI	578.11	59.43

1 wt.% C-CNT/PI	581.88	57.64
3 wt.% C-CNT/PI	590.05	57.99
5 wt.% C-CNT/PI	590.09	58.79
7 wt.% C-CNT/PI	595.49	58.93
9 wt.% C-CNT/PI	586.07	59.61

"PI" is the abbreviation of polyimide, "CNT" is the abbreviation of carbon nanotube, "C-CNT" is the abbreviation of carboxylated CNT.

Figure 11 shows the TMA results of PI films with various CNT and C-CNT contents, which reflect the dimensional stability of the films under heating. The temperature displacement curves are nearly linear because the test temperature is well below the glass transition temperature [57]. With the increase in filler content, the CTE of films showed a continuous downward trend and reached the lowest value at 9 wt.%. Because the deformation caused by thermal expansion at -10 °C to 150 °C was very small, the CTE values of CNT/PI film and C-CNT/PI film were very similar at the same content. Therefore, multiple measurements and statistical analyses were performed to obtain accurate results (Table 6). The increase in the CTE value of the film was found when CNT aggregates, so the C-CNT/PI film with the same content had a lower CTE value than the CNT/PI film. This result is consistent with the influence of CNT aggregation in the matrix on the CTE value of the material deduced from theoretical and numerical techniques [58,59]. When the filler content reached 9 wt.%, the CTE of CNT/PI and C-CNT/PI films decreased to 19.74 and 19.04 ppm/K, respectively, and decreased by 25.17% and 27.82% compared with pure PI films. At this time, C-CNT played a better role in reducing CTE because the C-CNT achieved better dispersibility and interfacial properties with the matrix due to the carboxylation modification. The CTE of a composite mainly depends on its structure, and each phase in the composite and the interaction between the phases together determine the CTE. The existence of CNT can be used as a physical barrier to interrupt the crystallization process, thus affecting the size of the lens and the structure of spherulites and ultimately leading to the increase in the amorphous area and the decrease in the crystallinity [22,60]. When the PI film has low crystalline morphology, it will reduce the CTE of PI and cause thermal expansion of the film in the thickness direction [61,62]. Meanwhile, C-CNT had more influence areas on the crystallinity of films (especially at low concentrations) due to its better dispersion in the film than CNT, which led to lower CTE of C-CNT/PI films than CNT/PI films. Moreover, the CTE of CNT and C-CNT is so low that they hardly deform during the experiment, and when the PI matrix is thermally expanded, they will limit the deformation of the matrix as reinforcements. CNT and C-CNT were uniformly dispersed in the PI matrix and had a solid interfacial bonding effect with the matrix, so the CTE of CNT/PI and C-CNT/

PI films can be effectively reduced.

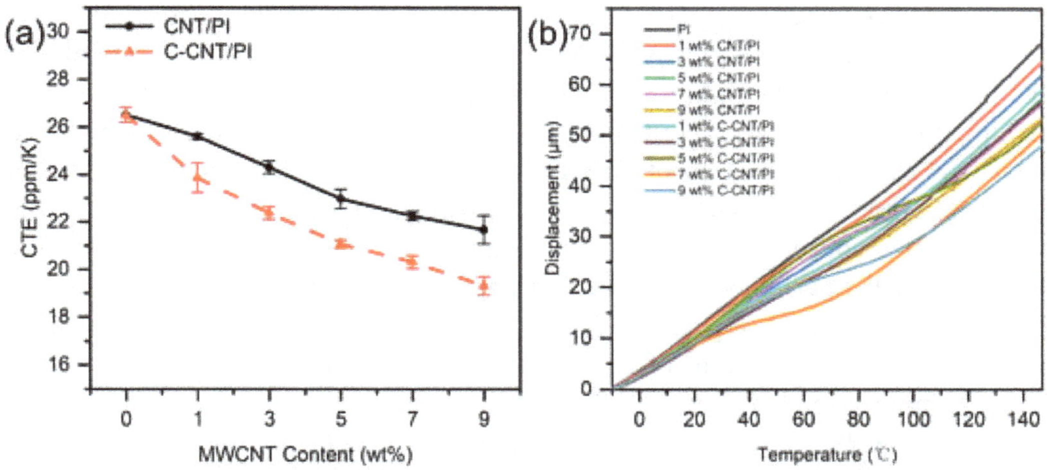

Figure 11. (a) Summary of the CTE values of PI films with different contents of CNT and C-CNT and (b) TMA curves of PI, CNT/PI and C-CNT/PI films.

Table 6. The CTE values of PI, CNT/PI and C-CNT/PI films.

Samples	CTE (ppm/K)	Samples	CTE (ppm/K)
PI	26.51 ± 0.32	-	-
1 wt.% CNT/PI	25.59 ± 0.12	1 wt.% C-CNT/PI	23.87 ± 0.62
3 wt.% CNT/PI	24.31 ± 0.26	3 wt.% C-CNT/PI	22.38 ± 0.27
5 wt.% CNT/PI	22.97 ± 0.41	5 wt.% C-CNT/PI	21.07 ± 0.17
7 wt.% CNT/PI	22.26 ± 0.18	7 wt.% C-CNT/PI	20.31 ± 0.27
9 wt.% CNT/PI	21.68 ± 0.60	9 wt.% C-CNT/PI	19.30 ± 0.38

"PI" is the abbreviation of polyimide, "CNT" is the abbreviation of carbon nanotube, "C-CNT" is the abbreviation of carboxylated CNT; The "±" symbol is used to connect the mean and standard deviation of the measurements.

3.4. Mechanical Properties of CNT/PI and C-CNT/PI Films

In order to investigate the effect of CNT and C-CNT in PI films, the mechanical properties of the films were further tested, and the experimental results are shown in **Figure 12**. According to the results in **Table 7**, both the tensile strength and Young's modulus of films increased up to a maximum value and then decreased with the content of the filler. The tensile strength and Young's modulus of CNT/PI and C-CNT/PI films reached the highest value at 5 wt.% filler content, and C-CNT had a better mechanical reinforcement effect on PI films. When the content of fillers reached 5 wt.%, the tensile strength of the CNT/PI and C-CNT/PI films increased to 95.61 and 97.52 MPa, respectively, which increased by about 21.50% and 23.93% compared with pure PI films (78.69 MPa). When the content of C-CNT was 5 wt.%, Young's modulus of C-CNT/PI film reached the highest value of 3.02 GPa, which was 30.82% higher than that of pure PI films (2.31 GPa). However, the turning point of Young's modulus of CNT/PI film was earlier (3 wt.%), which was only 14.35% higher than that of pure PI films. In particular, compared with CNT/PI films, the tensile strength and Young's modulus of C-CNT/PI films were higher when the filler content was the same. In **Figure** 7e,f, the 9 wt.% CNT/PI film surface had small visible particles formed by CNT agglomeration, while the 9 wt.% C-CNT/PI film did not have these particles. Therefore, the dispersion of C-CNT in the films was more uniform than that of CNT, and due to the existence of –COOH groups, both the interaction between C-CNT and PI matrix and the interface performance was improved. The dispersion of fillers in polymers and the interaction between fillers and polymers play a vital role in enhancing the mechanical properties of polymer composites; as a result, the mechanical properties of C-CNT/PI films are better improved. Similarly, when the filler content is too high, the dispersion of the filler in the film will be reduced, and agglomeration will occur. When CNT or C-CNT aggregates, the matrix is difficult to penetrate into it, and the CNT cluster lacks load transfer capability [**63**]. Meanwhile, there is a poor binding force between CNT clusters [**64**]; thus, the mechanical properties are mainly determined by the matrix. Therefore, at higher filler content (>5 wt.%), the mechanical properties of CNT/PI and C-CNT/PI films will decrease.

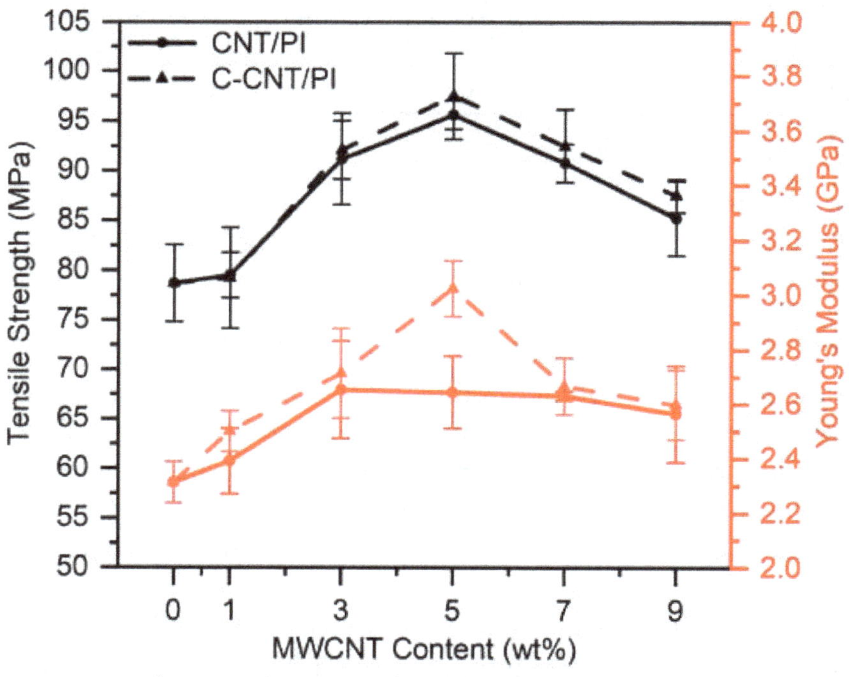

Figure 12. The tensile strength and tensile modulus of PI films with different contents of CNT and C-CNT.

Table 7. The mechanical properties of PI, CNT/PI and C-CNT/PI films.

Samples	Tensile Strength (MPa)	Tensile Modulus (GPa)
PI	78.69 ± 3.87	2.31 ± 0.08
1 wt.% CNT/PI	79.50 ± 2.27	2.39 ± 0.12
3 wt.% CNT/PI	91.18 ± 4.61	2.65 ± 0.18
5 wt.% CNT/PI	95.61 ± 1.42	2.64 ± 0.13
7 wt.% CNT/PI	90.80 ± 1.92	2.63 ± 0.02
9 wt.% CNT/PI	85.22 ± 3.72	2.56 ± 0.18
1 wt.% C-CNT/PI	79.21 ± 5.06	2.50 ± 0.07
3 wt.% C-CNT/PI	92.10 ± 2.93	2.71 ± 0.17
5 wt.% C-CNT/PI	97.52 ± 4.33	3.02 ± 0.10
7 wt.% C-CNT/PI	92.49 ± 3.68	2.67 ± 0.10
9 wt.% C-CNT/PI	87.50 ± 1.68	2.60 ± 0.13

"PI" is the abbreviation of polyimide, "CNT" is the abbreviation of carbon nanotube, "C-CNT" is the abbreviation of carboxylated CNT; The "±" symbol is used to connect the mean and standard deviation of the measurements.

4. CONCLUSIONS

In conclusion, carboxyl groups were formed on the surface of C-CNT after acid treatment, leading to better dispersion of C-CNT in solution and film than CNT and better interfacial adhesion between C-CNT and PI matrix. Thus C-CNT can better enhance the thermal and mechanical properties of PI films than CNT. C-CNT improved the thermal stability of the films, but high-temperature spots would appear in the films due to high C-CNT content (>7 wt.%), which reduced the thermal stability. The mechanical properties of the films were improved by adding C-CNT. However, it was difficult for C-CNT to be thoroughly wetted by the PI matrix at high content (>5 wt.%), which reduced the load transfer of the matrix to C-CNT, leading to the decline of the mechanical properties of the films. This paper studied the property advantages of C-CNT/PI films and the optimum content required for the highest performance, thus guiding the application of C-CNT/PI films in such fields as aerospace and electronic engineering.

5. CONCISE SUMMARY

Polyimide (PI) films with excellent heat resistance and outstanding mechanical properties have been widely researched in microelectronics and aerospace fields. However, most PI films can only be used under ordinary conditions due to their instability of dimension. The fabrication of multifunctional PI films for harsh conditions is still a challenge. Herein, flexible, low coefficient of thermal expansion (CTE) and improved mechanical properties films modified by carboxylated carbon nanotube (C-CNT) were fabricated. Acid treatment was adapted to adjust the surface characteristics by using a mixture of concentrated H_2SO_4/HNO_3 solution to introduce carboxyl groups on the surface and improve the interfacial performance between the CNT and matrix. Moreover, different C-CNT concentrations of 0, 1, 3, 5, 7, and 9 wt.% were synthesized to use for the PI film fabrication. The results demonstrated that the 9 wt.% and 5 wt.% C-CNT/PI films possessed the lowest CTE value and the highest mechanical properties. In addition, the thermal stability of the C-CNT/PI films was improved, making them promising applications in precise and harsh environments.

REFERENCES

1. Wang, K.; Amin, K.; An, Z.; Cai, Z.; Chen, H.; Chen, H.; Dong, Y.; Feng, X.; Fu, W.; Gu, J.; et al. Advanced functional polymer materials. Mater. Chem. Front. 2020, 4, 1803–1915. [Google Scholar] [CrossRef]
2. Ou, X.; Lu, X.; Chen, S.; Lu, Q. Thermal conductive hybrid polyimide with ultrahigh heat resistance, excellent mechanical properties and low coefficient of thermal expansion. Eur. Polym. J. 2020, 122, 109368. [Google Scholar] [CrossRef]
3. Ji, S.; Yang, J.; Zhao, J.; Hu, Y.; Gao, H. Study about mechanical property and machinability of polyimide. Polymers 2018, 10, 173. [Google Scholar] [CrossRef] [PubMed][Green Version]
4. Ruan, H.; Zhang, Q.; Liao, W.; Li, Y.; Huang, X.; Xu, X.; Lu, S. Enhancing tribological, mechanical, and thermal properties of polyimide composites by the synergistic effect between graphene and ionic liquid. Mater. Des. 2020, 189, 108527. [Google Scholar] [CrossRef]
5. Wang, B.; He, B.; Wang, Z.; Qi, S.; Zhang, D.; Tian, G.; Wu, D. Enhanced impact properties of hybrid composites reinforced by carbon fiber and polyimide fiber. Polymers 2021, 13, 2599. [Google Scholar] [CrossRef]
6. Chang, J.; Ge, Q.; Zhang, M.; Liu, W.; Cao, L.; Niu, H.; Sui, G.; Wu, D. Effect of pre-imidization on the structures and properties of polyimide fibers. RSC Adv. 2015, 5, 69555–69566. [Google Scholar] [CrossRef]
7. Zhang, M.; Niu, H.; Chang, J.; Ge, Q.; Cao, L.; Wu, D. High-performance fibers based on copolyimides containing benzimidazole and ether moieties: Molecular packing, morphology, hydrogen-bonding interactions and properties. Polym. Eng. Sci. 2015, 55, 2615–2625. [Google Scholar] [CrossRef]

8. Wang, Z.; Zhang, J.; Niu, H.; Wu, D.; Zhang, M.; Han, E.; Sheng, J.; Sun, X.; Fan, C. Influences of different imidization conditions on polyimide fiber properties and structure. J. Appl. Polym. Sci. 2021, 138, 51189. [Google Scholar] [CrossRef]

9. Wu, B.; Zhang, Y.; Yang, D.; Yang, Y.; Yu, Q.; Che, L.; Liu, J. Self-healing anti-atomic-oxygen phosphorus-containing polyimide film via molecular level incorporation of nanocage trisilanolphenyl POSS: Preparation and characterization. Polymers 2019, 11, 1013. [Google Scholar] [CrossRef] [PubMed][Green Version]

10. Chang, Z.; Sun, X.; Liao, Z.; Liu, Q.; Han, J. Des and preparation of polyimide/ TiO2@ MoS2 nanofibers by hydrothermal synthesis and their photocatalytic performance. Polymers 2022, 14, 3230. [Google Scholar] [CrossRef]

11. Yan, Y.; Mao, Y.; Li, B.; Zhou, P. Machinability of the thermoplastic polymers: PEEK, PI, and PMMA. Polymers 2020, 13, 69. [Google Scholar] [CrossRef] [PubMed]

12. Hornig, A.; Winkler, A.; Bauerfeind, E.; Gude, M.; Modler, N. Delamination behaviour of embedded polymeric sensor and actuator carrier layers in epoxy based CFRP laminates—A study of energy release rates. Polymers 2021, 13, 3926. [Google Scholar] [CrossRef] [PubMed]

13. Toshiyuki, T.; Maeda, S.; Tsukada, Y. Packaging substrate applications of ultralow CTE polyimide. J. Photopolym. Sci. Technol. 2012, 25, 359–364. [Google Scholar]

14. Tian, Y.; Luo, L.; Yang, Q.; Zhang, L.; Wang, M.; Wu, D.; Wang, X.; Liu, X. Construction of stable hydrogen bonds at high temperature for preparation of polyimide films with ultralow coefficient of thermal expansion and high Tg. Polymer 2020, 188, 122100. [Google Scholar] [CrossRef]

15. Zhuang, Y.; Seong, J.G.; Lee, Y.M. Polyimides containing aliphatic/alicyclic segments in the main chains. Prog. Polym. Sci. 2019, 92, 35–88. [Google Scholar] [CrossRef]

16. Huang, Y.; Shen, X.; Wang, Z.; Jin, K.; Lu, J.Q.; Wang, C. Zero thermal expansion polyarylamide film with reversible conformational change structure. Macromolecules 2018, 51, 8477–8485. [Google Scholar] [CrossRef]

17. Choi, J.Y.; Jin, S.W.; Kim, D.M.; Song, I.H.; Nam, K.N.; Park, H.J.; Chung, C.M. Enhancement of the mechanical properties of polyimide film by microwave irradiation. Polymers 2019, 11, 477. [Google Scholar] [CrossRef][Green Version]

18. Iijima, S. Helical microtubules of graphitic carbon. Nature 1991, 354, 56–58. [Google Scholar] [CrossRef]

19. Sholl, D.S.; Johnson, J.K. Making high-flux films with carbon nanotubes. Mater. Sci. 2006, 312, 1003–1004. [Google Scholar]

20. Ma, P.C.; Kim, J.-K.; Tang, B.Z. Functionalization of carbon nanotubes using a silane coupling agent. Carbon 2006, 44, 3232–3238. [Google Scholar] [CrossRef] [Green Version]

21. Xu, H.; Schubert, D.W. Electrical conductivity of polystyrene/poly (n-alkyl methacrylate) s/carbon nanotube ternary composite casting films. J. Polym. Res. 2020, 27, 153. [Google Scholar] [CrossRef]

22. Gissinger, J.R.; Pramanik, C.; Newcomb, B.; Kumar, S.; Heinz, H. Nanoscale structure-property relationships of polyacrylonitrile/CNT composites as a function of polymer crystallinity and CNT diameter. ACS Appl. Mater. Inter. 2018, 10, 1017–1027. [Google Scholar] [CrossRef] [PubMed]

23. Qin, S.; Cui, M.; Dai, Z.; Qiu, S.; Zhao, H.; Wang, L.; Zhang, A. Noncovalent functionalized graphene-filled polyimides with improved thermal, mechanical, and wear resistance properties. Tribol. Lett. 2018, 66, 69. [Google Scholar] [CrossRef]

24. Song, K.; Zhang, Y.; Meng, J.; Green, E.C.; Tajaddod, N.; Li, H.; Minus, M.L. Structural polymer-based carbon nanotube composite fibers: Understanding the processing-structure-performance relationship. Materials 2013, 6, 2543–2577. [Google Scholar] [CrossRef][Green Version]

25. Li, H.; Minus, M.L. On the formation of potential polymer-nanotube blends by liquid-solid phase separation. Polymer 2017, 131, 179–192. [Google Scholar] [CrossRef]

26. Coleman, J.N.; Khan, U.; Blau, W.J.; Gun'ko, Y.K. Small but strong: A review of the mechanical properties of carbon nanotube–polymer composites. Carbon 2006, 44, 1624–1652. [Google Scholar] [CrossRef]

27. Kim, C.; Yang, K. Electrochemical properties of carbon nanofiber web as an electrode for supercapacitor prepared by electrospinning. Appl. Phys. Lett. 2003, 83, 1216–1218. [Google Scholar] [CrossRef]

28. Dalton, A.B.; Collins, S.; Munoz, E.; Razal, J.M.; Ebron, V.H.; Ferraris, J.P.; Coleman, J.N.; Kim, B.G.; Baughman, R.H. Super-tough carbon-nanotube fibres. Nature 2003, 423, 703. [Google Scholar] [CrossRef]

29. Tanaike, O.; Hatori, H.; Yamada, Y.; Shiraishi, S.; Oya, A. Preparation and pore control of highly mesoporous carbon from defluorinated PTFE. Carbon 2003, 41, 1759–1764. [Google Scholar] [CrossRef]

30. Park, J.H.; Ko, J.M.; Park, O.O.; Kim, D.-W. Capacitance properties of graphite/polypyrrole composite electrode prepared by chemical polymerization of pyrrole on graphite fiber. J. Power Source 2002, 105, 20–25. [Google Scholar] [CrossRef]

31. Raunika, A.; Raj, S.A.; Jayakrishna, K.; Sultan, M. In carbon nanotube: A review on its mechanical properties and application in aerospace industry. IOP Conf. Ser. Mater. Sci. Eng. 2016, 270, 012027. [Google Scholar] [CrossRef][Green Version]

32. Jin, S.; Son, G.; Kim, Y.; Kim, C.-G. Enhanced durability of silanized multi-walled carbon nanotube/epoxy nanocomposites under simulated low earth orbit space environment. Compos. Sci. Technol. 2013, 87, 224–231. [Google Scholar] [CrossRef]

33. Ahmed, K.; Khosla, A.; Kawakami, M.; Furukawa, H. Poly ionic liquid-based nano composites for smart electro-mechanical devices. SPIE 2017, 10167, 97–103. [Google Scholar]

34. Ju, Y.-W.; Choi, G.-R.; Jung, H.-R.; Lee, W.-J. Electrochemical properties of electrospun PAN/MWCNT carbon nanofibers electrodes coated with polypyrrole. Electrochim. Acta 2008, 53, 5796–5803. [Google Scholar] [CrossRef]

35. Jiang, M.; Lin, D.; Jia, W.; Du, J.; Han, E.; Zhang, M.; Niu, H.; Wu, D. Preparation and properties of polyimide/carbon nanotube composite films with electromagnetic wave absorption performance. Polym. Eng. Sci. 2021, 61, 2691–2700. [Google Scholar] [CrossRef]

36. Monthioux, M.; Smith, B.W.; Burteaux, B.; Claye, A.; Fischer, J.E.; Luzzi, D.E. Sensitivity of single-wall carbon nanotubes to chemicalprocessing: An electron microscopy investigation. Carbon 2001, 39, 1251–1272. [Google Scholar] [CrossRef]

37. Rafiee, R.; Ghorbanhosseini, A. Investigating interaction between CNT and polymer using cohesive zone model. Polym. Compos. 2018, 39, 3903–3911. [Google Scholar] [CrossRef]

38. Kosynkin, D.V.; Higginbotham, A.L.; Sinitskii, A.; Lomeda, J.R.; Dimiev, A.; Price, B.K.; Tour, J.M. Longitudinal unzipping of carbon nanotubes to form graphene nanoribbons. Nature 2009, 458, 872–876. [Google Scholar] [CrossRef] [Green Version]

39. Lutsyk, P.M.; Shankar, P.; Rozhin, A.G.; Kulinich, S.A. Surface sensitivity of ultrasonically treated carbon nanotube network towards ammonia. Surf. Interfaces 2019, 17, 100363. [Google Scholar] [CrossRef]

40. Al Araimi, M.; Lutsyk, P.; Verbitsky, A.; Piryatinski, Y.; Shandura, M.; Rozhin, A. A dioxaborine cyanine dye as a photoluminescence probe for sensing carbon nanotubes. Beilstein. J. Nanotechnol. 2016, 7, 1991–1999. [Google Scholar] [CrossRef][Green Version]

41. Lutsyk, P.; Piryatinski, Y.; AlAraimi, M.; Arif, R.; Shandura, M.; Kachkovsky, O.; Verbitsky, A.; Rozhin, A. Emergence of additional visible-range

photoluminescence due to aggregation of cyanine dye: Astraphloxin on carbon nanotubes dispersed with anionic surfactant. J. Phys. Chem. C 2016, 120, 20378–20386. [Google Scholar] [CrossRef][Green Version]

42. Lutsyk, P.; Piryatinski, Y.; Shandura, M.; AlAraimi, M.; Tesa, M.; Arnaoutakis, G.E.; Melvin, A.A.; Kachkovsky, O.; Verbitsky, A.; Rozhin, A. Self-Assembly for Two Types of J-Aggregates: Cis-Isomers of Dye on the Carbon Nanotube Surface and Free Aggregates of Dye trans-Isomers. J. Phys. Chem. C 2019, 123, 19903–19911. [Google Scholar] [CrossRef]

43. Park, S.K.; Kim, S.H.; Hwang, J.T. Carboxylated multiwall carbon nanotube-reinforced thermotropic liquid crystalline polymer nanocomposites. J. Appl. Polym. Sci. 2008, 109, 388–396. [Google Scholar] [CrossRef]

44. Wu, H.-L.; Ma, C.-C.M.; Yang, Y.-T.; Kuan, H.-C.; Yang, C.-C.; Chiang, C.-L. Morphology, electrical resistance, electromagnetic interference shielding and mechanical properties of functionalized MWNT and poly (urea urethane) nanocomposites. J. Polym. Sci. Pol. Phys. 2006, 44, 1096–1105. [Google Scholar] [CrossRef]

45. Vast, L.; Philippin, G.; Destrée, A.; Moreau, N.; Fonseca, A.; Nagy, J.B.; Delhalle, J.; Mekhalif, Z. Chemical functionalization by a fluorinated trichlorosilane of multi-walled carbon nanotubes. Nanotechnology 2004, 15, 781–785. [Google Scholar] [CrossRef]

46. Sun, H.; Wang, T.; Xu, Y.; Gao, W.; Li, P.; Niu, Q.J. Fabrication of polyimide and functionalized multi-walled carbon nanotubes mixed matrix films by in-situ polymerization for CO2 separation. Sep. Purif. Technol. 2017, 177, 327–336. [Google Scholar] [CrossRef]

47. Zhang, Q.; Li, S.; Wang, C.; Chang, H.-C.; Guo, R. Carbon nanotube-based mixed-matrix films with supramolecularly engineered interface for enhanced gas separation performance. J. Film. Sci. 2020, 598, 117794. [Google Scholar]

48. Lee, J.-H.; Rhee, K.Y.; Park, S.J. Silane modification of carbon nanotubes and its effects on the material properties of carbon/CNT/epoxy three-phase composites. Compos. Part A Appl. Sci. Manuf. 2011, 42, 478–483. [Google Scholar] [CrossRef]

49. Mathkar, A.; Aichele, C.; Omole, I.; Singh, N.; Hashim, D.; Gullapalli, H.; Ajayan, P.M. Creating supersolvophobic nanocomposite materials. RSC Adv. 2013, 3, 4216–4220. [Google Scholar] [CrossRef]

50. Zhang, X.; Zhang, W.; Dai, J.; Sun, M.; Zhao, J.; Ji, L.; Chen, L.; Zeng, F.; Yang, F.; Huang, B. Carboxylated carbon nanotubes with high electrocatalytic activity for oxygen evolution in acidic conditions. InfoMat 2022, 4, 12273. [Google Scholar] [CrossRef]

51. Aroon, M.A.; Ismail, A.F.; Montazer-Rahmati, M.M.; Matsuura, T. Effect of chitosan as a functionalization agent on the performance and separation properties of polyimide/multi-walled carbon nanotubes mixed matrix flat sheet films. J. Membr. Sci. 2010, 364, 309–317. [Google Scholar] [CrossRef]

52. Holzinger, M.; Abraham, J.; Whelan, P.; Graupner, R.; Ley, L.; Hennrich, F.; Kappes, M.; Hirsch, A. Functionalization of single-walled carbon nanotubes with (R-) oxycarbonyl nitrenes. J. Am. Chem. Soc. 2003, 125, 8566–8580. [Google Scholar] [CrossRef] [PubMed]

53. Kim, S.; Lee, Y.-I.; Kim, D.-H.; Lee, K.-J.; Kim, B.-S.; Hussain, M.; Choa, Y.-H. Estimation of dispersion stability of UV/ozone treated multi-walled carbon nanotubes and their electrical properties. Carbon 2013, 51, 346–354. [Google Scholar] [CrossRef]

54. Yuan, W.; Che, J.; Chan-Park, M.B. A novel polyimide dispersing matrix for highly electrically conductive solution-cast carbon nanotube-based composite. Chem. Mater. 2011, 23, 4149–4157. [Google Scholar] [CrossRef]

55. Stephen, F.; Dinetz, E.J.B.; Raymond, L.; Wagner, A.W. Fountain III. A comparative study of the gaseous productsgenerated by thermal and ultra-violet laserpyrolyses of the polyimide PMDA-ODA. J. Anal. Appl. Pyrol. 2002, 63, 241–249. [Google Scholar]

56. Weng, T.-H.; Tseng, H.-H.; Wey, M.-Y. Preparation and characterization of multi-walled carbon nanotube/PBNPI nanocomposite film for H2/ CH4 separation. Int. J. Hydrog. Energy 2009, 34, 8707–8715. [Google Scholar] [CrossRef]

57. Tan, J.; Huang, J.; Liu, Y.; Ding, Q.; Zeng, Y.; Zhang, H.; Liu, Y.; Xiang, X. Novel high-barrier polyimide containing rigid planar dibenzofuran moiety in main chain. High Perform. Polym. 2017, 30, 539–548. [Google Scholar] [CrossRef]

58. Hassanzadeh-Aghdam, M.K.; Mahmoodi, M.J.; Ansari, R.; Mehdipour, H. Effects of adding CNTs on the thermo-mechanical characteristics of hybrid titanium nanocomposites. Mech. Mater. 2019, 131, 121–135. [Google Scholar] [CrossRef]

59. Pan, J.; Bian, L.C. Coefficients of Thermal Expansion for Composites with Agglomerated Carbon Nanotubes. IOP Conf. Ser. Mater. Sci. Eng. 2017, 281, 012043. [Google Scholar] [CrossRef]

60. Hamester, M.R.R.; Pietezak, D.F.; Dalmolin, C.; Becker, D. Influence of crystallinity and chain interactions on the electrical properties of polyamides/ carbon nanotubes nanocomposites. J. Appl. Polym. Sci. 2021, 138, 50817. [Google Scholar] [CrossRef]

61. Ishii, J.; Takata, A.; Oami, Y.; Yokota, R.; Vladimirov, L.; Hasegawa, M. Spontaneous molecular orientation of polyimides induced by thermal

imidization (6). Mechanism of negative in-plane CTE generation in non-stretched polyimide films. Eur. Polym. J. 2010, 46, 681–693. [Google Scholar] [CrossRef]

62. Pottiger, M.T.; Coburn, J.C.; Edman, J.R. The Effect of Orientation on Thermal Expansion Behavior in Polyimide Films. J. Polym. Sci. B Polym. Phys. 1994, 32, 825–837. [Google Scholar] [CrossRef]

63. Bakshi, S.R.; Agarwal, A. An analysis of the factors affecting strengthening in carbon nanotube reinforced aluminum composites. Carbon 2011, 49, 533–544. [Google Scholar] [CrossRef]

64. Deng, C.F.; Wang, D.Z.; Zhang, X.X.; Li, A.B. Processing and properties of carbon nanotubes reinforced aluminum composites. Mater. Sci. Eng. A 2007, 444, 138–145. [Google Scholar] [CrossRef]

CHAPTER-7

MINIMIZING THERMAL RESIDUAL STRAIN IN A METAL-CFRP-METAL HYBRID TUBE IS ACHIEVED BY EMPLOYING AN AXIAL PRELOAD TOOL, MONITORED VIA OPTICAL FIBER SENSORS

1. INTRODUCTION: AN OVERVIEW

In the pursuit of optimizing structural integrity and performance in composite materials, this study explores a novel approach to mitigate thermal residual strain within a metal-CFRP-metal hybrid tube. Employing an axial preload tool, the reduction of thermal residual strain is meticulously monitored through the utilization of state-of-the-art optical fiber sensors. This research not only addresses the challenges associated with thermal stresses in hybrid structures but also underscores the significance of real-time monitoring techniques for precise control and enhancement of material properties.

As a special hybrid structure, sandwich structure is formed by two thin, stiff, strong faces, such as metal or fiber-reinforced polymer (FRP) composites, with a lightweight core material, such as foam, honeycomb, FRP composites, balsa, etc. For meeting the high-quality needs of modern industry, it combines material science, functional design, intelligent sensing, and integrated manufacturing into an interdisciplinary concept [1,2]. The three-layered metal-FRP-metal sandwich structure has the potential for lightweight and high strength, vibration, and noise reduction in the automotive, rail transportation, marine, and aerospace industries. This metal-composite hybrid structure combines the superior durability of metals with the attractive properties of FRP composites, such as lightweight, high specific strength and stiffness, good damping capacity, and tailorable ability [**3**].

Many articles about metal-composite hybrid structures study theoretically and experimentally concerning connection performance between the metal parts and FRP composites [**1**,**4**], mechanical properties [**5**,**6**], impact resistance [**7**,**8**], durability [**9**,**10**], machinability [**11**,**12**], energy harvesting [**13**], etc. However, most of the research is conducted on plate and beam structures, and only a few papers refer to the circular tube with metal-FRP walls [**14**,**15**,**16**] and few papers are reported for square tube with metal-FRP-metal sandwich walls due to difficult fabrication. Meanwhile, there are a limited number of articles concerning the thermal residual strains/stresses in metal-composite hybrid structures. Therefore, this important issue still requires more understanding and knowledge, especially for complex structural components such as the metal-CFRP-metal hybrid tube.

The thermal residual strains/stresses in metal-composite hybrid structures are inevitably generated during the manufacturing process. The most important manufacturing process factors include differences in elastic properties and coefficients of the thermal expansion (CTE) of the FRP and metal layers, the cure cycle, and the tool-part interaction. The thermal residual strains/stresses cause several defects in hybrid structures such as transverse cracking and delamination, decrease the fatigue performance and dimensional accuracy, reduce the structure's strength and modulus, and affect the natural frequencies and flexural stiffness [**17**]. Various studies have reported the reduction of residual strain, including changing the material composition of the hybrid structure [**18**], modifying the curing cycle [**19**,**20**,**21**,**22**,**23**], using special tools [**24**,**25**,**26**], post-stretching [**27**], microwave curing process [**28**,**29**], etc. The axial preload tool proposed in this paper for residual strain reduction targets hybrid structures based on unidirectional CFRP composite. The measurement methods can be divided into three categories: non-destructive, semi-destructive, and destructive [**30**]. As a non-destructive measurement method, the FBG sensor is used to monitor residual strain development during cure processing because it is small in diameter, precise, stable, easy integration, and anti-interference [**31**]. In this paper, the MCMHT with sandwich walls based on steel skins and unidirectional CFRP core is proposed and fabricated. The axial preload tool is proposed to reduce the thermal residual strains/stresses during the co-cure bonding process by pre-compressing of the metal layers or pre-stretching of the CFRP layers. The analytical model for evaluation of thermal residual strains is proposed based on classical lamination theory. The thermal residual strains are measured in real-time by the embedded FBG sensors. The modal testing results show the intrinsic properties of the metal-composite hybrid structure can be modified by the proposed axial preload tool.

2. EXPERIMENTAL METHODS

2.1. Material and Structure of the MCMHT

As shown in **Figure 1**, the MCMHT, with dimensions of 50 mm × 50 mm × 400 mm, is composed of an internal steel square tube, a square layer of unidirectional CFRP core, and two orthogonal steel plates. The unidirectional CFRP core in this paper consists of 10 layers of USN 10000/T300 prepreg from the Weihai Guangwei composites company with dimensions of 460 mm × 1000 mm × 0.1 mm. The material of the steel square tube is AISI 1045 based on the American Iron and Steel Institute (AISI) grade system. Material properties of the unidirectional CFRP prepreg and the steel square tube are listed in **Table 1**.

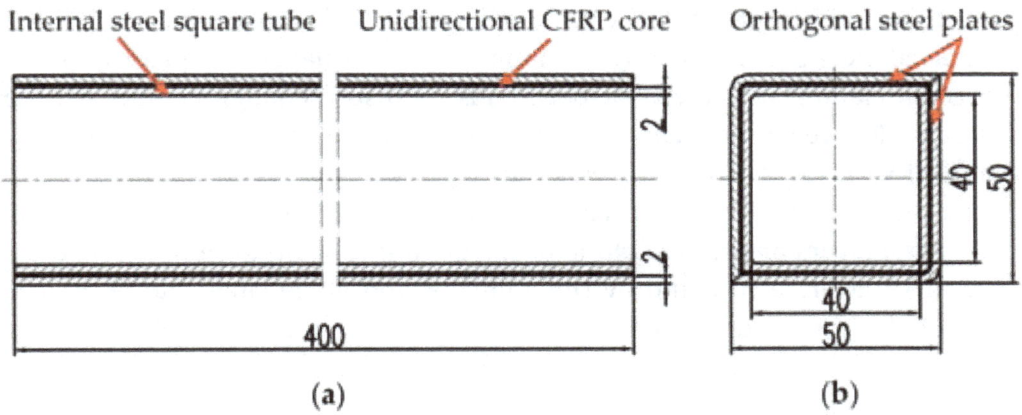

Figure 1. Structure and dimensions of the MCMHT: (a) main view; (b) left side view.

Table 1. Material properties of the unidirectional CFRP prepreg and the steel square tube.

Material Properties	USN 10000/T300 Prepreg	AISI 1045
Longitudinal modulus, E1 (GPa)	137	200
Transverse modulus, E2 (GPa)	9	200
Shear modulus, G12 (GPa)	3.78	80
Major Poisson's ratio, ν12	0.28	0.29
Longitudinal CTE, α1 (10-6/°C)	-0.5	11
Transverse CTE, α2 (10-6/°C)	27	11
Density (g/cm3)	1.76	7.85

2.2. Fabrication Method of the MCMHT with the Axial Preload Tool

The detailed fabrication processes of the MCMHT with the axial preload tool are as follows: (1) manufacturing the internal steel square tube and orthogonal steel plates, and the metal surfaces need to be roughened with wire wheels to increase the interfacial adhesive strength between the metal plate and the prepreg. Finally, metal surfaces are degreased and cleaned with acetone [**32,33**]. (2) Wrapping 5 unidirectional CFRP prepregs on the internal steel square tube by hand layup, and the fiber direction of the unidirectional prepreg is oriented in line with the axis of the internal steel square tube. (3) Placing the FBG strain sensor (FBGSS) in the middle of the face of the CFRP layers, and the FBG temperature sensor (FBGTS) placed nearby can be used for temperature compensation. The two optical fiber sensors are protected by a Teflon tube in the egress location. (4) Wrapping 5 other unidirectional CFRP prepregs, in the same manner as step 2, to make the total prepreg layers of 1 mm thickness. (5) Covering the wrapped internal steel square tube and sensors with the two orthogonal steel plates, which compose of an external steel square tube. (6) Clamping the external steel square tube with orthogonal clamps and screws at both ends to ensure enough contact between the steel-CFRP-steel sandwich walls for effective co-cure bonding. The MCMHT is assembled with $[St/0_{10}/St]$ symmetric tacking sequences. (7) Installing the axial preload tool to reduce the thermal residual strains/stresses by pre-compressing of the metal layers or pre-stretching of the CFRP layers. Because of the fastening of the CFRP layers at the end by the end caps and orthogonal clamps, and the design of the threaded screw motion and thrust bearing, the preload tool can apply compressive forces to the metal layers or tensile forces to the CFRP layers by rotating the handwheel. The assembly schematic of the MCMHT with the axial preload tool is shown in **Figure 2**. Two MCMHTs are fabricated in this paper, one of which is pre-stretched in the CFRP layers with 4 mm, as shown in **Figure 2**b.

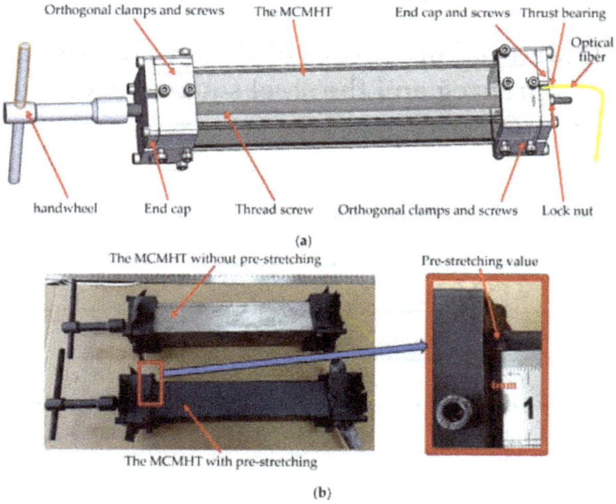

Figure 2. Structure of the MCMHT with the axial preload tool: (a) assembly schematic; (b) real fabricated MCMHTs.

The co-cure experimental setup for the MCMHT with the axial preload tool is shown in Figure 3. The two MCMHTs are put in the high-low temperature oven. A standard K-type thermocouple is fixed on the surface of each MCMHT with thermally conductive adhesive. Both signal wires of FBGs and thermocouples are fed through a specially reserved sealing hole in the oven wall and connected to the FBG interrogator and the thermocouple temperature indicator. The computer is used to record the Bragg wavelength shifts of FBGs by the cable connected to the FBG interrogator. The detailed experimental conditions are listed in Table 2.

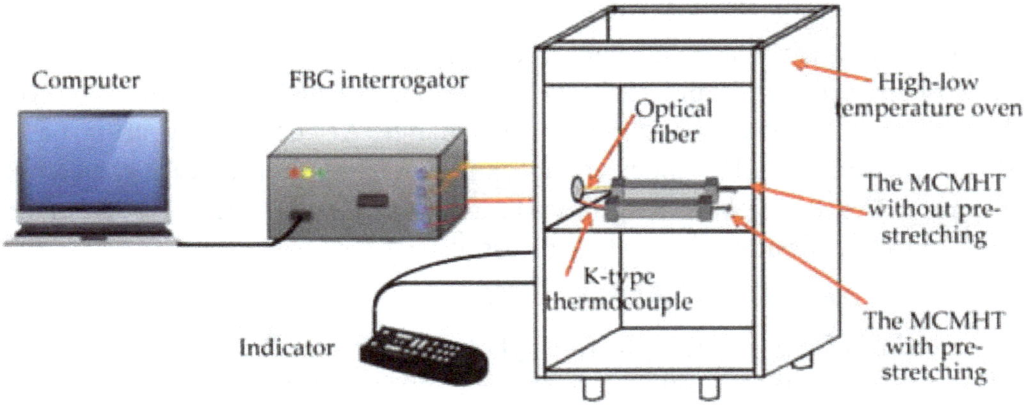

Figure 3. Co-cure experimental setup for the MCMHT with the axial preload tool.

Table 2. The detailed experimental conditions.

Experimental Condition	MCMHT without Pre-Stretching	MCMHT with Pre-Stretching
Material	Unidirectional CFRP prepreg and AISI 1045	
Curing equipment	The high-low temperature oven	
Sensor	FBGTS1 and FBGSS1	FBGTS2 and FBGSS2
Axial preload tool	Without pre-stretching	With pre-stretching of 4 mm

For the fabrication of the co-cure bonding of MCMHTs with the axial preload tool, the manufacturer's recommended cure cycle is used. The process is a typical curing cycle for thin CFRP/epoxy composite and is characterized by a heat-up ramp and dwell stages. The temperature is enhanced to the cure temperature (120 °C) in 1.5 h and held for 1.5 h. Finally, the MCMHTs are cooled to room temperature. During those stages, the adhesive bonding between the unidirectional CFRP prepregs and internal/external steel square tube is realized by the prepreg's resin. During the cooling stage, thermal residual strains/stresses appear due to the different CTE between the steel and the composite.

2.3. Measurement of Strains through Optical Fiber Sensors

2.3.1. Sensing Principle of FBG Sensor

The FBG is composed of a periodic distribution of the refractive index, which is made by ultraviolet exposure in the optical fiber core. When an incident broadband light passes through an FBG, a narrow-band light with a particular wavelength, called a Bragg wavelength, is reflected. The Bragg wavelength, λB, satisfies the Bragg scattering condition. It is expressed by the following equation [**34**]:

$$\lambda B{=}2{\cdot}neff{\cdot}\Lambda, \quad (1)$$

The value of the Bragg wavelength depends on the effective refractive index of the fiber core, $neff$, and the grating period, Λ. However, when the FBG is subjected to axial strain ε or temperature changes ΔT, both the grating period and the effective refractive index change, and then result in the Bragg wavelength shift, $\Delta\lambda B$. The Bragg wavelength variation which is sensitive to strain and temperature simultaneously can be expressed as:

$$\Delta\lambda B{=}(1{-}Pe)\varepsilon{+}\lambda B(\alpha f{+}\xi)\Delta T{=}K\varepsilon\varepsilon{+}KT\Delta T, \quad (2)$$

where $Pe, \alpha f, \xi$ are the effective photo-elastic coefficients, the thermal expansion coefficients, and the thermo-optic coefficients, respectively, and $K\varepsilon, KT$ are the strain sensitivity constants and the temperature sensitivity constants, respectively. The strain and temperature sensitivity constant of FBG sensors depend on the type of fibers. As Pe has a typical value of 0.22 for fused silica [35], $K\varepsilon$ is 1.2 pm/με in this paper without calibration for an FBG of a central wavelength of 1547 nm. However, KT requires a calibration procedure because FBG exhibits linear thermal-optic behavior only over a certain temperature range. The detailed calibration procedure can be seen in the next section.

According to Equation (2), it can be found that the changes of FBG wavelength are proportional to axial strain and temperature. As consequence, a single FBG cannot avoid strain-temperature cross-sensitivity, as both strain and temperature induce a Bragg wavelength shift. Several techniques to achieve such discrimination are available in the literature [36]. In this paper, two separate FBG sensors are embedded into a structure to avoid FBG cross-sensitivity. The FBGTS is a 10 mm long FBG encapsulated in a stainless-steel tube as shown in Figure 4. Considering a normal optical fiber with an outer diameter of 0.125 mm, the inner and outer diameters of the stainless-steel tube are 0.2 mm and 0.4 mm: as small as possible to avoid affecting hybrid structural integrity. As consequence, the FBGTS only has relations with the temperature change theoretically. Therefore, Equation (2) can be simplified as [37]:

$$\Delta\lambda B1{=}KT1\Delta T, \quad (3)$$

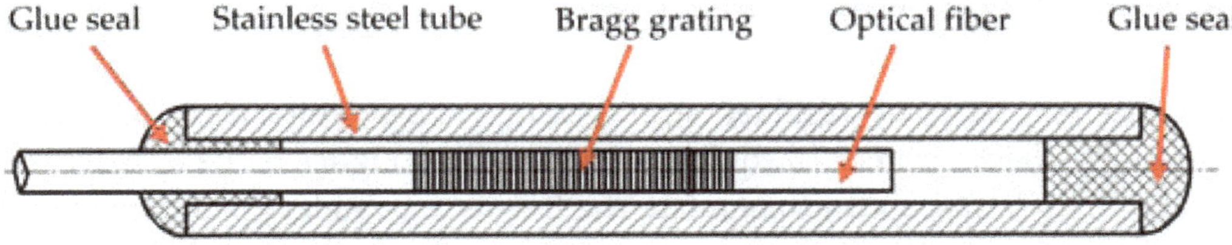

Figure 4. Encapsulated FBG for temperature measurement.

The FBGSS is a bare FBG with no treatment and is affected by axial strain and temperature. In this paper, the FBGTS is placed near to the FBGSS for the same temperature changes. Then the axial strain can be obtained from the measured wavelength shift by combining Equation (2) with Equation (3) [**37**].

$$\varepsilon = \Delta\lambda B/K\varepsilon - (\Delta\lambda B1KT)/(KT1K\varepsilon), \quad (4)$$

2.3.2. Temperature Calibration

Four FBGs (two FBGTSs and two FBGSSs) and a standard K-type thermocouple are fixed on an aluminum plate with a thermally conductive adhesive in an oven. It is ensured that the FBGs are in strain-free condition, so they respond to temperature change only. The Bragg wavelength shifts of FBGs are monitored by an FBG interrogator with a minimum resolution of 1 pm and a maximum sampling frequency of 4 kHz. The reliability of the encapsulated FBGTS must be confirmed before the calibration procedure. When the two FBGTSs are subjected to axial load at room temperature, no obvious wavelength shift is observed, so the two FBGTSs are considered as in strain-free condition. Both signal wires of FBGs and thermocouples are fed through the sealing strip of the oven door and connected to the FBG interrogator and the thermocouple temperature indicator. As the oven temperature is uniformly increased from 25 °C–200 °C, the results of the Bragg wavelength shifts are recorded by the computer. From the linear fitting results, the initial central wavelengths *KT* of the Four FBGs are listed in Table 3.

Table 3. The temperature sensitivity constants of the four FBGs.

FBG	Initial Central Wavelength/nm	*KT*/ pm/°Clh/s gh ih,a fcuk ikuh feyk,	Fitting Linear Correlation Coefficient
FBGTS1	1546.914	11.80	99.95%
FBGTS2	1546.860	11.84	99.91%
FBGSS1	1537.041	11.20	99.84%
FBGSS2	1546.936	11.07	99.82%

2.4. Experimental Setup of Modal Testing

Modal testing is performed to study the dynamic characteristics of the MCMHTs with different thermal residual strain states under vibrational excitation. According to the experimental equipment, as shown in **Figure 5**, the MCMHT with pre-stretching is suspended to emulate the free-free boundary. For comparison, the same experiment is done for the MCMHT without pre-stretching. A force transducer connected to the hammer is used to measure the force history of vibrations of the MCMHT caused by an impact hammer. With an accelerometer bonded on the surface, the acceleration response of the MCMHT is detected in a similar way. The excitation and response signals are subsequently acquired and analyzed by the LMS analysis system developed by Siemens company. Based on the obtained frequency response function (FRF), the modal parameters, including the first natural frequencies and damping ratio, can be processed by the modal analysis module. In order to obtain relatively accurate results, each MCMHT is measured with 7 excitation points along its length and each excitation point is applied 3 times to obtain the average FRF.

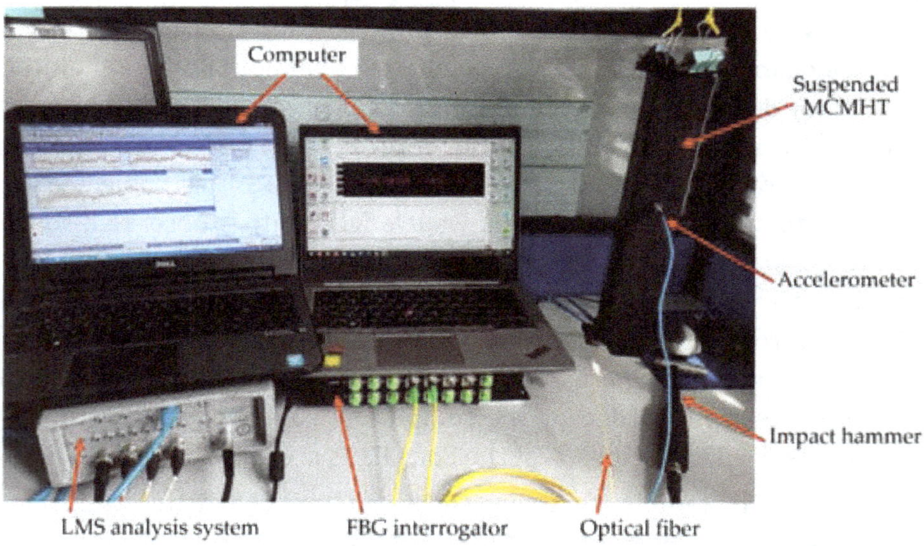

Figure 5. Experimental setup of modal testing.

3. EVALUATION OF THERMAL RESIDUAL STRAINS BASED ON CLASSICAL LAMINATION THEORY

The analytical model considers thermal residual strains produced only during the cooling phase stage, based on the following assumptions [**38**]: (1) the CFRP layers are plane stresses; (2) each lamina has a unique and linearly elastic deformation; (3) perfect bonding occurs between layers without gaps, debonding and other defects. Classical lamination theory is applied to predict the laminate properties of orthotropic continuous fiber laminated composites.

3.1. Material Properties of the Laminate with Arbitrary Lamina Orientation Angle

The stiffness and transformation matrices for predicting the engineering constants of the CFRP layers are expressed as follows [**39**]. The stiffness matrix $[Q]$ and transformation matrix $[T]$ are:

$$[Q] = \begin{bmatrix} \frac{E_1}{1-v_{12}v_{21}} & \frac{v_{12}E_2}{1-v_{12}v_{21}} & 0 \\ \frac{v_{21}E_1}{1-v_{12}v_{21}} & \frac{E_2}{1-v_{12}v_{21}} & 0 \\ 0 & 0 & G_{12} \end{bmatrix}, \quad (5)$$

$$[T] = \begin{bmatrix} \cos^2\theta & \sin^2\theta & 2\sin\theta\cos\theta \\ \sin^2\theta & \cos^2\theta & -2\sin\theta\cos\theta \\ -\sin\theta\cos\theta & \sin\theta\cos\theta & \cos^2\theta - \sin^2\theta \end{bmatrix}, \quad (6)$$

The stiffness for angled lamina is:

$$[Q] = [T]^{-1}[Q]\begin{bmatrix} 1 & 0 & 0 \\ 0 & 1 & 0 \\ 0 & 0 & 2 \end{bmatrix}[T], \quad (7)$$

where E_1, E_2, G_{12}, v_{12}, v_{21}, θ represent longitudinal Young's modulus, transverse Young's modulus, shear modulus, major Poisson's ratio, minor Poisson's ratio, and lamina orientation angle, respectively.

The extensional stiffnesses matrix $[A]$, strain-curvature coupling stiffness matrix $[B]$, and bending stiffness matrix $[D]$ for laminate are given by:

$$[A] = \sum_{k=1}^{N} (Q_{ij})_k (z_k - z_{k-1}) = \sum_{k=1}^{N} (Q_{ij})_k t_k, \quad (8)$$

$$[B] = \frac{1}{2}\sum_{k=1}^{N} (\overline{C}_{ij})_k (z_k^2 - z_{k-1}^2) = \sum_{k=1}^{N} (\overline{C}_{ij})_k t_k \overline{z}_k, \quad (9)$$

$$[D] = \frac{1}{3}\sum_{k=1}^{N} (\overline{C}_{ij})_k (z_k^3 - z_{k-1}^3) = \sum_{k=1}^{N} (\overline{C}_{ij})_k \left(t_k z_k^{-2} + \frac{t_k^3}{12}\right), \quad (10)$$

where zk and tk represent the vertical position of the kth lamina from the mid-plane and thickness of the kth lamina, respectively. N represents the total number of layers of the laminate. The subscript i,=1,2,...,6, whose meaning can be found in any textbook of composite mechanics.

For symmetric laminate, the effective longitudinal Young's modulus of the laminate Ex, the effective transverse Young's modulus of the laminate Ey, the effective laminate in-plane shear modulus Gxy, and the effective laminate longitudinal Poisson's ratio vxy, are defined as:

$$E_x = \frac{\sigma_x}{\varepsilon_x^0} = \frac{A_{11}A_{22} - A_{12}{}^2}{t\ A_{22}}, \quad (11)$$

$$E_y = \frac{\sigma_y}{\varepsilon_y^0} = \frac{A_{11}A_{22} - A_{12}{}^2}{t\ A_{11}}, \quad (12)$$

$$G_{xy} = \frac{\tau_{xy}}{\Upsilon_{xy}^0} = \frac{A_{66}}{t}, \quad (13)$$

$$\nu_{xy} = \frac{A_{12}}{A_{22}}, \quad (14)$$

$$\nu_{yx} = \frac{A_{12}}{A_{11}}, \quad (15)$$

where t represents the thickness of the laminate.

3.2. Strains in Metal-CFRP-Metal Hybrid Structure

Figure 6 is the evaluation of thermal residual strains in metal-CFRP-metal hybrid structure after the co-cure bonding process. **Figure 6**a represents that no strain appears in the dwell stage except for the pre-stretching strain ε_p. **Figure 6**b shows an ideal state where there is no interface interaction between the metal and CFRP layers. εtm and εtc represent the thermal strain in the metal layer and the CFRP layer, respectively. As shown in Figure 6c, εrm and εrc represent the thermal residual strain in the metal layer and the CFRP layer, respectively.

Based on the unique deformation, total strains of the two materials, which are composed of pre-stretching strain, thermal residual strain, and thermal strain, can be described as:

$$\varepsilon_m^r + \varepsilon_m^t = \varepsilon_p + \varepsilon_c^r + \varepsilon_c^t, \quad (16)$$

Taking into account the isotropic metal layer and the anisotropic CFRP layer, the above equation can be expressed as:

$$[\varepsilon_m] + [\alpha_m]\Delta T = [\varepsilon_p] + [\varepsilon_c] + [\alpha_c]\Delta T, \quad (17)$$

where

$$[\varepsilon_m] = \begin{bmatrix} \varepsilon_m \\ \varepsilon_m \end{bmatrix}, [\varepsilon_p] = \begin{bmatrix} \varepsilon_{p1} \\ \varepsilon_{p2} \end{bmatrix}, [\varepsilon_c] = \begin{bmatrix} \varepsilon_{c1} \\ \varepsilon_{c2} \end{bmatrix}, [\alpha_m] = \begin{bmatrix} \alpha_m \\ \alpha_m \end{bmatrix} \text{ and } [\alpha_c] = \begin{bmatrix} \alpha_1 \cos^2\theta + \alpha_2 \sin^2\theta \\ \alpha_1 \sin^2\theta + \alpha_2 \cos^2\theta \end{bmatrix}$$

In which εm, εc_1, c_2 represent the strains of the steel, and CFRP layer in the axial and transverse directions, respectively. εp_1,p_2 represent the pre-stretching strains of the CFRP layer in the axial and transverse directions, respectively. αm_1, c_1, αc_2 denote the CTEs of steel, and CFRP layers in the axial and transverse direction, respectively. ΔT denotes the temperature difference at different phases.

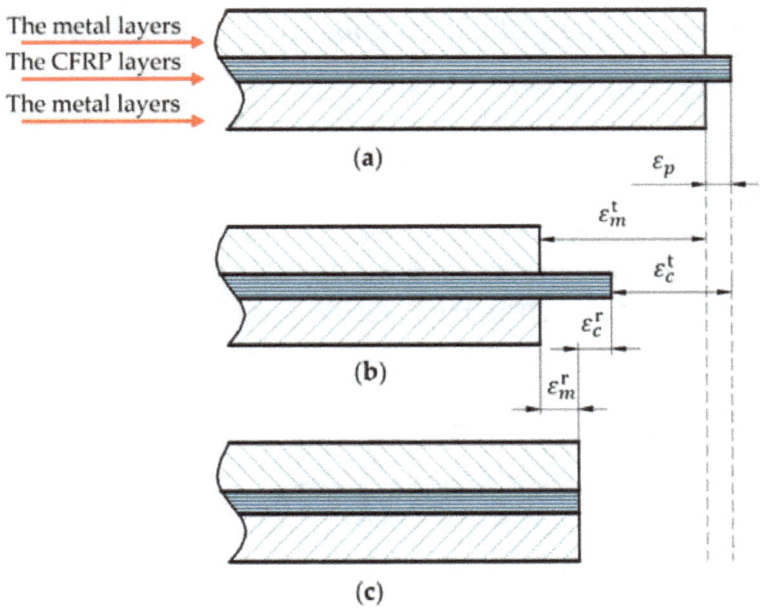

Figure 6. Schematic of thermal residual strains: (a) no strain state; (b) ideal state; (c) final state.

According to the force equilibrium equation from the mechanics of materials, residual strains in the MCMHT can be described as:

$$A_m E_m [\varepsilon_m] + A_c [C] [\varepsilon_c] = 0, \quad (18)$$

where $[C] = \begin{bmatrix} \frac{E_x}{1-\nu_{xy}\upsilon_{yx}} & \frac{\nu_{xy}E_y}{1-\nu_{xy}\upsilon_{yx}} \\ \frac{\upsilon_{yx}E_x}{1-\nu_{xy}\upsilon_{yx}} & \frac{E_y}{1-\nu_{xy}\upsilon_{yx}} \end{bmatrix}.$.

In which A_m and $A_{crepresent}$ the cross-sectional areas of steel and CFRP layers, respectively. Em and [C] represent Young's modulus of steel and the stiffness matrix of CFRP layers, respectively.

Then, from Equations (17) and (18), the thermal residual strain in the CFRP layers can be expressed by:

$$[\varepsilon_c] = [A]^{-1}([B]\,\Delta T - [\varepsilon_p]), \qquad (19)$$

Where,

$$[A] = \begin{bmatrix} 1 + \frac{E_x A_c}{E_m A_m (1-\nu_{xy}\upsilon_{yx})} & \frac{\nu_{xy} E_y A_c}{E_m A_m (1-\nu_{xy}\upsilon_{yx})} \\ \frac{\upsilon_{yx} E_x A_c}{E_m A_m (1-\nu_{xy}\upsilon_{yx})} & 1 + \frac{E_y A_c}{E_m A_m (1-\nu_{xy}\upsilon_{yx})} \end{bmatrix} \text{ and } [B] = [\alpha_s] - [\alpha_c].$$

4. RESULTS AND DISCUSSION

4.1. Comparison of Theoretical Calculation with Measurement by the FBGSS

The plot of temperature and strain history for the MCMHTs with and without pre-stretching during the co-cure bonding process is given in **Figure 7.** Simultaneous measurements of temperatures and strains during the co-cure bonding process are performed by the FBG sensors in real-time. It can be observed that the tendency of the temperature and strain measured for the MCMHTs with and without pre-stretching is consistent. The temperature measured by the FBGTS has good agreement with the cure cycle applied by the high-low temperature oven.

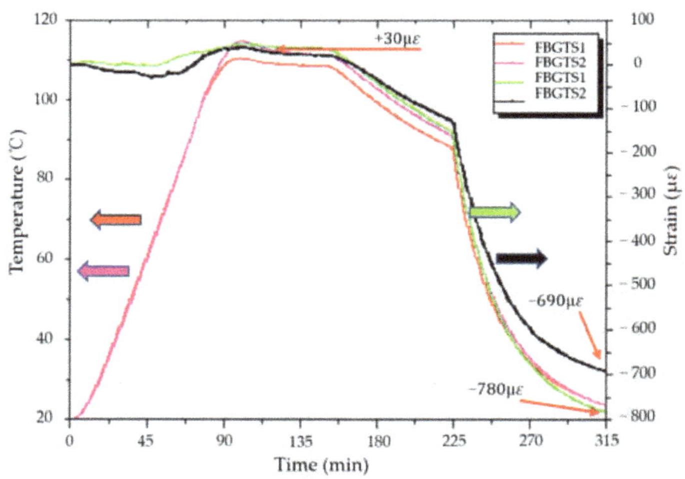

Figure 7. Temperature and strain history for the MCMHTs with and without pre-stretching during the co-cure bonding process.

The trend of strain history is also the same for the two MCMHTs, but their values are slightly different due to the pre-stretching strain. The residual strains of the CFRP layers transform from tensile strains to compressive strains during the dwell and cooling stages. As one of the MCMHTs is pre-stretched in the CFRP layers as shown in **Figure 2**b, the pre-stretching strain in this case is 100 με, taking into account the slippage between the CFRP layers and the axial preload tool. **Table**

4 reports the comparison of the residual strains for the two MCMHTs obtained by the FBGSS and theoretical calculation. Considering that the thermal tensile strain is 30 με, it is observed that the residual compression strains for the MCMHT reduce from 810 με to 720 με, obtained by the FBGSS through pre-stretching of the CFRP layers. Indeed, it can also be reduced by pre-compressing of the metal layer because of the interaction between the CFRP and the metal. The important point is that the axial preload tool proposed in this paper can be used either clockwise or counter-clockwise to achieve the pre-stretching of the CFRP layers or the pre-compressing of the metal layers. The measured results and theoretical calculations verify that the residual strains of the MCMHT can be modified by the axial preload tool. Furthermore, the residual strain can theoretically be eliminated by suitable preloading.

Table 4. Comparison of the residual compression strains for the two MCMHTs obtained by the FBGSS and theoretical calculation.

MCMHT	by the FBGSS/με	by Theoretical Calculation/με	Relative Difference
With pre-stretching	720	659.2	-8.4%
Without pre-stretching	810	759.2	-6.2%

4.2. The Dynamic Characteristics of the MCMHT

Acceleration FRFs of 7 excitation points for the MCMHTs with different thermal residual strain states are processed by the LMS analysis system as shown in **Figure 8.** Dynamic characteristics of the MCMHTs from the modal analysis module are given in **Table 5.** It is observed that the first natural frequency of the MCMHT with pre-stretching is 5.2% higher than that without pre-stretching. Nevertheless, the damping of the MCMHT with pre-stretching decrease by 11.5% compared to the MCMHT without pre-stretching. The modal testing results imply that the reduction of residual strain increases the natural frequency of the MCMHT, but is detrimental to the damping capability of the MCMHT.

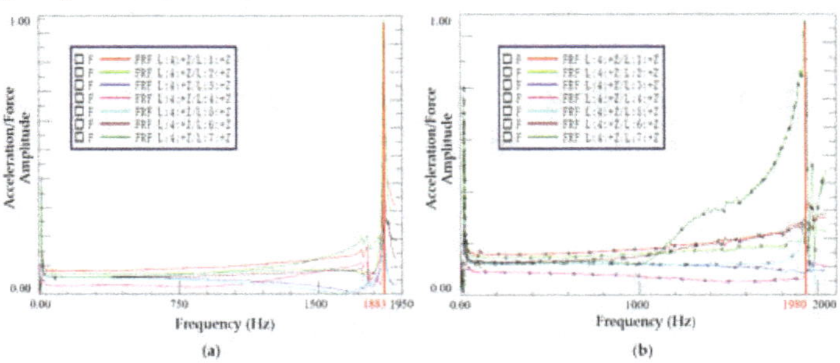

Figure 8. Acceleration FRFs of 7 excitation points for the MCMHTs: (a) without pre-stretching; (b) with pre-stretching.

Table 5. Dynamic characteristics of the MCMHTs from the modal analysis module.

Characteristics	without Pre-Stretching	with Pre-Stretching	Relative Difference
Natural frequency/Hz	1883	1980	5.2%
Damping ratio	0.96%	0.85%	-11.5%

5. CONCLUSIONS

The axial preload tool is proposed to reduce the thermal residual strains/stresses during the co-cure bonding process by pre-compressing of the metal layers or pre-stretching of the CFRP layers. Residual strain determination by embedded optical fiber sensors in metal-composite hybrid structures is presented. Thermal residual strain results obtained from FBG sensors are compared with those obtained from theoretical calculations. Moreover, the dynamic characteristics of the MCMHTs with different stress states are compared. For future work, the embedded FBG sensors can be used for real-time structural health monitoring of composite structures, based on the multiplexing capability of FBG sensing technology. For example, damage such as microcracks, delamination at interfaces, or crushing of polymer matrix can reduce stiffness overall. By the embedded FBG sensors array, it is possible to determine the initiation, size, and location of the damage. The following conclusions can be drawn from the results obtained:

- To reduce thermal residual strain, the proposed axial preload tool can apply compressive forces to the metal layers or tensile forces to the CFRP layers by rotating the handwheel. This shows the axial preload tool can change the strain state of the metal-composite hybrid structure;
- Thermal residual strain of the metal-composite hybrid structure obtained from embedded optical fiber sensors show good agreement with the theoretical calculation based on classic laminate theory;
- The modal testing results imply that the reduction of residual strain increases the natural frequency of the metal-composite hybrid structure, but is detrimental to its damping capability. This shows that the intrinsic properties of the metal-composite hybrid structure can be modified by the proposed axial preload tool.

6. CONCISE SUMMARY

Thermal residual strains/stresses cause several defects in hybrid structures and various studies have reported the reduction of residual strain. This paper describes a method for reducing thermal residual strains/stresses in metal-CFRP-metal hybrid tubes (MCMHT). The proposed axial preload tool provides two ways to reduce the thermal residual strains/stresses during the co-cure bonding process: pre-compressing of the metal layers and pre-stretching of the unidirectional carbon fiber reinforced polymer (CFRP) layers. An online measurement technique with embedded optical fiber Bragg grating (FBG) sensors is presented. Thermal residual strains are evaluated based on classical lamination theory with the assumption of plane stress. The theoretical calculations and measurement results agree well. Furthermore, the dynamic characteristics of the MCMHTs are tested. The results show that the reduction of residual strain increases the natural frequency of the MCMHT, but is detrimental to the damping capability of the MCMHT, which imply that the intrinsic properties of the metal-composite hybrid structure can be modified by the proposed axial preload tool.

REFERENCES

1. Sinmazçelik, T.; Avcu, E.; Bora, M.Ö.; Çoban, O. A review: Fibre metal laminates, background, bonding types and applied test methods. *Mater. Des.* **2011**, *32*, 3671–3685. **[Google Scholar]** **[CrossRef]**
2. Braga, D.F.O.; Tavares, S.M.O.; da Silva, L.F.M.; Moreira, P.M.G.P.; de Castro, P.M.S.T. Advanced design for lightweight structures: Review and prospects. *Prog. Aerosp. Sci.* **2014**, *69*, 29–39. **[Google Scholar]** **[CrossRef]**
3. Hollaway, L.C. A review of the present and future utilisation of FRP composites in the civil infrastructure with reference to their important in-service properties. *Constr. Build. Mater.* **2010**, *24*, 2419–2445. **[Google Scholar]** **[CrossRef]**
4. Pramanik, A.; Basak, A.K.; Dong, Y.; Sarker, P.K.; Uddin, M.S.; Littlefair, G.; Dixit, A.R.; Chattopadhyaya, S. Joining of carbon fibre reinforced polymer (CFRP) composites and aluminium alloys—A review. *Compos. Part A* **2017**, *101*, 1–29. **[Google Scholar]** **[CrossRef][Green Version]**
5. Nowak, T. Elastic-plastic behavior and failure analysis of selected Fiber Metal Laminates. *Compos. Struct.* **2018**, *183*, 450–456. **[Google Scholar]** **[CrossRef]**
6. Yao, Y.; Shi, P.; Chen, M.; Chen, G.; Gao, C.; Boisse, P.; Zhu, Y. Experimental and numerical study on Mode I and Mode II interfacial fracture toughness of co-cured steel-CFRP hybrid composites. *Int. J. Adhes. Adhes.* **2022**, *112*, 103030. **[Google Scholar]** **[CrossRef]**
7. Sun, G.; Chen, D.; Zhu, G.; Li, Q. Lightweight hybrid materials and structures for energy absorption: A state-of-the-art review and outlook. *Thin-Walled Struct.* **2022**, *172*, 108760. **[Google Scholar]** **[CrossRef]**
8. Düring, D.; Weiß, L.; Stefaniak, D.; Jordan, N.; Hühne, C. Low-velocity impact response of composite laminates with steel and elastomer protective layer. *Compos. Struct.* **2015**, *134*, 18–26. **[Google Scholar]** **[CrossRef]**

9. Alderliesten, R.C. Designing for damage tolerance in aerospace: A hybrid material technology. *Mater. Des.* **2015**, *66*, 421–428. **[Google Scholar] [CrossRef]**
10. Dadej, K.; Bieniaś, J.; Surowska, B. Residual fatigue life of carbon fibre aluminium laminates. *Int. J. Fatigue* **2017**, *100*, 94–104. **[Google Scholar] [CrossRef]**
11. Zafar, R.; Lihui, L.; Rongjing, Z. Analysis of hydro-mechanical deep drawing and the effects of cavity pressure on quality of simultaneously formed three-layer Al alloy parts. *Int. J. Adv. Manuf. Technol.* **2015**, *80*, 2117–2128. **[Google Scholar] [CrossRef]**
12. Kotik, H.G.; Perez Ipiña, J.E. Short-beam shear fatigue behavior of fiber metal laminate (Glare). *Int. J. Fatigue* **2017**, *95*, 236–242. **[Google Scholar] [CrossRef]**
13. Elsheikh, A. Bistable Morphing Composites for Energy-Harvesting Applications. *Polymers* **2022**, *14*, 1893. **[Google Scholar] [CrossRef] [PubMed]**
14. Kim, H.S.; Park, S.W.; Lee, D.G. Smart cure cycle with cooling and reheating for co-cure bonded steel/carbon epoxy composite hybrid structures for reducing thermal residual stress. *Compos. Part A* **2006**, *37*, 1708–1721. **[Google Scholar] [CrossRef]**
15. Kim, H.C.; Shin, D.K.; Lee, J.J. Characteristics of aluminum/CFRP short square hollow section beam under transverse quasi-static loading. *Compos. Part B Eng.* **2013**, *51*, 345–358. **[Google Scholar] [CrossRef]**
16. Zhou, Y.; Liu, X.; Xing, F.; Li, D.; Wang, Y.; Sui, L. Behavior and modeling of FRP-concrete-steel double-skin tubular columns made of full lightweight aggregate concrete. *Constr. Build. Mater.* **2017**, *139*, 52–63. **[Google Scholar] [CrossRef]**
17. Parlevliet, P.P.; Bersee, H.E.N.; Beukers, A. Residual stresses in thermoplastic composites—a study of the literature. Part III: Effects of thermal residual stresses. *Compos. Part A* **2007**, *38*, 1581–1596. **[Google Scholar] [CrossRef]**
18. Shokrieh, M.M.; Daneshvar, A.; Akbari, S. Reduction of thermal residual stresses of laminated polymer composites by addition of carbon nanotubes. *Mater. Des.* **2014**, *53*, 209–216. **[Google Scholar] [CrossRef]**
19. Agius, S.L.; Joosten, M.; Trippit, B.; Wang, C.H.; Hilditch, T. Rapidly cured epoxy/anhydride composites: Effect of residual stress on laminate shear strength. *Compos. Part A* **2016**, *90*, 125–136. **[Google Scholar] [CrossRef]**
20. White, S.R.; Hahn, H.T. Cure Cycle Optimization for the Reduction of Processing-Induced Residual Stresses in Composite Materials. *J. Compos. Mater.* **1993**, *27*, 1352–1378. **[Google Scholar] [CrossRef]**

21. Kim, H.-S.; Yoo, S.-H.; Chang, S.-H. In situ monitoring of the strain evolution and curing reaction of composite laminates to reduce the thermal residual stress using FBG sensor and dielectrometry. *Compos. Part B Eng.* **2013**, *44*, 446–452. **[Google Scholar] [CrossRef]**

22. Kim, S.S.; Murayama, H.; Kageyama, K.; Uzawa, K.; Kanai, M. Study on the curing process for carbon/epoxy composites to reduce thermal residual stress. *Compos. Part A* **2012**, *43*, 1197–1202. **[Google Scholar] [CrossRef]**

23. Prussak, R.; Stefaniak, D.; Kappel, E.; Hühne, C.; Sinapius, M. Smart cure cycles for fiber metal laminates using embedded fiber Bragg grating sensors. *Compos. Struct.* **2019**, *213*, 252–260. **[Google Scholar] [CrossRef]**

24. Hyun, D.; Lee, D.G. Manufacturing of co-cured composite aluminum shafts with compression during co-curing operation to reduce residual thermal stresses. *J. Compos. Mater.* **1998**, *32*, 1221–1241. **[Google Scholar] [CrossRef]**

25. Xue, J.; Wang, W.-X.; Takao, Y.; Matsubara, T. Reduction of thermal residual stress in carbon fiber aluminum laminates using a thermal expansion clamp. *Compos. Part A* **2011**, *42*, 986–992. **[Google Scholar] [CrossRef]**

26. Lee, D.G.; Kim, J.W.; Hwang, H.Y. Torsional Fatigue Characteristics of Aluminum–Composite Co-Cured Shafts with Axial Compressive Preload. *J. Compos. Mater.* **2004**, *38*, 737–756. **[Google Scholar] [CrossRef]**

27. Khan, S.U.; Alderliesten, R.C.; Benedictus, R. Post-stretching induced stress redistribution in Fibre Metal Laminates for increased fatigue crack growth resistance. *Compos. Sci. Technol.* **2009**, *69*, 396–405. **[Google Scholar] [CrossRef]**

28. Li, N.; Li, Y.; Hao, X.; Gao, J. A comparative experiment for the analysis of microwave and thermal process induced strains of carbon fiber/bismaleimide composite materials. *Compos. Sci. Technol.* **2015**, *106*, 15–19. **[Google Scholar] [CrossRef][Green Version]**

29. Li, N.; Li, Y.; Hang, X.; Gao, J. Analysis and optimization of temperature distribution in carbon fiber reinforced composite materials during microwave curing process. *J. Mater. Process. Technol.* **2014**, *214*, 544–550. **[Google Scholar] [CrossRef][Green Version]**

30. Wu, T.; Degener, S.; Tinkloh, S.; Liehr, A.; Zinn, W.; Nobre, J.P.; Tröster, T.; Niendorf, T. Characterization of residual stresses in fiber metal laminate interfaces—A combined approach applying hole-drilling method and energy-dispersive X-ray diffraction. *Compos. Struct.* **2022**, *299*, 116071. **[Google Scholar] [CrossRef]**

31. Parlevliet, P.P.; Bersee, H.E.N.; Beukers, A. Residual stresses in thermoplastic composites—A study of the literature—Part II: Experimental techniques. *Compos. Part A* **2007**, *38*, 651–665. **[Google Scholar] [CrossRef]**

32. Lee, D.-W.; Park, B.-J.; Park, S.-Y.; Choi, C.-H.; Song, J.-I. Fabrication of high-stiffness fiber-metal laminates and study of their behavior under low-velocity impact loadings. *Compos. Struct.* **2018**, *189*, 61–69. **[Google Scholar]** **[CrossRef]**

33. Harhash, M.; Sokolova, O.; Carradó,, A.; Palkowski, H. Mechanical properties and forming behaviour of laminated steel/polymer sandwich systems with local inlays—Part 1. *Compos. Struct.* **2014**, *118*, 112–120. **[Google Scholar]** **[CrossRef]**

34. Hill, K.O.; Meltz, G. Fiber Bragg Grating Technology Fundamentals and Overview. *J. Lightwave Technol.* **1997**, *15*, 1263–1276. **[Google Scholar]** **[CrossRef][Green Version]**

35. Kuang, K.S.C.; Zhang, L.; Cantwell, W.J.; Bennion, I. Process monitoring of aluminum-foam sandwich structures based on thermoplastic fibre–metal laminates using fibre Bragg gratings. *Compos. Sci. Technol.* **2005**, *65*, 669–676. **[Google Scholar]** **[CrossRef]**

36. Majumder, M.; Gangopadhyay, T.K.; Chakraborty, A.K.; Dasgupta, K.; Bhattacharya, D.K. Fibre Bragg gratings in structural health monitoring—Present status and applications. *Sens. Actuators A Phys.* **2008**, *147*, 150–164. **[Google Scholar]** **[CrossRef]**

37. Wang, Q.; Gao, L.; Wang, X.; Dong, Q. Numerical analysis and fiber Bragg grating monitoring of thermocuring processes of carbon fiber/epoxy laminates. *Polym. Test.* **2017**, *62*, 287–294. **[Google Scholar]** **[CrossRef]**

38. Xu, Y.; Li, H.; Yang, Y.; Hu, Y.; Tao, J. Determination of residual stresses in Ti/CFRP laminates after preparation using multiple methods. *Compos. Struct.* **2019**, *210*, 715–723. **[Google Scholar]** **[CrossRef]**

39. Gibson, R.F. *Principles of Composite Material Mechanics*, 4th ed.; Taylor & Francis Group: Oxford, UK, 2016; pp. 338–343. **[Google Scholar]**

CHAPTER-8

PROCESSES OF ELECTROSPUN POLYVINYLIDENE FLUORIDE-BASED NANOFIBERS, THEIR PIEZOELECTRIC PROPERTIES, AND SEVERAL FANTASTIC APPLICATIONS

1. INTRODUCTION: AN OVERVIEW

Embarking on the forefront of nanofiber technology, this exploration delves into the intricate processes involved in the electrospinning of polyvinylidene fluoride (PVDF)-based nanofibers. Beyond the complexity of fabrication, the study meticulously investigates the inherent piezoelectric properties of these nanofibers, unraveling their potential for transformative applications. Polyvinylidene fluoride, renowned for its remarkable piezoelectric characteristics, serves as a foundational material in this research, promising breakthroughs in various fields. From the intricacies of the electrospinning process to the dynamic world of piezoelectricity, this study sets the stage for a comprehensive understanding of PVDF-based nanofibers and their exciting prospects in the realm of innovative applications.

With the continuous progress and development of science and technology, piezoelectric materials can play an important role in the mutual conversion between mechanical force and electrical signals, and these materials have gradually attracted people's attention [1]. In 1880, P. Curie and J. Curie discovered the piezoelectric effect in quartz crystals [2]; the piezoelectric effect is divided into positive and negative piezoelectric effects. When a mechanical force is applied in a direction and deformed, polarization occurs inside, and opposite positive and negative charges are generated on the two opposite surfaces; a piezoelectric material can generate a built-in electric field under the action of mechanical stress. After removing the external force, the positive and negative charges disappear and return to the original state. On the contrary, there is also an inverse piezoelectric effect: the application of an external electric field can cause deformation of the material. After the electric field is removed, the deformation of the dielectric disappears, as shown in Figure 1. According to the principle of the piezoelectric effect, the material is able to convert the information of the structural deformation into an electrical signal [3]. This potential change can be captured by external detection equipment, which can be applied to energy harvesters [4,5], ultrasonic transducers [6,7], sensors [8], biomedicine [9,10,11], and nanogenerators [12,13]. Piezoelectric effects can also be used as a means of separating charges in photocatalysis [14].

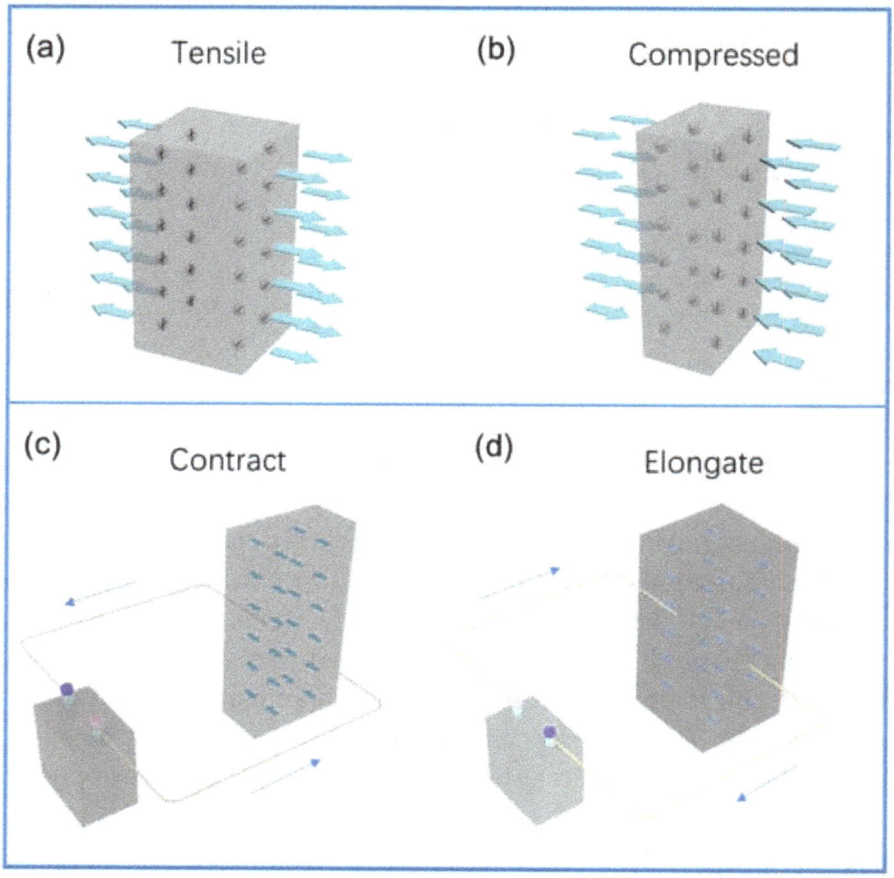

Figure 1. Schematic diagram of positive piezoelectric effect and negative piezoelectric effect. (a,b) show the positive piezoelectric effect exerted by compressive and tensile forces, respectively, the arrows represent tensile and compressive forces. (c,d) represent the negative piezoelectric effect, which causes the piezoelectric material to shrink and elongate, respectively, the arrow represents the direction of the current.

Piezoelectric materials can be divided into ceramic material and polymer material categories. Many lead-based and non-lead-based ceramic piezoelectric materials have been reported, such as lead metaniobate ($PbNb_2O_6$) [15], lead titanate ($PbTiO_3$) [16], barium titanate ($BaTiO_3$) [17,18], barium lead niobate (PBLN) [19], lithium tetraborate ($Li_2B_4O_7$) [20], lead zirconate titanate (PZT) [21], zinc oxide (ZnO) [22], and others. Although these piezoelectric materials have good piezoelectric properties, the flexibility of such materials is generally poor, which limits their application in flexible pressure sensors. PVDF is a semi-crystalline polymer composed of [CH_2=CF_2] monomer polymerization, its piezoelectric constant (20–28 pC/N) is the best among polymers [23], and it possesses good flexibility, piezoelectric properties, a dielectric constant, pyroelectric effect, ferroelectric effect, biocompatibility, and chemical stability [24]. There are five polymorphs of PVDF produced by different processing conditions, including α and δ (TGTG'), β (TTTT), and γ and ε (TTTGTTTG') phases [25]. Among them, α and ε phases are non-polar phases, while the β, γ, and δ phases are polar phases. The α phase is the most common phase in nature, but the β phase of PVDF exhibits the best piezoelectric properties. The melt crystallization of PVDF produces α phase [26,27,28,29]. In order to obtain better piezoelectric materials, β phase with good piezoelectric properties is needed.

Generally, the methods of converting the non-polar α phase in PVDF to the polar β phase include heat treatment [**30**], mechanical stretching [**31**], high-voltage electric field application [**32**], electrospinning [**33,34,35**], functional particles addition [**36,37,38**], and others. Compared with other methods, electrospinning is a simple and low-cost technology, which can produce fibers with diameters from micro-size to nano-size [**39,40,41**], and it can effectively convert the α phase of PVDF into β phase. Additionally, various functional particles are easily incorporated into the nanofibers by electrospinning, and functional particles/PVDF composite nanofibrous membranes with improved piezoelectric properties are easily constructed. NFs have many advantages; the piezoelectric constants of inorganic and polymeric piezoelectric materials are listed in **Table 1**. The piezoelectric properties of PBLN are the best among ceramic materials, while the piezoelectric properties of PVDF NFs are the best among polymers. The piezoelectric properties of polyacrylonitrile (PAN) and cellulose NFs are not excellent, but the good piezoelectric properties of composite fiber membranes can be obtained by adding other piezoelectric materials. Thus, these polymers are often used as substrates for piezoelectric membrane materials.

This review starts from the research on the preparation of NF piezoelectric materials by electrospinning, and introduces the preparation methods of various piezoelectric NFs and their advantages and disadvantages, including PVDF NFs, PVDF organic blend NFs, doped functional particles, PVDF composite NFs,

PVDF copolymer NFs, and other piezoelectric polymer NFs. The effects of the spinning process, addition of functional particles, and blending of components on the piezoelectric properties of materials—as well as the effect mechanism of the structure–activity relationship of materials on piezoelectric properties of materials—are mainly discussed. Finally, the applications of piezoelectric NFs in biomedicine, energy, and photocatalysis are introduced, and a summary and prospect are made.

Table 1. Piezoelectric constants and crystal structures of common piezoelectric materials.

Category	Materials	Piezoelectric Charge Constant, dij (pC/N)		Crystal Structure
		Bulk	Nanofiber	
Ceramic materials	PbNb2O6	d33 = 57 [42]	—	Orthorhombic (tungsten bronze)
	PbTiO3	d33 = 97 [43]	—	Tetragonal
	BaTiO3	d33 = 95 [44]	—	Tetragonal (perovskite)
	PBLN	d33 = 350 [45]	—	Orthorhombic
	Li2B4O7	d33 = 8.76 [46]	—	Tetragonal
	PZT	d33 = 223 [47]	—	Tetragonal (perovskite)
	ZnO	d33 = 12.3 [48]	—	Hexagonal (wurtzite)
Polymer materials	PVDF	d31 = 23 [49], d33 = 15 [50,51]	d33 = 57.6 [52]	α, β, γ, δ, σ-phase
	PLLA	d14 = 6.0 [49]	d33 = 3 ± 1 [53]	α, β, δ-phase
	PHBV	d33 = 0.43 [54]	d33 = 0.7 ± 0.5 [55]	α, β
	PAN	d31 = 0.6 [56]	d33 = 39 [57], 1.5 [58]	Zigzag
	Nylon-11	d33 = 6.5 [59], d31 = 14 [60]	—	α, α', β, β'
	Collagen	d14 = 12 [61], d33 = 2 [62]	d15 = 1 [63]	Triple helix
	Cellulose	d25 = 2.1 [64]	d33 = 31 [65]	Iα (triclinic), Iβ (monoclinic)
	Chitin	d33 = 9.49 [66]	—	α, β, γ-phase
	Chitosan	d33 = 4.4 [67], d31 = 10 [68]	—	Double helix

2. PREPARATION OF PVDF NFS BY ELECTROSPINNING

2.1. Introduction of Electrospinning Principle and Parameters

The electrospinning device is generally composed of a high-voltage power supply, injection pump, spinning head, and receiving plate [**69**]. The basic principle of electrostatic spinning is the viscosity of the polymer solution through Coulomb force stretching, where spinning polymer droplets (due to Coulomb and gravity forces greater than the surface tension) gradually change from hemispherical to conical, and from a cone pointed outward to a liquid, and the solvents evaporate quickly, creating fine fibers in the receiving plate, as shown in **Figure 2**, eventually forming nanometer fiber membranes [**70,71,72,73,74,75**]. NFs have excellent properties such as small diameter, large porosity, large specific surface area, and good mechanical energy along the fiber axis; the fiber morphology can be controlled by adjusting the parameter settings [**76,77**]. The influencing factors can be divided into three categories, including electrospinning parameters (voltage, needle diameter, feed rate, distance from needle to collector), solution properties (polymer molecular weight, concentration, solvent, viscosity, conductivity, surface tension), and environment parameters (temperature and relative humidity). The influence of different electrospinning parameters on the morphology and piezoelectric properties of PVDF NFs are analyzed in following sections.

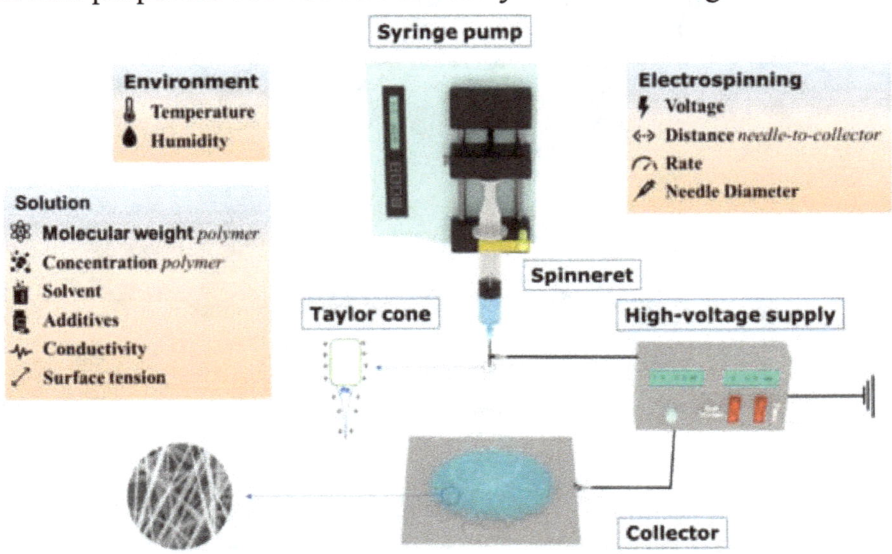

Figure 2. Electrospinning schematic diagram and influencing parameters.

The piezoelectric properties of PVDF are related to its strong dipole moment. Since the electronegativity of fluorine atoms is greater than that of hydrogen and carbon atoms, PVDF exhibits a large dipole moment. When the PVDF monomer polymerizes to form a crystal and produces a dipole, there is a net dipole moment in the β phase where the fluorine atoms are arranged on one side of the polymer chain, and the net dipole moment disappears in the α phase where the fluorine atoms are evenly distributed on both sides of the chain. There is also a weak net dipole moment in the γ phase [**78**]. Therefore, the β phase with the highest piezoelectric properties has attracted much attention. Many methods have been developed to obtain high-β-phase content. The electrospinning method is a simple and effective method to obtain the β phase; PVDF nanofiber membranes prepared by electrospinning have a higher content of β phase. Due to the stretching of the jet and its own electrical polarization during spinning, the non-polar α phase and weakly polar γ phase are transformed into polar β phase, which has strong piezoelectric properties. The molecular chain conformation of α phase, β phase, and γ phase is shown in **Figure 3.**

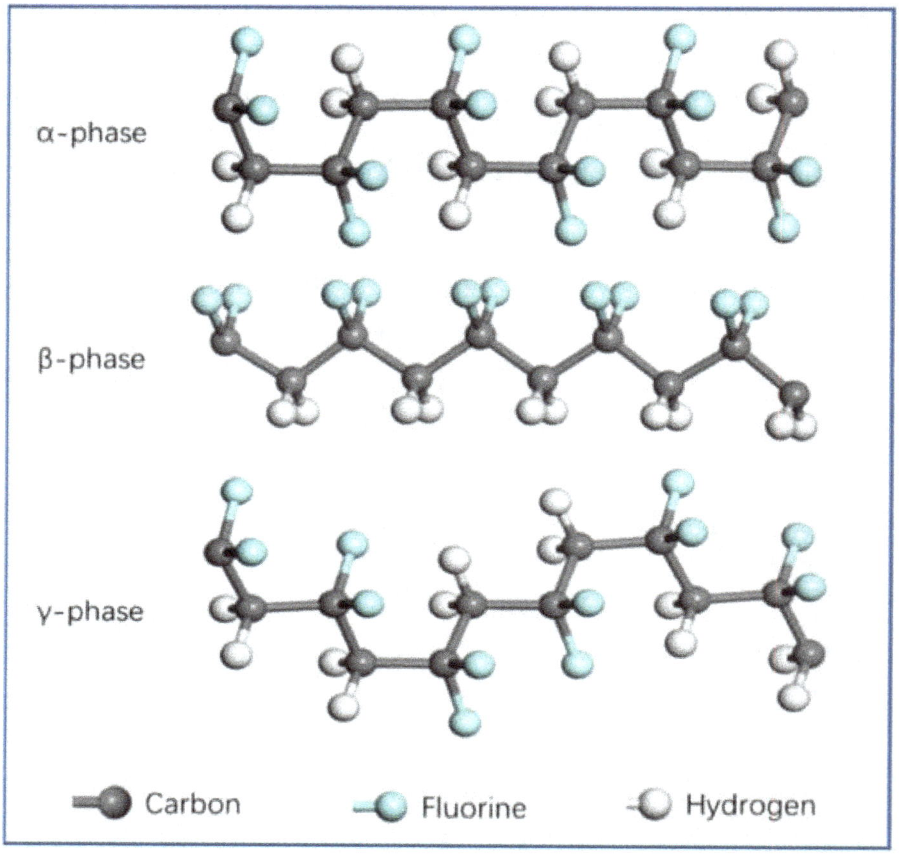

Figure 3. Molecular structure of the α, β, and γ phases of PVDF.

2.2. Effect of Electrospinning Parameters on Piezoelectric Properties of Pure PVDF

The morphology and properties of NFs are mainly determined by the following three aspects: process parameters, the nature of the polymer solution, and environmental parameters [**79,80,81,82,83**]. The influence of three parameters on the morphology and piezoelectric properties of PVDF is introduced in the following section.

2.2.1. Process Parameters

The process parameters included voltage, distance between receiving plate and spinning head, propulsion speed of syringe pump, and diameter of spinning head. During electrospinning, the effect of voltage on the structure and morphology of the fiber is enormous [**84**]; as the voltage increases, the diameter of the fiber decreases, and when the voltage increases to a certain value, the diameter of the fiber decreases and there is no more bead structure and no further increase in fiber diameter because the Taylor cone is smaller, and due to the increase in Coulomb force, the jet flow rate increases at the same propulsion speed [**85**].

Zaarour et al. [86] studied the influence of process parameters on the surface morphology of PVDF. Dimethylformamide (DMF), acetone (ACE), and DMF/ACE were used as solvents at voltages of 6, 12, 18, and 24 kV, and other amounts were kept constant. Finally, the average diameter of the fibers was measured, and it was found that the diameter of the fibers gradually decreased when the voltage was gradually increased. When the voltage is changed from 18 kV to 24 kV, the fiber diameter increases. In addition, the surface morphology of the fibers is different. When the voltage gradually increases, the large surface aperture becomes elliptical and the grooves become more numerous because the Taylor cone is stretched more rapidly, and the roughness decreases because it is less affected by the buckling instability, as shown in Figure 4(1). In addition, the PVDF fiber under high voltage is more uniform, and the crystallinity of PVDF is improved [87]. In this process, the polymer solution is stretched to produce β phase due to the effect of the electric field. Singh et al. [88] studied the influence of process parameters on the formation of β phase in PVDF. Using DMF/ACE as a solvent, the voltage was determined as a single variable, and electrospinning was carried out at voltages of 10, 12, 15, 18, 22.5, and 27 kV. These voltage values corresponded to 0.67, 0.8, 1.0, 1.2, and 1.2 kV, respectively. Electric fields of 1.5 and 1.8 kV/cm were used. The β phase in the samples was determined by infrared spectroscopy, and it was concluded that the content of β phase in the fibers obtained from 10 kV to 18 kV voltage increased gradually (up to 79%), while the content of β phase in the fibers obtained from increasing the voltage continued to decrease, as shown in Figure 4(2). This is because with the continuous increase in voltage (greater than 18 kV), the flow rate also gradually increases, the evaporation time of the solvent is shortened, and the effective stretching of the NFs is reduced. The effect of voltage on the crystalline phase transition was also confirmed in the work of Nugraha et al. [87], and the change in β phase is a key factor affecting the piezoelectric properties of PVDF.

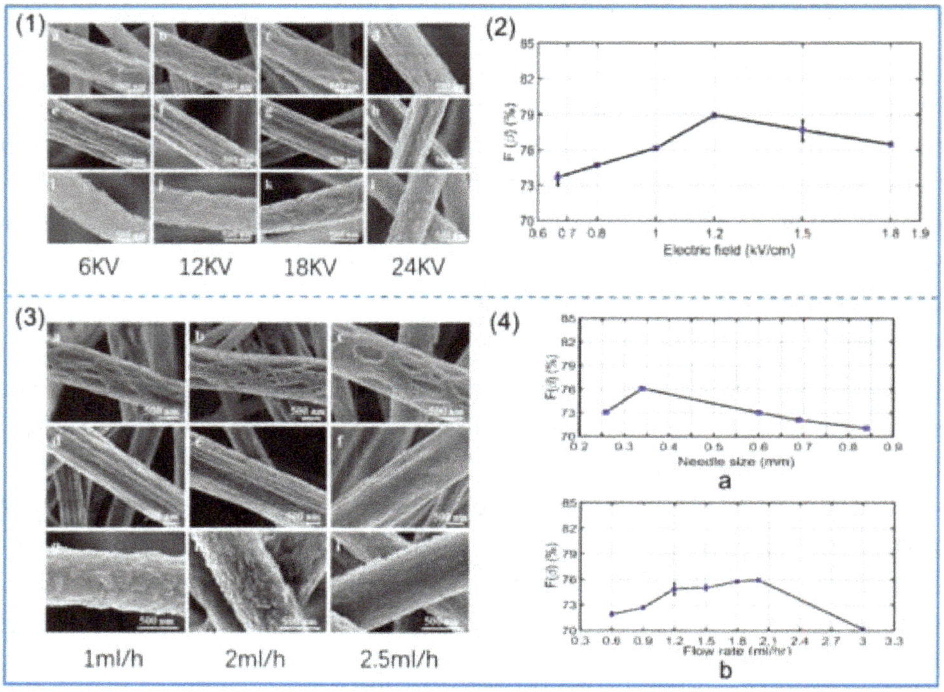

Figure 4. (1) SEM images of surface morphology of electrospun PVDF fibers under different applied voltage levels (a,e,i) 6 kV; (b,f,j) 12 kV; (c,g,k) 18 kV; (d,h,l) 24 kV. Reprinted with permission from Ref. [86], copyright 2019, IOP Publishing. (2) curves of percentage of β phase in PVDF NFs and voltage values of 10, 12, 15, 18, 22.5, and 27 kV. Reprinted with permission from Ref. [88], copyright 2021, Polymer. (3) SEM images of surface morphology of electrospun PVDF fibers at different flow rates (a,d,g) 1 mL/h; (b,e,h) 2 mL/h; (c,f,i) 2.5 mL/h. Reprinted with permission from Ref. [86], copyright 2019, IOP Publishing. (4) (a) Relationship between the proportion of β phase of PVDF NFs and the size of the needle eye (flow rate 2.0 mL/h). (b) Relationship between the percentage of β phase and the flow rate of solution (needle diameter 0.6 mm). Reprinted with permission from Ref. [88], copyright 2021, Polymer.

The distance between the receiving plate and the spinning head mainly affects the evaporation of the solvent, and because the distance affects the electric field strength, it affects the stretching of the polymer. Usually, this distance is set to 15 cm [**89**], and the larger the distance, the finer the diameter of the fiber [**90,91**]. When the spacing is shortened to very small, uneven fibers are obtained [**92**], and large-diameter fibers with many defects are produced [**93**]. The content of α phase decreases with a decrease in this distance, and the content of β phase increases with a decrease in this distance [**94**]. However, this parameter has little effect on the fiber [**95**]. Generally, the distance between the receiving plate and the spinning head often affects the fiber morphology and performance together with parameters such as voltage and flow rate.

The propulsion speed of the injection pump directly affects the flow rate of the polymer solution and maintaining a stable flow rate is a key factor in the production of homogeneous fibers; faster flow rates are not conducive to fiber formation and may result in coarse fibers or even unstable jets that can damage fiber formation [**96**]. Relatively coarse fibers may even produce unstable jet damage formation, which is unfavourable to solvent evaporation; velocities exceeding the critical value would produce microspheres [**85**]. However, the lower flow rate also affects the formation of Taylor cones. At a low flow rate, more elliptical large holes are formed on the fiber surface. At a high flow rate, the number of holes is reduced, and jet instability is weakened due to the increase in flow rate, resulting in a smoother surface, as shown in **Figure 4**(3). Flow rate also has a great influence on the formation of β phase. Singh et al. [**88**] studied the effect of flow rate change on the formation of β phase and found that the content of β phase also increased when the flow rate gradually increased from 0.6 mL/h to 2.0 mL/h; then, as they continued to increase the flow rate, the content of β phase began to decrease, which was when the flow rate was small. Because the sheer force of the needle wall leads to the instability of the solution flow, the solution flow tends to be stable when the flow rate increases. However, when the flow rate continues to increase, the stretching of the fiber is incomplete due to the delayed evaporation of the solvent, and the content of β phase is reduced. The diameter of the spinning head can also have an effect on the formation of PVDF β phase, and a decrease in the diameter of the needle can change the number of solution droplets that flow out of the needle. When the diameter of the needle becomes smaller, the polymer solution drops from the needle shrink, and the increase in surface tension affects the jet, thus affecting surface morphology and piezoelectric properties, as shown in **Figure 4**(4).

2.2.2. Polymer Solution Parameters

These parameters are mainly determined by the properties of the polymer solution, usually including the molecular weight of PVDF, the concentration of PVDF solution, the solvent, the solution conductivity, and the surface tension. The molecular weight of PVDF affects the viscosity of the solution. When PVDF with a higher molecular weight is added to the solvent, a solution with higher viscosity is produced. A solution with high viscosity cannot smoothly pass through the needle tip and spin normally. In addition, an increase in viscosity gradually changes the bead shape from spherical to spindle until smooth fibers are obtained [**97**]. Magniez et al. [**98**] found that higher-molecular-weight PVDF can produce more β phase, and this result was also demonstrated in the study of Zaarour et al. [**99**]. When the molecular weight increases, the viscosity increases, leading to a longer spinning time, the evaporation time of the solvent becomes longer, and the stretching time of the polymer solution increases, so the β-phase content increases. Therefore, in order to obtain a large amount of β phase, the molecular weight can be increased, but pinholes can be blocked due to increased viscosity, so it should not be too high. Although increasing the diameter of the needle prevents clogging, the fiber diameter increases. Surface morphology and piezoelectric properties are shown in **Figure 5**(1,2).

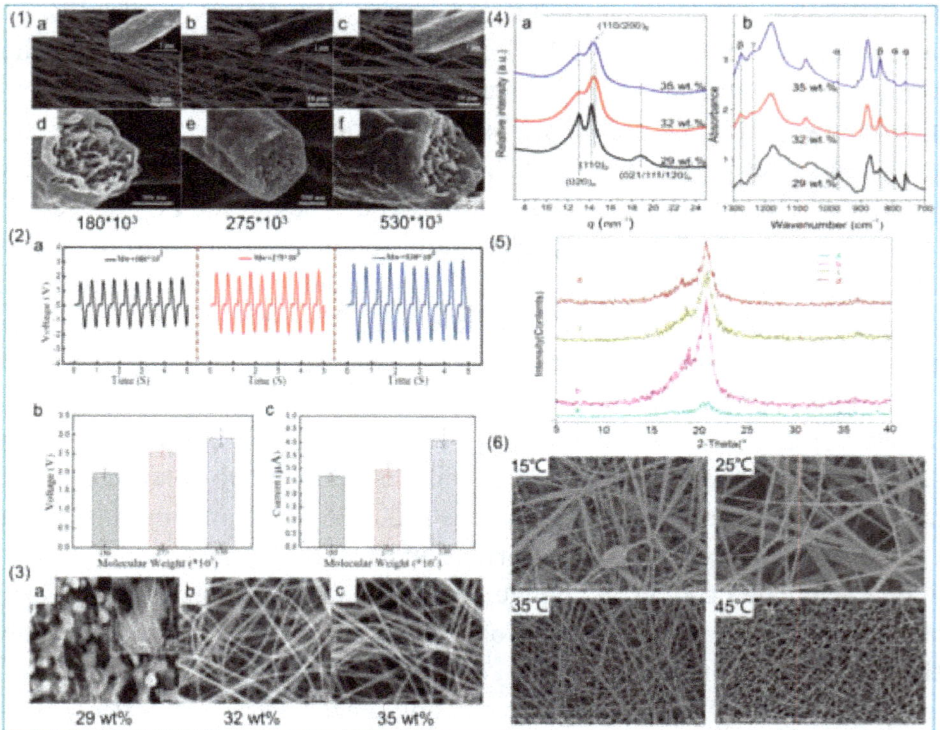

Figure 5. (1) SEM images of electrospun PVDF NFs with different molecular weights (MWs) (a,d) 180 × 10³; (b,e) 275 × 10³; (c,f) 530 × 10³. Reprinted with permission from Ref. [99], copyright 2019, Taylor & Francis. (2) Piezoelectric effect of PVDF NFs under different MWs. (a) Voltage output. (b) Statistical results of average voltage output. (c) Statistical results of average current output. Reprinted with permission from Ref. [99], copyright 2019, Taylor & Francis. (3) SEM images of PVDF NFs with different concentrations (a) 29 wt%; (b) 32 wt%; (c) 35 wt%. Reprinted with permission from Ref. [102], copyright 2011, Polymer. (4) WAXD (a) and FTIR spectra of PVDF fibrous membranes (b). Reprinted with permission from Ref. [102], copyright 2011, Polymer. (5) XRD patterns of PVDF NFs at ambient temperature (a) 15 °C (b) 25 °C (c) 35 °C (d) 45 °C. Reprinted with permission from Ref. [103], copyright 2008, Walter De Gruyter GmbH. (6) FE-SEM images of PVDF NFs prepared at different ambient temperatures. Reprinted with permission from Ref. [103], copyright 2008, Walter De Gruyter GmbH.

The concentration of PVDF is generally in the range of 10–25%, and different concentrations of PVDF solutions produce different fiber morphologies [**100**]. Lower concentrations of PVDF solution produce polymer particles or nanoparticles when passing high-voltage electricity. This technology is the electrospray method rather than electrospinning, and when the concentration increases to a certain value, continuous NFs are produced. This is electrospinning technology, which produces smooth and continuous NFs in the optimal concentration range [**101**]. Similarly, the concentration of PVDF also affects the viscosity, and the stretching effect of the electric field on the polymer is more obvious when the viscosity increases. When the concentration reaches the best value, the formed fiber morphology is smooth and uniform and the content of β phase is the highest. When the concentration exceeds this value, the tensile effect is reduced due to the increase in viscosity, and the content of β phase is reduced. This can be proven from the study of Zhong et al. [**102**]. Fiber surface morphology and β-phase content are shown in **Figure 5**(3,4).

When using polar solvents to dissolve PVDF such as dimethylacetamide (DMAc), DMF, etc. [**104**], in the case of the same temperature and pressure, the saturated vapour pressure of these solvents is relatively low, so ACE with higher saturated vapour pressure is also selected as a solvent to dissolve PVDF for electrospinning. In this way, due to the higher vapour pressure of the solution, the complete evaporation of the solvent in the process of electrospinning can obtain smoother fibers and reduce the formation of beaded structure, which is conducive to the formation of β phase. Generally, the mixed solution of DMF and ACE with a ratio of 6:4 is used as the solvent [**100**].

The conductivity of the solution mainly depends on the polymer type, the solvent, and the added electrolyte. The conductivity of a polymer solution can be changed by adding a small amount of salt or electrolyte to the solution. When the charge density of the jet increases as the conductivity increases, the jet is stretched more fully, resulting in a smoother fiber [**101**]. Surface tension is an important factor affecting fiber properties: reducing surface tension can produce smoother fibers, and adding surfactants to the solution can reduce surface tension.

2.2.3. Environmental Parameters

Environmental parameters can be divided into two parameters: ambient temperature and relative humidity. An increase in ambient temperature leads to faster solvent evaporation, which reduces the surface tension and viscosity of the solution. Zaarour et al. [**105**] found that with an increase in temperature, the diameter of the fiber becomes smaller and the surface becomes rough. This is because DMF with low vapour pressure hinders the evaporation of ACE with high vapour pressure when a DMF/ACE mixed solution is used as a solvent. The content of β phase from 20 °C to 60 °C was studied, and it was found that the content of β phase was the highest at 20 °C, and the content of β phase decreased with the increase in temperature. However, in the study of Huang et al. [**103**], it was found that the content of β phase was the highest at 25 °C. The increase in temperature led to the decrease in surface tension and viscosity, thus affecting the stretching of the solution, which was the main reason for the highest content of β phase being at 25 °C. Surface topography and XRD spectra are shown in **Figure 5**(5,6).

Relative humidity also affects the evaporation of solvent. When the humidity increases, the evaporation of solvent is incomplete, which affects the surface morphology of fibers. Some studies have found that when the relative humidity increases from 2% to 62%, the content of β phase increases from 55% to 73.06% [**106**]. Higher relative humidity leads to increased water on the surface of the jet, which prolongs the stretching time, and the nucleation and growth time of the β phase is long, which increases the content of the β phase.

2.3. Effect of Adding Functional Particles on Piezoelectric Properties of PVDF NFs

The piezoelectric properties of PVDF NFs can be improved by adding various additives. Various nanomaterials can be easily added into PVDF solutions by electrospinning. The various additives can be divided into inorganic piezoelectric particles, metal–organic framework (MOF), conductive particles, organic additives, etc. Table 2 lists the influence factors of some additives, the content, the average diameter of the obtained fiber membranes, the peak voltage, the elastic modulus, and their related applications.

Table 2. The influence of Various additives to PVDF.

Category	Material	Influence	Addition (wt%)	Average Diameter (μm)	Peak Voltage (V)	Modulus of Elasticity (MP)	Application
Inorganic piezoelectric material	PZT [107]	Enhanced the mechanical and piezoelectric properties of the fiber membrane	20.0	370	—	2053	Piezoelectric pressure sensor
	BaTiO3 [108]	Enhanced piezoelectric performance	16.0	0.200	0.48	—	Smart textiles and biomedical devices
	ZnO [109]	Improved the crystal structure and improved the electrical output	15.0	0.757	1.10	—	Self-powering of microelectronics clothing
	G-ZnO [110]	Reduced average diameter, increased the β phase, and improved piezoelectric performance	1.0	0.087	0.84	—	Wearable piezoelectric nanogenerators
Conductive particles	AgNO3 [111]	Conductivity increased	0.3	0.201	2.00	4047	Nanoelectronics, and energy harvesting
	FeCl3·6H2O [111]	Conductivity increased, increased the β phase	0.8	0.254	4.80	5621	
	Modified graphene [111]	Conductivity increased, increased the β phase, improved the mechanical properties	1.0	0.296	1.80	6583	
	Carbon nanotubes [112]	Increased specific surface area, increased the β-phase content, and improved the piezoelectric properties	18	0.138	—	—	Efficient sound absorbers
	BNNs [113]	Enhanced the triboelectric effect	2.0	0.007	0.25	—	Human motion detection

Category	Material	Influence	Addition (wt%)	Average Diameter (μm)	Peak Voltage (V)	Modulus of Elasticity (MP)	Application
MOF	Uio-66(Zr) [114]	Increased the β phase, its own piezoelectricity	5.0	0.053	0.57	1100	Flexible energy convertors, biomedical monitoring
	[CdI2-INH=CMe2] [115]	Increased the β phase, controllable porous structure	1.0	—	12.00	—	Real-time wearable medical devices
Organic additives	DDAC [116]	Increased the β phase	8.0	0.09	—	—	Artificial intelligence

2.3.1. Inorganic Piezoelectric Material

As the most typical inorganic piezoelectric material, PZT is very suitable as an additive to improve the piezoelectric properties of PVDF. Adding 20 wt% PZT to PVDF for electrospinning can obtain a fiber film with the best piezoelectric properties [107]. However, because lead-based piezoelectric materials are harmful to human health and are not conducive to recovery and carry the possibility of harm to the environment, research on lead-free piezoelectric materials is now a hot spot. A study on PVDF NFs with different concentrations of BaTiO3 found that the piezoelectric output voltage of PVDF NFs containing 16 wt% BaTiO3 was 1.7 times that of pure PVDF under the same test environment [108]. A similar conclusion can also be obtained in the study of Dashtizad et al. [117]: the content of β phase in PVDF NFs with 0.8 wt% BaTiO3 is the highest (90%), as shown in Figure 6(1), and silver particles are added by ultraviolet irradiation of AgNO3 aqueous solution containing BaTiO3. By controlling the number of silver ions added and the irradiation time, this method can increase the content of β phase and the piezoelectric properties of PVDF, as shown in Figure 6(2). The results showed that the output voltage of PVDF NFs with ZnO nanoparticles was increased by 1.1 V compared with pure PVDF NFs [109]. This was also confirmed in the study of Li et al. [118], in which nanofiber membranes prepared by electrospinning technology after mixing PVDF with ZnO nanoparticles showed smaller fiber diameters and increased output voltage. In addition, modified ZnO can also improve the piezoelectric properties of PVDF NFs. Graphene has excellent optical, electrical, and mechanical properties; combining it with other materials can improve various physical and chemical properties [119,120]. Hasanzadeh et al. [110] synthesized graphene–zinc oxide nanocomposite (G–ZnO) by the hydrothermal method and added it to PVDF for electrospinning, and used XRD and TFIR to characterize the β phase of the NFs. It was found that the β-phase content of the composite was higher, and the piezoelectric output performance of this composite was better.

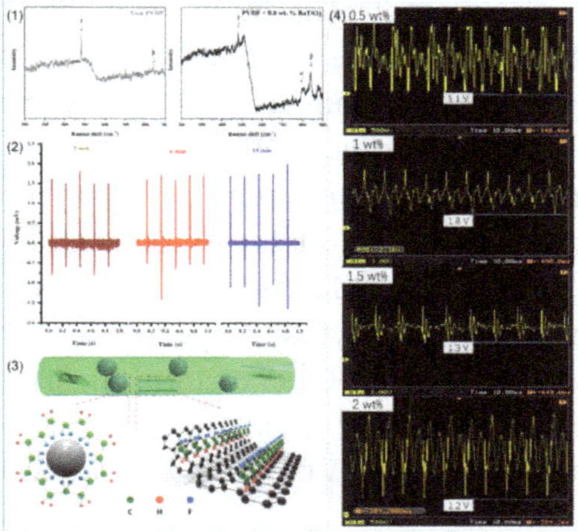

Figure 6. (1) Raman spectroscopy results of pure PVDF and PVDF containing 0.8 wt% $BaTiO_3$. Reprinted with permission from Ref. [117], copyright 2021, Elsevier. (2) Voltage curve of PVDF-$BatiO_3$-Ag composite as a function of time. Reprinted with permission from Ref. [117], copyright 2021, Elsevier. (3) Schematic diagram of modified nanographene to promote β-phase generation. Reprinted with permission from Ref. [111], copyright 2021, MDPI. (4) the piezoelectric signals of the fiber membrane when different amounts of modified nanographene were added. Reprinted with permission from Ref. [111], copyright 2021, MDPI.

2.3.2. Conductive Particles

The addition of conductive particles can also improve the piezoelectric properties of PVDF NFs. Because the addition of conductive particles increases the surface charge density of the jet during electrospinning, the bending whipping effect of the jet is enhanced, the fiber is better stretched, and the content of β phase is relatively improved. Li et al. [111] studied the piezoelectric properties of electrospun composite NFs with the addition of AgNO3, FeCl3·6H2O, and modified graphene to PVDF. The relevant data are shown in Table 2. The results showed that the piezoelectric signal of the composite fiber membrane with 0.3 wt% AgNO3 was the highest when the number of silver ions was low, and the piezoelectric signal of the composite fiber membrane with 0.3 wt% AgNO3 was the highest when the number of silver ions was low. As the solution conductivity increases, its level jet whip dynamic instability is improved, and because its content is more than 0.3 wt%, the fluid resistance due to the increase in surface tension increases, and as a result of the Ag^+ and NO_3^- gathered, results in an uneven surface charge releasing; additionally, dynamic instability increases, causing a whip against solution stretching. The content of β phase decreases instead, so high $AgNO_3$ concentrations are not conducive to the improvement of piezoelectric properties of PVDF composite NFs. $FeCl_3{\cdot}6H_2O$ was also added to PVDF and its piezoelectric properties were studied. With the increase in $FeCl_3{\cdot}6H_2O$ content, the piezoelectric output voltage of the fiber membrane reached its highest (4.8 V) when the content reached 0.8 wt%. Moreover, this addition also affects the β nucleation of PVDF, since the specific interactions near the Fe/PVDF interface can induce PVDF β nucleation [121], and the strong hydrogen bond interaction formed between the water molecules in $FeCl_3{\cdot}6H_2O$ and polar CF_2 may be the driving factor for β nucleation. After adding modified graphene, due to the van der Waals force between carbon atoms, they are easy to collect. In order to make it more evenly dispersed into the PVDF solution, rare earth element La is added to form La–C bonds to make it stable, and the diameter of PVDF fibers after adding modified graphene is smaller, and the conductivity is improved. Among them, in the PVDF with the addition of 1 wt% modified graphene, β-phase content was the highest and the piezoelectric output voltage was the highest, as shown in Figure 6(4). In addition to promoting jet stretching by increasing solution conductivity, adding nanographene can also promote the formation of β phase; the surface of nanographene in the interface interaction forms β phase of PVDF, as shown in Figure 6(3). It was also found that the addition of unmodified nanographene affected the movement of PVDF chains due to the aggregation effect, resulting in a lower β-phase content than pure PVDF. Adding carbon nanotubes (CNTs) can also improve the sensor performance of fibrous PVDF membranes, and piezoelectric properties of the PVDF with CNTs added for electrostatic spinning used a sound-absorbing device. The results found that the specific surface area of a fiber membrane improved, increasing the acoustic contact and improving the crystallinity of β phase; making it in the low frequency area of the sound waves can also result in good absorption [112]. In another study, CNTs were added to PVDF/potassium sodium niobate (KNN). The results showed that the addition of 0.1% CNTs could significantly improve the piezoelectric performance, with an output voltage of 23.24 V, a current of 9 μA, and a power density of 52.29 $\mu W/cm^2$. Since the addition of CNTs improves the electrical conductivity of the solution, the jet stretches better, resulting in more β phase [122]; this can also be confirmed in the work of Hehata et al. [123]. Boron nitride nanosheets (BNNs) can also be added to PVDF to improve the piezoelectric properties of PVDF [113]. The results show that the piezoelectric properties of PVDF/BNNs composite nanofiber film are improved, and the triboelectric effect between the metal electrode and the composite fiber film also helps to improve the electrical output.

2.3.3. Other Additives

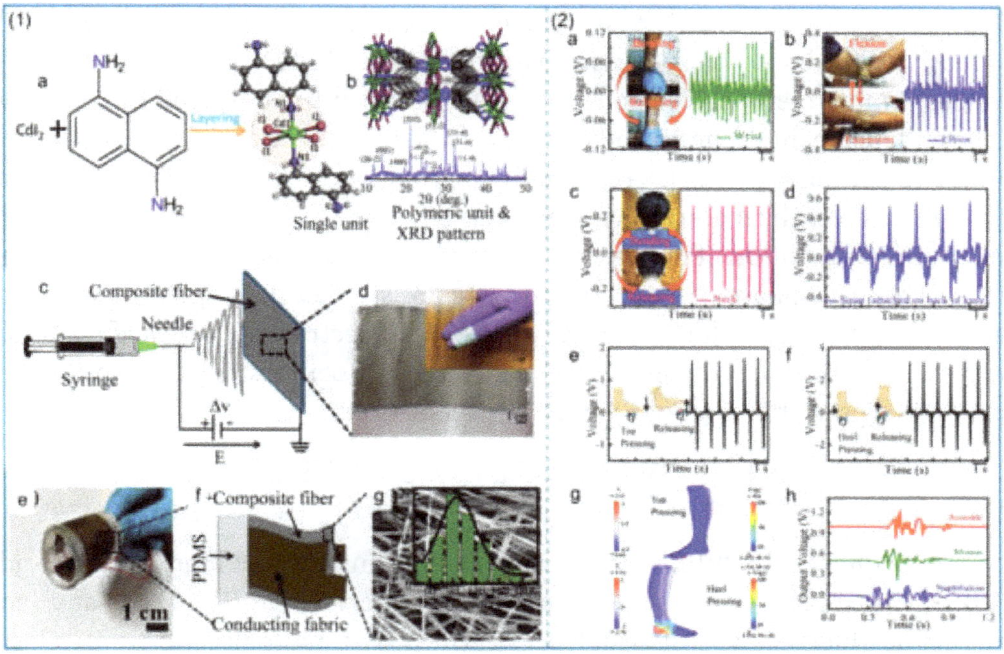

Figure 7. (1) The macroscopic and microscopic structure and state diagram of the material, (a) the schematic diagram of the synthesis route of two-dimensional MOF (CdI_2-NAP) and its unit structure. (b) MOF structure (top), XRD (bottom). (c,d) electrospinning. (e,f) photos of C-PNG. (g) FE-SEM images of composite NFs, inset shows a histogram of fiber diameter distribution. Reprinted with permission from Ref. [126], copyright 2021, American Chemical Society. (2) Analysis of piezoelectric properties during human movement, (a) voltage responses of C-PNGs attached to a wrist, (b) elbow, (c) neck, (d) knee, (e) toe, and (f) heel movements, the arrows represent pressure and release. (g) Simulation based on FEM. (h) Voltage response curves of C-PNGs attached to the throat during "acoustic," "motor," and "naphthene" vocations. Reprinted with permission from Ref. [126], copyright 2021, American Chemical Society.

In addition, MOF particles can be added to enhance the piezoelectric properties of PVDF. Generally speaking, we hope to obtain composite fiber membranes with better performance by increasing the loading of MOF. However, when the MOF content is too high, it makes electrospinning difficult to carry out, so the performance of the fiber membrane decreases [124]. Moghadam et al. [114] found that the NFs produced by adding zirconium-based MOF to PVDF can greatly improve its piezoelectric properties. The β-phase content of PVDF with Zr-based MOF increased by 16%, and the output voltage of PVDF with 5 wt% MOF was found to be the highest when it was used for human arterial pulse detection. In addition, Zr-based MOF can also be used for SO2 detection, and the results show that it has high sensitivity and excellent flexibility [125]. Roy et al. [126] incorporated 2D MOF into PVDF, and the structure of MOF is shown in Figure 7(1). The composite fiber-based piezoelectric nanogenerator (C-PNG) was prepared by the electrospinning method. The coupling effect between the MOF and PVDF NFs led to the enhancement of the piezoelectric performance of the composite. The output voltage can reach 6 V—the corresponding motion test results are shown in Figure 7(2)—which gives the material good applicability in the field of low-frequency noise detection. The addition of organic additives to the PVDF solution can also enhance the electroactive phase of PVDF. Xue et al. [116] studied the effect of dioctadecyl dimethyl ammonium chloride (DDAC) on the piezoelectric properties of PVDF, and the results showed that the β-phase content of PVDF composite NFs with 4% DDAC was 39.1% higher than that of PVDF NFs without DDAC.

3. PREPARATION OF PVDF COPOLYMER NFS BY ELECTROSPINNING

The copolymers of PVDF are mainly poly (vinylidene fluoride-trifluoride) (PVDF-TRFE), poly (vinylidene fluoride-tetrafluoride) (PVDF-TFE), and poly (vinylidene fluoride-hexafluoride) (PVDF-HFP). The piezoelectric properties of these polymers, as well as their biocompatibility and flexibility, are very excellent. Zhang et al. [127] made two kinds of piezoelectric nanogenerators (PNG) based on PVDF and PVDF-TRFE, and studied the mechanical and piezoelectric properties of PVDF and PVDF-TRFE NFs as well as the crystallinity of β phase. The structure and XRD analysis are shown in Figure 8(1), and the piezoelectric performance test of the piezoelectric nanogenerator is shown in Figure 8(2). The results show that the mechanical properties and piezoelectric properties of PVDF-TRFE are better than those of pure PVDF. For the piezoelectric nanogenerator made of PVDF-TRFE, its β-phase content is higher and the output voltage is higher. In addition, compared with the performance of electrospinning PVDF and PVDF-TRFE with polyethylene oxide (PEO) and LiCl, the β-phase content of PVDF-TRFE with PEO is higher than that of PVDF with the same addition [128], and the addition of LiCl improves the conductivity of the solution and also leads to an increase in the β-phase content. The fabricated PVDF-TRFE/PEO composite nanofiber membrane can achieve an instantaneous power of 40.7 μW/cm2 under a mechanical force of 1.58 N.

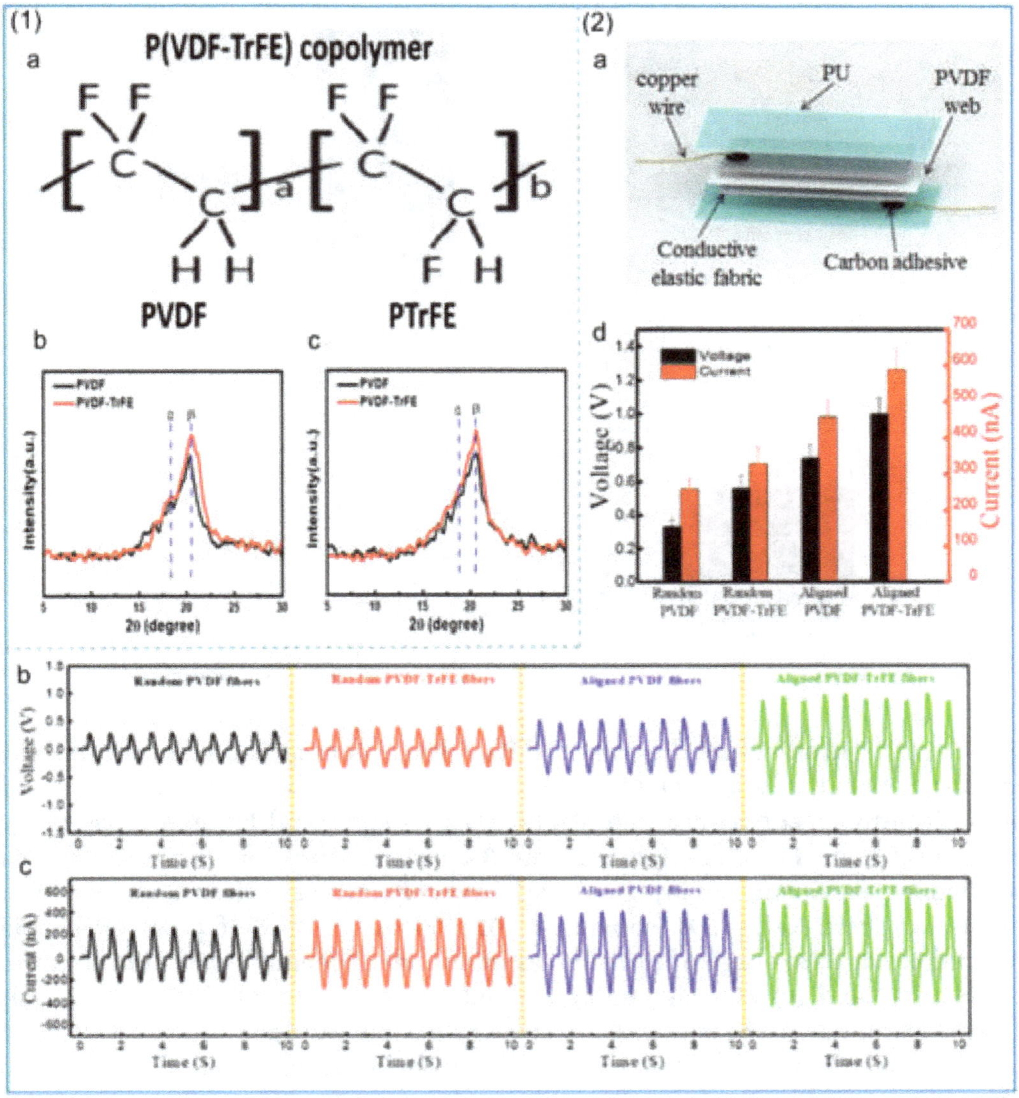

Figure 8. **(1) Analysis of PVDF and PVDF–TRFE NFs, and (a) Molecular structure of PVDF–TRFE. (b) XRD patterns of randomly oriented PVDF and PVDF–TRFE NFs. (c) XRD patterns of PVDF and PVDF–TRFE fibers arranged in order. Reprinted with permission from Ref. [127], copyright 2020, SAGE Publications Ltd. STM. (2) PNG analysis, (a) Schematic diagram of the structure of PNG. Statistical results of (b) output voltage, (c) output current, (d) average voltage, and current generated by PNG during folding and release at 90° bending. Reprinted with permission from Ref. [127], copyright 2020, SAGE Publications Ltd. STM.**

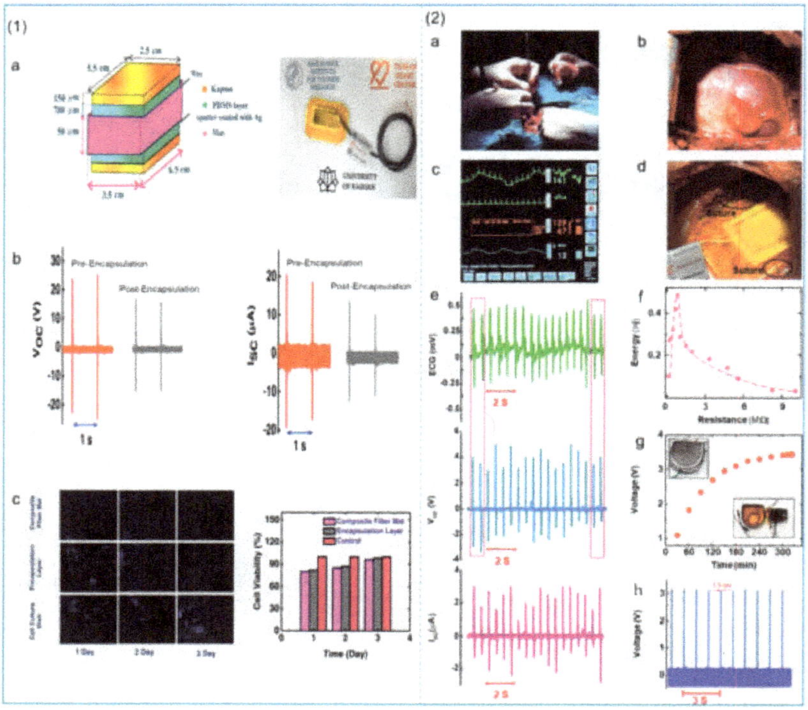

Figure 9. (1) PNG structure diagram and its voltage, current, and biocompatibility test, (a) Schematic diagram and photo of PNG. (b) Open-circuit voltage and short-circuit current before and after PNG packaging. (c) Fluorescence images and viability of MEF cells after 1, 2, and 3 days of culture on composite fibrous membranes, encapsulated layers of PNG, and cell culture dishes. Reprinted with permission from Ref. [129], copyright 2021, Elsevier. (2) Self-powered cardiac pacemakers and (a) surgical procedures in animal experiments. (b) Photographs of the heart and (c) ECG signals and blood pressure recorded simultaneously. (d) PNG stitched to the epicardium facing the left outdoor lateral wall (e) In vivo ECG signal, open-circuit voltage, and short-circuit current of PNG. (f) Energy generated by PNG. (g) Charging curve of a 100 μF capacitor charged by PNG. (h) Pacing pulses generated by a PNG-powered pacemaker. Reprinted with permission from Ref. [129], copyright 2021, Elsevier.

The addition of other particles can also improve the piezoelectric properties of PVDF–TRFE. ZnO and reduced graphene oxide (rGO) were added to PVDF–TRFE using electrospinning technology to prepare NFs and fabricate PNG for self-powered cardiac pacemakers [129]. The mass ratio of ZnO to rGO is 9:1. The PNG structure and its voltage, current, and biocompatibility test are shown in Figure 9(1). The results show that the PVDF nanofiber membrane containing 0.1 wt% ZnO/rGO reaches the maximum output power of 138 ± 2.82 μW/cm2. An electrical energy of 0.487 μJ can be collected from each heartbeat, as shown in Figure 9(2), for surgical pictures and related tests, which is a high collection rate. The high-speed roller speed during the electrospinning of PVDF–HFP can produce nanofiber films with smaller fiber diameters and more β-phase content [130]. A similar conclusion can be obtained in the study of Conte et al. [131], where the piezoelectric properties of PVDF–HFP fiber membranes can be enhanced by optimizing processing conditions such as tensile rate and temperature. Wu et al. [132], by incorporating barium calcium zirconia titanate (BCZT) into PVDF–HFP and using electrospinning technology to fabricate NFs, significantly improved its piezoelectric properties, and the maximum power density reached 161.7 Mw/cm2. The PVDF–HFP/ZnO NF piezoelectric sensor fabricated by the addition of ZnO has excellent performance and an ultra-high sensitivity of 1.9 V/kPa, which is due to the combination effect between the electric dipole inside the fiber and the asymmetric lattice of ZnO to improve the piezoelectric performance [133]. Pinnamma et al. [134] fabricated ZnO and TiO2 containing PVDF/PVDF–HFP core–shell-structured NFs based on electrospinning, which produced output voltages of up to 14 V and had a dielectric constant that was five times higher than that of pure polymers.

4. PREPARATION OF OTHER PIEZOELECTRIC NFS BY ELECTROSPINNING

In addition to PVDF, there are many other electrospun piezoelectric fiber membranes, such as poly-l-lactic acid (PLLA), which is a biodegradable polymer derived from plant alkalis, poly (3-hydroxybutyric acid) (PHB), which is also a biodegradable polymer, PAN, Nylon-11, collagen, cellulose, chitin, chitosan derived from chitin, etc. A subset of spinnable piezoelectric polymers and their applications are listed in **Figure 10.**

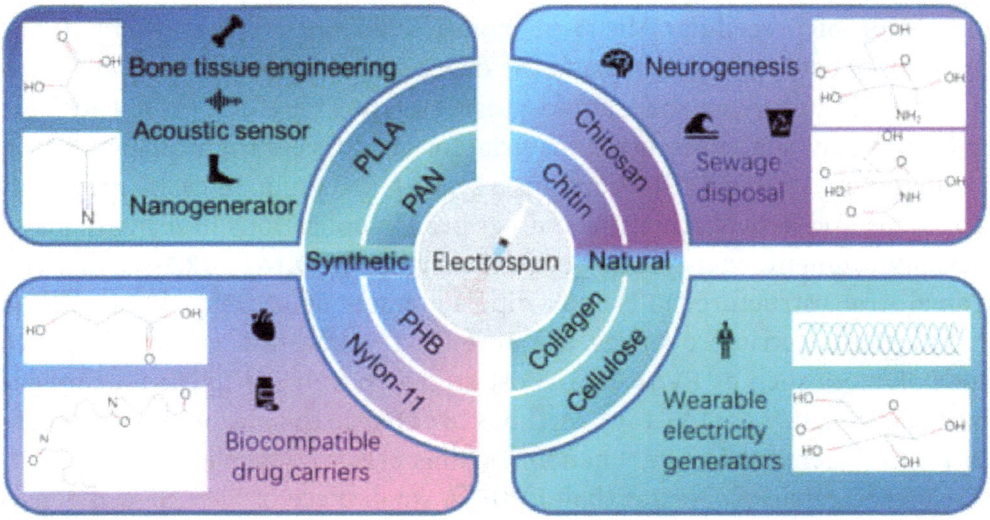

Figure 10. Structures and applications of other piezoelectric NFs prepared by electrospinning.

4.1. Preparation of Synthetic Piezoelectric Polymers by Electrospinning

PLLA has good biocompatibility, which can be proved by the FDA's approval of PLLA injection for human tissue in 2004, and has a good prospect in the application of some implantable devices. Compared to PVDF, the polarization direction of PLLA in the piezoelectric domain is not the same, and the polarization direction is parallel to the plane where the shear stress is applied, leading to the shear piezoelectric property of the polymer. This is different from PVDF fiber membranes that lack a shear strain response, and PLLA can be more effectively applied in shear piezoelectrics [**135**]. PLLA NFs prepared by electrospinning have different effects on their piezoelectric properties under different surface morphologies and different heat treatments [**136**]. Fiber diameter and heat treatment have a great impact on the transverse and longitudinal output voltage of PLLA NFs, which affects the differentiation of stem cells. Orthogonal and shear piezoelectric properties affect neurogenesis and osteogenesis, respectively, and the improvement of piezoelectric properties enhances the regeneration process accordingly. Zhao et al. [**137**] also prepared PLLA NFs that can reach a current of 8 pA and a voltage of 20 mV, and prepared a blood pulse sensor, which can produce an output current of 2 pA and can effectively detect blood pulses. The addition of hydroxyapatite (HAP) to PLLA can also enhance the related ability of PLLA in cell cultures, where electrospinning is performed to prepare NFs [**138**]. These fiber matrices have good performance as osteocyte culture substrates. The results showed that pure PLLA NFs did not release substances that were toxic to cells. Addition of 0.5% (w/v) HAP particles to PLLA fibrous membrane scaffolds not only promoted the attachment and proliferation of precellular Mc3t3-e1 mouse osteoblasts, but also increased the expression of osteocalcin mRNA and the degree of mineralization of cells after 14 and 21 days of culture on the scaffolds, respectively. Therefore, PLLA/HAP fibrous membranes may be the preferred material for bone tissue engineering. PAN is an amorphous vinyl polymer that contains a cyano (–CN) group in each repeat unit. It has a zigzag and helical structure and has broad applications in many fields such as the environment, energy, and drug delivery [**139,140,141,142,143**]. The zigzag has an all-trans structure (TTT) and a dipole moment of 3.5 Debye [**144**]. Although PAN has piezoelectric properties, the piezoelectric constant is much lower than other polymers such as PVDF [**56**]. Most studies have used PAN as a carrier for piezoelectric materials [**145,146**]. Methods to enhance the piezoelectric properties of PAN have been studied. PAN nanofiber films prepared by electrospinning have better piezoelectric properties than PVDF. A small piece of PAN NF can generate an output voltage of 2.0 V and an output current of 1.1 μA. The residual charge on PAN NFs also contributes to energy conversion [**147**]. This can also be confirmed in the study by Street et al. [**58**]. Among the PAN NFs with ZnO addition, the pressure-sensing ability and energy harvesting efficiency of ZnO/PAN composite NFs are 2.7 times higher than those of pure PAN NFs [**146**]. Using carbonized PAN/ barium strontium titanate (BTO) NFs to make the sensor, the pressure-sensing sensitivity was increased by more than 2.4 times when BTO nanoparticles were added to the PAN nanofiber film due to the synergistic effect of piezoelectric and triboelectric effects [**148**]. Airagi et al. [**149**] added CuO to PAN to prepare an electrospun fiber nanogenerator, and found that the piezoelectric performance of the nanogenerator increased with an increase in CuO addition. The main reason for this is that CuO provides a conductive path for PAN, which makes the charge on the electrode increase and the energy collection efficiency increase. The PAN nanofiber film containing 0.5% CuO had an output voltage of 5 V and an output current of 0.172 μA, and the power density of the nanogenerator was 0.215 μW/ cm^2. In situ polarization and stretching via an electrospinning process causes the –CN groups of PAN to rearrange on both sides of the polymer chain, increasing the number of electroactive conformations. PAN NFs can also be used for acoustic detection. Poon et al. [**150**] studied the ability of PAN NFs in acoustic detection and the ability of electrospun PAN NFs to accurately detect low- and medium-frequency sounds (100–600 Hz) at a medium sound pressure level (60–95 dB), which covers the main sound spectrum in our daily activities. Compared with acoustic sensors made from PVDF NFs under the same conditions, the PAN device has a wider response bandwidth, greater sensitivity, and higher fidelity, which indicates that PAN plays an important role in acoustic detection of NFs.

PHB is a biodegradable material that has good biocompatibility. Based on the sustainable development strategy, it can be used to replace polypropylene and polyethylene. However, the piezoelectric coefficient of PHB is very low. Jiang et al. [**151**] prepared PHB NFs with multi-walled CNTs (MWCNTs) by the electrospinning method. Compared with pure PHB nanofiber scaffolds, PHB/MWCNTs composite nanofiber scaffolds contain higher β-shaped crystal content. The piezoelectric charge constant and induced potential measured on the surface of the nanofiber scaffold indicate that the addition of MWCNTs and thermal stretching treatment to the PHB nanofiber matrix helps to improve the piezoelectric properties of PHB. Chhernozem et al. [**152**] prepared PHB NFs with rGO. The surface potential distribution of PHB–rGO fibers was more uniform, and the average voltage increased from 33 ± 29 mV to 314 ± 31 mV compared with the pure PHB nanofiber membrane. Further increases in rGO content lead to increased α crystal deformation and prevents zigzag chain formation, which results in decreased crystallinity and pressure sensitivity of PHB scaffolds. The degradable electrospun composite scaffolds of polyaniline (PANI) and PHB have better piezoelectric performances. The doping of piezoelectric polymer PHB and conductive PANI can significantly improve the piezoelectric charge coefficient and surface potential of pure PHB scaffolds. Compared with pure PHB scaffolds, the surface potential under cyclic mechanical stress at a frequency of 4 Hz is increased by 4.2 times and 3.5 times, respectively [**153**]. Karpov et al. [**154**] also reported the piezoelectric response of PHB NFs and studied the piezoelectric properties of polycaprolactone (PCL), PHB, and PHB–PANI, and found that PCL had the worst piezoelectric properties and PHB–PANI had the best piezoelectric properties.

Nylon (polyamide) is a general term for thermoplastic resin containing repetitive unit amide group (-[NHCO]-) in the molecular backbone, among which Nylon-11 of odd nylon has better piezoelectric properties, and there are four crystalline phases of Nylon-11, namely α, α', γ, δ'. Only the δ' phase has better piezoelectric properties. This is because α and α' phases of Nylon-11 do not have piezoelectric properties due to different grain orientation and also have strong hydrogen bonds that cannot be electrically polarized. The γ phase has a very weak polarity, so the piezoelectric properties are very poor [**155**]. The δ' phase has randomly distributed hydrogen bonds, so dipoles are generated when electrically polarized. However, it is difficult to generate the piezoelectric δ' phase by traditional production methods of Nylon-11 fibers. The cooling rate determines the crystal structure of Nylon-11, and it has been found that the faster the cooling, the higher the content of δ' phase [**156**]. The high-vapor-pressure and low-boiling-point solvents trifluoroacetic acid (TFA) and ACE were used for electrospinning. The reason why ACE is added is that TFA forms hydrogen bonds with amide groups on the Nylon-11 chain, some of them are left in the Nylon-11 fibers when they evaporate, and the hydrogen bond between ACE and TFA is stronger. ACE removes TFA from the Nylon-11 as it evaporates. Anwar et al. [**157**] found that Nylon-11 NFs prepared by electrospinning with TFA and ACE as solvents could generate an output voltage of 6 V under the action of mechanical force.

4.2. Preparation of Natural Piezoelectric Polymers by Electrospinning

Although natural polymers are popular in applications associated with medical products [**158,159,160,161,162**], there are also some natural piezoelectric polymers in addition to the above-mentioned polymers, such as collagen, cellulose, chitin, chitin-derived chitosan, etc. These polymers are extracted from organisms in nature, so their biocompatibility is very good and their toxicity is very weak.

Collagen comprises the main amino acid glycine, proline, hydroxyproline, and alanine; these amino acids have piezoelectric properties [**163**]. Collagen is the main component of connective tissue, accounting for nearly a quarter of total protein in human body, three-fourths of the human body skin dry weight, more than 90% of human tendon and corneal tissue, and nearly 80% of organic matter in bone [**164**]. Twenty-nine collagen types have been identified, of which type I collagen in particular may be a good candidate for biomedical applications, such as scaffolds for wound dressings and tissue engineering. Rho et al. [**165**] prepared a collagen nanofiber matrix by the electrospinning method for wound dressing. This nanofiber has the characteristics of wide pore size distribution, high porosity, good mechanical strength, and a high surface area-to-volume ratio, which is conducive to cell attachment, growth, and proliferation. They found that NFs coated with type I collagen promoted adhesion and spreading of human keratinocytes.

Cellulose is a kind of polysaccharide. Constituting the basic unit of cellulose is glucose, which can be found in plants, but due to its low piezoelectric constant, cellulose will, generally, be broken after being mixed with piezoelectric materials during the pouring process in the film production of nanogenerators, for example, the combination of cellulose and MoS2 can greatly improve mechanical strength and piezoelectric properties [65]. There are not many studies on the preparation of cellulose by electrospinning technology because cellulose has poor spinnability. Usually, cellulose derivatives such as cellulose acetate are used for electrospinning to prepare NFs, and then hydrolyzed to obtain cellulose NFs.

Chitin is a polysaccharide substance extracted from the shells of arthropods, the cell walls of fungi, and the wings of butterflies. Its structure is very similar to that of cellulose, which is a polymer of six-carbon sugar. The basic unit of chitin is acetylglucosamine. Chitin exists in three crystal forms, namely α, β, and γ, among which α has piezoelectric properties [166]. Treet et al. [167] demonstrated that chitin NFs prepared by electrospinning have higher crystallinity and piezoelectric properties. The crystallinity and piezoelectric properties of chitin can be adjusted by controlling the parameters of electrospinning, which is helpful for the adsorption of heavy metals [168]. This means chitin has good application prospects in sewage treatment due to its convenient access.

Chitosan is the product of N-deacetylation of chitin, and is similar to chitin and cellulose in structure. The monomer is glucosamine, which is the only basic polysaccharide in the group of natural polysaccharides due to its free amino group [**169**]. Chitosan has great potential for drug release. Dexamethasone (DEX) was loaded onto chitosan nanoparticles and mixed with a PCL solution, then ZnO nanorods were mixed with the PVDF solution; mixing two kinds of solution using a double-electrospinning method allowed for the preparation of compound electrospun NF membranes. A DEX controlled-release scaffold was fabricated to induce the osteogenic differentiation of bone tissue-engineered mouse bone marrow mesenchymal stem cells (MSCS); this method of controlled release of DEX was proven to be beneficial to osteogenic differentiation [**170**].

5. APPLICATION OF ELECTROSPUN PIEZOELECTRIC NF MEMBRANES

Because of their unique piezoelectric properties, good biocompatibility, and flexibility, piezoelectric fiber membranes in biomedicine, water treatment, piezoelectric photocatalysis, energy collection, nanogenerators, and air particle adsorption have very broad application prospects, as shown in **Figure 11**. In recent years, researchers have done a lot of related work in these areas [**171,172,173,174**].

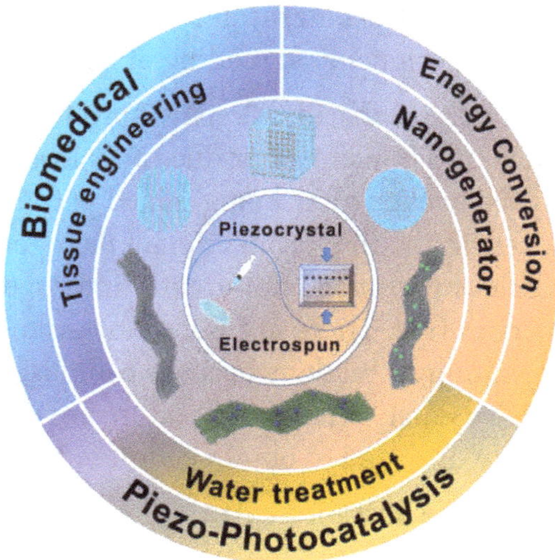

Figure 11. Application of electrospun piezoelectric NFs.

5.1. Biomedical Applications

Compared with ceramic piezoelectric materials, many piezoelectric polymer materials have very good biocompatibility, and electrospun fibers have broad prospects in the fields of biomedicine and ergonomics. Since piezoelectric materials can generate electrical signals and electrical stimulation, they can be applied to tissue engineering scaffolds. Piezoelectric fiber membranes can also be used in biosensors, drug delivery platforms, and so on. Due to the self-electric function of piezoelectric materials, they can be used to replace implantable devices or wearable sensors that require battery replacement.

5.1.1. Scaffold for Tissue Engineering

Tissue engineering scaffold materials refer to materials that can combine with living tissue cells and implant different tissues of organisms according to the function of specific replacement tissues. In order to enable seed cells to proliferate and differentiate, it is necessary to provide a cell scaffold composed of biological materials, which is equivalent to an artificial extracellular matrix. The preparation of tissue engineering scaffolds by electrospinning has great application prospects [**175,176,177**], for example, charge generation on the surface of PVDF–TRFE NF scaffolds induced neural stem cells to differentiate into multiple neuronal, oligodendrocytic, and astrocytic phenotypes simultaneously under underwater acoustic driving [**178**]. Compared to traditional biochemically mediated differentiation, the 3D neuron–glial interface induced by mechanical–electrical stimulation results in enhanced interactions between cellular components, leading to superior neural connectivity and function.

Electrospun scaffolds have great advantages over other scaffolds. These scaffolds generate a fiber network similar to the morphology of an extracellular matrix in vivo to support cell growth, proliferation and differentiation. PVDF and its derivative PVDF–TRFE NFs show relatively high piezoelectric constant values when synthesized by electrospinning [**179,180**]. Ins et al. [**181**] found that PVDF NF scaffolds can enhance neuronal cell alignment and neural synapses, which is due to the piezoelectric effect induced by cell contractility, which in turn affects cell behaviour.

In addition to its use in neural tissue engineering, piezoelectric polymer materials also show promising applications in wound healing. Polyurethane/PVDF composite NFs were prepared by electrospinning for wound healing experiments in rats. The results showed that electrospinning technology had transformed the α phase of PVDF into β phase, and under the piezoelectric effect caused by mechanical deformation, the activity of fibroblasts in vitro and in vivo was improved [**182**]. In bone tissue engineering, Amaraju et al. [**183**] found that MSCS showed increased chondrogenesis and lower piezoelectric properties when cultured on electrospun PVDF–TRFE scaffolds compared to cells cultured on heat-treated PVDF–TRFE scaffolds. Similarly, PLLA and collagen are considered suitable piezoelectric materials for bone regeneration: their shear piezoelectricity is particularly relevant to the structure of collagen, the major organic component in bone. The use of PLLA as a bone substitute is advantageous, and PLLA implants can promote bone growth through piezoelectric effects [**184**]. PLLA NFs can stimulate cell differentiation and cell migration. A study by Schofer et al. [**185**] demonstrated that PLLA nanofiber scaffolds can promote cell migration and, thus, can achieve high cell density. However, they lack sufficient osteogenic stimulation to allow further differentiation of these cells, a problem that can be overcome by the incorporation of recombinant human bone morphogenetic protein-2 (rhBMP-2) into PLLA NFs. A portion of mouse skull resection was divided into four groups (group I: unfilled as negative control, group II: implanted with bovine cavernosum as positive control, group III: implanted with PLLA nanofiber scaffolds, and group IV: implanted with PLLA/rhBMP-2 nanofiber scaffolds). Morphometric measurements after 12 weeks are shown in **Figure 12**(1). After implantation of PLLA/rhBMP-2 nanofiber scaffolds, approximately 30% of the defect sites were filled with hard calluses after 4 weeks, which was significantly higher than what was observed in all other treatments. During the experiment, callus formation increased to 45% in the PLLA/rhBMP-2 group after 12 weeks, which was significantly higher than that in the negative control and PLLA groups, but there was no significant difference between the PLLA/rhBMP-2 group and the bovine spongiosum group after 12 weeks, although the average relative bone formation difference was about 20%, as shown in **Figure 12**(2).

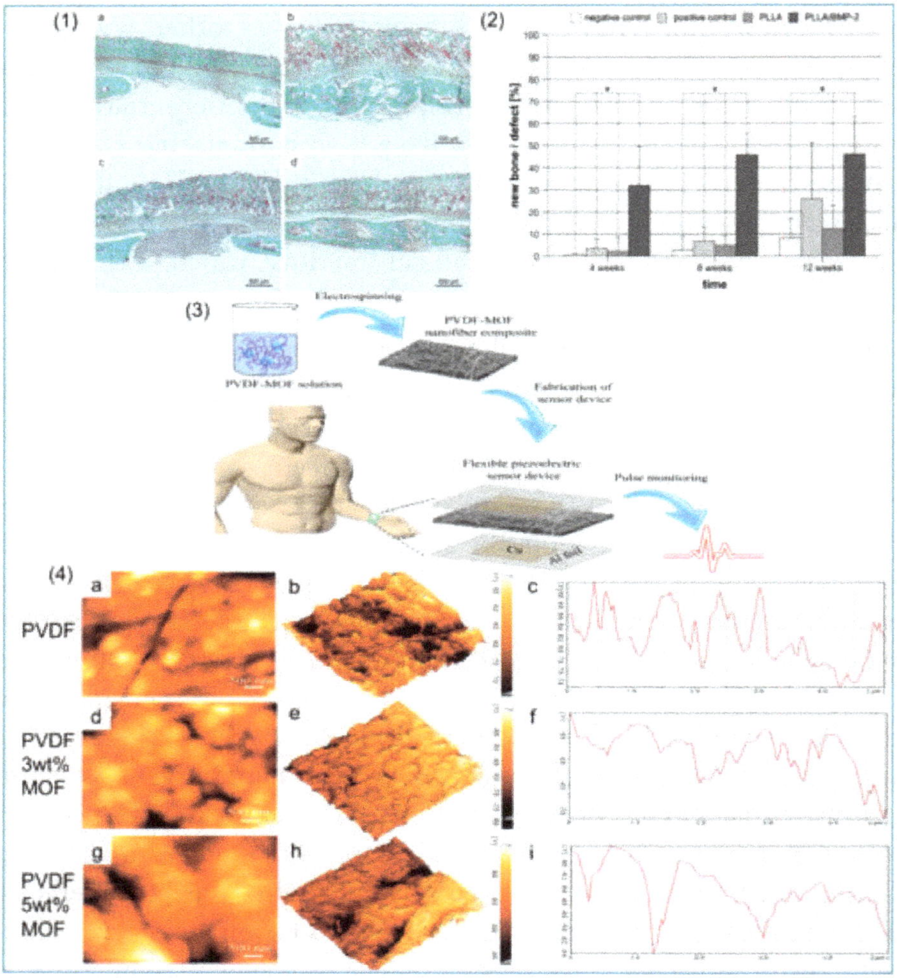

Figure 12. (1) Histological morphology at week 12: (a) negative control. (b) Positive control. (c) PLLA. (d) PLLA/rhBMP-2. Reprinted with permission from Ref. [185], copyright 2011, Public Library of Science. (2) The relationship between new bone formation and the whole defect area, * p < 0.05. Reprinted with permission from Ref. [185], copyright 2011, Public Library of Science. (3) Schematic diagram of arterial pulse-sensing device. Reprinted with permission from Ref. [114], copyright 2020, American Chemical Society. (4) Surface characteristics and piezoelectric response of PVDF NFs. (a,d,g) and (b,e,h) atomic force microscopy. (c,f,i) Voltage curves showing the piezoelectric response of the NFs. Reprinted with permission from Ref. [114], copyright 2020, American Chemical Society.

Electrospinning techniques that combine mechanical stretching with spontaneous electrical polarization have generated unprecedented research enthusiasm in the preparation of fibrous polymers with piezoelectric properties. However, due to some disadvantages of piezoelectric polymers as tissue scaffolds and possible rejection in transplantation, systematic models for clinical application are still lacking.

5.1.2. Biosensor and Drug Release Carrier

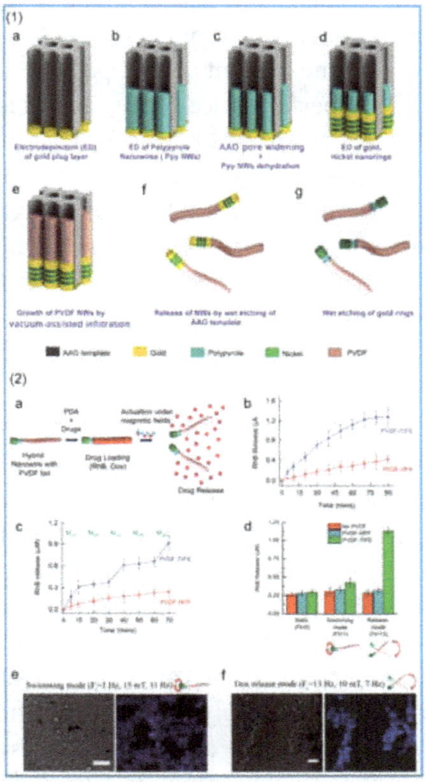

Figure 13. (1) Fabrication process of nano-electric eel, (a) Au sputtering and electrodeposition of Au plug layer in an AAO membrane, (b) electrodeposition of Ppy NWs, followed by (c) their dehydration and the pore widening of AAO template to allow for (d) sequential electrodeposition of Ni and Au nanoring segments. (e) Vacuum infiltration to form PVDF NWs, and RIE etching. Wet etching of (f) AAO template and (g) etching of gold nanoring segments to obtain released hybrid nano-electric eel. Reprinted with permission from Ref. [190], copyright 2019, John Wiley & Sons, Ltd. (2) Controlled drug release of mixed nano-electric eel. (a) Schematic representation of the magnetic drug release from mixed nano-electric eel. (b) Continuous release of RhB at 10 mT and 7 Hz. (c) The amount of RhB released with and without magnetic field. (d) Release plots of RhB in NWs without PVDF, with PVDF–HFP, and with PVDF–TRFE, bright-field and fluorescence images of cancer cells cultured with Adriamycin-coated PVDF–TRFE NWs in (e) swimming mode and (f) drug release mode. Reprinted with permission from Ref. [190], copyright 2019, John Wiley & Sons, Ltd.

The polymer composite NFs prepared by the electrospinning method have micron-sized piezoelectric fiber meshes and have a good piezoelectric response. Due to the tensile and self-polarization of the fibers, the NFs produced by this process have good porosity and enhanced ferroelectric and piezoelectric properties. This has promising applications in self-powered implantable devices and electric clothing, especially small biosensors, which have very sensitive mechanical signal detection capabilities [**186**]. There is a lot of research on small self-powered devices for medicine. Moghadam et al. [**114**] developed a novel arterial pulse-sensing device using a microporous zirconium-based MOF embedded in PVDF NFs, as shown in **Figure 12**(3). By varying the content of MOF particles, the β crystal phase of PVDF is changed, and the piezoelectric performance of the sensor is controlled. In this device, the β-phase content of PVDF reaches up to 75%, the output voltage can reach 568 ± 76 mV, and the piezoelectric conversion sensitivity can reach 118 mV/N; its piezoelectric response and surface properties are shown in **Figure 12**(4). PVDF NFs can also be used in wearable flexible sensors that can monitor movement signals in the human body. Mahanty et al. [**187**] used the electrospinning method to prepare composite NFs by mixing MWCNTs and PVDF, and fabricated a skin sensor. The electronic skin sensor was installed on different parts of the human body, such as the wrist and arm muscles, and could generate a voltage of 5 V to monitor the physiological signals of the human body. The device can not only be used as an electronic skin pressure sensor, but can also be used in self-powered portable electronics.

The treatment of chronic diseases often requires a systemic drug release. Through the sustained release of drug concentration in the body for a long time, maintaining a certain level, we can accelerate the local regeneration or achieve the goal of continuous treatment for localized disease. This is a big challenge for conventional drug release systems [**188**]. Recent studies have found that the preparation of drug carriers by electrospinning is an excellent solution [**189**]. Piezoelectric materials can act as a switch for drug release. Due to its unique piezoelectric properties, the potential difference can affect the release of controlled drugs by generating opposite charges on opposite surfaces through the piezoelectric effect. Muhtaq et al. [**190**] designed a nanorobot as a drug delivery vehicle. The vehicle was a soft composite nanorobot that mimicked an electric eel, resembled a knifefish with a slender cylindrical body, and generated electricity as it moves. First, a gold coating is electrodeposited inside the anodized aluminum oxide (AAO) template, as shown in **Figure 13**(1a), to avoid electrolyte leakage during subsequent fabrication steps. Then, polypyrrole (PPy) NWs are electrodeposited, as shown in **Figure 13**(1b). Subsequently, electrodeposited PPy NWs contract radially in AAO pores, and then the template is reamed using a diluted NaOH solution. These two steps introduce a void between the PPy NWs and AAO pore walls, as shown in **Figure 13**(1c), and electrodeposit Ni/Au nanofragments around the PPy NWs, as shown in **Figure 13**(1d). Next, a solution containing the PVDF-based copolymer is filled into the template pores by vacuum permeation, and the template is subsequently dried and annealed to form NWs, shown in **Figure 13**(1e). Finally, the AAO template is wet etched to release hybrid nanoparticles, as shown in **Figure 13**(1f), and then the gold nanoring segment is etched to obtain the monomer hybrid nano-electric eel, as shown in **Figure 13**(1g). Incorporation of doxorubicin and rhodamine B (RhB) into the nanorobot's tail portion by adsorption allows on-demand release from the tail surface under optimally tuned magnetic fields (10 mT and 7 Hz). Its drug release and related tests are shown in **Figure 13**(2e), we can see multiple nuclei (blue) without the presence of the drug Adriamycin, while in f, we can clearly see nuclei (blue) and released Adriamycin (green) in multiple locations.

5.2. Piezo-Photocatalysis

Due to the existence of dipole moments in piezoelectric materials, mechanical force causes the two opposite surfaces of the material to produce opposite charges to form a built-in electric field and generate a potential difference. This potential change changes the electronic energy level and drives the electrochemical reaction. For example, in photocatalytic water treatment, if the piezoelectric potential generated by the piezoelectric material is greater than the potential required for the REDOX reaction, the conduction band (CB) of the piezoelectric material is reduced below the highest-occupied molecular orbital (HOMO) of the solution in contact with it, and the oxidation reaction occurs when electrons are transferred from the HOMO to the CB. However, the valence band (VB) moves above the lowest-unoccupied orbital (LUMO), and the reduction reaction occurs when electrons are transferred from the VB to the LUMO, as shown in **Figure 14 [191]**. The preparation of piezoelectric photocatalyst nanofiber membranes by the electrospinning method not only improves catalytic activity but also solves the shortcomings of the catalyst that are not easy to recover.

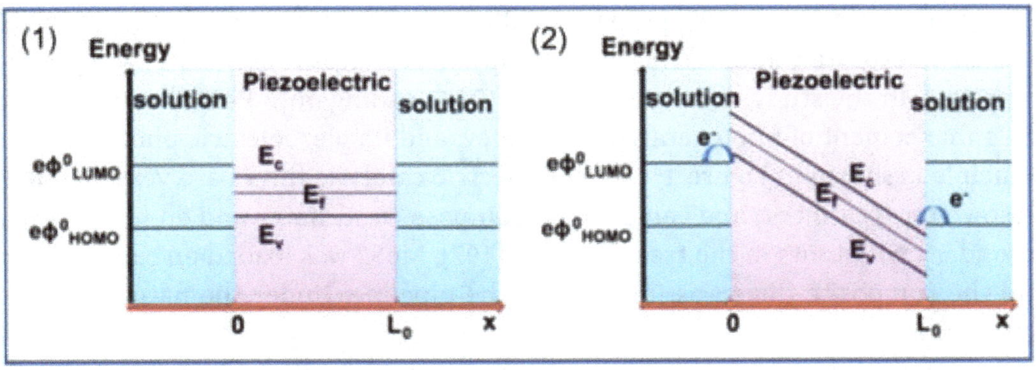

Figure 14. Band-structure changes before (1) and after (2) strain, and the energy shift on the occupied and unoccupied states. Reprinted with permission from Ref. [191], copyright 2015, Elsevier.

Combining piezoelectric materials with photocatalysts with piezoelectric properties can result in composites that outperform both. Ma et al. [192] used electrospinning technology to prepare composite NFs of molybdenum disulfide (MoS2) nanosheets and PVDF, and their piezoelectric catalytic activity was confirmed by degradation of the antibiotic oxytetracycline. The electrostatic interaction between MoS2 nanosheets and –CH2 groups of PVDF promoted the transition from α phase to β phase. A schematic diagram of β-phase formation in MoS2/PVDF is shown in Figure 15(2), that is, MoS2 nanosheets induce β nucleation of PVDF and increase the piezoelectric potential, and Figure 15(3) illustrates the basic principle of their piezoelectric photocatalysis. In addition, MoS2 nanosheets are exposed to the surface of the NFs, which act as additional active sites to enhance the piezoelectric catalytic activity. In the study of enhancing photocatalytic activity, the use of piezoelectric polymer nanofibers as composite reinforcers is an effective strategy. Because most of the light catalysts produce electrons and holes before migrating to the catalyst surface recombination, the carrier recombination rate is higher, its photocatalytic performance is affected, and the built-in electric field generated by the piezoelectric materials can effectively prevent the electrons and holes in the compound to improve the catalytic performance. Titanium dioxide (TiO2) is a simple and stable photocatalyst. The catalytic performance of the composite fiber membrane is enhanced by loading TiO2 onto PVDF to prepare bilayer hollow NFs [193]. In another study, TiO2 was loaded on the surface of PbTiO3, a single-crystal fiber was prepared by the secondary hydrothermal method, and a PN junction was formed at the TiO2/PbTiO3 interface. Due to the piezoelectric properties of PbTiO3, under the action of ultrasonic waves, a built-in electric field was generated in the crystal, which promoted the carrier separation and improved the catalytic efficiency of TiO2. The catalytic efficiency of the composite fiber for methylene orange was increased by 60% compared with that of TiO2 [194]. Urairaj et al. [195] prepared TiO2-loaded PVDF NFs and, similarly, found that the photocatalytic efficiency for degradation of organic pollutants was greatly improved. In the study of Ding et al. [196], TiO2 loading onto PAN also resulted in the enhancement of photocatalytic efficiency, and its piezoelectric photocatalytic principle is shown in Figure 15(4). Piezoelectric catalytic filters are very useful in wastewater treatment, and can easily decompose wastewater and do not produce secondary pollutants in the treated water [197]. MoS2 was loaded on carbon fiber, and the composite fiber was installed in the pipeline. Under the natural flow of wastewater, the fiber was bent by mechanical force and generated piezoelectric polarization, thus enhancing the catalytic performance at the interface between MoS2 and carbon fiber, producing hydroxyl radicals (·OH) and an electrochemical reaction to decompose organic matter in wastewater. An amount of 1 L of degraded 10 ppm methyl blue was completely decomposed within 40 min. Ingh et al. [198] calcinated the prepared ZnO/PAN NFs to obtain mesoporous ZnO NFs. When the naphthalene and anthracene dyes were degraded, the contact area between ZnO and them was larger and the reaction rate was faster, which effectively improved the photocatalytic efficiency.

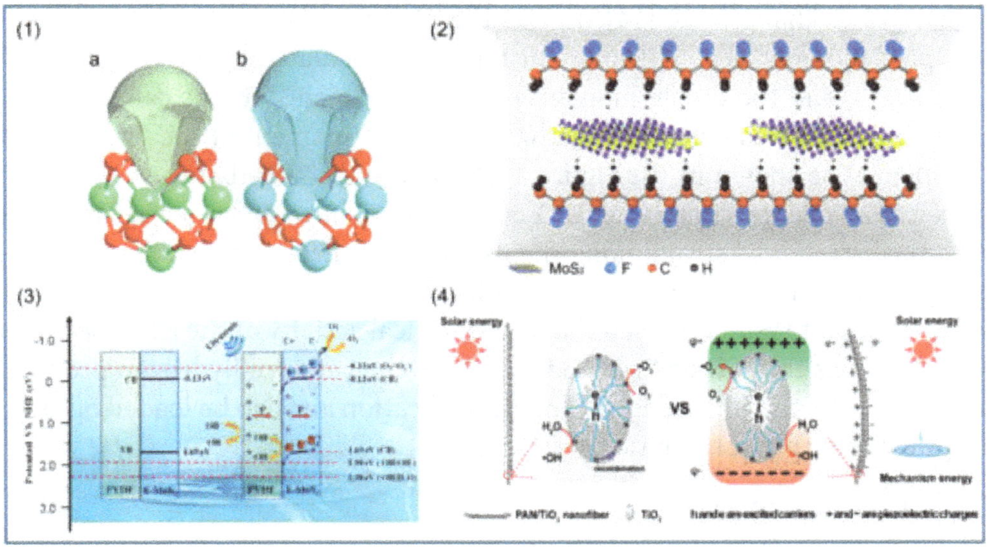

Figure 15. ELF basin analysis of the optimized geometry of (1) (a) Zr-oxo and (b) HF-oxo clusters. Reprinted with permission from Ref. [14], copyright 2021, John Wiley & Sons, Ltd. (2) Schematic diagram of β-phase formation in MoS2/PVDF. Reprinted with permission from Ref. [192], copyright 2021, Elsevier. (3) the mechanism of interfacial charge transfer process under the influence of band structure and band bending. Reprinted with permission from Ref. [192], copyright 2021, Elsevier. (4) Piezoelectric photocatalysis principle of TiO2/PVDF. Reprinted with permission from Ref. [196], copyright 2022, Elsevier.

It is promising to enhance the piezoelectric properties of polymer NFs by adjusting the electrospinning process parameters to control the nanofiber morphology and electroactive phase content, which is very promising in piezoelectric catalytic applications. Many catalysts have good catalytic performance [199]. In particular, piezoelectric polymer NFs have good biocompatibility, non-toxicity, and flexibility, which makes them outperform many ceramic materials for certain environmental remediation. NaNbO3 is a typical single-component catalyst. Under the irradiation of ultraviolet light, NaNbO3 nanorods produce electrons and holes and then produce hydroxyl radicals and superoxide radicals to degrade organic matter. When ultrasonic vibration gives NaNbO3 mechanical strain, it produces polarization due to piezoelectric effect. It is important to reduce the recombination of photogenerated charge carriers and improve the degradation efficiency [200]. In order to combine the functions of each component more effectively, researchers also constructed a core–shell-structured piezoelectric photocatalyst to prepare a PZT/TiO2 composite structure, and synthesized a core-shell-configuration piezoelectric catalyst nanoparticle by coating TiO2 with PZT microspheres. PZT can generate a built-in electric field to effectively prevent the electron–hole pair recombination generated by TiO2, thus improving the catalytic capacity [201]. Zhang et al. [14] studied the photocatalytic activity of two MOFs. The dipole moment of Zr/HF–oxo clusters was calculated by density functional theory (DFT), and the dipole moment of the HF–oxo cluster was calculated to be 2.940 Debye. This was much higher than the dipole moment of the Zr–oxo cluster (1.739 Debye) as shown in the Figure 15(1), so the distance between the positive and negative charge centers of UIO-66-NH2 (Hf) shows a larger change under mechanical stretch or strain along the asymmetric direction, leading to stronger spontaneous polarization and a larger piezoelectric response. The piezoelectric performance test results showed that the electrical activity of the HF-based UO-66-NH2 is 2.2 times that of the Zr-based, and MOF can catalyze hydrogen production under both light and ultrasonic mechanical vibration and can be loaded on the NFs to improve their photocatalytic performance.

5.3. Energy Conversion

In recent years, with the gradual depletion of primary energy such as traditional fossil fuels and the improvement of environmental awareness, the demand for clean energy and renewable energy has gradually increased. Compared with solar energy, tidal energy, wind energy, geothermal energy, and other energy sources that require large-scale energy conversion devices, the micro-electromechanical conversion system using piezoelectric materials can be more convenient for some applications and can also collect mechanical energy and convert it into electric energy utilizing small mechanical forces such as human kinematics and micro-vibrations.

Recently, the electrostatic spinning technique used to prepare NFs used in generators has become a hot spot. Because the electrostatic spinning technology can improve the piezoelectric properties of materials, Zhu et al. [**202**] use the electrostatic spinning technique to prepare PLLA NFs; this method of preparation of the nanometer carbon fiber's main chain of electric dipole polarization components can be along the alignment direction. Moreover, the piezoelectric properties of NFs can be affected by controlling the electrospinning parameters, and the piezoelectric properties of polymers can also be improved by adding nanoparticles and surfactants. Sun et al. [**146**] loaded ZnO nanorods (ZnO NRs) onto PAN to form ZnO/PAN composite NFs, and fabricated PNG to drive the electric heating plate, as shown in **Figure 16**(1). Moreover, the NFs' piezoelectric sensitivity and energy harvesting efficiency are about 2.7 times higher than those of pure PAN, and their output power is twice that of ZnO/PVDF composite NFs. This excellent performance is attributed to the improved zigzag conformation in PAN NFs by the addition of ZnO NRs. In the medical field, this piezoelectric composite nanofiber membrane can be used as a power supply device for small medical devices [**129**]. ZnO/rGO/PVDF–TRFE composite NFs can be used in battery-free cardiac pacemakers. The piezoelectric properties were also increased by the addition of nanoparticles ZnO and rGO. The hydrogen atoms in ZnO, rGO, and PVDF–TRFE interact with fluorine atoms so that the fluorine atoms are arranged in one direction in the main chain to form β phase (TTTT). The pacemaker has demonstrated its excellent performance in animal experiments, and it can collect more energy per heartbeat than human heart pacing. Sultana et al. [**203**] studied the addition of methyl lead ammonium bromide (MAPbBr) to PVDF in electrospinning to make PNG, and the relevant tests are shown in **Figure 16**(2). The results show that the content of □ phase in PVDF is increased to 91%, and the PNG has high piezoelectric sensitivity and energy harvesting efficiency.

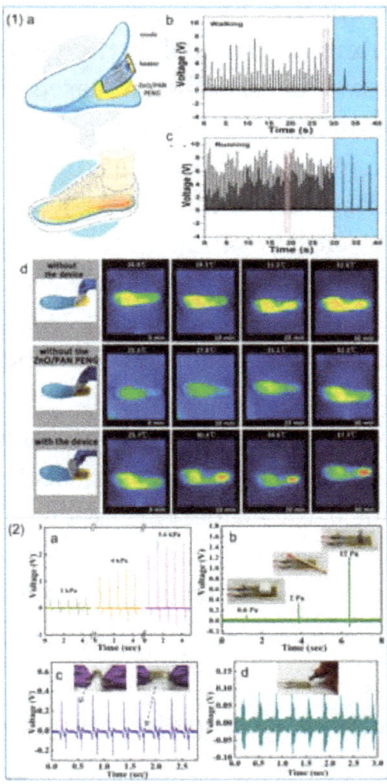

Figure 16. (1) The electric energy made by PNG is used to drive the electric heating plate. (a) Schematic diagram of self-heating insole. Digital photo and infrared image of the output voltage generated by PNG during (b) walking and (c) running, the blue inset is a partial enlargement of the red box, (d) insole. Reprinted with permission from Ref. [146], copyright 2020, American Chemical Society. (2) (a) Plot of voltage response as a function of time for different applied pressures under finger impingement. (b) Output voltage response of PNG with polystyrene, matchstick, and weight placed at the top. Piezo-response plots for loading and unloading cycles under (c) bending and (d) blower pressure. Reprinted with permission from Ref. [203], copyright 2018, Elsevier.

Human kinematics is the main source of piezoelectric energy collected by NFs, which show great potential in self-powered wearable devices, converting mechanical energy into electrical signals through walking, joint bending, sound vibration, and other means. It is promising to prepare these piezoelectric materials into flexible fabrics or wearable items for use in human life. Particularly, some raw natural materials [**204,205,206,207**] and nanostructures [**208,209,210**] may play roles in the developments of these piezoelectric materials.

6. CONCLUSIONS

Electrospinning technology is a simple and efficient technology for preparing nanofiber films. Various properties of nanofiber films can be well-controlled by adjusting spinning parameters. In particular, the content of piezoelectric phase in piezoelectric polymers can be greatly increased by stretching and electric polarization of the jet, so as to improve their piezoelectric properties. Piezoelectric NFs prepared by electrospinning have better flexibility and biocompatibility than traditional ceramic piezoelectric materials. By adding different additives to piezoelectric materials with electrospinning technology to produce piezoelectric properties of nanometer fiber membranes, we observe increases in piezoelectric performance improvement factors, including piezoelectric phase content. Additives themselves have piezoelectric properties and by changing the solution properties in electrospinning, NFs' piezoelectric properties also change. Many piezoelectric polymer NFs, especially those containing fluorine polymers with good biocompatibility, also have very good piezoelectric constants, but they are biodegradable and their renewable property is lower, so it is necessary to research some natural polymers. By electrospinning, parameters optimized the synthesis of natural polymers and enhanced piezoelectric properties.

- The high-voltage electric field applied during electrospinning has a significant effect on the in situ polarization and induced piezoelectric phase formation, so this method has a broad prospect in the preparation of piezoelectric materials;
- By studying ways to change the piezoelectric properties of piezoelectric polymers, it is expected that the piezoelectric properties of piezoelectric polymers can be continuously improved to meet the needs of various applications;
- Piezoelectric NFs can convert mechanical energy from various small vibrations into electric energy. Because of their good flexibility, piezoelectric NFs have been widely studied in wearable electronic devices;

- Due to their good biocompatibility, piezoelectric polymers have broad application prospects in the biomedical field, such as in power supply for tissue engineering scaffolds, cardiac pacemakers, and other small internal medical devices, which is promising for clinical treatment after a lot of research;
- Piezoelectric NF membranes prepared by electrospinning also have important applications in the field of piezoelectric photocatalysis. The piezoelectric effect can generate a built-in electric field inside the material to inhibit the composition of photogenerated charge carriers, improve the carrier life, and enhance the catalytic efficiency. Some single-component piezoelectric photocatalysts are known to have excellent catalytic performance. It is expected that piezoelectric photocatalysts with better catalytic performance can be obtained by combining them with electrospinning;
- Electrospinning is moving forward rapidly from single-fluid processes to coaxial, triaxial, side-by-side, and tri-layer side-by-side processes for generating all kinds of complicated nanostructures [**211,212,213,214,215**]. With PVDF as a filament-forming polymeric matrix, more and more excellent piezoelectric nanomaterials are predicted to be witnessed in future.

7. CONCISE SUMMARY

Since the third scientific and technological revolution, electronic information technology has developed rapidly, and piezoelectric materials that can convert mechanical energy into electrical energy have become a research hotspot. Among them, piezoelectric polymers are widely used in various fields such as water treatment, biomedicine, and flexible sensors due to their good flexibility and weak toxicity. However, compared with ceramic piezoelectric materials, the piezoelectric properties of polymers are poor, so it is very important to improve the piezoelectric properties of polymers. Electrospinning technology can improve the piezoelectric properties of piezoelectric polymers by adjusting electrospinning parameters to control the piezoelectrically active phase transition of polymers. In addition, the prepared nanofibrous membrane is also a good substrate for supporting piezoelectric functional particles, which can also effectively improve the piezoelectric properties of polymers by doping particles. This paper reviews the piezoelectric properties of various electrospun piezoelectric polymer membranes, especially polyvinylidene fluoride (PVDF)-based electrospun nanofibrous membranes (NFs). Additionally, this chapter introduces the various methods for increasing piezoelectric properties from the perspective of structure and species. Finally, the applications of NFs in the fields of biology, energy, and photocatalysis are discussed, and the future research directions and development are prospected.

REFERENCES

1. Mason, W.P. Piezoelectricity, its history and applications. *J. Acoust. Soc. Am.* **1981**, *70*, 1561–1566. **[Google Scholar] [CrossRef]**

2. Curie, P.; Curie, J. Développement par compression de l'électricité polaire dans les cristaux hémièdres à faces inclinées. *Bull. Minéral.* **1880**, *3*, 90–93. **[Google Scholar] [CrossRef]**

3. Sirohi, J.; Chopra, I. Fundamental understanding of piezoelectric strain sensors. *J. Intell. Mater. Syst. Struct.* **2000**, *11*, 246–257. **[Google Scholar] [CrossRef]**

4. Bowen, C.R.; Kim, H.A.; Weaver, P.M.; Dunn, S. Piezoelectric and ferroelectric materials and structures for energy harvesting applications. *Energy Environ. Sci.* **2014**, *7*, 25–44. **[Google Scholar] [CrossRef][Green Version]**

5. Chang, J.; Dommer, M.; Chang, C.; Lin, L. Piezoelectric nanofibers for energy scavenging applications. *Nano Energy* **2012**, *1*, 356–371. **[Google Scholar] [CrossRef]**

6. Zheng, T.; Zhang, Y.; Ke, Q.; Wu, H.; Heng, L.W.; Xiao, D.; Zhu, J.; Pennycook, S.J.; Yao, K.; Wu, J. High-performance potassium sodium niobate piezoceramics for ultrasonic transducer. *Nano Energy* **2020**, *70*, 104559. **[Google Scholar] [CrossRef]**

7. Liu, P.; Hu, Y.; Geng, B.; Xu, D. Investigation on novel embedded piezoelectric ultrasonic transducers for corrosion monitoring of reinforced concrete. *Smart Mater. Struct.* **2019**, *28*, 115041. **[Google Scholar] [CrossRef]**

8. Hu, Y.; Wang, Z.L. Recent progress in piezoelectric nanogenerators as a sustainable power source in self-powered systems and active sensors. *Nano Energy* **2015**, *14*, 3–14. **[Google Scholar] [CrossRef][Green Version]**

9. Surmenev, R.A.; Orlova, T.; Chernozem, R.V.; Ivanova, A.A.; Bartasyte, A.; Mathur, S.; Surmeneva, M.A. Hybrid lead-free polymer-based nanocomposites with improved piezoelectric response for biomedical energy-harvesting applications: A review. *Nano Energy* **2019**, *62*, 475–506. **[Google Scholar] [CrossRef][Green Version]**

10. Motamedi, A.S.; Mirzadeh, H.; Hajiesmaeilbaigi, F.; Bagheri-Khoulenjani, S.; Shokrgozar, M.A. Piezoelectric electrospun nanocomposite comprising Au NPs/PVDF for nerve tissue engineering. *J. Biomed. Mater. Res. Part A* **2017**, *105*, 1984–1993. **[Google Scholar] [CrossRef]**

11. Hwang, G.T.; Kim, Y.; Lee, J.H.; Oh, S.; Jeong, C.K.; Park, D.Y.; Ryu, J.; Kwon, H.; Lee, S.G.; Joung, B.; et al. Self-powered deep brain stimulation via a flexible PIMNT energy harvester. *Energy Environ. Sci.* **2015**, *8*, 2677–2684. **[Google Scholar] [CrossRef]**

12. Hu, D.; Yao, M.; Fan, Y.; Ma, C.; Fan, M.; Liu, M. Strategies to achieve high performance piezoelectric nanogenerators. *Nano Energy* **2019**, *55*, 288–304. **[Google Scholar] [CrossRef]**

13. Wang, X.; Song, J.; Liu, J.; Wang Zhong, L. Direct-current nanogenerator driven by ultrasonic waves. *Science* **2007**, *316*, 102–105. **[Google Scholar] [CrossRef] [Green Version]**

14. Zhang, C.; Lei, D.; Xie, C.; Hang, X.; He, C.; Jiang, H.-L. Piezo-photocatalysis over metal–organic frameworks: Promoting photocatalytic activity by piezoelectric effect. *Adv. Mater.* **2021**, *33*, 2106308. **[Google Scholar] [CrossRef]**

15. Yu, H.X.; Liu, Y.F.; Deng, C.C.; Xia, M.T.; Zhang, X.K.; Zhang, L.Y.; Shui, M.; Shu, J. Lithium storage behaviors of PbNb2O6 in rechargeable batteries. *Ceram Int.* **2021**, *47*, 26732–26737. **[Google Scholar] [CrossRef]**

16. Jaffe, B.; Roth, R.S.; Marzullo, S. Piezoelectric properties of lead zirconate-lead titanate solid-solution ceramics. *J. Appl. Phys.* **1954**, *25*, 809–810. **[Google Scholar] [CrossRef]**

17. Shrout, T.R.; Zhang, S.J. Lead-free piezoelectric ceramics: Alternatives for PZT? *J. Electroceram.* **2007**, *19*, 113–126. **[Google Scholar] [CrossRef]**

18. Chang, Y.; Yin, B.; Qiu, Y.; Zhang, H.; Lei, J.; Zhao, Y.; Luo, Y.; Hu, L. ZnO nanorods array/BaTiO3 coating layer composite structure nanogenerator. *J. Mater. Sci. Mater. Electron.* **2016**, *27*, 3773–3777. **[Google Scholar] [CrossRef]**

19. Yokosuka, M. A new transparent ferroelectric ceramic, lanthanum modified lead-barium metaniobate (PBLN). *Jpn. J. Appl. Phys.* **1977**, *16*, 379–380. **[Google Scholar] [CrossRef]**

20. Sysoev, A.M.; Parshikov, S.A.; Zaitsev, A.I.; Zamkov, A.V.; Aleksandrov, K.S. Piezoelectric properties of glassceramic based on lithium tetraborate. *Ferroelectrics* **1996**, *186*, 277–280. **[Google Scholar] [CrossRef]**

21. Haertling, G.H. Ferroelectric ceramics: History and technology. *J. Am. Ceram. Soc.* **1999**, *82*, 797–818. **[Google Scholar] [CrossRef]**

22. Ponnamma, D.; Cabibihan, J.-J.; Rajan, M.; Pethaiah, S.S.; Deshmukh, K.; Gogoi, J.P.; Pasha, S.K.K.; Ahamed, M.B.; Krishnegowda, J.; Chandrashekar, B.N.; et al. Synthesis, optimization and applications of ZnO/polymer nanocomposites. *Mater. Sci. Eng. C* **2019**, *98*, 1210–1240. **[Google Scholar] [CrossRef]**

23. Xin, Y.; Sun, H.; Tian, H.; Guo, C.; Li, X.; Wang, S.; Wang, C. The use of polyvinylidene fluoride (PVDF) films as sensors for vibration measurement: A brief review. *Ferroelectrics* **2016**, *502*, 28–42. **[Google Scholar] [CrossRef]**

24. Ribeiro, C.; Costa, C.M.; Correia, D.M.; Nunes-Pereira, J.; Oliveira, J.; Martins, P.; Gonçalves, R.; Cardoso, V.F.; Lanceros-Méndez, S. Electroactive poly(vinylidene

fluoride)-based structures for advanced applications. *Nat. Protoc.* **2018,** *13,* 681–704. **[Google Scholar] [CrossRef]**

25. Lovinger, A.J. Poly(vinylidene fluoride). In *Devel Cryst Polum-1*; Springer: Berlin/Heidelberg, Germany, 1982. **[Google Scholar]**
26. Sajkiewicz, P.; Wasiak, A.; Gocłowski, Z. Phase transitions during stretching of poly(vinylidene fluoride). *Eur. Polym. J.* **1999,** *35,* 423–429. **[Google Scholar] [CrossRef]**
27. Broadhurst, M.G.; Davis, G.T.; McKinney, J.E.; Collins, R.E. Piezoelectricity and pyroelectricity in polyvinylidene fluoride—A model. *J. Appl. Phys.* **1978,** *49,* 4992–4997. **[Google Scholar] [CrossRef]**
28. Gregorio, R.; Ueno, E.M. Effect of crystalline phase, orientation and temperature on the dielectric properties of poly (vinylidene fluoride) (PVDF). *J. Mater. Sci.* **1999,** *34,* 4489–4500. **[Google Scholar] [CrossRef]**
29. Lanceros-Méndez, S.; Mano, J.F.; Costa, A.M.; Schmidt, V.H. FTIR and DSC studies of mechanically deformed β-PVDF films. *J. Macromol. Sci. B* **2001,** *40,* 517–527. **[Google Scholar] [CrossRef][Green Version]**
30. Wang, J.; Li, H.; Liu, J.; Duan, Y.; Jiang, S.; Yan, S. On the α → β transition of carbon-coated highly oriented PVDF ultrathin film induced by melt recrystallization. *J. Am. Chem. Soc.* **2003,** *125,* 1496–1497. **[Google Scholar] [CrossRef] [PubMed]**
31. Lando, J.B.; Olf, H.G.; Peterlin, A. Nuclear magnetic resonance and X-ray determination of the structure of poly(vinylidene fluoride). *J. Polym. Sci. A Polym. Chem.* **1966,** *4,* 941–951. **[Google Scholar] [CrossRef]**
32. Davis, G.T.; McKinney, J.E.; Broadhurst, M.G.; Roth, S.C. Electric-field-induced phase changes in poly(vinylidene fluoride). *J. Appl. Phys.* **1978,** *49,* 4998–5002. **[Google Scholar] [CrossRef]**
33. Li, D.; Xia, Y. Electrospinning of nanofibers: Reinventing the wheel? *Adv. Mater.* **2004,** *16,* 1151–1170. **[Google Scholar] [CrossRef]**
34. Lei, T.P.; Yu, L.K.; Zheng, G.F.; Wang, L.Y.; Wu, D.Z.; Sun, D.H. Electrospinning-induced preferred dipole orientation in PVDF fibers. *J. Mater. Sci.* **2015,** *50,* 4342–4347. **[Google Scholar] [CrossRef]**
35. Ruan, L.; Yao, X.; Chang, Y.; Zhou, L.; Qin, G.; Zhang, X. Properties and applications of the β phase poly(vinylidene fluoride). *Polymers* **2018,** *10,* 228. **[Google Scholar] [CrossRef][Green Version]**
36. Xin, Y.; Qi, X.; Tian, H.; Guo, C.; Li, X.; Lin, J.; Wang, C. Full-fiber piezoelectric sensor by straight PVDF/nanoclay nanofibers. *Mater. Lett.* **2016,** *164,* 136–139. **[Google Scholar] [CrossRef]**

37. Ahn, Y.; Lim, J.Y.; Hong, S.M.; Lee, J.; Ha, J.; Choi, H.J.; Seo, Y. Enhanced piezoelectric properties of electrospun poly(vinylidene fluoride)/multiwalled carbon nanotube composites due to high β-phase formation in poly(vinylidene fluoride). *J. Phys. Chem. C* **2013**, *117*, 11791–11799. **[Google Scholar] [CrossRef]**

38. Fu, R.F.; Chen, S.; Lin, Y.; Zhang, S.H.; Jiang, J.; Li, Q.B.; Gu, Y.C. Improved piezoelectric properties of electrospun poly(vinylidene fluoride) fibers blended with cellulose nanocrystals. *Mater. Lett.* **2017**, *187*, 86–88. **[Google Scholar] [CrossRef]**

39. Afifi, A.M.; Nakano, S.; Yamane, H.; Kimura, Y. Electrospinning of continuous aligning yarns with a 'funnel' target. *Macromol. Mater. Eng.* **2010**, *295*, 660–665. **[Google Scholar] [CrossRef]**

40. Sill, T.J.; von Recum, H.A. Electrospinning: Applications in drug delivery and tissue engineering. *Biomaterials* **2008**, *29*, 1989–2006. **[Google Scholar] [CrossRef]**

41. Jiang, S.; Chen, Y.; Duan, G.; Mei, C.; Greiner, A.; Agarwal, S. Electrospun nanofiber reinforced composites: A review. *Polym. Chem.* **2018**, *9*, 2685–2720. **[Google Scholar] [CrossRef]**

42. Sahini, M.G.; Grande, T.; Fraygola, B.; Biancoli, A.; Damjanovic, D.; Setter, N. Solid solutions of lead metaniobate—Stabilization of the ferroelectric polymorph and the effect on the lattice parameters, dielectric, ferroelectric, and piezoelectric properties. *J. Am. Ceram. Soc.* **2014**, *97*, 220–227. **[Google Scholar] [CrossRef]**

43. Morita, T. Piezoelectric materials synthesized by the hydrothermal method and their applications. *Materials* **2010**, *3*, 5236–5245. **[Google Scholar] [CrossRef] [PubMed][Green Version]**

44. Nagata, K.; Kiyota, T. Piezoelectric properties of low coercive-field BaTiO3/ ceramics and its application. *Jpn. J. Appl. Phys.* **1989**, *28*, 98. **[Google Scholar] [CrossRef]**

45. Acharya, V.V.N.; Bhanumathi, A.; Uchino, K.; Prasad, N.S. Investigation on modified lead barium niobate to optimize the piezoelectric properties. *Ferroelectr. Lett.* **2000**, *27*, 7–10. **[Google Scholar] [CrossRef]**

46. Jung, H.-R.; Jin, B.-M.; Cha, J.-W.; Kim, J.-N. Piezoelectric and elastic properties of Li_2B4O_7 single crystal. *Mater. Lett.* **1997**, *30*, 41–45. **[Google Scholar] [CrossRef]**

47. Nguyen, M.D.; Dekkers, M.; Vu, H.N.; Rijnders, G. Film-thickness and composition dependence of epitaxial thin-film PZT-based mass-sensors. *Sens. Actuators A Phys.* **2013**, *199*, 98–105. **[Google Scholar] [CrossRef]**

48. Kobiakov, I.B. Elastic, piezoelectric and dielectric properties of ZnO and CdS single crystals in a wide range of temperatures. *Solid State Commun.* **1980**, *35*, 305–310. **[Google Scholar] [CrossRef]**

49. Ando, M.; Kawamura, H.; Kitada, H.; Sekimoto, Y.; Inoue, T.; Tajitsu, Y. Pressure-sensitive touch panel based on piezoelectric poly(l-lactic acid) film. *Jpn. J. Appl. Phys.* **2013**, *52*, 09KD17. **[Google Scholar] [CrossRef]**

50. Shuai, C.J.; Liu, G.F.; Yang, Y.W.; Qi, F.W.; Peng, S.P.; Yang, W.J.; He, C.X.; Wang, G.Y.; Qian, G.W. A strawberry-like Ag-decorated barium titanate enhances piezoelectric and antibacterial activities of polymer scaffold. *Nano Energy.* **2020**, *70*, 104825. **[Google Scholar] [CrossRef]**

51. Varadan, V.K.; Vinoy, K.J.; Gopalakrishnan, S. *Smart Material Systems and MEMS: Design and Development Methodologies*; John Wiley Sons Ltd.: New York, NY, USA, 2006; pp. 43–62. **[Google Scholar] [CrossRef]**

52. Pu, J.; Yan, X.; Jiang, Y.; Chang, C.; Lin, L. Piezoelectric actuation of direct-write electrospun fibers. *Sens. Actuators A Phys.* **2010**, *164*, 131–136. **[Google Scholar] [CrossRef]**

53. Sultana, A.; Ghosh, S.K.; Sencadas, V.; Zheng, T.; Higgins, M.J.; Middya, T.R.; Mandal, D. Human skin interactive self-powered wearable piezoelectric bio-e-skin by electrospun poly-l-lactic acid nanofibers for non-invasive physiological signal monitoring. *J. Mater. Chem. B* **2017**, *5*, 7352–7359. **[Google Scholar] [CrossRef] [PubMed][Green Version]**

54. Jacob, J.; More, N.; Mounika, C.; Gondaliya, P.; Kalia, K.; Kapusetti, G. Smart piezoelectric nanohybrid of poly(3-hydroxybutyrate-co-3-hydroxyvalerate) and barium titanate for stimulated cartilage regeneration. *ACS Appl. Bio. Mater.* **2019**, *2*, 4922–4931. **[Google Scholar] [CrossRef] [PubMed]**

55. Chernozem, R.V.; Surmeneva, M.A.; Shkarina, S.N.; Loza, K.; Epple, M.; Ulbricht, M.; Cecilia, A.; Krause, B.; Baumbach, T.; Abalymov, A.A.; et al. Piezoelectric 3-D fibrous poly(3-hydroxybutyrate)-based scaffolds ultrasound-mineralized with calcium carbonate for bone tissue engineering: Inorganic phase formation, osteoblast cell adhesion, and proliferation. *ACS Appl. Mater. Inter.* **2019**, *11*, 19522–19533. **[Google Scholar] [CrossRef] [PubMed]**

56. Ueda, H.; Carr, S.H. Piezoelectricity in polyacrylonitrile. *Polym. J.* **1984**, *16*, 661–667. **[Google Scholar] [CrossRef][Green Version]**

57. Yu, S.; Milam-Guerrero, J.; Tai, Y.; Yang, S.; Choi, Y.Y.; Nam, J.; Myung, N.V. Maximizing polyacrylonitrile nanofiber piezoelectric properties through the optimization of electrospinning and post-thermal treatment processes. *ACS Appl. Polym. Mater.* **2022**, *4*, 635–644. **[Google Scholar] [CrossRef]**

58. Street, R.M.; Minagawa, M.; Vengrenyuk, A.; Schauer, C.L. Piezoelectric electrospun polyacrylonitrile with various tacticities. *J. Appl. Polym. Sci.* **2019**, *136*, 47530. **[Google Scholar]** **[CrossRef]**

59. Meurisch, K.; Gojdka, B.; Strunskus, T.; Zaporojtchenko, V.; Faupel, F. Vapour phase deposition of highly crystalline self-poled piezoelectric nylon-11. *J. Phys. D Appl. Phys.* **2012**, *45*, 055304. **[Google Scholar]** **[CrossRef]**

60. Takase, Y.; Lee, J.W.; Scheinbeim, J.I.; Newman, B.A. High-temperature characteristics of nylon-11 and nylon-7 piezoelectrics. *Macromolecules* **1991**, *24*, 6644–6652. **[Google Scholar]** **[CrossRef]**

61. Denning, D.; Kilpatrick, J.I.; Fukada, E.; Zhang, N.; Habelitz, S.; Fertala, A.; Gilchrist, M.D.; Zhang, Y.; Tofail, S.A.M.; Rodriguez, B.J. Piezoelectric tensor of collagen fibrils determined at the nanoscale. *ACS Biomater. Sci. Eng.* **2017**, *3*, 929–935. **[Google Scholar]** **[CrossRef]**

62. Fukada, E. History and recent progress in piezoelectric polymers. *IEEE Trans. Ultrason.* **2000**, *47*, 1277–1290. **[Google Scholar]** **[CrossRef]**

63. Minary-Jolandan, M.; Yu, M.-F. Nanoscale characterization of isolated individual type I collagen fibrils: Polarization and piezoelectricity. *Nanotechnology* **2009**, *20*, 085706. **[Google Scholar]** **[CrossRef]** **[PubMed]**

64. Csoka, L.; Hoeger, I.C.; Rojas, O.J.; Peszlen, I.; Pawlak, J.J.; Peralta, P.N. Piezoelectric effect of cellulose nanocrystals thin films. *ACS Macro. Lett.* **2012**, *1*, 867–870. **[Google Scholar]** **[CrossRef]** **[PubMed]**

65. Wu, T.; Song, Y.; Shi, Z.; Liu, D.; Chen, S.; Xiong, C.; Yang, Q. High-performance nanogenerators based on flexible cellulose nanofibril/MoS2 nanosheet composite piezoelectric films for energy harvesting. *Nano Energy* **2021**, *80*, 105541. **[Google Scholar]** **[CrossRef]**

66. Hoque, N.A.; Thakur, P.; Biswas, P.; Saikh, M.M.; Roy, S.; Bagchi, B.; Das, S.; Ray, P.P. Biowaste crab shell-extracted chitin nanofiber-based superior piezoelectric nanogenerator. *J. Mater. Chem. A* **2018**, *6*, 13848–13858. **[Google Scholar]** **[CrossRef]**

67. Praveen, E.; Murugan, S.; Jayakumar, K. Investigations on the existence of piezoelectric property of a bio-polymer—Chitosan and its application in vibration sensors. *RSC Adv.* **2017**, *7*, 35490–35495. **[Google Scholar]** **[CrossRef]** **[Green Version]**

68. Amran, A.; Ahmad, F.B.; Akmal, M.H.M.; Ralib, A.A.M.; Bin Suhaimi, M.I. Biosynthesis of thin film derived from microbial chitosan for piezoelectric application. *Mater. Today Commun.* **2021**, *29*, 102919. **[Google Scholar]** **[CrossRef]**

69. Ji, Y.; Song, W.; Xu, L.; Yu, D.-G.; Annie Bligh, S.W. A review on electrospun poly(amino acid) nanofibers and their applications of hemostasis and wound healing. *Biomolecules* **2022**, *12*, 794. **[Google Scholar] [CrossRef]**

70. Zhao, K.; Kang, S.-X.; Yang, Y.-Y.; Yu, D.-G. Electrospun functional nanofiber membrane for antibiotic removal in water: Review. *Polymers* **2021**, *13*, 226. **[Google Scholar] [CrossRef]**

71. Jiang, W.L.; Zhao, P.; Song, W.L.; Wang, M.L.; Yu, D.G. Electrospun zein/ polyoxyethylene core-sheath ultrathin fibers and their antibacterial food packaging applications. *Biomolecules* **2022**, *12*, 1110. **[Google Scholar] [CrossRef]**

72. Liu, X.K.; Zhang, M.X.; Song, W.L.; Zhang, Y.; Yu, D.G.; Liu, Y.B. Electrospun core (HPMC-acetaminophen)-shell (PVP-sucralose) nanohybrids for rapid drug delivery. *Gels* **2022**, *8*, 357. **[Google Scholar] [CrossRef]**

73. Liu, H.R.; Jiang, W.L.; Yang, Z.L.; Chen, X.R.; Yu, D.G.; Shao, J. Hybrid films prepared from a combination of electrospinning and casting for offering a dual-phase drug release. *Polymers* **2022**, *14*, 2132. **[Google Scholar] [CrossRef] [PubMed]**

74. Xu, L.; Liu, Y.A.; Zhou, W.H.; Yu, D.G. Electrospun medical sutures for wound healing: A review. *Polymers* **2022**, *14*, 1637. **[Google Scholar] [CrossRef]**

75. Liu, Y.B.; Chen, X.H.; Gao, Y.H.; Yu, D.G.; Liu, P. Elaborate design of shell component for manipulating the sustained release behavior from core-shell nanofibres. *J. Nanobio. Technol.* **2022**, *20*, 244. **[Google Scholar] [CrossRef] [PubMed]**

76. Du, Y.T.; Zhang, X.Y.; Liu, P.; Yu, D.G.; Ge, R.L. Electrospun nanofiber-based glucose sensors for glucose detection. *Front. Chem.* **2022**, *10*, 944428. **[Google Scholar] [CrossRef] [PubMed]**

77. Huang, C.; Xu, X.; Fu, J.; Yu, D.-G.; Liu, Y. Recent progress in electrospun polyacrylonitrile nanofiber-based wound dressing. *Polymers* **2022**, *14*, 3266. **[Google Scholar] [CrossRef]**

78. Giannetti, E. Semi-crystalline fluorinated polymers. *Polym. Int.* **2001**, *50*, 10–26. **[Google Scholar] [CrossRef]**

79. Zhou, Y.; Wang, M.; Yan, C.; Liu, H.; Yu, D.-G. Advances in the application of electrospun drug-loaded nanofibers in the treatment of oral ulcers. *Biomolecules* **2022**, *12*, 1254. **[Google Scholar] [CrossRef]**

80. Zhao, K.; Lu, Z.H.; Zhao, P.; Kang, S.X.; Yang, Y.Y.; Yu, D.G. Modified tri-axial electrospun functional core-shell nanofibrous membranes for natural photodegradation of antibiotics. *Chem. Eng. J.* **2021**, *425*, 131455. **[Google Scholar] [CrossRef]**

81. Liu, Y.B.; Chen, X.H.; Gao, Y.H.; Liu, Y.Y.; Yu, D.G.; Liu, P. Electrospun core-sheath nanofibers with variable shell thickness for modifying curcumin release to achieve a better antibacterial performance. *Biomolecules* **2022**, *12*, 1057. **[Google Scholar]** **[CrossRef]** **[PubMed]**

82. Ning, T.B.; Zhou, Y.J.; Xu, H.X.; Guo, S.R.; Wang, K.; Yu, D.G. Orodispersible membranes from a modified coaxial electrospinning for fast dissolution of diclofenac sodium. *Membranes* **2021**, *11*, 802. **[Google Scholar]** **[CrossRef]**

83. Sivan, M.; Madheswaran, D.; Valtera, J.; Kostakova, E.K.; Lukas, D. Alternating current electrospinning: The impacts of various high-voltage signal shapes and frequencies on the spinnability and productivity of polycaprolactone nanofibers. *Mater. Des.* **2022**, *213*, 110308. **[Google Scholar]** **[CrossRef]**

84. Okutan, N.; Terzi, P.; Altay, F. Affecting parameters on electrospinning process and characterization of electrospun gelatin nanofibers. *Food Hydrocolloid* **2014**, *39*, 19–26. **[Google Scholar]** **[CrossRef]**

85. Haider, A.; Haider, S.; Kang, I.K. A comprehensive review summarizing the effect of electrospinning parameters and potential applications of nanofibers in biomedical and biotechnology. *Arab. J. Chem.* **2018**, *11*, 1165–1188. **[Google Scholar]** **[CrossRef]**

86. Zaarour, B.; Zhu, L.; Jin, X. Maneuvering the secondary surface morphology of electrospun poly (vinylidene fluoride) nanofibers by controlling the processing parameters. *Mater. Res. Express* **2019**, *7*, 015008. **[Google Scholar]** **[CrossRef]**

87. Nugraha, A.S.; Chou, C.C.; Yu, P.H.; Lin, K.-L. Effects of applied voltage on the morphology and phases of electrospun poly(vinylidene difluoride) nanofibers. *Polym. Int.* **2022**, *71*, 1176–1183. **[Google Scholar]** **[CrossRef]**

88. Singh, R.K.; Lye, S.W.; Miao, J. Holistic investigation of the electrospinning parameters for high percentage of □-phase in PVDF nanofibers. *Polymer* **2021**, *214*, 123366. **[Google Scholar]** **[CrossRef]**

89. Matabola, K.P.; Moutloali, R.M. The influence of electrospinning parameters on the morphology and diameter of poly(vinyledene fluoride) nanofibers-effect of sodium chloride. *J. Mater. Sci.* **2013**, *48*, 5475–5482. **[Google Scholar]** **[CrossRef]**

90. Doshi, J.; Reneker, D.H. Electrospinning process and applications of electrospun fibers. *J. Electrostat.* **1995**, *35*, 151–160. **[Google Scholar]** **[CrossRef]**

91. Jaeger, R.; Bergshoef, M.M.; Batlle, C.M.I.; Schönherr, H.; Julius Vancso, G. Electrospinning of ultra-thin polymer fibers. *Macromol. Symp.* **1998**, *127*, 141–150. **[Google Scholar]** **[CrossRef]**

92. Megelski, S.; Stephens, J.S.; Chase, D.B.; Rabolt, J.F. Micro-and nanostructured surface morphology on electrospun polymer fibers. *Macromolecules* **2002**, *35*, 8456–8466. **[Google Scholar] [CrossRef]**

93. Wang, T.; Kumar, S. Electrospinning of polyacrylonitrile nanofibers. *J. Appl. Polym. Sci.* **2006**, *102*, 1023–1029. **[Google Scholar] [CrossRef]**

94. Zheng, J.; He, A.; Li, J.; Han, C.C. Polymorphism control of poly(vinylidene fluoride) through electrospinning. *Macromol. Rapid Commun.* **2007**, *28*, 2159–2162. **[Google Scholar] [CrossRef]**

95. Motamedi, A.S.; Mirzadeh, H.; Hajiesmaeilbaigi, F.; Bagheri-Khoulenjani, S.; Shokrgozar, M. Effect of electrospinning parameters on morphological properties of PVDF nanofibrous scaffolds. *Prog. Biomater.* **2017**, *6*, 113–123. **[Google Scholar] [CrossRef]**

96. Zulfikar, M.A.; Afrianingsih, I.; Nasir, M.; Alni, A. Effect of processing parameters on the morphology of PVDF electrospun nanofiber. *J. Phys. Confer. Ser.* **2018**, *987*, 012011. **[Google Scholar] [CrossRef]**

97. Senador, A.E.; Shaw, M.T.; Mather, P.T. Electrospinning of polymeric nanofibers: Analysis of jet formation. *MRS Online Proc. Libr.* **2001**, *661*, KK59. **[Google Scholar] [CrossRef]**

98. Magniez, K.; De Lavigne, C.; Fox, B.L. The effects of molecular weight and polymorphism on the fracture and thermo-mechanical properties of a carbon-fibre composite modified by electrospun poly (vinylidene fluoride) membranes. *Polymer* **2010**, *51*, 2585–2596. **[Google Scholar] [CrossRef]**

99. Zaarour, B.; Zhu, L.; Jin, X.Y. Controlling the surface structure, mechanical properties, crystallinity, and piezoelectric properties of electrospun PVDF nanofibers by maneuvering molecular weight. *Soft Mater.* **2019**, *17*, 181–189. **[Google Scholar] [CrossRef]**

100. Gee, S.; Johnson, B.; Smith, A.L. Optimizing electrospinning parameters for piezoelectric PVDF nanofiber membranes. *J. Membr. Sci.* **2018**, *563*, 804–812. **[Google Scholar] [CrossRef]**

101. Zhu, G.; Zhao, L.Y.; Zhu, L.T.; Deng, X.Y.; Chen, W.L. Effect of experimental parameters on nanofiber diameter from electrospinning with wire electrodes. *IOP Conf. Ser. Mater. Sci. Eng.* **2017**, *230*, 012043. **[Google Scholar] [CrossRef]**

102. Zhong, G.; Zhang, L.; Su, R.; Wang, K.; Fong, H.; Zhu, L. Understanding polymorphism formation in electrospun fibers of immiscible Poly(vinylidene fluoride) blends. *Polymer* **2011**, *52*, 2228–2237. **[Google Scholar] [CrossRef]**

103. Huang, F.; Wei, Q.; Wang, J.; Cai, Y.; Huang, Y. Effect of temperature on structure, morphology and crystallinity of PVDF nanofibers via electrospinning. *E-Polymers* **2008**, *8*, 1–8. **[Google Scholar] [CrossRef]**

104. Zaarour, B.; Zhang, W.; Zhu, L.; Jin, X.Y.; Huang, C. Maneuvering surface structures of polyvinylidene fluoride nanofibers by controlling solvent systems and polymer concentration. *Text. Res. J.* **2018**, *89*, 2406–2422. **[Google Scholar] [CrossRef]**

105. Zaarour, B.; Zhu, L.; Huang, C.; Jin, X. Controlling the secondary surface morphology of electrospun PVDF nanofibers by regulating the solvent and relative humidity. *Nanoscale Res. Lett.* **2018**, *13*, 285. **[Google Scholar] [CrossRef] [PubMed][Green Version]**

106. Zaarour, B.; Zhu, L.; Huang, C.; Jin, X. Fabrication of a polyvinylidene fluoride cactus-like nanofiber through one-step electrospinning. *RSC Adv.* **2018**, *8*, 42353–42360. **[Google Scholar] [CrossRef][Green Version]**

107. Yun, J.S.; Park, C.K.; Jeong, Y.H.; Cho, J.H.; Paik, J.-H.; Yoon, S.H.; Hwang, K.-R. The fabrication and characterization of piezoelectric PZT/PVDF electrospun nanofiber composites. *Nanomater. Nanotechnol.* **2016**, *6*, 20. **[Google Scholar] [CrossRef][Green Version]**

108. Lee, C.; Wood, D.; Edmondson, D.; Yao, D.; Erickson, A.E.; Tsao, C.T.; Revia, R.A.; Kim, H.; Zhang, M. Electrospun uniaxially-aligned composite nanofibers as highly-efficient piezoelectric material. *Ceram. Int.* **2016**, *42*, 2734–2740. **[Google Scholar] [CrossRef]**

109. Sorayani Bafqi, M.S.; Bagherzadeh, R.; Latifi, M. Fabrication of composite PVDF-ZnO nanofiber mats by electrospinning for energy scavenging application with enhanced efficiency. *J. Polym. Res.* **2015**, *22*, 130. **[Google Scholar] [CrossRef]**

110. Hasanzadeh, M.; Ghahhari, M.R.; Bidoki, S.M. Enhanced piezoelectric performance of PVDF-based electrospun nanofibers by utilizing in situ synthesized graphene-ZnO nanocomposites. *J. Mater. Sci. Mater. Electron.* **2021**, *32*, 15789–15800. **[Google Scholar] [CrossRef]**

111. Li, C.; Wang, H.; Yan, X.; Chen, H.; Fu, Y.; Meng, Q. Enhancement research on piezoelectric performance of electrospun PVDF fiber membranes with inorganic reinforced materials. *Coatings* **2021**, *11*, 1495. **[Google Scholar] [CrossRef]**

112. Wu, C.M.; Chou, M.H. Polymorphism, piezoelectricity and sound absorption of electrospun PVDF membranes with and without carbon nanotubes. *Compos. Sci. Technol.* **2016**, *127*, 127–133. **[Google Scholar] [CrossRef]**

113. Zhang, J.; Wang, H.; Blanloeuil, P.; Li, G.; Sha, Z.; Wang, D.; Lei, W.; Boyer, C.; Yu, Y.; Tian, R.; et al. Enhancing the triboelectricity of stretchable electrospun piezoelectric polyvinylidene fluoride/boron nitride nanosheets composite nanofibers. *Compos. Commun.* **2020**, *22*, 100535. **[Google Scholar] [CrossRef]**

114. Moghadam, B.H.; Hasanzadeh, M.; Simchi, A. Self-Powered Wearable Piezoelectric Sensors Based on Polymer Nanofiber–Metal–Organic Framework Nanoparticle Composites for Arterial Pulse Monitoring. *ACS Appl. Nano Mater.* **2020**, *3*, 8742–8752. **[Google Scholar] [CrossRef]**

115. Roy, K.; Jana, S.; Ghosh, S.K.; Mahanty, B.; Mallick, Z.; Sarkar, S.; Sinha, C.; Mandal, D. Three-dimensional MOF-assisted self-polarized ferroelectret: An effective autopowered remote healthcare monitoring approach. *Langmuir* **2020**, *36*, 11477–11489. **[Google Scholar] [CrossRef]**

116. Xue, W.; Lv, C.; Jing, Y.; Chen, F.; Fu, Q. Fabrication of electrospun PVDF nanofibers with higher content of polar β phase and smaller diameter by adding a small amount of dioctadecyl dimethyl ammonium chloride. *Chin. J. Polym. Sci.* **2017**, *35*, 992–1000. **[Google Scholar] [CrossRef]**

117. Dashtizad, S.; Alizadeh, P.; Yourdkhani, A. Improving piezoelectric properties of PVDF fibers by compositing with BaTiO3-Ag particles prepared by sol-gel method and photochemical reaction. *J. Alloys Compd.* **2021**, *883*, 160810. **[Google Scholar] [CrossRef]**

118. Li, G.-Y.; Zhang, H.-D.; Guo, K.; Ma, X.-S.; Long, Y.-Z. Fabrication and piezoelectric-pyroelectric properties of electrospun PVDF/ZnO composite fibers. *Mater. Res. Express* **2020**, *7*, 095502. **[Google Scholar] [CrossRef]**

119. Khan, M.; Tiehu, L.; Hussain, A.; Raza, A.; Zada, A.; Alei, D.; Khan, A.R.; Ali, R.; Hussain, H.; Hussain, J.; et al. Physiochemical evaluations, mechanical attenuations and thermal stability of graphene nanosheets and functionalized nanodiamonds loaded pitch derived carbon foam composites. *Diam. Relat. Mater.* **2022**, *126*, 109077. **[Google Scholar] [CrossRef]**

120. Raza, J.; Hamid, A.; Khan, M.; Hussain, F.; Zada, A.; Tiehu, L.; Ali, A.; Fazil, P.; Wahab, Z. Preparation and comparative evaluation of PVC/PbO and PVC/PbO/graphite based conductive nanocomposites. *Z. Für Phys. Chem.* **2022**. **[Google Scholar] [CrossRef]**

121. Martins, P.; Costa, C.M.; Benelmekki, M.; Botelho, G.; Lanceros-Mendez, S. On the origin of the electroactive poly(vinylidene fluoride) □-phase nucleation by ferrite nanoparticles via surface electrostatic interactions. *CrystEngComm* **2012**, *14*, 2807–2811. **[Google Scholar] [CrossRef]**

122. Bairagi, S.; Ali, S.W. Investigating the role of carbon nanotubes (CNTs) in the piezoelectric performance of a PVDF/KNN-based electrospun nanogenerator. *Soft Matter* **2020**, *16*, 4876–4886. **[Google Scholar] [CrossRef]**

123. Shehata, N.; Elnabawy, E.; Abdelkader, M.; Hassanin, A.H.; Salah, M.; Nair, R.; Ahmad Bhat, S. Static-aligned piezoelectric poly (vinylidene fluoride) electrospun nanofibers/MWCNT composite membrane: Facile

method. *Polymers* **2018**, *10*, 965. **[Google Scholar] [CrossRef] [PubMed] [Green Version]**

124. Liu, Y.A.; Lv, H.; Liu, Y.; Gao, Y.M.; Kim, H.Y.; Ouyang, Y.M.; Yu, D.G. Progresses on electrospun metal-organic frameworks nanofibers and their wastewater treatment applications. *Mater. Today Chem.* **2022**, *25*, 100974. **[Google Scholar] [CrossRef]**

125. Zhang, X.; Zhai, Z.; Wang, J.; Hao, X.; Sun, Y.; Yu, S.; Lin, X.; Qin, Y.; Li, C. Zr-MOF combined with nanofibers as an efficient and flexible capacitive sensor for detecting SO_2. *ChemNanoMat* **2021**, *7*, 1117–1124. **[Google Scholar] [CrossRef]**

126. Roy, K.; Jana, S.; Mallick, Z.; Ghosh, S.K.; Dutta, B.; Sarkar, S.; Sinha, C.; Mandal, D. Two-dimensional MOF modulated fiber nanogenerator for effective acoustoelectric conversion and human motion detection. *Langmuir* **2021**, *37*, 7107–7117. **[Google Scholar] [CrossRef] [PubMed]**

127. Zhang, W.; Zaarour, B.; Zhu, L.; Huang, C.; Xu, B.; Jin, X. A comparative study of electrospun polyvinylidene fluoride and poly(vinylidenefluoride-co-trifluoroethylene) fiber webs: Mechanical properties, crystallinity, and piezoelectric properties. *J. Eng. Fibers Fabr.* **2020**, *15*, 1558925020939290. **[Google Scholar] [CrossRef]**

128. Diaz Sanchez, F.J.; Chung, M.; Waqas, M.; Koutsos, V.; Smith, S.; Radacsi, N. Sponge-like piezoelectric micro- and nanofiber structures for mechanical energy harvesting. *Nano Energy* **2022**, *98*, 107286. **[Google Scholar] [CrossRef]**

129. Azimi, S.; Golabchi, A.; Nekookar, A.; Rabbani, S.; Amiri, M.H.; Asadi, K.; Abolhasani, M.M. Self-powered cardiac pacemaker by piezoelectric polymer nanogenerator implant. *Nano Energy* **2021**, *83*, 105781. **[Google Scholar] [CrossRef]**

130. Maurya, A.K.; Mias, E.; Schoeller, J.; Collings, I.E.; Rossi, R.M.; Dommann, A.; Neels, A. Understanding multiscale structure–property correlations in PVDF-HFP electrospun fiber membranes by SAXS and WAXS. *Nanoscale Adv.* **2022**, *4*, 491–501. **[Google Scholar] [CrossRef]**

131. Conte, A.A.; Shirvani, K.; Hones, H.; Wildgoose, A.; Xue, Y.; Najjar, R.; Hu, X.; Xue, W.; Beachley, V.Z. Effects of post-draw processing on the structure and functional properties of electrospun PVDF-HFP nanofibers. *Polymer* **2019**, *171*, 192–200. **[Google Scholar] [CrossRef]**

132. Wu, Y.; Qu, J.; Daoud, W.A.; Wang, L.; Qi, T. Flexible composite-nanofiber based piezo-triboelectric nanogenerators for wearable electronics. *J. Mater. Chem. A* **2019**, *7*, 13347–13355. **[Google Scholar] [CrossRef]**

133. Li, G.-Y.; Li, J.; Li, Z.-J.; Zhang, Y.-P.; Zhang, X.; Wang, Z.-J.; Han, W.-P.; Sun, B.; Long, Y.-Z.; Zhang, H.-D. Hierarchical PVDF-HFP/ZnO composite

nanofiber–based highly sensitive piezoelectric sensor for wireless workout monitoring. *Adv. Compos. Hybrid Mater.* **2021**, *5*, 766–775. **[Google Scholar]** **[CrossRef]**

134. Ponnamma, D.; Chamakh, M.M.; Alahzm, A.M.; Salim, N.; Hameed, N.; AlMaadeed, M.A. Core-shell nanofibers of polyvinylidene fluoride-based nanocomposites as piezoelectric nanogenerators. *Polymers* **2020**, *12*, 2344. **[Google Scholar]** **[CrossRef]** **[PubMed]**
135. Chorsi, M.T.; Curry, E.J.; Chorsi, H.T.; Das, R.; Baroody, J.; Purohit, P.K.; Ilies, H.; Nguyen, T.D. Piezoelectric biomaterials for sensors and actuators. *Adv. Mater.* **2019**, *31*, 1802084. **[Google Scholar]** **[CrossRef][Green Version]**
136. Tai, Y.; Yang, S.; Yu, S.; Banerjee, A.; Myung, N.V.; Nam, J. Modulation of piezoelectric properties in electrospun PLLA nanofibers for application-specific self-powered stem cell culture platforms. *Nano Energy* **2021**, *89*, 106444. **[Google Scholar]** **[CrossRef]**
137. Zhao, G.; Huang, B.; Zhang, J.; Wang, A.; Ren, K.; Wang, Z.L. Electrospun poly(l-lactic acid) nanofibers for nanogenerator and diagnostic sensor applications. *Macromol. Mater. Eng.* **2017**, *302*, 1600476. **[Google Scholar]** **[CrossRef]**
138. Chuenjitkuntaworn, B.; Supaphol, P.; Pavasant, P.; Damrongsri, D. Electrospun poly(l-lactic acid)/hydroxyapatite composite fibrous scaffolds for bone tissue engineering. *Polym. Int.* **2010**, *59*, 227–235. **[Google Scholar]** **[CrossRef]**
139. Lv, H.; Guo, S.R.; Zhang, G.Y.; He, W.L.; Wu, Y.H.; Yu, D.G. Electrospun structural hybrids of acyclovir-polyacrylonitrile at acyclovir for modifying drug release. *Polymers* **2021**, *13*, 4286. **[Google Scholar]** **[CrossRef]** **[PubMed]**
140. Xu, X.; Lv, H.; Zhang, M.; Wang, M.; Zhou, Y.; Liu, Y.; Yu, D.G. Recent progresses of electrospun nanofibers and their applications in treating heavy metal wastewater. *Front. Chem. Sci. Eng.* **2022**, *17*, 1–34. **[Google Scholar]**
141. Zhang, X.L.; Guo, S.Q.; Qin, Y.; Li, C.J. Functional electrospun nanocomposites for efficient oxygen reduction reaction. *J. Chem. Res. Chin. Univ.* **2021**, *37*, 379–393. **[Google Scholar]** **[CrossRef]**
142. Lv, H.; Zhang, M.; Wang, P.; Xu, X.; Liu, Y.; Yu, D.-G. Ingenious construction of Ni(DMG)2/TiO2-decorated porous nanofibers for the highly efficient photodegradation of pollutants in water. *Colloids Surf. A Physicochem. Eng. Asp.* **2022**, *650*, 129561. **[Google Scholar]** **[CrossRef]**
143. Cao, X.; Chen, W.; Zhao, P.; Yang, Y.; Yu, D.G. Electrospun porous nanofibers: Pore-forming mechanisms and photocatalytic degradation applications. *Polymers* **2022**, *14*, 3990. **[Google Scholar]** **[CrossRef]**

144. Minagawa, M.; Miyano, K.; Takahashi, M.; Yoshii, F. Infrared characteristic absorptionbandsofhighlyisotacticpoly(acrylonitrile). *Macromolecules* **1988**, *21*, 2387–2391. **[Google Scholar] [CrossRef]**

145. Yuan, L.; Fan, W.; Yang, X.; Ge, S.; Xia, C.; Foong, S.Y.; Liew, R.K.; Wang, S.; Van Le, Q.; Lam, S.S. Piezoelectric PAN/BaTiO3 nanofiber membranes sensor for structural health monitoring of real-time damage detection in composite. *Compos. Commun.* **2021**, *25*, 100680. **[Google Scholar] [CrossRef]**

146. Sun, Y.; Liu, Y.; Zheng, Y.; Li, Z.; Fan, J.; Wang, L.; Liu, X.; Liu, J.; Shou, W. Enhanced energy harvesting ability of ZnO/PAN hybrid piezoelectric nanogenerators. *ACS Appl. Mater. Interfaces* **2020**, *12*, 54936–54945. **[Google Scholar] [CrossRef]**

147. Wang, W.; Zheng, Y.; Jin, X.; Sun, Y.; Lu, B.; Wang, H.; Fang, J.; Shao, H.; Lin, T. Unexpectedly high piezoelectricity of electrospun polyacrylonitrile nanofiber membranes. *Nano Energy* **2019**, *56*, 588–594. **[Google Scholar] [CrossRef]**

148. Zhao, G.; Zhang, X.; Cui, X.; Wang, S.; Liu, Z.; Deng, L.; Qi, A.; Qiao, X.; Li, L.; Pan, C.; et al. Piezoelectric polyacrylonitrile nanofiber film-based dual-function self-powered flexible sensor. *ACS Appl. Mater. Interfaces* **2018**, *10*, 15855–15863. **[Google Scholar] [CrossRef] [PubMed]**

149. Bairagi, S.; Chowdhury, A.; Banerjee, S.; Thakre, A.; Saini, A.; Ali, S.W. Investigating the role of copper oxide (CuO) nanorods in designing flexible piezoelectric nanogenerator composed of polyacrylonitrile (PAN) electrospun web-based fibrous material. *J. Mater. Sci. Mater. Electron.* **2022**, *33*, 13152–13165. **[Google Scholar] [CrossRef]**

150. Peng, L.; Jin, X.; Niu, J.; Wang, W.; Wang, H.; Shao, H.; Lang, C.; Lin, T. High-precision detection of ordinary sound by electrospun polyacrylonitrile nanofibers. *J. Mater. Chem. C* **2021**, *9*, 3477–3485. **[Google Scholar] [CrossRef]**

151. Cai, Z.; Xiong, P.; He, S.; Zhu, C. Improved piezoelectric performances of highly orientated poly(β-hydroxybutyrate) electrospun nanofiber membrane scaffold blended with multiwalled carbon nanotubes. *Mater. Lett.* **2019**, *240*, 213–216. **[Google Scholar] [CrossRef]**

152. Chernozem, R.V.; Romanyuk, K.N.; Grubova, I.; Chernozem, P.V.; Surmeneva, M.A.; Mukhortova, Y.R.; Wilhelm, M.; Ludwig, T.; Mathur, S.; Kholkin, A.L.; et al. Enhanced piezoresponse and surface electric potential of hybrid biodegradable polyhydroxybutyrate scaffolds functionalized with reduced graphene oxide for tissue engineering. *Nano Energy* **2021**, *89*, 106473. **[Google Scholar] [CrossRef]**

153. Chernozem, R.V.; Surmeneva, M.A.; Surmenev, R.A. Hybrid biodegradable scaffolds of piezoelectric polyhydroxybutyrate and conductive polyaniline:

Piezocharge constants and electric potential study. *Mater. Lett.* **2018,** *220*, 257–260. **[Google Scholar] [CrossRef]**

154. Karpov, T.E.; Muslimov, A.R.; Zyuzin, M.V.; Peltek, O.O.; Sergeev, I.S.; Goncharenko, A.A.; Surmenev, R.A.; Timin, A.S. Multifunctional scaffolds based on piezoelectric electrospun fibers modified with biocompatible drug carriers for regenerative medicine. *J. Phys. Confer. Ser.* **2020,** *1461*, 012060. **[Google Scholar] [CrossRef][Green Version]**

155. Nair, S.S.; Ramesh, C.; Tashiro, K. Crystalline phases in Nylon-11: studies using HTWAXS and HTFTIR. *Macromolecules* **2006,** *39*, 2841–2848. **[Google Scholar] [CrossRef]**

156. Chocinski-Arnault, L.; Gaudefroy, V.; Gacougnolle, J.L.; Rivière, A. Memory effect and crystalline structure in polyamide 11. *J. Macromol. Sci. B* **2002,** *41*, 777–785. **[Google Scholar] [CrossRef]**

157. Anwar, S.; Hassanpour Amiri, M.; Jiang, S.; Abolhasani, M.M.; Rocha, P.R.F.; Asadi, K. Piezoelectric Nylon-11 Fibers for Electronic Textiles, Energy Harvesting and Sensing. *Adv. Funct. Mater.* **2021,** *31*, 2004326. **[Google Scholar] [CrossRef]**

158. Zhang, L.Y.; He, G.; Yu, Y.; Zhang, Y.; Li, X.; Wang, S.G. Design of biocompatible chitosan/polyaniline/laponite hydrogel with Photothermal Conversion Capability. *Biomolecules* **2022,** *12*, 1089. **[Google Scholar] [CrossRef]**

159. Yu, D.-G.; Zhao, P. The key elements for biomolecules to biomaterials and to bioapplications. *Biomolecules* **2022,** *12*, 1234. **[Google Scholar] [CrossRef] [PubMed]**

160. Zhao, P.; Chen, W.; Feng, Z.; Liu, Y.; Liu, P.; Xie, Y.; Yu, D.-G. Electrospun nanofibers for periodontal treatment: A recent progress. *Int. J. Nanomed.* **2022,** *17*, 4137–4162. **[Google Scholar] [CrossRef]**

161. Huang, C.X.; Dong, J.; Zhang, Y.Y.; Chai, S.L.; Wang, X.C.; Kang, S.X.; Yu, D.G.; Wang, P.; Jiang, Q. Gold nanoparticles-loaded polyvinylpyrrolidone/ethylcellulose coaxial electrospun nanofibers with enhanced osteogenic capability for bone tissue regeneration. *Mater. Design* **2021,** *212*, 110240. **[Google Scholar] [CrossRef]**

162. Chen, W.; Zhao, P.; Yang, Y.; Yu, D.-G. Electrospun beads-on-the-string nanoproducts: Preparation and drug delivery application. *Curr. Drug Deliv.* **2022,** *19*, 991–1000. **[Google Scholar] [CrossRef]**

163. Guerin, S.; Syed, T.A.M.; Thompson, D. Deconstructing collagen piezoelectricity using alanine-hydroxyproline-glycine building blocks. *Nanoscale* **2018,** *10*, 9653–9663. **[Google Scholar] [CrossRef] [PubMed]**

164. Viguet-Carrin, S.; Garnero, P.; Delmas, P.D. The role of collagen in bone strength. *Osteoporos. Int.* **2006**, *17*, 319–336. **[Google Scholar] [CrossRef] [PubMed]**

165. Rho, K.S.; Jeong, L.; Lee, G.; Seo, B.-M.; Park, Y.J.; Hong, S.-D.; Roh, S.; Cho, J.J.; Park, W.H.; Min, B.-M. Electrospinning of collagen nanofibers: Effects on the behavior of normal human keratinocytes and early-stage wound healing. *Biomaterials* **2006**, *27*, 1452–1461. **[Google Scholar] [CrossRef] [PubMed]**

166. Binetti, V.R.; Schiffman, J.D.; Leaffer, O.D.; Spanier, J.E.; Schauer, C.L. The natural transparency and piezoelectric response of the Greta oto butterfly wing. *Integr. Biol.* **2009**, *1*, 324–329. **[Google Scholar] [CrossRef] [PubMed] [Green Version]**

167. Street, R.M.; Huseynova, T.; Xu, X.; Chandrasekaran, P.; Han, L.; Shih, W.Y.; Shih, W.-H.; Schauer, C.L. Variable piezoelectricity of electrospun chitin. *Carbohyd. Polym.* **2018**, *195*, 218–224. **[Google Scholar] [CrossRef] [PubMed]**

168. Jaworska, M.; Sakurai, K.; Gaudon, P.; Guibal, E. Influence of chitosan characteristics on polymer properties. I: Crystallographic properties. *Polym. Int.* **2003**, *52*, 198–205. **[Google Scholar] [CrossRef]**

169. Ahmed, S.; Arshad, T.; Zada, A.; Afzal, A.; Khan, M.; Hussain, A.; Hassan, M.; Ali, M.; Xu, S. Preparation and Characterization of a Novel Sulfonated Titanium Oxide Incorporated Chitosan Nanocomposite Membranes for Fuel Cell Application. *Membranes* **2021**, *11*, 450. **[Google Scholar] [CrossRef] [PubMed]**

170. FotouhiArdakani, F.; Mohammadi, M.; Mashayekhan, S. ZnO-incorporated polyvinylidene fluoride/poly(ε-caprolactone) nanocomposite scaffold with controlled release of dexamethasone for bone tissue engineering. *Appl. Phys. A-Mater.* **2022**, *128*, 654. **[Google Scholar] [CrossRef]**

171. Wang, Y.B.; Zhu, M.M.; Wei, X.D.; Yu, J.Y.; Li, Z.L.; Ding, B. A dual-mode electronic skin textile for pressure and temperature sensing. *Chem. Eng. J.* **2021**, *425*, 130599. **[Google Scholar] [CrossRef]**

172. Zhu, M.M.; Li, J.L.; Yu, J.Y.; Li, Z.L.; Ding, B. Superstable and intrinsically self-healing fibrous membrane with bionic confined protective structure for breathable electronic skin. *Angew. Chem.* **2022**, *134*, e202200226. **[Google Scholar] [CrossRef]**

173. Cai, J.Y.; Du, M.J.; Li, Z.L. Flexible temperature sensors constructed with fiber materials. *Adv. Mater. Technol.* **2022**, *7*, 2101182. **[Google Scholar] [CrossRef]**

174. Hamid, A.; Khan, M.; Hussain, F.; Zada, A.; Li, T.; Alei, D.; Ali, A. Synthesis and physiochemical performances of PVC-sodium polyacrylate and PVC-sodium polyacrylate-graphite composite polymer membrane. *Z. Für Phys. Chem.* **2021**, *235*, 1791–1810. **[Google Scholar] [CrossRef]**

175. Ngadiman, N.H.A.; Noordin, M.Y.; Idris, A.; Kurniawan, D. A review of evolution of electrospun tissue engineering scaffold: From two dimensions to three dimensions. *Proc. Inst. Mech. Eng. H J. Eng. Med.* **2017**, *231*, 597–616. **[Google Scholar] [CrossRef] [PubMed]**

176. Lin, W.; Chen, M.; Qu, T.; Li, J.; Man, Y. Three-dimensional electrospun nanofibrous scaffolds for bone tissue engineering. *J. Biomed. Mater. Res. B* **2020**, *108*, 1311–1321. **[Google Scholar] [CrossRef] [PubMed]**

177. Hu, X.; Wang, G. Research progress of electrospun nanofibers scaffold in nerve tissue engineering. *Zhongguo Xiu Fu Chong Jian Wai Ke Za Zhi Zhongguo Xiufu Chongjian Waike Zazhi Chin. J. Reparative Reconstr. Surg.* **2010**, *24*, 1133–1137. **[Google Scholar]**

178. Tai, Y.; Ico, G.; Low, K.; Liu, J.; Jariwala, T.; Garcia-Viramontes, D.; Lee, K.H.; Myung, N.V.; Park, B.H.; Nam, J. Formation of 3D self-organized neuron-glial interface derived from neural stem cells via mechano-electrical stimulation. *Adv. Healthc. Mater.* **2021**, *10*, 2100806. **[Google Scholar] [CrossRef]**

179. Ico, G.; Myung, A.; Kim, B.S.; Myung, N.V.; Nam, J. Transformative piezoelectric enhancement of P(VDF-TrFE) synergistically driven by nanoscale dimensional reduction and thermal treatment. *Nanoscale* **2018**, *10*, 2894–2901. **[Google Scholar] [CrossRef][Green Version]**

180. Persano, L.; Dagdeviren, C.; Su, Y.; Zhang, Y.; Girardo, S.; Pisignano, D.; Huang, Y.; Rogers, J.A. High performance piezoelectric devices based on aligned arrays of nanofibers of poly(vinylidenefluoride-co-trifluoroethylene). *Nat. Commun.* **2013**, *4*, 1633. **[Google Scholar] [CrossRef][Green Version]**

181. Lins, L.C.; Wianny, F.; Livi, S.; Dehay, C.; Duchet-Rumeau, J.; Gérard, J.-F. Effect of polyvinylidene fluoride electrospun fiber orientation on neural stem cell differentiation. *J. Biomed. Mater. Res. B* **2017**, *105*, 2376–2393. **[Google Scholar] [CrossRef] [PubMed]**

182. Guo, H.-F.; Li, Z.-S.; Dong, S.-W.; Chen, W.-J.; Deng, L.; Wang, Y.-F.; Ying, D.-J. Piezoelectric PU/PVDF electrospun scaffolds for wound healing applications. *Colloids Surf. B* **2012**, *96*, 29–36. **[Google Scholar] [CrossRef] [PubMed]**

183. Damaraju, S.M.; Shen, Y.; Elele, E.; Khusid, B.; Eshghinejad, A.; Li, J.; Jaffe, M.; Arinzeh, T.L. Three-dimensional piezoelectric fibrous scaffolds selectively promote mesenchymal stem cell differentiation. *Biomaterials* **2017**, *149*, 51–62. **[Google Scholar] [CrossRef] [PubMed]**

184. Ikada, Y.; Shikinami, Y.; Hara, Y.; Tagawa, M.; Fukada, E. Enhancement of bone formation by drawn poly(L-lactide). *J. Biomed. Mater. Res.* **1996**, *30*, 553–558. **[Google Scholar] [CrossRef]**

185. Schofer, M.D.; Roessler, P.P.; Schaefer, J.; Theisen, C.; Schlimme, S.; Heverhagen, J.T.; Voelker, M.; Dersch, R.; Agarwal, S.; Fuchs-Winkelmann, S.; et al. Electrospun PLLA nanofiber scaffolds and their use in combination with BMP-2 for reconstruction of bone defects. *PLoS ONE* **2011**, *6*, e25462. **[Google Scholar] [CrossRef][Green Version]**

186. Azimi, B.; Milazzo, M.; Lazzeri, A.; Berrettini, S.; Uddin, M.J.; Qin, Z.; Buehler, M.J.; Danti, S. Electrospinning piezoelectric fibers for biocompatible devices. *Adv. Healthc. Mater.* **2020**, *9*, 1901287. **[Google Scholar] [CrossRef]**

187. Mahanty, B.; Maity, K.; Sarkar, S.; Mandal, D. Human skin interactive self-powered piezoelectric e-skin based on PVDF/MWCNT electrospun nanofibers for non-invasive health care monitoring. *Mater. Today Proceed.* **2020**, *21*, 1964–1968. **[Google Scholar] [CrossRef]**

188. Rambhia, K.J.; Ma, P.X. Controlled drug release for tissue engineering. *J. Control. Release* **2015**, *219*, 119–128. **[Google Scholar] [CrossRef][Green Version]**

189. Zhang, Y.; Kim, I.; Lu, Y.; Xu, Y.; Yu, D.-G.; Song, W. Intelligent poly(l-histidine)-based nanovehicles for controlled drug delivery. *J. Control. Release* **2022**, *349*, 963–982. **[Google Scholar] [CrossRef]**

190. Mushtaq, F.; Torlakcik, H.; Hoop, M.; Jang, B.; Carlson, F.; Grunow, T.; Läubli, N.; Ferreira, A.; Chen, X.-Z.; Nelson, B.J.; et al. Motile piezoelectric nanoeels for targeted drug delivery. *Adv. Funct. Mater.* **2019**, *29*, 1808135. **[Google Scholar] [CrossRef]**

191. Starr, M.B.; Wang, X. Coupling of piezoelectric effect with electrochemical processes. *Nano Energy* **2015**, *14*, 296–311. **[Google Scholar] [CrossRef][Green Version]**

192. Ma, W.; Yao, B.; Zhang, W.; He, Y.; Yu, Y.; Niu, J. Fabrication of PVDF-based piezocatalytic active membrane with enhanced oxytetracycline degradation efficiency through embedding few-layer E-MoS2 nanosheets. *Chem. Eng. J.* **2021**, *415*, 129000. **[Google Scholar] [CrossRef]**

193. Paredes, L.; Murgolo, S.; Dzinun, H.; Dzarfan Othman, M.H.; Ismail, A.F.; Carballa, M.; Mascolo, G. Application of immobilized TiO2 on PVDF dual layer hollow fibre membrane to improve the photocatalytic removal of pharmaceuticals in different water matrices. *Appl. Catal. B Environ.* **2019**, *240*, 9–18. **[Google Scholar] [CrossRef]**

194. Bai, Y.; Zhao, J.; Li, Y.; Lv, Z.; Lu, K. Preparation and photocatalytic performance of TiO2/PbTiO3 fiber composite enhanced by external force induced piezoelectric field. *J. Am. Ceram. Soc.* **2019**, *102*, 5415–5423. **[Google Scholar] [CrossRef]**

195. Durairaj, A.; Ramasundaram, S.; Sakthivel, T.; Ramanathan, S.; Rahaman, A.; Kim, B.; Vasanthkumar, S. Air bubbles induced piezophotocatalytic degradation of organic pollutants using nanofibrous poly(vinylidene fluoride)-titanium dioxide hybrid. *Appl. Surf. Sci.* **2019**, *493*, 1268–1277. **[Google Scholar]** **[CrossRef]**

196. Ding, D.; Li, Z.; Yu, S.; Yang, B.; Yin, Y.; Zan, L.; Myung, N.V. Piezo-photocatalytic flexible PAN/TiO2 composite nanofibers for environmental remediation. *Sci. Total Environ.* **2022**, *824*, 153790. **[Google Scholar]** **[CrossRef]**

197. Lee, J.-T.; Lin, M.-C.; Wu, J.M. High-efficiency cycling piezo-degradation of organic pollutants over three liters using MoS2/carbon fiber piezocatalytic filter. *Nano Energy* **2022**, *98*, 107280. **[Google Scholar]** **[CrossRef]**

198. Singh, P.; Mondal, K.; Sharma, A. Reusable electrospun mesoporous ZnO nanofiber mats for photocatalytic degradation of polycyclic aromatic hydrocarbon dyes in wastewater. *J. Colloid Interface Sci.* **2013**, *394*, 208–215. **[Google Scholar]** **[CrossRef]**

199. Khan, M.; Tiehu, L.; Zaidi, S.B.A.; Javed, E.; Hussain, A.; Hayat, A.; Zada, A.; Alei, D.; Ullah, A. Synergistic effect of nanodiamond and titanium oxide nanoparticles on the mechanical, thermal and electrical properties of pitch-derived carbon foam composites. *Polym. Int.* **2021**, *70*, 1733–1740. **[Google Scholar]** **[CrossRef]**

200. Singh, S.; Khare, N. Coupling of piezoelectric, semiconducting and photoexcitation properties in $NaNbO_3$ nanostructures for controlling electrical transport: Realizing an efficient piezo-photoanode and piezo-photocatalyst. *Nano Energy* **2017**, *38*, 335–341. **[Google Scholar]** **[CrossRef]**

201. Feng, Y.; Li, H.; Ling, L.; Yan, S.; Pan, D.; Ge, H.; Li, H.; Bian, Z. Enhanced photocatalytic degradation performance by fluid-induced piezoelectric field. *Environ. Sci. Technol.* **2018**, *52*, 7842–7848. **[Google Scholar]** **[CrossRef]**

202. Zhu, J.; Jia, L.; Huang, R. Electrospinning poly(l-lactic acid) piezoelectric ordered porous nanofibers for strain sensing and energy harvesting. *J. Mater. Sci. Mater. Electron.* **2017**, *28*, 12080–12085. **[Google Scholar]** **[CrossRef]**

203. Sultana, A.; Alam, M.M.; Sadhukhan, P.; Ghorai, U.K.; Das, S.; Middya, T.R.; Mandal, D. Organo-lead halide perovskite regulated green light emitting poly(vinylidene fluoride) electrospun nanofiber mat and its potential utility for ambient mechanical energy harvesting application. *Nano Energy* **2018**, *49*, 380–392. **[Google Scholar]** **[CrossRef]**

204. Shen, Y.H.; Yu, X.; Cui, J.; Yu, F.; Liu, M.Y.; Chen, Y.J.; Wu, J.L.; Sun, B.B.; Mo, X.M. Development of Biodegradable Polymeric Stents for the Treatment of Cardiovascular Diseases. *Biomolecules* **2022**, *12*, 1245. **[Google Scholar]** **[CrossRef]** **[PubMed]**

205. Liu, Y.; Wang, X.; Wu, Q.; Pei, W.; Teo, M.J.; Chen, Z.S.; Huang, C. Application of lignin and lignin-based composites in different tissue engineering fields. *Int. J. Biol. Macromol.* **2022**, *222*, 994–1006. **[Google Scholar] [CrossRef] [PubMed]**

206. Huang, C.; Peng, Z.; Li, J.; Li, X.; Jiang, X.; Dong, Y. Unlocking the role of lignin for preparing the lignin-based wood adhesive: A review. *Ind. Crops Prod.* **2022**, *187*, 115388. **[Google Scholar] [CrossRef]**

207. Deng, J.P.; Song, Q.; Liu, S.Y.; Pei, W.H.; Wang, P.; Zheng, L.M.; Huang, C.X.; Ma, M.G.; Jiang, Q.; Zhang, K. Advanced applications of cellulose-based composites in fighting bone diseases. *Compos. B Eng.* **2022**, *245*, 110221. **[Google Scholar] [CrossRef]**

208. Huang, C.; Yu, Y.; Li, Z.; Yan, B.; Pei, W.; Wu, H. The preparation technology and application of xylo-oligosaccharide as prebiotics in different fields: A review. *Front. Nutr.* **2022**, *9*, 996811. **[Google Scholar] [CrossRef]**

209. Xie, D.C.; Zhou, X.; Xiao, B.; Duan, L.; Zhu, Z.H. Mucus-Penetrating Silk Fibroin-Based Nanotherapeutics for Efficient Treatment of Ulcerative Colitis. *Biomolecules* **2022**, *12*, 1263. **[Google Scholar] [CrossRef]**

210. Tang, Y.X.; Varyambath, A.; Ding, Y.C.; Chen, B.L.; Huang, X.Y.; Zhang, Y.; Yu, D.G.; Kim, I.; Song, W.L. Porous organic polymers for drug delivery: Hierarchical pore structures, variable morphologies, and biological properties. *Biomater. Sci.* **2022**, *10*, 5369–5390. **[Google Scholar] [CrossRef]**

211. Kang, S.X.; Hou, S.C.; Chen, X.W.; Yu, D.G.; Wang, L.; Li, X.Y.; Williams, G.R. Energy-saving electrospinning with a concentric teflon-core rod spinneret to create medicated nanofibers. *Polymers* **2020**, *12*, 2421. **[Google Scholar] [CrossRef]**

212. Jiang, W.L.; Zhang, X.Y.; Liu, P.; Zhang, Y.; Song, W.L.; Yu, D.G.; Lu, X.H. Electrospun healthcare nanofibers from medicinal liquor of Phellinus igniarius. *Adv. Compos. Hybrid Mater.* **2022**, 5. **[Google Scholar] [CrossRef]**

213. Wang, M.L.; Yu, D.G.; Williams, G.R.; Bligh, S.W.A. Co-loading of inorganic nanoparticles and natural oil in the electrospun janus nanofibers for a synergetic antibacterial effect. *Pharmaceutics* **2022**, *14*, 1208. **[Google Scholar] [CrossRef] [PubMed]**

214. Liu, H.; Wang, H.B.; Lu, X.H.; Murugadoss, V.; Huang, M.N.; Yang, H.S.; Wan, F.X.; Yu, D.G.; Guo, Z.H. Electrospun structural nanohybrids combining three composites for fast helicide delivery. *Adv. Compos. Hybrid Mater.* **2022**, *5*, 1017–1029. **[Google Scholar] [CrossRef]**

215. Wang, M.L.; Hou, J.S.; Yu, D.G.; Li, S.Y.; Zhu, J.W.; Chen, Z.Z. Electrospun tri-layer nanodepots for sustained release of acyclovir. *J. Alloys Compd.* **2020**, *846*, 156471. **[Google Scholar] [CrossRef]**

CHAPTER-9

SUPERCAPACITORS WITH A FIBER-SHAPED STRUCTURE COMPOSED OF GRAPHENE, FEATURING A POROUS CORE-SHEATH DESIGN, ARE ENHANCED THROUGH NITROGEN DOPING

1. INTRODUCTION: AN OVERVIEW

In the realm of energy storage devices, graphene-based materials have emerged as promising candidates for their exceptional conductivity and structural versatility. Among these, nitrogen-doped porous core-sheath graphene fiber-shaped supercapacitors have garnered significant attention. This innovative design, incorporating nitrogen doping and a porous core-sheath architecture, aims to elevate the performance of supercapacitors, enhancing their energy storage capabilities and overall efficiency. In this context, the integration of nitrogen doping and a fiber-shaped structure presents a novel approach to advancing the state-of-the-art in energy storage technology.

With the emerging electronic textiles, the demand for rechargeable energy storage technologies is urgently needed [**1**,**2**]. Fiber-based flexible supercapacitors have the characteristics of lightweight, good flexibility, high power density, great cycle stability, and excellent charge–discharge performance [**3**,**4**,**5**,**6**,**7**]. Various types of fiber-shaped supercapacitors have been widely studied in recent years.

Owing to the remarkable electrical conductivity, higher tensile strength, and controllable structure [**8**,**9**,**10**], graphene fibers are considered as promising electrode materials for flexible fiber-shaped supercapacitors. However, the graphene nanosheets are easily aggregated and often closely packed in graphene fibers which hinder ion and electron transport in electrodes due to strong π–π interaction between graphene interlayers, resulting in limited specific surface area and thus power and energy density [**11,12,13,14**]. Therefore, it is necessary to prevent the restacking of graphene nanosheets, and to activate the graphene nanosheets for efficient utilization for energy storage [**15**]. For example, Zhang et al. [**16**] made atomic-level modification efforts to limit the restacking and improve the accessibility with the electrolytes to facilitate the ion and electron transport. Tian et al. [**17**] introduced heteroatom groups in the graphene surfaces is one of the most promising methods. Nitrogen atoms share a comparable atomic size with carbon atoms but have higher electronegativity than that of carbon [**18**], showing promise in n-type doping of carbon materials. The nitrogen doping leads to enhanced electrochemical properties in organic, aqueous solution, and ionic liquid electrolytes, which may open a new avenue for the research on the chemical doping of carbon-based materials and their electrochemical device applications [**19**]. N doping can enhance the surface polarity and improve surface wettability of graphene-based materials [**20,21**].The nitrogenous active site can efficiently facilitate charge transfer and enhance the electrochemical activity in graphene-based materials [**22,23,24**]. One method for preparing nitrogen-doped graphene fibers is by uniformly mixing nitrogen-containing substance with graphene oxide solution for wet spinning with further carbonization. For example, Ding et al. [**25**] obtained nitrogen-doped graphene fiber by wet spinning blended pyridine and graphene oxide solution. Guan et al. [**26**] used the microfluidic method to obtain nitrogen-doped graphene fibers by injecting a mixture of graphene oxide and urea solution into microtubes with further thermal carbonization. However, the nitrogen source may destroy the structure of graphene fibers. Currently, there are limited studies to achieve both porous structures and superior electrochemical performances.

In this work, nitrogen-doped graphene fiber electrodes by dip-coating wet-spun graphene oxide fibers with slurry containing small-sized graphene oxide and pyrrole monomers as nitrogen sources and thermal carbonization are designed. The influence of mass ratio of different components on the electrochemical performance of graphene fiber electrodes was systematically investigated. The graphene fibers were successfully doped with nitrogen atoms. Moreover, the nitrogen doping resulted in 159.7% enhancement of area specific capacitance (38.3 mF cm^{-2}) and 159.5% enhancement of energy density 3.40 μWh cm^{-2}), compared with that of the pure graphene fibers. NSG@GF FSSC also demonstrated ultralong cycling life and good conductivity.

2. MATERIALS AND METHODS

2.1. Materials

The graphite oxide powder was bought from the Sixth Element Materials Technology Co., Ltd., Changzhou, China. Polyvinyl alcohol (PVA, 99%) was purchased from Sigma-Aldrich, Shanghai, China. Pyrrole (PY), ferric chloride ($FeCl_3$, 98%), and sulfuric acid (H_2SO_4, 98 wt.%) were purchased from Chemical Reagent Co., Ltd., Shanghai, China. Deionized (DI) water was made with Master-Q15 (resistivity~18.3 MΩ cm), Shanghai, China.

2.2. Preparation of Small-Sized Graphene Oxide (SGO)

A certain amount of graphite oxide was dispersed in abundant DI water and subsequently ultrasonicated by a Biosafer with a power of 312 W for 1 h to form graphene oxide (GO) solution. To remove the unexfoliated graphite oxide, the above dispersion was centrifuged at 3000 rpm for 15 min. Then the supernatant was centrifugated at 8000 rpm for 15 min to remove the small-sized graphene oxide, and condensed to serve as the solution for wet spinning graphene oxide fibers.

The small-sized graphene oxide (SGO) (~150 nm in diameter) was prepared by breaking graphene oxide into smaller pieces with high-power ultrasonication for 20 h. Then the SGO dispersion was centrifuged at 3000 rpm for 30 min to obtain SGO with suitable sizes. Lastly the SGO dispersion was thermally concentrated to a suitable concentration (15 mg mL^{-1}) for further wet spinning.

2.3. Preparation of GF

The graphene oxide fiber (GOF) was wet spun by the method described in our earlier work [27]. A total of 15 mg mL-1 GO dispersion was injected into an ethanol/ DI water (1:3 v/v) coagulation bath containing 5 wt.% CaCl2 through a spinneret (inner diameter of 0.3 mm) at room temperature. By mounting onto a syringe pump (LSP02-1B), the continuous wet-spun GOFs were drawn out from the coagulation and washed by DI water and ethanol. The GOFs were then wound onto a winder and dried under tension by infrared drying at 100 °C for 4 h.

2.4. Preparation of SG@GFs

Small-sized graphene-based fibers (SG@GFs) were fabricated by the method as described below. First, the as-prepared GOFs were immersed in the SGO solution, and dried and solidified at 80 °C. The resultant fibers are denoted as SGO@GOF. SG@GF was prepared by pre-oxidized SGO@GOFs with multiple temperature steps at 120 °C, 150 °C, and 180 °C for 1 h. Then, they were thermally annealed at 800 °C for 3 h under nitrogen protection to form SG@GFs.

2.5. Preparation of NSG@GFs

Core-sheath nitrogen-doped graphene-based fibers (NSG@GFs) were prepared by the following two-step method. The core fiber is GOF, the sheath was fabricated of polypyrrole in SGO. The PPY/SGO@GOFs were fabricated by the "dip-coating" method. First, a certain amount of pyrrole was mixed into SGO slurry. Then, the mixture was homogenized with a high shear dispersion device (FLUKO, FA25-D) at 10,000 rpm for 15 min and cooled in an ice bath for 3 h. Then GOF were dipped in the above mixture for sheath coating, and then immersed in FeCl3 solution to polymerize pyrrole monomers in the SGO sheath. Then, PPY/SGO@GOF was finally obtained after drying at 60°C for 1 h.

NSG@GFs were prepared by pre-oxidizing PPY/SGO@GOFs with multiple temperature steps at 120°C, 150°C, and 180°C for 1 h, respectively. The pre-oxided PPY/SGO@GOFs were thermally annealed by a high-temperature vacuum tubular furnace (OTF-1200X) under nitrogen (N2) purge with a heating rate of 2.5°C/min. Then the core-sheath NSG@GF was produced at 800°C for 3 h, towards the final required NSG@GFs.

2.6. Assembly of NSG@GF-Based FSSCs

The NSG@GF-based FSSCs in aqueous electrolyte were prepared with the following method. A total of 1 g PVA powder was dissolved in 10 mL DI water at 90 °C. The PVA solution was vigorously stirred for 2 h until fully dissolved. Subsequently, 0.98 g H2SO4 (98 wt.%) was added and stirred at room temperature to prepare PVA/H2SO4 gel electrolyte. To fabricate quasi-solid-state fiber-shaped supercapacitor, two fiber electrodes with length of ~7 mm were placed in parallel on a flexible polyethylene terephthalate (PET) substrate with an electrode spacing of ~2 mm. Each end of the fibers was connected to polished copper foils by silver paste. Then assembled NSG@GF FSSCs were solidified for 24 h at room temperature to make the electrolyte sufficiently infiltrate into fiber electrodes.

2.7. Characterization

Scanning electron microscopy (SEM, HITACHI, TM3000, and SU5000) was used to characterize the morphological feature of GF, SG@GF, and NSG@GF samples. The diameters of GF, SG@GF, and NSG@GF were measured with polarizing optical microscope (NIKON ECLIPSE LV100POL). The microstructure and compositional element distribution of the NSG@GF were further investigated by using transmission electron microscopy (TEM, JEM-2100) coupled with energy dispersive spectrometer (EDS) mapping. Element valence states and contents were preceded on an X-ray photoelectron spectroscopy (XPS, ESCALAB250Xi). Fourier transform infrared spectroscopic (FTIR) measurements were conducted on a Nicolet NEXUS-670; where the resolution is 4 cm^{-1}, the scanning wave number range is 4000–400 cm^{-1}, to characterize the chemical structure of GOF, GF, SG@GF, and NSG@GF, where KBr was used to mix with samples to prepare thin films for FTIR measurements.

Electrochemical tests were carried out on an electrochemical workstation (CHI 660E, CH Instruments Inc., Bee Cave, TX, USA) with a two-electrode configuration for analysis of cyclic voltammetry (CV), galvanostatic charge/discharge (GCD) and electrochemical impedance spectroscopy (EIS, 0.01 Hz to 100 kHz). The electrical conductivity (σ) was measured with four-wire resistivity measurement method at room temperature.

3. RESULTS AND DISCUSSION

Figure 1 shows the fabrication process of NSG@GF. The preparation of NSG@GF was performed by combining wet spinning and dip-coating with subsequent thermal carbonization. The core GOF was wet spun by using concentrated LGO dispersion. At the same time, pyrrole monomers were added into aqueous SGOs solution. FeCl3 solution was then carefully dropped into the above mixture to produce PPy@SGO slurry. The slurry was then dip-coated onto GOF due to the hydrophilic characteristics between them. Finally, thermal carbonization was conducted to form the designed core-sheath porous graphene-based composite fiber, NSG@GF. For electrochemical measurements, NSG@GF FSSC was assembled by immersing two pieces of NSG@GF into PVA/H2SO4 gel electrolyte and aligning them in parallel and dried under ambient condition. SG@GF was fabricated by the as-prepared GOF immersed in the SGO solution with subsequent thermal carbonization.

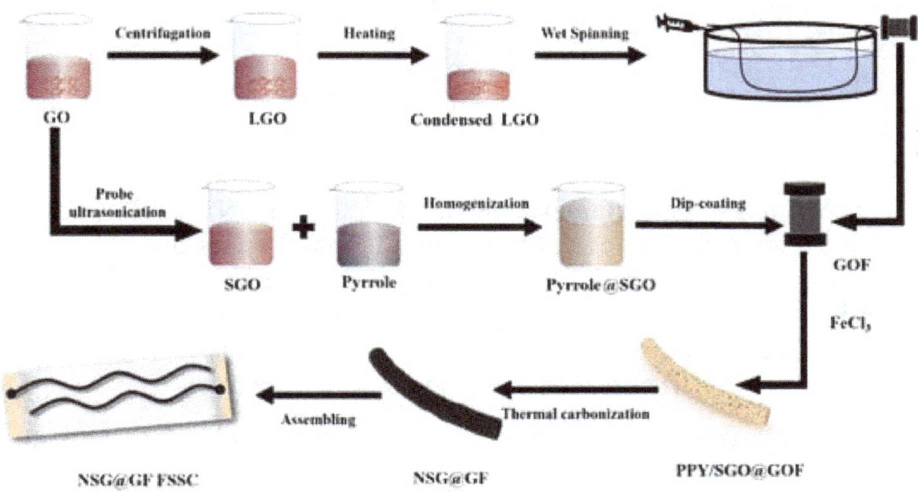

Figure 1. Schematic illustration for the preparation of PPY/SGO@GOF and NSG@GF FSSC.

Figure 2 shows the SEM images of GF, SG@GF, and NSG@GF. Figure 2a exhibits the surface morphology of GF, showing aligned graphene sheets along the fiber axis, which is attributed to the drawing force during the wet spinning process.

The fiber surface shows a typical wrinkled morphology as seen in Figure 2b. Figure 2c shows the cross section of GF, which does appear round, probably due to the irregular shrinkage caused by the rapid evaporation of water.

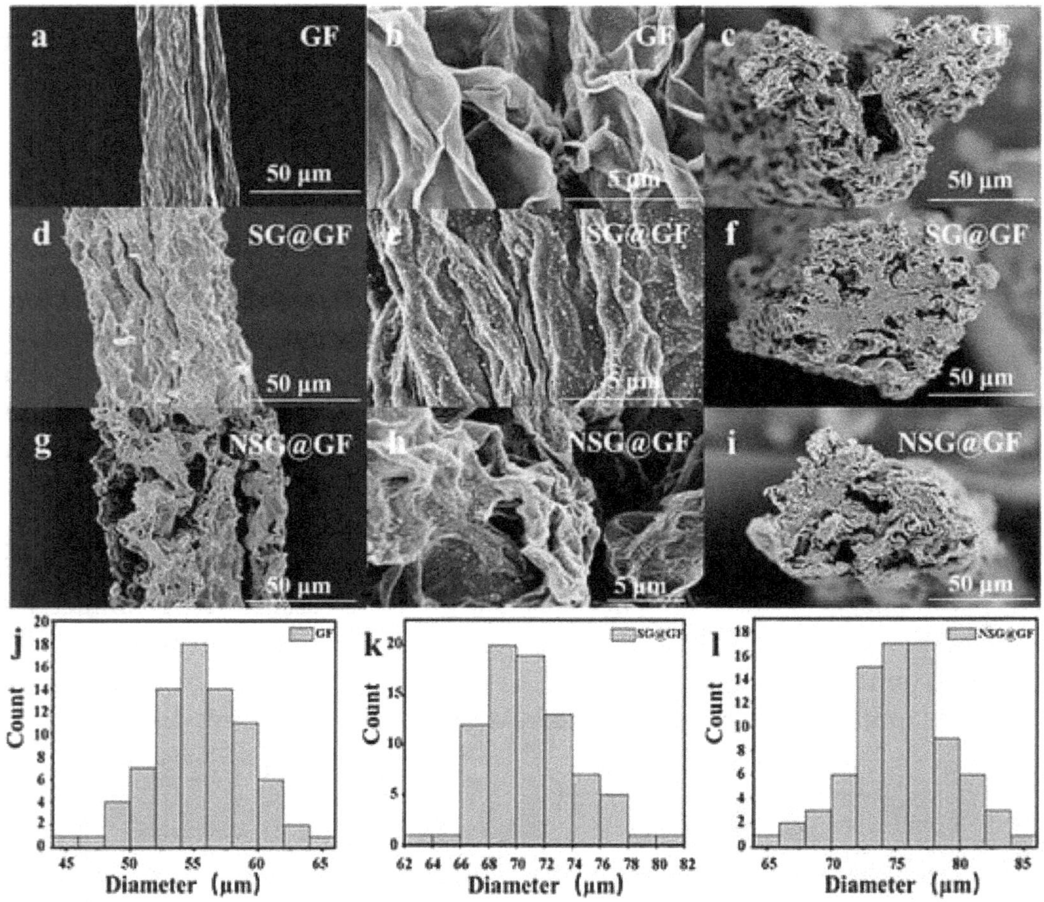

Figure 2. SEM images and diameter of GF, SG@GF, and NSG@GF. (a,b) Surface image of as-prepared GF. (c) Cross-sectional image of as-prepared GF. (d,e) Surface image of as-prepared SG@GF. (f) Cross-sectional image of as-prepared SG@GF. (g,h) Surface image of as-prepared NSG@GF. (i) Cross-sectional image of as-prepared NSG@GF. Diameter distributions of GF (j), SG@GF (k), and NSG@GF (l) obtained by optical microscopy.

In comparison, the surface of SG@GF shows less wrinkled morphology along the fiber axis as seen in **Figure** 2d,e. However, it possesses a slightly round cross-section and apparent interface between core and sheath layers as seen in **Figure** 2f. **Figure** 2j shows the distribution of GF diameters with an average diameter of ~55 μm. While, in **Figure** 2k, SG@GF shows the average diameter of ~72 μm. The average diameter of GF and SG@GF was compared, and it was indicated that SGO solution was successfully coated onto GOF.

It can be clearly seen from **Figure 2**g,h that nitrogen-doping method involves porous structure [**27**] on the surface of NSG@GF, which may be able to increase the specific surface area of NSG@GF. In **Figure 2**i, NSG@GF shows a blurred interface between the core and the sheath, which should be attributed to the existence of C-C cross-link between the graphene layers in both core and sheath [**28**,**29**]. The mean diameter distribution of NSG@GF is 75 μm as seen in **Figure 2**l.

To further identify the nitrogen doping, EDS was conducted to characterize the element distribution of NSG@GF. As seen in **Figure 3**a–d, carbon (C), nitrogen (N), and oxygen (O) elements are evenly distributed on the NSG@GF, indicating the successful nitrogen doping.

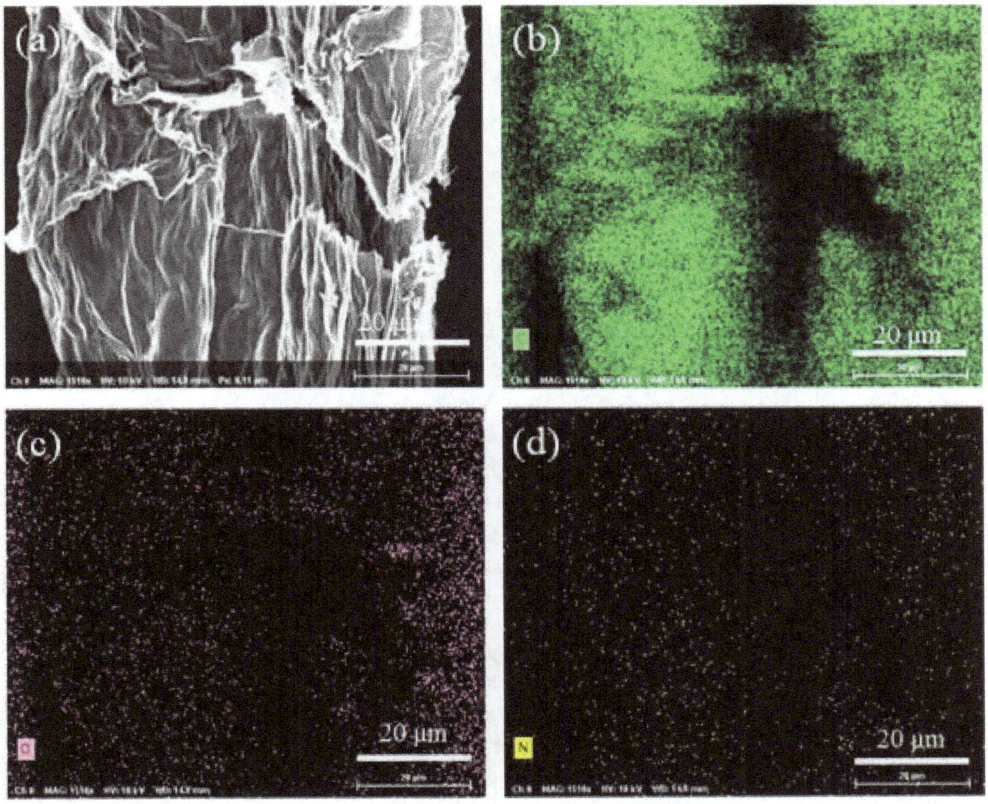

Figure 3. (a) EDX images of NSG@GF and (b–d) element distribution of carbon, oxygen, and nitrogen.

To quantitively characterize the valency and nitrogen doping level, we conducted XPS characterization for NSG@GF. Figure 4 shows N1s spectrum which can be deconvoluted into three different peaks. There are three types of N species in NSG@GF [30,31], which are pyridinic-N (N-6) [32], pyrrolic-N (N-5) [33], and quaternary N (N-Q) [34] locating at 398.4 eV, 399.4 eV, and 401.1 eV, respectively. Different nitrogen doping configurations have different effects. N-6 is easy to go through redox reaction and provide pseudo-capacitance because of its lower energy band values [33]. N-5 forms a five-membered ring structure [35]. N-5 mainly exists on the edge of the graphene nanosheets which can provide additional electrochemical active sites to enhance pseudo-capacitance [36]. N-Q is doped within the graphitic basal plane, which can improve the electrical conductivity and enhance fast charge/discharge [37]. It can be seen from Figure 4 that the N atoms in NSG@GF mostly exist in the form of N-Q and N-6. Hence, the nitrogen-doping method is believed to show large impact on the improvement of electrochemical performance due to the pseudo-capacitive and highly electrical conductive contributions [38].

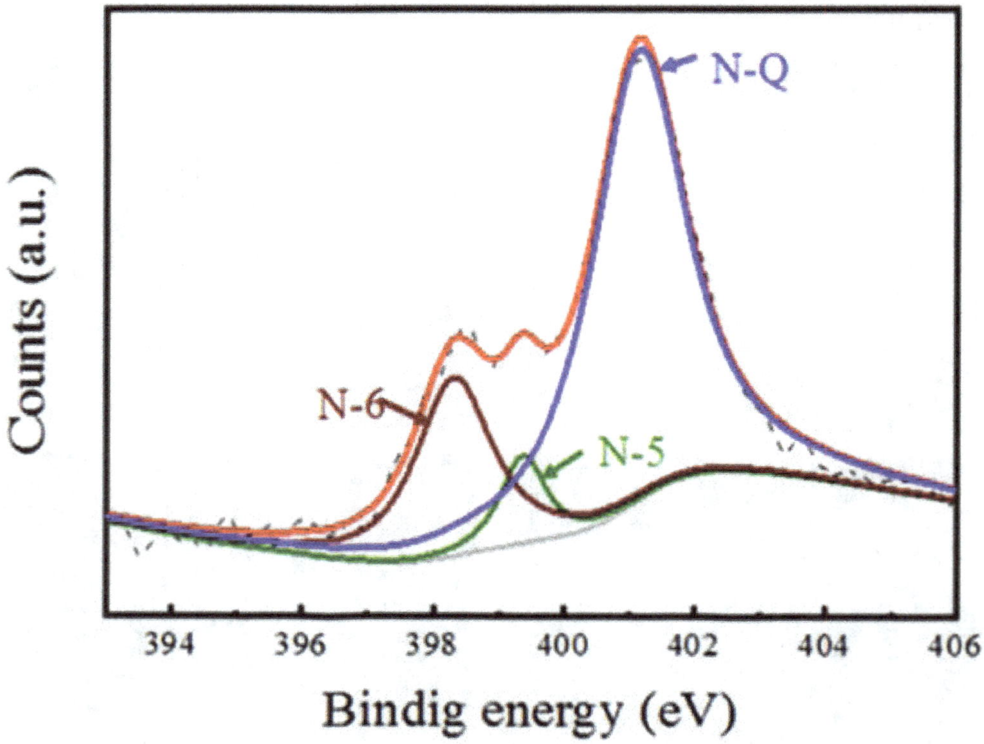

Figure 4. XPS spectra for N1s of NSG@GF.

Furthermore, Figure 5 shows the FTIR patterns of the GO, GOF, SG@GF, and NSG@GF. Distinct peaks are visible at 3483 cm−1, 1714 cm−1, and 1400 cm−1, which originate from -OH, C=O, and the C-O bond, respectively. this indicates that the surfaces of GF, GOF, and SG@GF contain abundant oxygen-contained groups [36]. However, the absorption peaks of -OH, C=O, and C-O become smaller or even disappear after thermal carbonization. Moreover, NSG@GF shows additional characteristic peaks at 2922 cm−1 (-CH), 1632 cm−1 (C=C), and 1115 cm−1 (C-N), confirming that NSG@GF is heavily reduced and doped with nitrogen atoms, which has good agreement with XPS results.

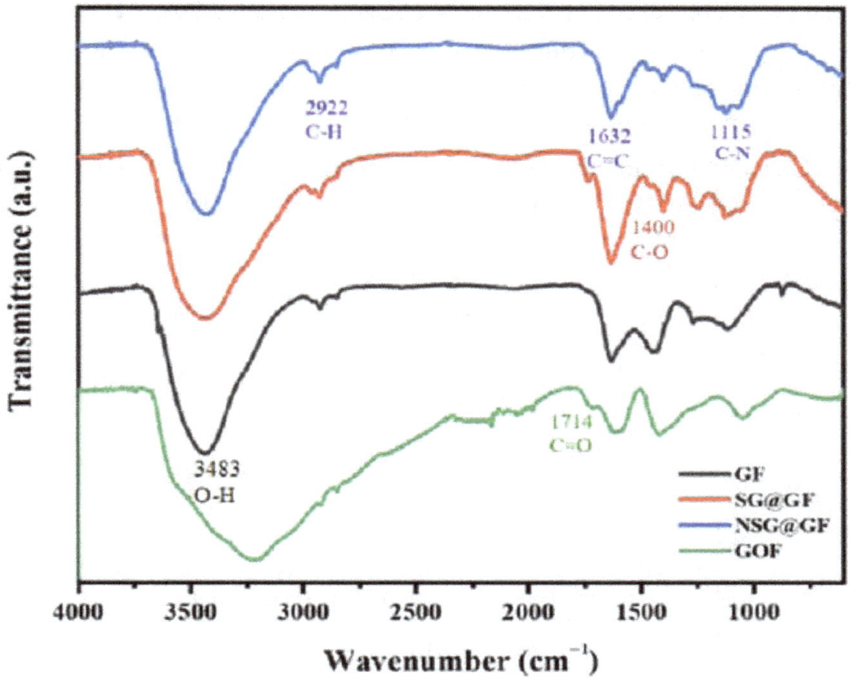

Figure 5. FTIR spectra of GOF, GF, SG@GF, and NSG@GF.

It is seen from Figure 6 that the electrical conductivity of GF, SG@GF, and NSG@GF is 81.79 S cm−1, 47.96 S cm−1, and 155 S cm−1, respectively. The decrease in the conductivity of the SG@GF is due to the increase in the diameter and the internal resistance of the fibers, due to the stacking of the GO layer. In contrast, the increase in the conductivity of the fiber after the nitrogen doping coating is due to the improvement of the electrical conductivity structure of the fiber [39].

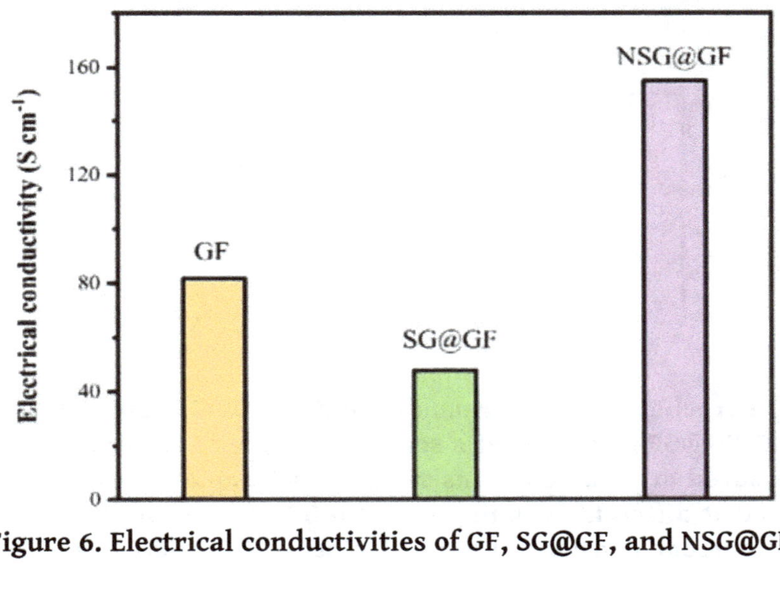

Figure 6. Electrical conductivities of GF, SG@GF, and NSG@GF.

The electrochemical properties of GF, SG@GF, and NSG@GF were summarized in Figure 7. Figure 7a shows the CV curves at 5 mVs-1 for GF and SG@GF, of which the shapes are nearly rectangular [40], indicating typical double-layer capacitance behaviour [41]. The curve area of SG@GF is smaller than that of GF, indicating that pure SG coating deteriorates the electrochemical performance, which may be caused by dense SG coating and poor electrical conductivity.

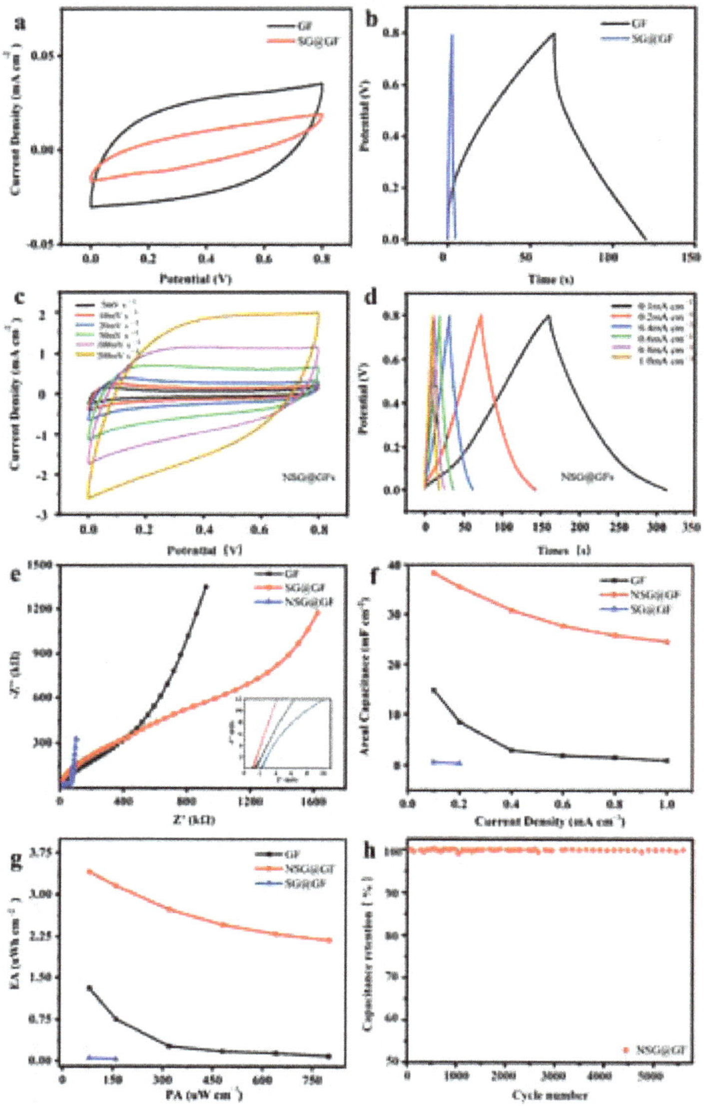

Figure 7. Electrochemical performances of GF, SG@GF, and NSG@GF. (a) CV curves of GF, SG@GF measured at a scan rate of 5 mV s-1. (b) GCD curves of GF, SG@GF measured at a current density of 0.1 mA cm-2. (c) CV curves of NSG@GFs measured at different scanning rate. (d) GCD curves of NSG@GFs measured at different current density. (e) Nyquist plots of GF, SG@GF, and NSG@GF. (f)

Specific capacitances (CA) of GF, SG@GF, and NSG@GF based on GCD test and measured current density. (g) The Ragone plots of the FSSCs for GF, SG@GF, and NSG@GF graphene-based fiber electrodes. (h) Cycle life of NSG@GF.

The GCD curves of GF and SG@GF electrode were measured at a current density of 0.1 mA cm^{-2} (**Figure 7**b). Compared with that of GF (~320 mV), SG@GF has a large voltage drop (IR drop) of 820 mV, indicating the high internal resistance in SG@GF. Moreover, the discharge time for SG@GF is 2.4 s, which is much smaller than that of pure GF. These results confirm the poor structure of direct coating small graphene sheets on GF. Hence, further modification to the graphene-based composite fiber electrodes is needed for better electrochemical performance.

In Figure 7c, all CV curves of NSG@GFs at different scanning rates demonstrate a rectangular-like shape but with broad peaks, indicating rapid current response to voltage scanning, and suggesting typical electrical double-layer capacitor behaviour [3]. However, the broad peak in these curves is considered to be redox reaction during scanning. It is believed to be caused by the nitrogen doping which may have provided extra pseudo-capacitance for enhanced electrochemical performance. The CV curves at 5 mVs−1 for NSG@GF show the smallest CV curve area compared with the other scanning rates. More importantly, it is obvious that the CV curve area of NSG@GFs are much larger than those of GF and SG@GF electrodes (Figure 7a), confirming that the introduction of nitrogen doping can significantly enhance the electrochemical performance. The CV curves at 200 mVs−1 for NSG@GF shows the largest enclosed area of its CV curve.

Furthermore, the discharging curves (Figure 7d) are not symmetric to its corresponding charge counterpart even at low current density (0.1 mA cm−2), indicating the existence of pseudo-capacitance from redox reactions on the shell surface with nitrogen-doped areas in graphene. Compared with the data for GF and SG@GF (Figure 7b), NSG@GF possesses much longer discharging time (~153 s), suggesting much higher energy storage capability.

Figure 7e shows the Nyquist plots of GF, SG@GF, and NSG@GF electrodes. Since electron transfer limited process can be shown by a high frequency, the diffusion process is reflected by a low frequency [42]. In the low frequency region, the NSG@GF has a higher slope, which indicates that the nitrogen-doped fiber has a better charge storage capacity [36]. However, the slope of the pure SG@GF is very small, even lower than that of GF, which indicates that its electrochemical performance is deteriorated [43,44]. Equivalent series resistance (ESR) of GF, SG@GF, and NSG@GF is 1.323 kΩ, 0.9699 kΩ, and 2.076 kΩ. The effect on equivalent series resistance of GF, SG@GF, and NSG@GF is in accordance with the trend in electrical conductivity.

Based on these GCD curves, we calculated their specific capacitances with respect to the current density from 0.1 to 1 mA cm−2 (Figure 7f). It shows that the areal specific capacitance of GF at 0.1 mA cm−2 is 14.75 mF cm−2; however, SG@GF only has a specific capacitance of 0.6 mF cm−2. After nitrogen doping, the areal specific capacitance of NSG@GF is significantly improved to 38.3 mF cm−2, which is believed to be ascribed to the synergetic effect of nitrogen doping and porous structure. Ragone plots for graphene-based fiber supercapacitor is shown in Figure 7g. The SG@GF had the lowest EA (0.05 μWh cm−2), lower than that of GF (1.31 μWh cm−2), but the EA of NSG@GF was improved to 3.40 μWh cm−2 after nitrogen doping.

As observed in Figure 7h, NSG@GF-based FSSC exhibits good cycling performance with the capacitance retention rate of 98.2% over 5000 cycles, illustrating its superior cyclic stability with a long cycle life.

4. CONCLUSIONS

In summary, we report a novel strategy to fabricate high-performance FSSCs assembled by nitrogen-doped core-sheath graphene fibers incorporated by slurry containing small-sized graphene oxide and pyrrole monomers serve as nitrogen sources. NSG@GF FSSCs display high electrical conductivity and excellent electrochemical performance. Moreover, the as-assembled FSSCs exhibit good capacitance and cyclic stability. The facile nitrogen-doped method, unique core-sheath graphene-shaped structure, and superior electrochemical properties endow a new avenue in the fields of fiber-shaped energy storage devices.

5. CONCISE SUMMARY

In this study, a strategy to fabricate nitrogen-doped porous core-sheath graphene fibers with the incorporation of polypyrrole-induced nitrogen doping and graphene oxide for porous architecture in sheath is reported. Polypyrrole/graphene oxide was introduced onto wet-spun graphene oxide fibers by dip-coating. Nitrogen-doped core-sheath graphene-based fibers (NSG@GFs) were obtained with subsequently thermally carbonized polypyrrole/small-sized graphene oxide and graphene oxide fiber slurry (PPY/SGO@GOF). Both nitrogen doping and small-sized graphene sheets can improve the utilization of graphene layers in graphene-based fiber electrode by preventing stacking of the graphene sheets. Enhanced electrochemical performance is achieved due to the introduced pseudo-capacitance and enhanced electrical double-layered capacitance. The specific capacitance (38.3 mF cm−2) of NSG@GF is 2.6 times of that of pure graphene fiber. The energy density of NSG@GF reaches 3.40 μWh cm−2 after nitrogen doping, which is 2.59 times of that of as-prepared one. Moreover, Nitrogen-doped graphene fiber-based supercapacitor (NSG@GF FSSC) exhibits good conductivity (155 S cm−1) and cycle stability (98.2% capacitance retention after 5000 cycles at 0.1 mA cm−2).

REFERENCES

1. Lee, J.W.; Hall, A.S.; Kim, J.-D. A Facile and Template-Free Hydrothermal Synthesis of Mn3O4 Nanorods on Graphene Sheets for Supercapacitor Electrodes with Long Cycle Stability. Chem. Mater. A Publ. Am. Chem. Soc. 2012, 24, 1158–1164. [Google Scholar] [CrossRef]

2. Wang, T.; Xu, Y.; Shi, B.; Gao, S.; Meng, G.; Huang, K. Novel activated N-doped hollow microporous carbon nanospheres from pyrrole-based hyper-crosslinking polystyrene for supercapacitors. React. Funct. Polym. 2019, 143, 104326. [Google Scholar] [CrossRef]

3. Meng, J.; Nie, W.; Zhang, K.; Xu, F.; Ding, X.; Wang, S.; Qiu, Y. Enhancing Electrochemical Performance of Graphene Fiber-Based Supercapacitors by Plasma Treatment. ACS Appl. Mater. Interfaces 2018, 10, 13652–13659. [Google Scholar] [CrossRef]

4. Lu, C.; Meng, J.; Zhang, J.; Chen, X.; Du, M.; Chen, Y.; Hou, C.; Wang, J.; Ju, A.; Wang, X.; et al. Three-Dimensional Hierarchically Porous Graphene Fiber-Shaped Supercapacitors with High Specific Capacitance and Rate Capability. ACS Appl. Mater. Interfaces 2019, 11, 25205–25217. [Google Scholar] [CrossRef]

5. Wu, G.; Yang, Z.; Zhang, Z.; Ji, B.; Hou, C.; Li, Y.; Jia, W.; Zhang, Q.; Wang, H. High performance stretchable fibrous supercapacitors and flexible strain sensors based on CNTs/MXene-TPU hybrid fibers. Electrochim. Acta 2021, 395, 139141. [Google Scholar] [CrossRef]

6. Gopi, C.V.M.; Vinodh, R.; Sambasivam, S.; Obaidat, I.M.; Kim, H.-J. Recent progress of advanced energy storage materials for flexible and wearable supercapacitor: From design and development to applications. J. Energy Storage 2020, 27, 101035. [Google Scholar]

7. Wei, W.; Aifang, Y.; Junyi, Z.; Lin, W.Z. Recent Progress of Functional Fiber and Textile Triboelectric Nanogenerators: Towards Electricity Power Generation and Intelligent Sensing. Adv. Fiber Mater. 2021, 3, 394–412. [Google Scholar]

8. Simon, P.; Gogotsi, Y. Perspectives for electrochemical capacitors and related devices. Nat. Mater. 2020, 19, 1151–1163. [Google Scholar] [CrossRef] [PubMed]

9. You, R.; Liu, Y.Q.; Hao, Y.L.; Han, D.D.; Zhang, Y.L.; You, Z. Laser Fabrication of Graphene-Based Flexible Electronics. Adv. Mater. 2020, 32, 1901981. [Google Scholar] [CrossRef]

10. Xu, D.; Xuan, C.; Li, X.; Luo, Z.; Wang, Z.; Tang, T.; Wen, J.; Li, M.; Xiao, J. Novel helical carbon nanotubes-embedded reduced graphene oxide in three-dimensional architecture for high-performance flexible supercapacitors. Electrochim. Acta 2020, 339, 135912. [Google Scholar] [CrossRef]

11. Zhu, Y.; Stoller, M.D.; Cai, W.; Velamakanni, A.; Piner, R.D.; Chen, D.; Ruoff, R.S. Exfoliation of graphite oxide in propylene carbonate and thermal reduction of the resulting graphene oxide platelets. ACS Nano 2010, 4, 1227–1233. [Google Scholar] [CrossRef] [PubMed]
12. Wang, W.; Xu, G.; Cui, X.T.; Sheng, G.; Luo, X. Enhanced catalytic and dopamine sensing properties of electrochemically reduced conducting polymer nanocomposite doped with pure graphene oxide, Biosensors & Bioelectronics: The International Journal for the Professional Involved with Research. Technol. Appl. Biosensers Relat. Devices 2014, 58, 153–156. [Google Scholar]
13. Gao, W.; Singh, N.; Song, L.; Liu, Z.; Reddy, A.L.M.; Ci, L.; Vajtai, R.; Zhang, Q.; Wei, B.; Ajayan, P.M. Direct laser writing of micro-supercapacitors on hydrated graphite oxide films. Nat. Nanotechnol. 2011, 6, 496–500. [Google Scholar] [CrossRef] [PubMed]
14. Mankge, N.S.; Madito, M.J.; Hlongwa, N.W.; Kuvarega, A.T. Review of electrochemical production of doped graphene for energy storage applications. J. Energy Storage 2022, 46, 103527. [Google Scholar] [CrossRef]
15. Li, Z.; Lin, J.; Li, B.; Yu, C.; Wang, H.; Li, Q. Construction of heteroatom-doped and three-dimensional graphene materials for the applications in supercapacitors: A review. J. Energy Storage 2021, 44, 103437. [Google Scholar] [CrossRef]
16. Ding, D.; Ma, L.; Li, X.; Liu, Z.; Hui, L.; Zhang, F.; Zhao, Y. Porous Carbon Material Derived from Steam-Exploded Poplar for Supercapacitor: Insights into Synergistic Effect of KOH and Urea on the Structure and Electrochemical Properties. Materials 2022, 15, 2741. [Google Scholar] [CrossRef]
17. Tian, W.; Gao, Q.; Tan, Y.; Zhang, Y.; Xu, J.; Li, Z.; Yang, K.; Zhu, L.; Liu, Z. Three-dimensional functionalized graphenes with systematical control over the interconnected pores and surface functional groups for high energy performance supercapacitors. Carbon 2015, 85, 351–362. [Google Scholar] [CrossRef]
18. Du, Y.; Liu, L.; Xiang, Y.; Zhang, Q. Enhanced electrochemical capacitance and oil-absorbability of N-doped graphene aerogel by using amino-functionalized silica as template and doping agent. J. Power Sources 2018, 379, 240–248. [Google Scholar] [CrossRef]
19. Abbas, Q.; Raza, R.; Shabbir, I.; Olabi, A.G. Heteroatom doped high porosity carbon nanomaterials as electrodes for energy storage in electrochemical capacitors: A review. J. Sci. Adv. Mater. Devices 2019, 4, 341–352. [Google Scholar] [CrossRef]

20. Wang, K.; Wang, J.; Wu, Y.; Zhao, S.; Wang, Z.; Wang, S. Nitrogen-doped graphene prepared by a millisecond photo-thermal process and its applications. Org. Electron. 2018, 56, 221–231. [Google Scholar] [CrossRef]

21. Babel, K.; Jurewicz, K. KOH activated carbon fabrics as supercapacitor material. J. Phys. Chem. Solids 2004, 65, 275–280. [Google Scholar] [CrossRef]

22. Xu, B.; Yue, S.; Sui, Z.; Zhang, X.; Hou, S.; Cao, G.; Yang, Y. What is the choice for supercapacitors: Graphene or graphene oxide? Energy Environ. Sci. EES 2011, 4, 2826–2830. [Google Scholar] [CrossRef]

23. Jeong, H.M.; Lee, J.W.; Shin, W.H.; Choi, Y.J.; Shin, H.J.; Kang, J.K.; Choi, J.W. Nitrogen-doped graphene for high-performance ultracapacitors and the importance of nitrogen-doped sites at basal planes. Nano Lett. 2011, 11, 2472–2477. [Google Scholar] [CrossRef] [PubMed]

24. Wang, Q.; Zhong, T.; Wang, Z. Plasma-Engineered N-CoOx Nanowire Array as a Bifunctional Electrode for Supercapacitor and Electrocatalysis. Nanomaterials 2022, 12, 2984. [Google Scholar] [CrossRef]

25. Ding, X.; Bai, J.; Xu, T.; Li, C.; Zhang, H.-M.; Qu, L. A novel nitrogen-doped graphene fiber microelectrode with ultrahigh sensitivity for the detection of dopamine. Electrochem. Commun. 2016, 72, 122–125. [Google Scholar] [CrossRef]

26. Wu, G.; Tan, P.; Wu, X.; Peng, L.; Cheng, H.; Wang, C.-F.; Chen, W.; Yu, Z.; Chen, S. High-Performance Wearable Micro-Supercapacitors Based on Microfluidic-Directed Nitrogen-Doped Graphene Fiber Electrodes. Adv. Funct. Mater. 2017, 27, 1702493. [Google Scholar] [CrossRef]

27. Zheng, X.; Zhang, K.; Yao, L.; Qiu, Y.; Wang, S. Hierarchically porous sheath–core graphene-based fiber-shaped supercapacitors with high energy density. J. Mater. Chem. A 2018, 6, 896–907. [Google Scholar] [CrossRef]

28. Worsley, M.A.; Pauzauskie, P.J.; Olson, T.Y.; Biener, J.; Satcher, J.H., Jr.; Baumann, T.F. Synthesis of graphene aerogel with high electrical conductivity. J. Am. Chem. Soc. 2010, 132, 14067–14069. [Google Scholar] [CrossRef]

29. Cheng, H.; Xue, H.; Zhao, G.; Hong, C.; Zhang, X. Preparation, characterization, and properties of graphene-based composite aerogels via in situ polymerization and three-dimensional self-assembly from graphene oxide solution. RSC Adv. 2016, 6, 78538–78547. [Google Scholar] [CrossRef]

30. Xu, C.; Chen, J.; Li, S.; Gu, Q.; Wang, D.; Jiang, C.; Liu, Y. N-doped honeycomb-like porous carbon derived from biomass as an efficient carbocatalyst for H2S selective oxidation. J. Hazard. Mater. 2021, 403, 123806. [Google Scholar] [CrossRef]

31. Sun, Y.; Zhang, G.; Xu, Y.; Zhang, R. Catalytic performance of dioxide reforming of methane over Co/AC-N catalysts: Effect of nitrogen doping content and calcination temperature. Int. J. Hydrog. Energy 2019, 44, 16424–16435. [Google Scholar] [CrossRef]

32. Zhang, X.; Zhang, G.; Qin, X.; Liu, J.; Li, G.; Xu, Y.; Lv, Y. Catalytic performance of CH4–CO2 reforming over metal free nitrogen-doped biomass carbon catalysts: Effect of different preparation methods. Int. J. Hydrog. Energy 2021, 46, 31586–31597. [Google Scholar] [CrossRef]

33. Li, J.; Han, K.; Wang, D.; Teng, Z.; Cao, Y.; Qi, J.; Li, M.; Wang, M. Fabrication of high performance structural N-doped hierarchical porous carbon for supercapacitors. Carbon 2020, 164, 42–50. [Google Scholar] [CrossRef]

34. Cai, X.; Wang, Q.; Liu, Y. Hybrid of Polyoxometalate-Based Ionic Salt and N-Doped Carbon toward Reductant-Free Aerobic Hydroxylation of Benzene to Phenol. ACS Sustain. Chem. Eng. 2016, 4, 4986–4996. [Google Scholar] [CrossRef]

35. Mehetre, S.S.; Maktedar, S.S.; Singh, M. Understanding the mechanism of surface modification through enhanced thermal and electrochemical stabilities of N-doped graphene oxide. Appl. Surf. Sci. A J. Devoted Prop. Interfaces Relat. Synth. Behav. Mater. 2016, 366, 514–522. [Google Scholar] [CrossRef]

36. Han, F.; Jing, W.; Wu, Q.; Tian, B.; Lin, Q.; Wang, C.; Zhao, L.; Liu, J.; Sun, Y.; Jiang, Z. Nitrogen-doped graphene fiber electrodes with optimal micro-/meso-/macro-porosity ratios for high-performance flexible supercapacitors. J. Power Sources 2022, 520, 230866. [Google Scholar] [CrossRef]

37. Hulicova-Jurcakova, D.; Seredych, M.; Lu, G.Q.; Bandosz, T.J. Combined Effect Of Nitrogen- And Oxygen-Containing Functional Groups of Microporous Activated Carbon on its Electrochemical Performance in Supercapacitors. Adv. Funct. Mater. 2009, 19, 438–447. [Google Scholar] [CrossRef]

38. Dumont, J.H.; Martinez, U.; Artyushkova, K.; Purdy, G.M.; Dattelbaum, A.M.; Zelenay, P.; Mohite, A.; Atanassov, P.; Gupta, G. Nitrogen-doped graphene oxide electrocatalysts for the oxygen reduction reaction. ACS Appl. Nano Mater. 2019, 2, 1675–1682. [Google Scholar] [CrossRef]

39. Zhao, X.; Dong, H.; Xiao, Y.; Hu, H.; Cai, Y.; Liang, Y.; Sun, L.; Liu, Y.; Zheng, M. Three-dimensional Nitrogen-doped graphene as binder-free electrode materials for supercapacitors with high volumetric capacitance and the synergistic effect between nitrogen configuration and supercapacitive performance. Electrochim. Acta 2016, 218, 32–40. [Google Scholar] [CrossRef]

40. Obodo, R.M.; Onah, E.O.; Nsude, H.E.; Agbogu, A.; Nwanya, A.C.; Ahmad, I.; Zhao, T.; Ejikeme, P.M.; Maaza, M.; Ezema, F.I. Performance evaluation of graphene oxide based Co3O4@ GO, MnO2@ GO and Co3O4/MnO2@ GO electrodes for supercapacitors. Electroanalysis 2020, 32, 2786–2794. [Google Scholar] [CrossRef]

41. Nwanya, A.C.; Ndipingwi, M.M.; Ikpo, C.O.; Obodo, R.M.; Nwanya, S.C.; Botha, S.; Ezema, F.I.; Iwuoha, E.I.; Maaza, M. Zea mays lea silk extract mediated synthesis of nickel oxide nanoparticles as positive electrode material for asymmetric supercabattery. J. Alloys Compd. Interdiscip. J. Mater. Sci. Solid State Chem. Phys. 2020, 822, 153581. [Google Scholar] [CrossRef]

42. Wu, H.; Yu, Y.; Gao, W.; Gao, A.; Qasim, A.M.; Zhang, F.; Wang, J.; Ding, K.; Wu, G.; Chu, P.K. Nickel plasma modification of graphene for high-performance non-enzymatic glucose sensing. Sens. Actuators B Chem. 2017, 251, 842–850. [Google Scholar] [CrossRef]

43. Zheng, Y.; Lia, Z.; Xu, J.; Wang, T.; Liu, X.; Duan, X.; Ma, Y.; Zhou, Y.; Pei, C. Multi-channeled hierarchical porous carbon incorporated Co3O4 nanopillar arrays as 3D binder-free electrode for high performance supercapacitors. Nano Energy 2016, 20, 94–107. [Google Scholar] [CrossRef]

44. Iessa, K.H.S.; Zhang, Y.; Zhang, G.; Xiao, F.; Wang, S. Conductive porous sponge-like ionic liquid-graphene assembly decorated with nanosized polyaniline as active electrode material for supercapacitor. J. Power Sources 2016, 302, 92–97. [Google Scholar] [CrossRef]

CHAPTER-10

IMPROVING THE PIEZOELECTRIC CHARACTERISTICS OF FLEXIBLE NANOFIBROUS MEMBRANES THROUGH HIERARCHICAL STRUCTURES AND NANOPARTICLES

1. INTRODUCTION: AN OVERVIEW

In the pursuit of flexible and efficient piezoelectric materials, the integration of hierarchical structures and nanoparticles represents a compelling avenue for enhancing their properties. Piezoelectricity, the ability of certain materials to generate an electric charge in response to mechanical stress, holds immense potential for various applications, including energy harvesting and sensor technologies. This paper explores the innovative approach of improving the piezoelectric properties of flexible nanofibrous membranes through the incorporation of hierarchical structures and nanoparticles. By delving into the synergistic effects of these enhancements, we aim to unveil a deeper understanding of how the combined influence of hierarchical architecture and nanoparticle integration contributes to the overall performance and functionality of piezoelectric nanofibrous materials

With the coming era of smart wearable devices, there is an increasing need for self-powered wearable electronics, and the advancement of wearable electronics is largely dependent on the power output of power sources [**1**,**2**]. The occurrence of piezoelectric nanogenerators (PENGs) has solved the long-standing problem of energy sources for wearable devices [**3**]. As essential components of PENGs, piezoelectric polymers can effectively convert mechanical energy into electric energy [**4**]. In general, piezoelectric polymers possess reasonable piezoelectricity, high mechanical flexibility, sensitivity to voltage change, and low impedance, and have wide applications in gas, liquid, and biological sensors [**5**]. According to their physical structure, piezoelectric polymers can be classified into semicrystalline or amorphous polymers [**6**]. Poly(vinylidene fluoride) (PVDF) is a well-known semicrystalline piezoelectric polymer with strong piezoelectricity in its stretched and polarized forms. Many studies have shown that PVDF nanofibrous membranes with piezoelectric output can be obtained by electrospinning [**7,8**]. In addition, PAN, as a commonly used amorphous polymer, has demonstrated excellent piezoelectric properties. Wang et al. found that electrospun PAN nanofibers can generate stronger piezoelectricity even than PVDF nanofibers due to their high content of planar Sawtooth PAN conformation [**9**]. Unfortunately, it is still a challenge to further substantially improve the piezoelectricity of polymer-based nanomaterials.

In order to solve the above issue, some researchers who are interested in the micro/nanostructured design of piezoelectric composite materials, such as porous membranes [**10**], nanopillar arrays [**11**], and trigonal line-shaped and pyramid-shaped membranes [**12**], aim to simultaneously obtain high electrical performance and flexibility in nanocomposites. For example, Mu et al. used two piezoelectric polymers to achieve a much higher acoustoelectric than their individual counterpart by increasing the density and generating a synergistic piezoelectric effect [**13**]. Hence, the fabrication of hybrid micro-nanofibers with excellent piezoelectric properties and large differences in fiber diameters by selecting two polymers is an essential method.

Another research direction in this field is focused on adding particle doping onto piezoelectric polymers, which can utilize various properties of inorganic nanomaterials to improve the piezoelectric resistance of nanofibrous PENGs. The most commonly used nanoparticles include ZnO [**14**], ZrO [**15**], graphene oxide (GO) [**16**], Al_2O_3 [**17**], TiO_2 [**18**], carbon nanotubes [**19**], and others. Among these, ZnO is a kind of semiconductor material, and the asymmetric central crystal of ZnO and its structure give it piezoelectric properties. Singh et al. reported that ZnO nanorods can significantly enhance the piezoelectric properties of PVDF nanofibrous membranes [**20**]. Sun et al. incorporated ZnO nanorods into PAN nanofibers with improvements in energy-harvesting ability of about 2.7 times [**5**]. It is obvious that enriching hierarchical structures by integrating organic-inorganic materials is an effective and simple approach to increasing the piezoelectric properties of PENGs to a large extent.

In this paper, PAN-PAN (P-PAN), PAN-PVDF (P-PVDF), and PAN-PVA (P-PVA) nanofibrous membranes were prepared by changing the types of polymers or concentrations in order to regulate the hierarchical structure. After comparing the piezoelectric and physical properties, the P-PVDF nanofibrous membrane was selected as the optimal outcome for the in situ growth of ZnO nanorods. The P-PVDF/ZnO membrane shows higher piezoelectric properties and fracture stress based on the hierarchical structure, and can lighten nine commercial LEDs. Thus, this approach offers a new strategy for the enhancement of the piezoelectric properties of PENGs, with potential applications in flexible wearable electronics, health monitoring, and signal sensing.

2. MATERIALS AND METHODS

2.1. Materials

All chemicals used in the experiments were of analytical grade. PVDF (HSV900, Mw~1,000,000) was obtained from Arkema, France. PAN (number average molecular weight, Mw~15,000) was purchased from Shanghai Chenqi Chemical Technology Co., Ltd., Shanghai, China. PVA (degree of polymerization: 1700; degree of hydrolysis: 88%) was provided by Aladdin Biochemical Technology Co., Ltd., Shanghai, China. N, N-dimethylformamide (DMF), hexamethylenetetramine (C6H12N4), zinc nitrate hexahydrate (Zn(NO3)2·6H2O), isopropyl alcohol (IPA), and ammonia hydroxide (NH3·H2O) were purchased from Chengdu Colon Chemicals Co., Ltd., Chengdu, China. Zinc acetate (Zn(AC)2) was obtained from Chengdu Aikeda Chemical Reagent Co., Ltd., Chengdu, China. Acetone was obtained from Sinopharm Chemical Reagent Co., Ltd., Shanghai, China.

2.2. Preparation of Nanofibrous Membranes

First, the electrospinning solutions were prepared by dissolving PAN (10% w/v), PAN (6% w/v) in DMF, PVA (10% w/v) in DI water, and PVDF (14% w/v) in the mixture of DMF and acetone (volume ratio 7:3). The solutions were stirred at room temperature for an appropriate time. Next, nanofibrous membranes were prepared with electrospinning equipment. The corresponding solutions were fed into a 10-milliliter plastic syringe controlled by a syringe pump. The process of producing nanofibers is shown in Figure 1a. A high voltage of 16 KV was applied and electrospinning of two spinning solutions simultaneously was performed at an appropriate flow rate from 0.2 mL·h−1 to 0.6 mL·h−1, with a spinning distance of 13 cm. Nanofibers were collected on a grounded collector for 3 h for the uniform thickness.

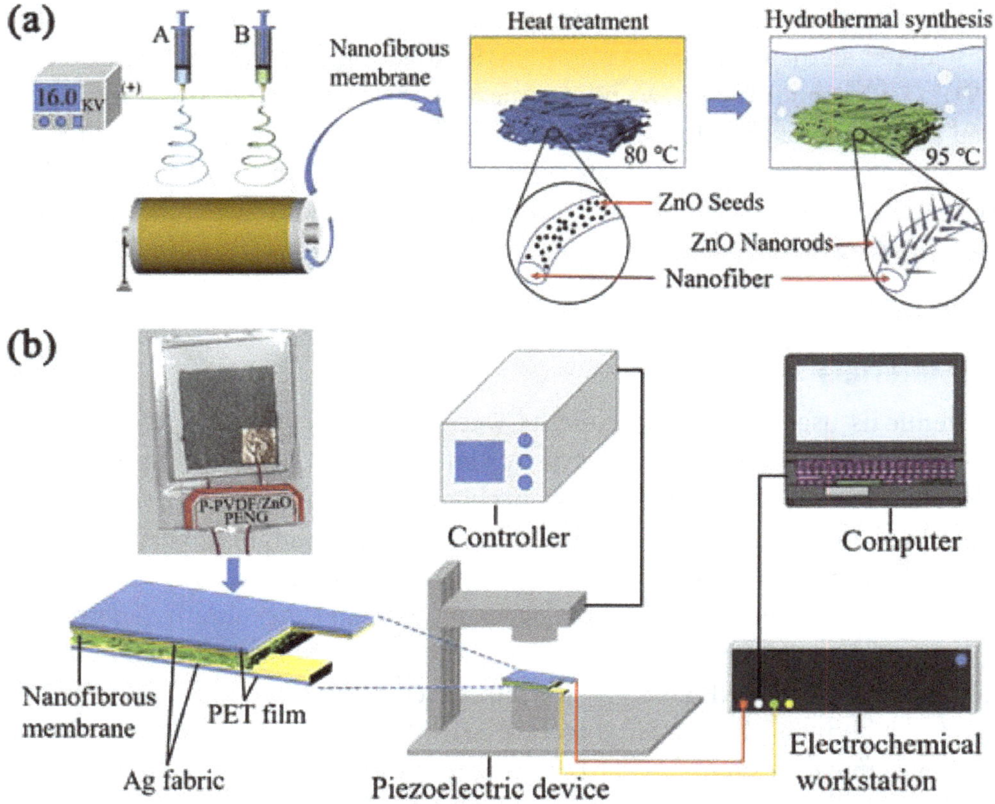

Figure 1. Schematic diagram of the experimental setup. (a) Nanofibers fabricated by double-nozzle electrospinning. (b) The real photograph of the P-PVDF/ZnO PENG and schematic illustration of the determination of piezoelectric properties.

2.3. Preparation of P-PVDF/ZnO Nanofibrous Membrane

Zn (AC)2 was dissolved in IPA (50 mL) and stirred vigorously at 60 °C for 20 min to obtain a seed solution (10 mM). The nanofibrous membranes were immersed in the seed solution for 20 min and then cured at 80 °C for 30 min. This process was repeated three times. All samples prepared with the ZnO seed layer were used for the hydrothermal growth in the next process.

ZnO nanorod arrays were prepared by low-temperature hydrothermal synthesis technology. The standard growth solution consisted of Zn(NO3)2·6H2O (30 mM), C6H12N4 (10 mM), and NH3·H2O (5 mL) and deionized (DI) water (100 mL) with stirring at room temperature for 3 min. The mixed solution and P-PVDF nanofibrous membranes were transferred into a 100-milliliter stainless-steel autoclave and maintained at different temperatures (85, 95, and 105 °C) for 3 h. Finally, the samples with ZnO nanorods were rinsed with a large amount of running DI water, ultrasonically cleaned for 30 min to remove the residue, and then dried in air to obtain P-PVDF/ZnO nanofibrous membranes. ZnO growth solutions with five concentrations were prepared, based on the standard solution as 100% and corresponding composition dilutions from 80% to 20%, which were labeled as shown in Table 1.

Table 1. The sample formulation of growth solutions with the different parameters.

Sample	$Zn(NO_3)^2 \cdot 6H_2O$ (mM)	$C_6H_{12}N_4$ (mM)	$NH_3 \cdot H_2O$ (mL)	DI Water (mL)
100%	30	10	5	100
80%	24	8	4	100
60%	18	6	3	100
40%	12	4	2	100
20%	6	2	1	100

2.4. Fabrication of PENGs

Figure 1b shows the real photograph of the P-PVDF/ZnO PENG and schematically illustrates the structure of piezoelectric device and test setup. All nanofibrous membranes obtained were cut into small pieces with an effective working area of 40 × 40 mm^2. Next, the silver fabric (30 × 30 mm^2) used for electrodes was placed on both sides of the nanofibrous membranes, and conductive copper tapes were attached to these two sliver fabric electrodes to collect the output voltage signals. To protect from any external mechanical damage, the polyethylene terephthalate (PET) membrane was further covered on the PENGs to make an encapsulation using a laminator.

2.5. Characterizations

The surface morphology of the nanofibrous membranes was characterized by scanning electron microscopy (SEM, Quanta-450-FEG + X-MAX50, FEI, Switzerland) and the elements of P-PVDF/ZnO were tested with energy-dispersive spectroscopy (EDS). It is worth noting that all samples should be sputter-coated with gold before surface-morphology observation. Fourier-transform infrared spectroscopy (FTIR) experiments were carried out in attenuated-total-reflectance (ATR) mode, and the spectra of nanofibrous membranes were recorded with a Fourier-transform spectrometer (Spotlight 400, Perkin Elmer, MA, USA) in the range of 650 to 4000 cm−1 with resolution of 4 cm−1. The composition and crystallographic structures were confirmed by X-ray diffraction (XRD, D8 Advance, Bruker, Germany) with Cu Kα radiation operated at 40 kV and 30 mA and the sample was scanned in the 2θ range of 15–70°. Mechanical properties of all membranes cut into 50 mm (length) × 10 mm (width) were measured by a universal testing device (UTM5205X, Shenzhen,

China) with stretching rate of 10 mm/min at room temperature until the sample was broken. The final results were the average value of at least 5 measurements.

A self-built device which consisted of a piezoelectric device, a computer, a controller, and an electrochemistry workstation (Keithley 6514, Tektronix, OR, USA) was used for measuring the piezoelectric properties (as shown in **Figure 1**b). The real-time voltage output of the PENG with dimensions of 30 mm × 30 mm was recorded by the electrochemistry workstation and computer. During testing, the amplitude and frequency of compressive impacts were controlled by the controller. The PENG was repeatedly tapped through piezoelectric device at a fixed frequency of 2 Hz and amplitude of 10 mm.

3. RESULTS AND DISCUSSION

3.1. Morphologies of Different Hierarchies of Nanofibers

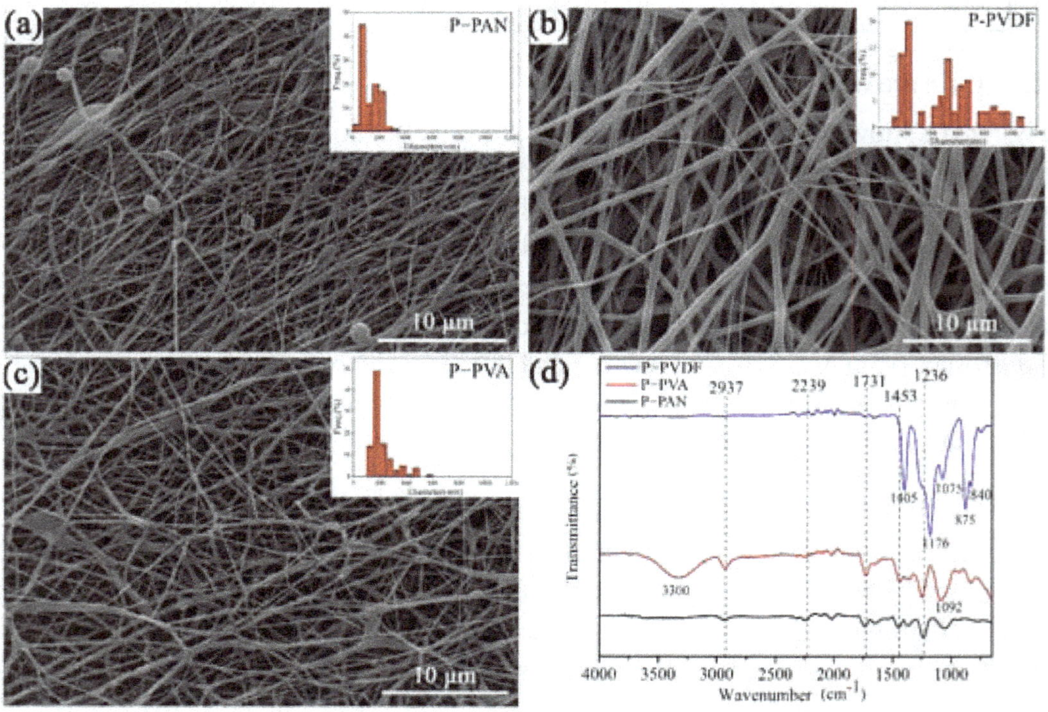

Figure 2. SEM and fiber-diameter distribution of (a) P-PAN, (b) P-PVDF, and (c) P-PVA and (d) FTIR spectra of different micro-nanofibers.

According to **Figure 2**a–c, different morphologies and sizes of electrospun nanofibers were clearly presented due to the distinctive polymer composition of the spinning solutions. All three nanofibers showed one thin part and one thick part of the diameter distribution, but different morphologies. In **Figure 2**a, two distinctive diameters of P-PAN nanofibers, whose average diameters were 120 nm and 230 nm, respectively, are shown, and discontinuous beads can be observed. The beads were caused by the low viscosity of the spinning liquid with low concentration. For the P-PVDF nanofibers in **Figure** 2b, a more obviously hierarchical structure appeared, with average diameters of 210 nm and 640 nm. In addition, all the nanofibers were smooth and uniform, indicating the integration of PAN and PVDF. Hierarchically structured P-PVA nanofibers with diameters 140 nm and 270 nm also occurred. Because the deionized water was the solvent of the PVA, adhesion between the nanofibers and the rough surface was observed.

3.2. FTIR Analysis of Different Hierarchies of Nanofibers

Figure 2d presents the FTIR spectra of the three different nanofibers. The P-PAN membrane exhibited characteristic absorption peaks from the PAN near the wave numbers of 1236 cm−1, 1375 cm−1, 1453 cm−1, 1731 cm−1, 2239 cm−1, and 2937 cm−1. Of these, the characteristic peaks near 1375 and 2937 cm−1 were related to the bending vibration and the contraction vibration of CH2, respectively; the characteristic peak at 2239 cm−1 corresponds to the rocking vibration of the cyan (C≡N); the characteristic absorption intensity near 1731 cm−1 was attributed to carbonyl (C=O) [21]. In addition, the characteristic peak at 1453 cm−1 was related to the in-plane bending of CH. The absorption peak at 1236 cm−1 corresponds to the 31- helical conformation of the identical sequence in the PAN molecular structure [22,23]. For the P-PVA, approximately 3300 cm−1, 1421 cm−1, and 1092 cm−1 correspond to the stretching-vibration peak of hydroxyl (-OH), the bending-vibration peak of CH-OH stretching, and the stretching-vibration-absorption peak of the C-O stretching, respectively [24], which belongs to the characteristic peaks of the PVA. It can be clearly observed that the characteristic peak of the PVDF appeared in the FTIR spectra of the P-PVDF. The characteristic absorption peaks at 765 cm−1, 875 cm−1, 1075 cm−1, 1176 cm−1, and 1405 cm−1 were produced by the α crystal form of the PVDF, and the β crystal has prominent characteristic absorption peaks at 840 cm−1 [25]. Furthermore, the characteristic peaks of the PAN can be seen in all three curves, which indicates the co-existence of two polymers in the nanofibers by double-nozzle electrospinning, as in the results of the SEM.

3.3. Piezoelectric and Mechanical Properties of Different Hierarchical Nanofibers

The results from the piezoelectric test of the P-PAN, P-PVDF, and P-PVA nanofibrous membranes are shown in **Figure** 3a–c, respectively. It is worth noting that all nanofibrous membranes can generate voltage output, but the peak voltage of the P-PVDF PENG can reach 7.2 V, mainly due to the molecular structure of PVDF β crystal phase [**26**]. The peak voltage of the P-PAN PENG was 3.4 V, about half of that of the P-PVDF PENG, because PAN owns the high dipole moment and 3^1-helical [**27,28**]. Further, the output voltage of the P-PVA PENG was only 1.2 V due to the low piezoelectricity of the PVA.

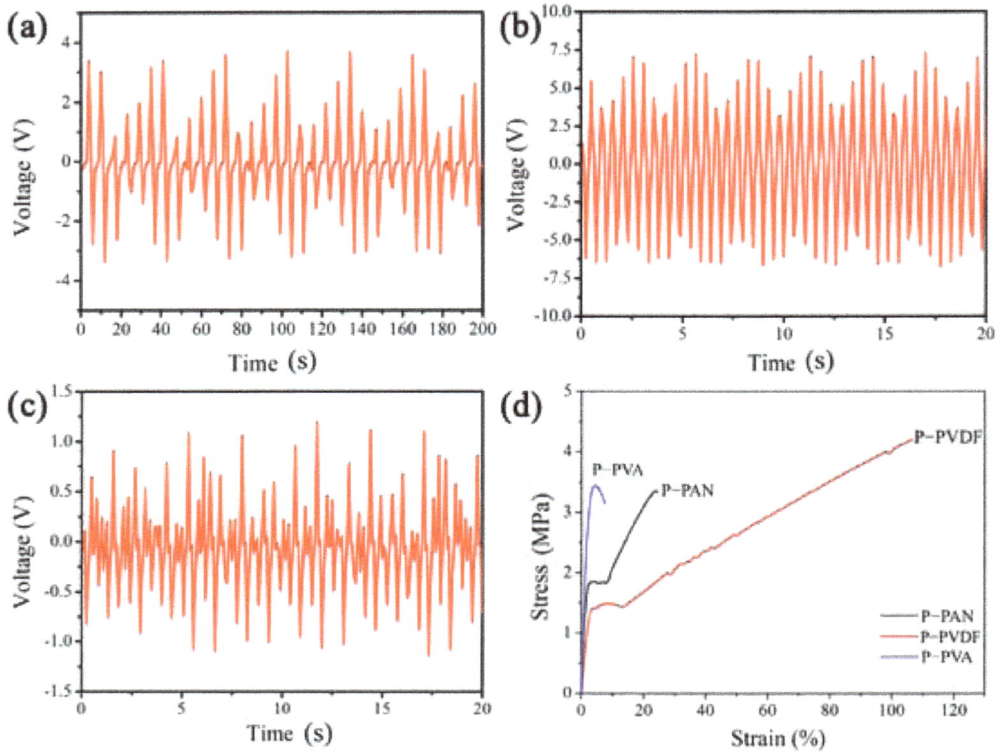

Figure 3. Output voltage generated by the nanofibrous membranes (a) P-PAN, (b) P-PVDF, and (c) P-PVA. (d) Comparison of tensile properties of P-PAN, P-PVDF, and P-PVA nanofibrous membranes.

From the tensile-test curves of the three nanofibrous membranes shown in **Figure 3**d, it can be observed that the fracture stress of the P-PVA and P-PAN nanofibrous membranes was nearly the same, at 3.4 MPa and 3.5 MPa, respectively, while the P-PVDF nanofibrous membrane reached 4.3 MPa. This can be attributed to the excellent uniformity, fineness, and lack of defects of P-PVDF nanofibers. The above results show that the piezoelectric output and mechanical properties of P-PVDF are optimal. Therefore, the P-PVDF was selected to explore the further enhancement of piezoelectric properties through the in situ growth of ZnO nanorods.

3.4. Morphologies of P-PVDF/ZnO Nanofibrous Membranes

Figure 4 shows the SEM image of the P-PVDF/ZnO generated by the ZnO growth solutions with five different concentrations during the hydrothermal reaction. It can be easily observed that the ZnO-growth-solution concentration had a critical effect on the coating and morphology of the ZnO. In the 100%-growth solution in Figure 4a, the ZnO was nanorod-shaped and perfectly embedded into the P-PVDF nanofibers. When the concentration was 80%, the nanorod shape became a nanoneedled structure and the amount of ZnO decreased, as shown in Figure 4b. The reduction in ZnO-growth-solution concentration to 60% led to the scattering of a few nanoclusters over the nanofibers, which can be seen in Figure 4c. Furthermore, no obvious nanorods grew on the surfaces of the P-PVDF nanofibers shown in Figure 4d–e, resulting from the low concentration of the ZnO growth solution. Furthermore, the element distribution of the P-PVDF/ZnO nanofibers was verified via EDS from a single nanofiber (Figure 4f), which demonstrates the successful growth of the ZnO on the P-PVDF nanofibers.

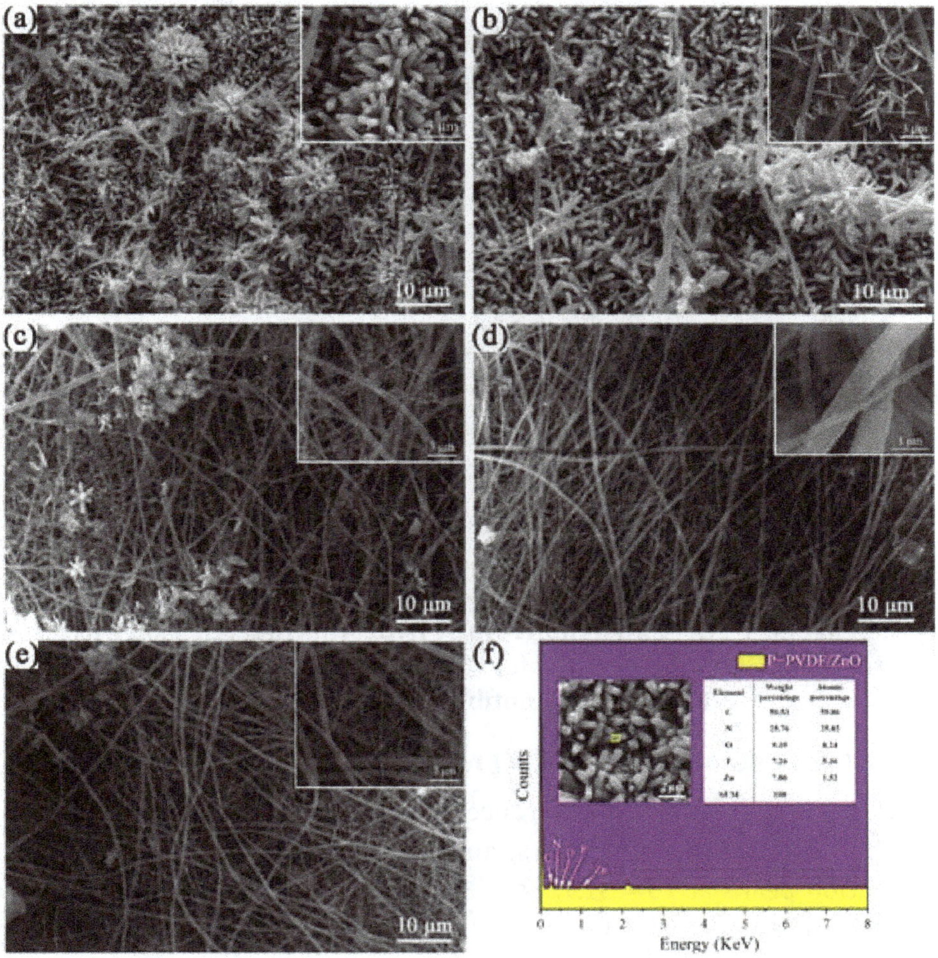

Figure 4. SEM of P-PVDF/ZnO nanofibrous membranes by hydrothermal reaction with different growth solutions: (a) 100%, (b) 80%, (c) 60%, (d) 40%, and (e) 20%. (f) The element distribution of P-PVDF/ZnO nanofibers via EDS.

On the other hand, the growth temperature of ZnO also plays an essential role in the formation of hierarchically structured nanofibers. **Figure** 5a–c shows the difference in the morphologies of the P-PVDF/ZnO nanofibers at 100% ZnO growth solution with temperatures of 85°C, 95°C, and 105°C. A few nanoneedled structures can be seen at 85 °C, because the temperature was too low to favour the growth of ZnO nanorods. However, after the temperature was increased to 95°C, a large quantity of ZnO nanorods completely coated the nanofibers, tightly forming the nanocomposite with the P-PVDF nanofibers. When the temperature reached 105°C, the coverage of the ZnO nanorods dramatically dropped, and some smooth nanofibers were exposed. Hence, this is a simple way to further regulate the hierarchical structure of P-PVDF/ZnO nanofibers by controlling the growth temperature.

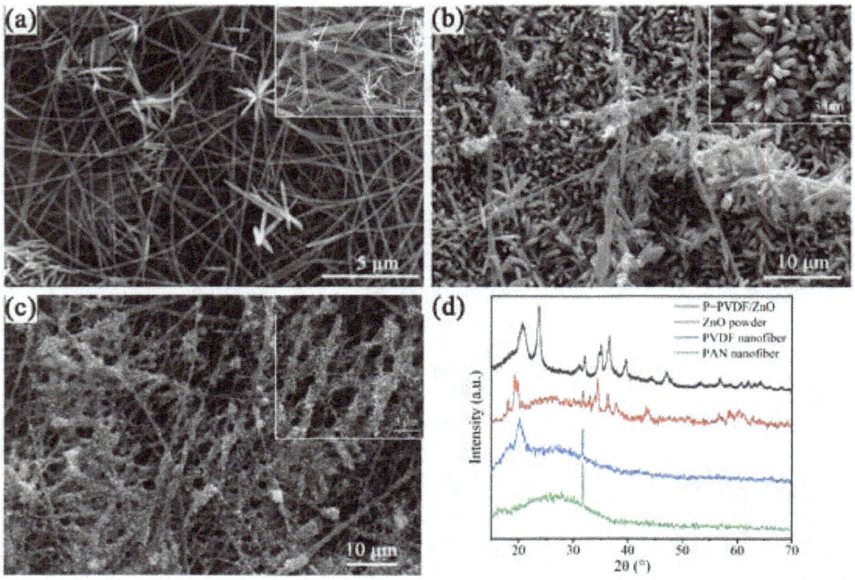

Figure 5. SEM of P-PVDF/ZnO nanofibers by hydrothermal reaction with different temperatures: (a) 85°C, (b) 95°C and (c) 105°C. (d) XRD patterns of PAN, PVDF, and P-PVDF/ZnO nanofibers and ZnO powder.

3.5. XRD Analysis of P-PVDF/ZnO Nanofibers

The XRD patterns prove the difference in the molecular conformation of the PAN, PVDF, P-PVDF/ZnO nanofibers, and ZnO powder. As shown in **Figure 5**d, the diffraction peaks at 2θ = 31.7, 34.4, 36.2, 47.5, 56.6, 62.8, 66.1, 67.9, and 69.0° correspond to the (100), (002), (101), (102), (110), (103), (200), (112), and (201) crystal faces of the ZnO hexagonal wurtzite structure, respectively [**29**]. The diffraction peaks at 2θ = 18.4 and 26.7° correspond to the crystallization peaks of the PVDF [**30**]. As suggested by the XRD pattern of the PVDF nanofiber, the diffraction peak around 20.5° can be attributed to the (110) and (200) planes of the β-phase content of the PVDF [**31**]. The XRD spectrum of the PAN consisted of characteristic peaks at 2θ = 16.8 and 17.0° [**27**]. The diffraction peaks of the PAN, PVDF, and ZnO powder appeared in the XRD patterns of the P-PVDF/ZnO. Hence, the XRD patterns also demonstrated the growth of the ZnO nanorods on the P-PVDF nanofibers.

3.6. Piezoelectric and Mechanical Properties of P-PVDF/ZnO Nanofibers

In order to further enhance the piezoelectric properties of the P-PVDF nanofibrous membrane, typical ZnO nanorods were hydrothermally grown on the surfaces of the P-PVDF nanofibers, since ZnO nanoparticles have a relative high surface activity. **Figure 6a**,b shows that the P-PVDF/ZnO PENG had the highest piezoelectric performance, yielding a peak voltage of 16.0 V. Compared with those of the PENGs without the ZnO nanorods, the output voltage of the P-PVDF/ZnO PENG was significantly increased, at least doubling. There are several reasons for this phenomenon. Firstly, ZnO nanorods can contribute to the formation of □-phase crystallization of PVDF and the zigzag conformation of PAN molecular chains [**5,15**], resulting in the enhancement of piezoelectric properties of P-PVDF/ZnO PENG. Secondly, P-PVDF/ZnO PENGs with hierarchical structures produce greater deformation under compressive force, resulting in a larger voltage output. Finally, PVDF, PAN, and ZnO have intrinsic piezoelectric properties, creating synergistic piezoelectric effects.

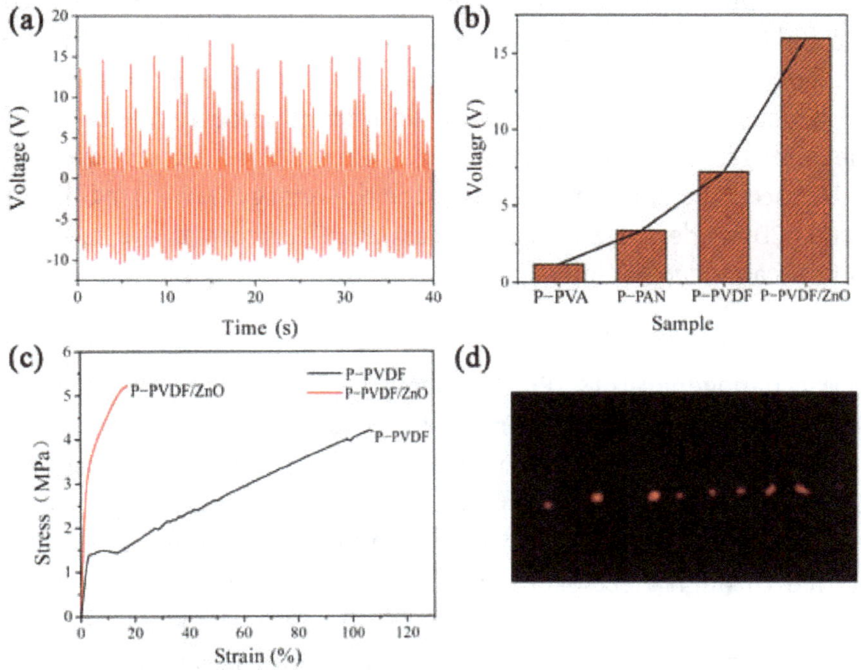

Figure 6. (a) Voltage output of P-PVDF/ZnO PENG. (b) Changes in voltage output with different PENGs. (c) Comparison of tensile stress between P-PVDF and P-PVDF/ZnO nanofibrous membranes. (d) Nine commercial LEDs were lit by tapping P-PVDF/ZnO PENG (working area, 9 cm^2; impact frequency 2 Hz).

Furthermore, the mechanical properties were also improved by adding the ZnO nanorods. As shown in **Figure 6**c, the tensile strength of the P-PVDF/ZnO nanofibrous membrane was 20% higher than that of the P-PVDF nanofibrous membrane, but the breaking elongation of the P-PVDF/ZnO nanofibrous membrane was sharply reduced to about one sixth of that of the P-PVDF nanofibrous membrane, indicating the change from toughness to rigidity. This was due to the deformation of the molecules of the polymers at 95 °C during the hydrothermal reaction and the reinforcement of the ZnO nanorods fully embedded as nanofillers. Notably, the PENG could be used in energy-harvesting applications, such as flexible electronics, by transferring mechanical energy to electrical energy. As shown in **Figure 6**d, the energy generated under 2 Hz lasting pressure was usable and nine commercial LEDs were powered.

4. CONCLUSIONS

In summary, hierarchically structured nanofibers can be easily fabricated by the use of two polymers in electrospinning and by in situ growth of nanoparticles in hydrothermal reactions. The SEM images clearly show the two-part diameter distribution of the nanofibers, as well as the full embedment of the ZnO nanorods as reinforcement fillers. The results of the FTIR and XRD confirmed every component in the nanocomposites. Furthermore, owing to the synergistic piezoelectric effect of organic and inorganic piezoelectric materials, namely the typical PVDF, the PAN polymers, and the hierarchically arranged ZnO nanorods, both the mechanical and the piezoelectric performance were significantly boosted, showing 5.25 MPa of tensile strength and 16.0 V of voltage output. Further, the designed PENG with the hierarchical structure was used for the energy harvesting, lighting nine commercial LEDs under 2 Hz of lasting pressure. Therefore, the high-performance flexible PENG can be potentially applied to wearable electronics.

5. CONCISE SUMMARY

Piezoelectric nanogenerators (PENGs) show superiority in self-powered energy converters and wearable electronics. However, the low power output and ineffective transformation of mechanical energy into electric energy l limit the role of PENGs in energy conversion and storage devices, especially in fiber-based wearable electronics. Here, a PAN-PVDF/ZnO PENG with a hierarchical structure was designed through electrospinning and a hydrothermal reaction. Compared with other polymer nanofibers, the PAN-PVDF/ZnO nanocomposites not only showed two distinctive diameter distributions of uniform nanofibers, but also the complete coverage and embedment of ZnO nanorods, which brought about major improvements in both mechanical and piezoelectric properties. Additionally, a simple but effective method to integrate the inorganic nanoparticles into different polymers and regulate the hierarchical structure by altering the types of polymers, concentrations of spinning solutions, and growth conditions of nanoparticles is presented. Further, the designed P-PVDF/ZnO PENG was demonstrated as an energy generator to successfully power nine commercial LEDs. Thus, this approach reveals the critical role of hierarchical structures and processing technology in the development of high-performance piezoelectric nanomaterials.

REFERENCES

1. Zaszczynska, A.; Gradys, A.; Sajkiewicz, P. Progress in the Applications of Smart Piezoelectric Materials for Medical Devices. *Polymers* **2020**, *12*, 2754. **[Google Scholar] [CrossRef] [PubMed]**
2. Wang, Z.; Tan, L.; Pan, X.; Liu, G.; He, Y.; Jin, W.; Li, M.; Hu, Y.; Gu, H. Self-Powered Viscosity and Pressure Sensing in Microfluidic Systems Based on the Piezoelectric Energy Harvesting of Flowing Droplets. *ACS Appl. Mater. Inter.* **2017**, *9*, 28586–28595. **[Google Scholar] [CrossRef] [PubMed]**
3. Kaczmarek, H.; Krolikowski, B.; Chylinska, M.; Klimiec, E.; Bajer, D. Piezoelectric Films Based on Polyethylene Modified by Aluminosilicate Filler. *Polymers* **2019**, *11*, 1345. **[Google Scholar] [CrossRef] [PubMed][Green Version]**
4. Tamate, R.; Mizutani Akimoto, A.; Yoshida, R. Recent Advances in Self-Oscillating Polymer Material Systems. *Chem. Rec.* **2016**, *16*, 1852–1867. **[Google Scholar] [CrossRef] [PubMed]**
5. Sun, Y.; Liu, Y.; Zheng, Y.; Li, Z.; Fan, J.; Wang, L.; Liu, X.; Liu, J.; Shou, W. Enhanced Energy Harvesting Ability of ZnO/PAN Hybrid Piezoelectric Nanogenerators. *ACS Appl. Mater. Inter.* **2020**, *12*, 54936–54945. **[Google Scholar] [CrossRef] [PubMed]**
6. Kaczmarek, H.; Królikowski, B.; Klimiec, E.; Chylińska, M.; Bajer, D. Advances in the study of piezoelectric polymers. *Russ. Chem. Rev.* **2019**, *88*, 749–774. **[Google Scholar] [CrossRef]**
7. Zhang, D.; Zhang, X.; Li, X.; Wang, H.; Sang, X.; Zhu, G.; Yeung, Y. Enhanced piezoelectric performance of PVDF/BiCl3/ZnO nanofiber-based piezoelectric nanogenerator. *Eur. Polym. J.* **2022**, *166*, 110956. **[Google Scholar] [CrossRef]**
8. Lu, L.; Ding, W.; Liu, J.; Yang, B. Flexible PVDF based piezoelectric nanogenerators. *Nano Energy* **2020**, *78*, 105251. **[Google Scholar] [CrossRef]**
9. Wang, W.; Zheng, Y.; Jin, X.; Sun, Y.; Lu, B.; Wang, H.; Fang, J.; Shao, H.; Lin, T. Unexpectedly high piezoelectricity of electrospun polyacrylonitrile nanofiber membranes. *Nano Energy* **2019**, *56*, 588–594. **[Google Scholar] [CrossRef]**
10. Mao, Y.; Zhao, P.; McConohy, G.; Yang, H.; Tong, Y.; Wang, X. Sponge-Like Piezoelectric Polymer Films for Scalable and Integratable Nanogenerators and Self-Powered Electronic Systems. *Adv. Energy Mater.* **2014**, 4, 1301624. **[Google Scholar] [CrossRef]**
11. Chen, X.; Shao, J.; An, N.; Li, X.; Tian, H.; Xu, C.; Ding, Y. Self-powered flexible pressure sensors with vertically well-aligned piezoelectric nanowire arrays

for monitoring vital signs. *J. Mater. Chem. C* **2015**, *3*, 11806–11814. **[Google Scholar] [CrossRef]**

12. Hu, X.; Jiang, Y.; Ma, Z.; He, Q.; He, Y.; Zhou, T.; Zhang, D. Highly Sensitive P(VDF-TrFE)/BTO Nanofiber-Based Pressure Sensor with Dense Stress Concentration Microstructures. *ACS Appl. Polym. Mater.* **2020**, *2*, 4399–4404. **[Google Scholar] [CrossRef]**
13. Mu, J.; Xian, S.; Yu, J.; Zhao, J.; Song, J.; Li, Z.; Hou, X.; Chou, X.; He, J. Synergistic Enhancement Properties of a Flexible Integrated PAN/PVDF Piezoelectric Sensor for Human Posture Recognition. *Nanomaterials* **2022**, *12*, 1155. **[Google Scholar] [CrossRef] [PubMed]**
14. Que, M.; Lin, C.; Sun, J.; Chen, L.; Sun, X.; Sun, Y. Progress in ZnO Nanosensors. *Sensors* **2021**, *21*, 5502. **[Google Scholar] [CrossRef]**
15. Naik, R.; S, M.; Chavan, S. Piezoelectric property investigation on PVDF/ZrO2/ZnO nanocomposite for energy harvesting application. *Eng. Res. Express* **2021**, *3*, 025003. **[Google Scholar] [CrossRef]**
16. Zeyrek Ongun, M.; Oguzlar, S.; Doluel, E.C.; Kartal, U.; Yurddaskal, M. Enhancement of piezoelectric energy-harvesting capacity of electrospun □-PVDF nanogenerators by adding GO and rGO. *J. Mater. Sci. Mater. El.* **2019**, *31*, 1960–1968. **[Google Scholar] [CrossRef]**
17. Yu, Z.; Chen, X.; Su, Y.; Lian, H.; Lu, J.; Zhou, J.; Liu, P. Hot-press sintering $K_{0.5}Na_{0.5}NbO_3$–0.5 mol% Al_2O_3 ceramics with enhanced ferroelectric and piezoelectric properties. *J. Mater. Sci.* **2019**, *54*, 13457–13466. **[Google Scholar] [CrossRef]**
18. Zhu, J.; Sun, H.; Xu, Y.; Liu, T.; Hou, T.; Liu, L.; Li, Y.; Lin, T.; Xin, Y. Preparation of PVDF/TiO2 nanofibers with enhanced piezoelectric properties for geophone applications. *Smart Mater. Struct.* **2019**, *28*, 085006. **[Google Scholar] [CrossRef]**
19. Wu, C.M.; Chou, M.H.; Zeng, W.Y. Piezoelectric Response of Aligned Electrospun Polyvinylidene Fluoride/Carbon Nanotube Nanofibrous Membranes. *Nanomaterials* **2018**, *8*, 420. **[Google Scholar] [CrossRef][Green Version]**
20. Singh, H.H.; Khare, N. Flexible ZnO-PVDF/PTFE based piezo-tribo hybrid nanogenerator. *Nano Energy* **2018**, *51*, 216–222. **[Google Scholar] [CrossRef]**
21. Jin, S.Y.; Kim, M.H.; Jeong, Y.G.; Yoon, Y.I.; Park, W.H. Effect of alkaline hydrolysis on cyclization reaction of PAN nanofibers. *Mater. Design* **2017**, *124*, 69–77. **[Google Scholar] [CrossRef]**
22. Rizzo, P.A.; Auriemma, F.; Guerra, G.; Petraccone, V.; Corradini, P. Conformational Disorder in the Pseudohexagonal Form of Atactic

Polyacrylonitrile. *Macromolecules* **1996**, *29*, 8852–8861. **[Google Scholar]** **[CrossRef]**

23. Grobelny, J.; Sokól, M.; Turska, E. A study of conformation, configuration and phase structure of polyacrylonitrile and their mutual dependence by means of WAXS and 1H BL-n.m.r. *Polymer* **1984**, *25*, 1415–1418. **[Google Scholar]** **[CrossRef]**
24. Khan, M.Q.; Kharaghani, D.; Ullah, S.; Waqas, M.; Abbasi, A.M.R.; Saito, Y.; Zhu, C.; Kim, I.S. Self-Cleaning Properties of Electrospun PVA/TiO_2 and PVA/ZnO Nanofibers Composites. *Nanomaterials* **2018**, *8*, 644. **[Google Scholar]** **[CrossRef]** **[PubMed][Green Version]**
25. Yang, T.; Pan, H.; Tian, G.; Zhang, B.; Xiong, D.; Gao, Y.; Yan, C.; Chu, X.; Chen, N.; Zhong, S.; et al. Hierarchically structured PVDF/ZnO core-shell nanofibers for self-powered physiological monitoring electronics. *Nano Energy* **2020**, *72*, 104706. **[Google Scholar]** **[CrossRef]**
26. Zhu, G.D.; Zeng, Z.G.; Zhang, L.; Yan, X.J. Piezoelectricity in β-phase PVDF crystals: A molecular simulation study. *Comput. Mater. Sci.* **2008**, *44*, 224–229. **[Google Scholar]** **[CrossRef]**
27. Wang, Z.Y.; Su, K.H.; Fan, H.Q.; Wen, Z.Y. Possible reasons that piezoelectricity has not been found in bulk polymer of polyvinylidene cyanide. *Polymer* **2008**, *49*, 2542–2547. **[Google Scholar]** **[CrossRef]**
28. Minagawa, M.; Miyano, K.; Takahashi, M. Infrared characteristic absorption bands of highly isotactic poly(acrylonitrile). *Macromolecules* **1988**, *21*, 2387–2391. **[Google Scholar]** **[CrossRef]**
29. Arul Hency Sheela, J.; Lakshmanan, S.; Manikandan, A.; Arul Antony, S. Structural, Morphological and Optical Properties of ZnO, ZnO:Ni^{2+} and ZnO:Co^{2+} Nanostructures by Hydrothermal Process and Their Photocatalytic Activity. *J. Inorg. Organomet. P.* **2018**, *28*, 2388–2398. **[Google Scholar]** **[CrossRef]**
30. Mahato, P.K.; Seal, A.; Garain, S.; Sen, S. Effect of fabrication technique on the crystalline phase and electrical properties of PVDF films. *Mater. Sci-Poland* **2015**, *33*, 157–162. **[Google Scholar]** **[CrossRef]**
31. Kanik M, A.O.; Sen, H.S.; Durgun, E.; Bayindir, M. Spontaneous High Piezoelectricity in Poly(vinylidene fluoride) Nanoribbons Produced by Iterative Thermal Size Reduction Technique. *ACS Nano* **2014**, *8*, 9311–9323. **[Google Scholar]** **[CrossRef]**

CHAPTER-11

ACCURATE MANAGEMENT OF PROTON EXCHANGE MEMBRANE FABRICATION USING DIRECT ELECTROSTATIC DEPOSITION

1. INTRODUCTION: AN OVERVIEW

In the realm of advanced materials for energy conversion and storage, proton exchange membranes (PEMs) play a pivotal role, particularly in fuel cell technology. The precise control of PEM preparation is a critical aspect in optimizing their performance and durability. One promising avenue for achieving this control is through the utilization of direct electrostatic deposition techniques. This innovative approach offers a means to intricately manage the fabrication process, allowing for the creation of PEMs with tailored structures and properties.

The use of fossil fuels brings about tremendous problems for resources and the environment, such as greenhouse effects, acid rain, ozone depletion, etc. [**1**,**2**]. Among them, greenhouse effects have caused concern around the world due to their serious effect on the environment and climate. Global decarbonization is of great importance, and China has put forward its carbon-neutral strategy [**3**]. Therefore, the research and development of new energy conversion devices will be vigorously promoted. However, it is difficult to apply renewable energy sources (such as solar energy and wind energy) continuously and stably due to their instability and intermittence during generation [**4**]. To tackle this issue, the employment of electrochemical energy storage systems, especially direct methanol fuel cells (DMFCs), has received wide attention throughout the world [**5**,**6**,**7**,**8**]. In the future, they will play a major role in improving energy efficiency and reducing fossil fuels. In DMFC components, proton exchange membranes (PEMs) act as proton-conductive mediums for protons as well as barriers for the passage of electrons and fuels between the anode and cathode components [**9**,**10**,**11**,**12**]. PEM is one of the key components which can directly affect the performance of DMFCs. Perfluorinated sulfonic acid resin, such as Nafion from Dupont, has been widely used as the PEM in DMFC because of its excellent chemical stability, and good mechanical strength derived from the hydrophobic PTFE backbone. Furthermore, the ionic domains formed between the hydrophilic–SO_3H in the side chain and the hydrophobic PTFE backbone in the Nafion structure could provide good proton conductivity (≥0.1 S cm^{-1}), which ensures its practical applications [**13**]. However, some drawbacks, such as high cost, high methanol permeability, and low proton conductivity under low humidity conditions, drastically limit the widespread commercial application of Nafion in fuel cells.

Recently, nanocomposites have raised a lot of research interest in the preparation of PEMs due to the significant improvement in performance based on the nature of nanomaterials, including nanoparticles, nanowires, nanofibers, nanosheets, etc. [**14,15**]. Among these nanomaterials, silica has attracted the greatest interest due to its high specific surface area and convenient surface modification. Much literature has proved that the addition of inorganic silica to PEM polymers can improve thermal stability, proton conductivity as well as methanol resistance [**16**]. Zhao et al. [**17**] prepared composite PEMs by doping amino-functionalized mesoporous silica (AMS) with SP/IL (N-ethylimidazole trifluoromethanesulfonate and found that amino-functionalized mesoporous silica contributed to the proton transfer due to large lumen channels and acid–base pairs between $-NH_2$ and $-SO_3H$. The prepared composite membrane with AMS reached a high proton conductivity of 1.494 mS/cm under anhydrous conditions at 200 °C, which is four times that of the composite membrane with pure silica. Mahdavi et al. [**18**] presented a novel nanocomposite PEM containing sulfonated polysulfone, metal–organic frameworks and silica nanoparticles. The combination of silica nanoparticles and MOFs in a matrix can act as proton hopping sites to enhance the transport efficiency of protons. Results showed that the prepared PEMs containing 5% nanoparticles demonstrated a high proton conductivity of 17 mS/cm at 70 °C and a maximum power density of 40.80 mW/cm^2. These experiments proved that functional silica is of great significance to the performance of PEMs.

Nowadays, the strategies and techniques for the preparation of PEM mainly include recasting or blending [**19,20,21**], hot-pressing [**22,23**] and impregnation [**24,25,26,27**]. Recasting is a simple and low-cost membrane formation method that can offer easy optimization of the processing parameters. The primary requirement for this method is to have the materials well-dissolved in the solvent to ensure the solution is uniform and homogeneous. Hot-pressing is a method for preparing PEMs by means of the difference in melting temperatures of poly-materials. The dense membrane can be prepared by this method only at high temperatures and pressure. Ballengee [**28**] prepared composite PEMs via hot pressing (127 °C and

15,000 psi) and annealing (from 130 °C to 250 °C). In this process, melted Nafion flowed into the void space between the polyphenylsulfone nanofibers to create a fully dense membrane structure. As the name suggests, the impregnation method refers to incorporating porous materials in a polymer matrix to form a dense membrane [**29**]. Similar to hot-pressing, the impregnated membrane is prepared by filling voids of porous materials with the polymer matrix solution. The nanofiber composite membranes are frequently prepared using this route. These methods have their own advantages; however, the precise control of the preparation process and the preparation of ultrathin composite membranes still remain major challenges for them. However, the membrane size and thickness can only be controlled by the volume of the casting solution roughly.

Recently, the direct membrane formation method has been reported for simplifying and optimizing the fabrication process of MEAs. Klingele et al. [30] directly deposited a Nafion® dispersion onto gas diffusion electrodes with catalyst layers as membrane layers, and then pressed two electrodes together with the membrane layers facing each other. This approach constructed the relatively thinner PEM in MEAs to strongly decrease the contact resistance of the membrane and the proton conducting phase of the catalyst layer. Their directly deposited MEAs demonstrated a high power density up to 4.07 W/cm2 under H2/O2 single cell performance test. Breitwieser et al. [31] presented a novel method of MEA preparation by combining scalable deposition and electrospinning to achieve the manufacturing of MEAs with a controlled 3D design; the fabricated composite membranes showed an ultra-thin thickness of 12 μm. These studies demonstrated deposition can improve the freedom degrees of complex MEAs design.

In this work, we present a novel membrane preparation technology of direct electrostatic deposition (DED), where the membrane is directly prepared on a substrate via electrostatic spraying, which is similar to the electrospinning technology. In this process, the polymer solution was sprayed onto a substrate via a spinneret which has a hole diameter of 0.1 mm. With the solution solidified layer by layer under high-temperature treatment of a substrate, the thickness of the membrane increased at a very slow rate; then the robust and continuous membrane formed. Depending on the increase in thickness on the nanometer scale, the thickness of the membrane can be controlled precisely and simply by spraying time and spraying rate. Besides, the size of the membrane can be precisely controlled by the operation track. Considering the weak proton conductivity of pure SiO_2 nanospheres, amino groups ($-NH_2$) were introduced on their surface to improve compatibility and conduction. More importantly, amino groups in nanomaterials and acid groups in the matrix can form acid–base pairs to accelerate proton transfer. Therefore, we introduced amino-modified SiO_2 nanoparticles (SiO_2-NH_2) into Nafion to prepare the hybrid PEM by DED. The schematic workflow of the preparation of the PEM is shown in **Figure 1**. Furthermore, the DMFC single cell performance of the as-prepared membrane and Nafion membrane was investigated.

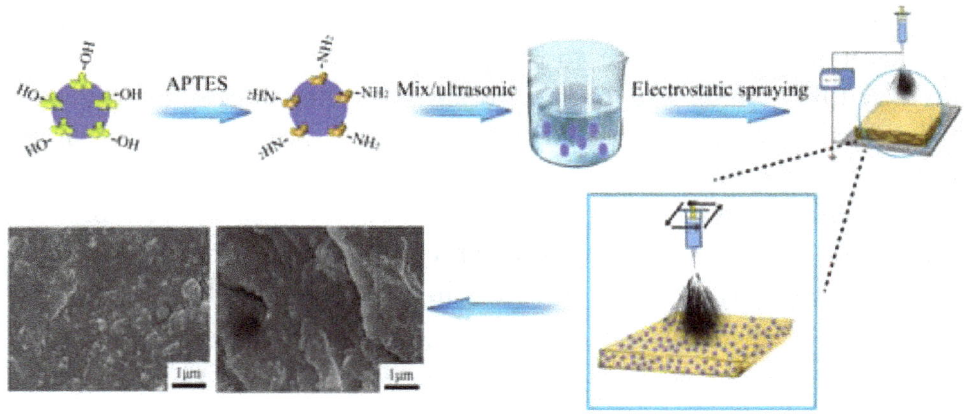

Figure 1. The schematic workflow of the preparation of the PEM.

2. MATERIALS AND METHODS

2.1. Preparation of the Hybrid Membrane

The detailed synthetic method of SiO2–NH2 referred to the published literature [32]. Ethanol was chosen as the solvent due to its low boiling point advantage. A certain amount of SiO2–NH2 nanoparticles and Nafion solution (5%) were successively dispersed in ethanol to obtain a silica/Nafion suspension. Herein, the percent of silica and Nafion in suspension was 0.15% and 2.5%, respectively, and the total fraction of SiO2–NH2 in the final membrane without solvent was approximately 5.7%. Then the final suspension with Nafion and SiO2–NH2 underwent ultrasonic treatment for 2 h to break the aggregates. For electrospraying, an electrostatic painting instrument equipped with a solution extrusion device, liquid injection needle, voltage system and heating metal collector was used. The process parameters of hybrid membrane preparation were 1 kV voltage, a tip collector distance of 2 cm, an operation track of 5 cm × 5 cm, a spray rate of 0.1 and 0.15 mL/min and a collector temperature of 75 °C. To compare the single cell performances in DMFC, the prepared hybrid membrane with a thickness of 30 μm was prepared and designed as Nafion/SiO2--NH2 in this work. All membranes were impregnated in 2 M H2SO4 for 12 h and washed with deionized water until neutralized.

2.2. Characterization

Scanning electron microscopy (SEM, Hitachi S-4800) and transmission electron microscopy (TEM, JEM 2200FS) were used to observe the morphologies of samples. Energy-dispersive X-ray spectra (EDS) mapping and X-ray photoelectron spectroscopy (XPS) were used to examine the composition of SiO_2–NH_2. Wide-angle X-ray diffractometry (XRD) and small-angle X-ray scattering (SAXS) measurements were carried out using an X-ray diffractometer (Rigaku SmartLab SE, Japan) and an Anton Paar SAXS system (SAXS ess mc2, Austria), respectively.

Proton conductivity (σ) was measured by AC impedance spectroscopy using an electrochemical workstation under a heated water bath. The frequency range from 0.1 to 105 Hz σ was calculated using the following equation:

$$\sigma = L/(R \cdot A) \quad (1)$$

where L, R, and A are the electrode distance, the impedance, and the membrane cross-sectional area, respectively.

The methanol permeability was measured via a diffusion cell containing two glass compartments sandwiching the test sample. The methanol permeability was calculated through the following equation:

$$DK = \frac{L \cdot V_B \cdot C_{B(t)}}{A \cdot C_{A(t-t_0)}} \quad (2)$$

where DK is the methanol permeability; L, A, and VB correspond to the thickness of the membrane, the effective area, and the volume of the water side, respectively; CA and CB are the concentration of methanol (M) in the A side and B side, which can be monitored by gas chromatography (Agilent 7820); *t-t0* is the test time.

The MEA was prepared by (i) spraying the anode catalyst (PtRu/C, Pt:Ru = 1:1, Johnson Matthey) and cathode catalyst (Pt/C, 60% Pt, Johnson Matthey, London, UK) on the PEM layer (2 cm × 2 cm), and both the catalyst loading was 1 mg/cm^2; (ii) Sandwiching the above membrane with gas diffusion layers and hot pressing at 100 °C. The DMFC performances of membrane electrode assemblies (MEAs) with different membranes were characterized by polarization curves in a fuel cell testing station (Model TEID160-1NBNNS, Arbin Inc., College Station, TX, USA) at 40 °C. The aqueous methanol (2 M) and oxygen were fed to the anode and cathode at 2 mL/min and 500 mL/min, respectively.

3. RESULTS AND DISCUSSION

3.1. Characterization of SiO_2–NH_2

SEM, TEM-EDS mapping and XPS tests were used to characterize the morphology and elemental composition of SiO2–NH2. It could be seen in Figure 2a that the SiO2–NH2 we synthesized showed a well-defined spherical appearance, and possessed a rough surface caused by the aggregation of –NH2. As shown in Figure 2b, N, O and

Si elements are uniformly distributed in the nanoparticles, which demonstrates the successful synthesis of the SiO2–NH2. In addition, the peaks of O (1s), N (1s), C (1s), Si (2s) and Si (2p) shown in Figure 2c could further prove the successful preparation of SiO2–NH2. Furthermore, the elemental analysis of SiO2–NH2 by XPS confirmed the Nitrogen percentage of 1.45%, which corresponds to 1.66% of–NH2.

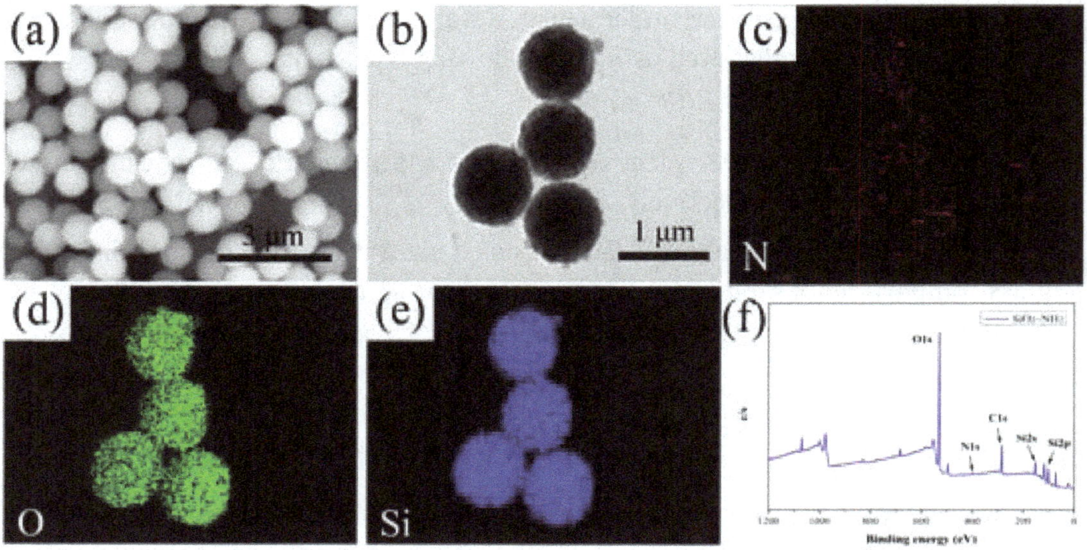

Figure 2. (a) SEM, (b–e) TEM-EDS mapping and (f) XPS images of SiO_2–NH_2.

3.2. Characterization of Nafion/SiO_2–NH_2

The realization of DED via electrostatic spraying mainly depends on the electric force and high-temperature solidification. The membrane fabrication process can be divided into two stages. In the first stage, the surface tension and viscoelastic force of membrane solution are overcome by the electric force, and then spraying type jets are formed and deposited on the collector. Different from the electrospinning process, solvent evaporation during spraying is extremely slow due to the short distance and relatively low voltage. In the second stage, the solution deposited on the collector solidifies to the membrane rapidly because of the high temperature.

The thickness of the composite membranes changed with the spraying time and spraying rate as shown in **Figure 3**. The thickness of the membrane shows a linearly increasing trend with the increased spraying time. Moreover, the thickness of the membrane prepared by a spraying rate of 0.15 mL/min is larger than that of the membrane with a spraying rate of 0.1 mL/min. In particular, the error bars of membrane thickness is quite small. Therefore, a membrane with a certain thickness and size can be prepared on a large scale using DED. The above phenomenon shows that the thickness of the membrane prepared by DED can be precisely controlled by spraying time and spraying rate.

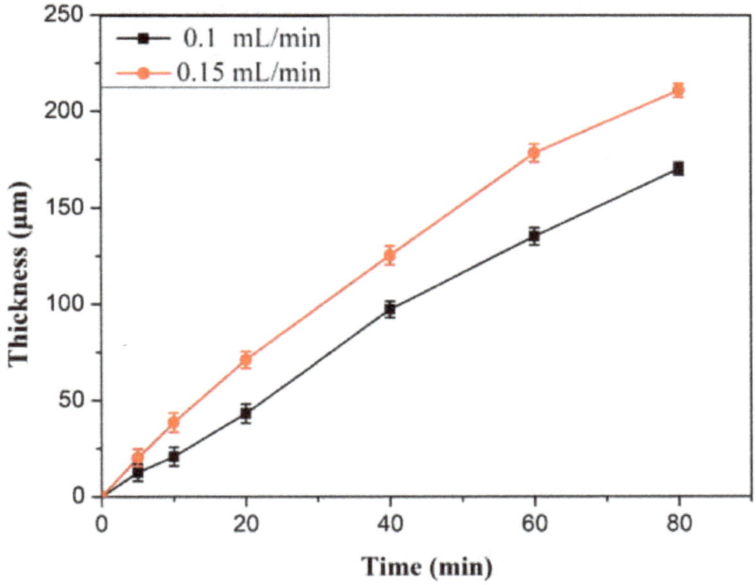

Figure 3. The thickness-time relationship at spraying rates of 0.1 and 0.15 mL/min.

Morphology of the Nafion/SiO2–NH2: SEM images of the surface and cross-sectional hybrid membranes at different magnifications are shown in Figure 4. As shown in Figure 4, the surface and cross-section of Nafion/SiO2–NH2 is compact, and no significant crack is shown in the membrane. Furthermore, the SiO2–NH2 nanoparticles could be clearly observed at both their surface and cross-section. This result revealed the good dispersion of SiO2–NH2 in the Nafion matrix.

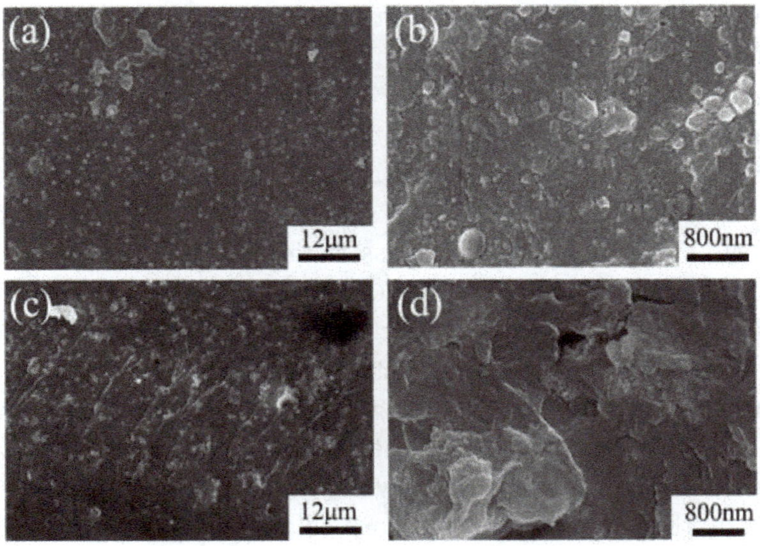

Figure 4. (a,b) the surface and (c,d) cross-sectional SEM images of hybrid membranes at different magnifications.

Figure 5a shows the XRD patterns of Nafion, Nafion/SiO2–NH2 and SiO2–NH2. All the samples showed amorphous peaks, indicating amorphous characteristics. Comparing the XRD patterns of Nafion/SiO2–NH2 with Nafion, a new broad peak appeared for the composite membranes at 24°. This result is caused by the redistribution of SiO2–NH2 in the Nafion matrix and reveals the good compatibility of these components.

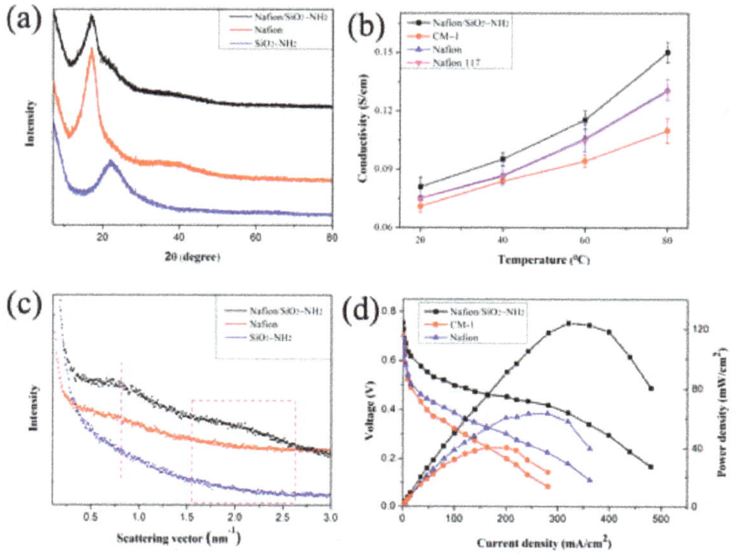

Figure 5. (a) XRD curves of Nafion, Nafion/SiO_2–NH_2 and SiO_2–NH_2; (b) Proton conductivity curve, (c) SAXS curves and (d) DMFC performance (operated at 40°C and 100% RH) of Nafion/SiO_2–NH_2, Nafion and CM^{-1}.

The proton conductivity of commercial Nafion 117, pure Nafion and hybrid Nafion membrane prepared by DED is shown in Figure 5b. Pure Nafion exhibited similar proton conductivity with Nafion 117 indicating the processing reliability of DED in membrane formation. The compared membrane containing the same content of SiO2–NH2 and Nafion matrix are prepared by the casting method and named CM$_{-1}$. It is interesting that CM-1 showed lower proton conductivity than Nafion/SiO_2–NH2 and Nafion. Nevertheless, Nafion/SiO_2–NH_2 exhibited the highest proton conductivity of 0.15 S/cm at 80 °C. This difference in proton conductivity originated from the different microstructure; better distribution of SiO_2–NH_2 in a hybrid membrane could bridge ionic clusters in the membrane to form continuous proton transferred channels [33]. During the casting process, nanospheres tend to be distributed on one side of the membrane, affected by gravity. However, high temperature facilitated the micro-volume polymer solution spinneret from solidification on the collector and then formed a layer-by-layer membrane with a homogeneous nanocomposite structure. This conclusion could be verified by the results of cross-sectional SEM images of hybrid membranes. To better verify the above explanation, SAXS of all membranes were characterized (Figure 5c). Nafion/SiO2–NH2 showed an obvious matrix segment peak and ionomer peak at lower and higher q, respectively. However, the peaks of CM-1 and pure Nafion were not obvious. Based on Bragg's law, the distance between neighboring ionic clusters in Nafion/SiO2–NH2 was smaller than in other membranes [34]. Such an observation is also consistent with the proton conductivity results.

Table 1 shows the methanol permeability of Nafion, Nafion/SiO2–NH2, and CM-1. Compared with the Nafion membrane, Nafion/SiO2–NH2 exhibited lower methanol permeability, indicating that the introduction of SiO2 improves the methanol barrier properties. However, the methanol permeability of CM-1 is lower compared to Nafion/SiO2–NH2, probably because of the reunion distribution of inorganic particles on one side of the membrane to form methanol barrier layers. This result is consistent with the proton conductivity results.

Table 1. The methanol permeability of Nafion, Nafion/SiO_2–NH_2, and CM^{-1}.

Samples	Methanol Permeability (10^{-7} cm^2 s^{-1})
Nafion	17.5
Nafion/SiO_2–NH_2	9.8
CM^{-1}	9.1

The polarization and performance curves of passive DMFCs based on pure Nafion and hybrid Nafion/SiO_2–NH_2 membranes were collected at 40°C and 100% RH and are shown in Figure 5c. It can be seen from Figure 5c that the DMFC performance of Nafion/SiO2–NH2 is enhanced compared to that of Nafion. Particularly, Nafion/SiO2–NH2 had a maximum power density output of 124.01 mW/cm2. However, Nafion and CM-1 only showed maximum power density values of 63.50 and 40.60 mW/cm2, respectively. This result is likely due to the following aspects: (i) the ultrathin Nafion/SiO2–NH2 can transport protons effectively through the membrane; (ii) the well-distributed inorganic silica may improve the water retention and methanol permeability of Nafion; (iii) the more ionic clusters in Nafion/SiO2–NH2 can provide massive proton transfer sites.

Some published works related to inorganic/organic hybrid membranes were cited for comparison with proton conductivity and power density, as shown in Table 2. The proton conductivity and power density for Nafion/SiO2–NH2 showed a competitive overall performance than other membranes, verifying that DED is a good application prospect in PEM preparation.

Table 2. Comparison of proton conductivity and power density with other reported PEMs.

PEMs	Proton Conductivity (S/cm)	Power Density (mW/cm²)	Ref.
PSU/mMOF/Si-SO3H	0.017 (70 °C, 100% RH)	40.8 (70 °C)	[18]
SPEEK/TiNFs-1.0	0.037 (80 °C, 100% RH)	431.5 (60 °C)	[35]
SPEEK/S-SiO2/MOF-5	0.00369 (30 °C, 100% RH)	NA	[36]
Nafion/SPES/SiO2-3%	0.23 (80 °C, 100% RH)	77.22 (80 °C)	[37]
Nafion/SiO2–NH2	0.15 (80 °C, 100% RH)	124.01 (40 °C)	This work

4. CONCLUSIONS

A novel approach to precisely control the fabrication of PEMs for DMFCs operating was presented in this work. Nafion/SiO_2–NH_2 was directly formed on a metal collector enabling the fast, simple and precise fabrication of 30 μm thin composite membranes. Nafion/SiO2–NH2 showed a maximum power density of 124.01 mW/cm2 at 40 °C and 100% RH, which was 95.29% higher than that of Nafion. The results proved that the DED can be a potential method for the precise production of cost-effective and ultrathin membranes.

5. CONCISE SUMMARY

In this work, we reported a novel preparation method for a proton exchange membrane (PEM) named, the direct electrostatic deposition method. In theory, any required thickness and size of PEM can be precisely controlled via this method. By direct electrostatic spraying of Nafion solution containing amino modified SiO2 nanoparticles onto a metal collector, a hybrid membrane of 30 μm thickness was fabricated. The DMFC assembled with a prepared ultrathin membrane showed a maximum power density of 124.01 mW/cm2 at 40 °C and 100% RH, which was 95.29% higher than that of Nafion. This membrane formation method provides potential benefits for the preparation of ultrathin PEMs.

REFERENCES

1. Elwan, H.; Mamlouk, M.; Scott, K. A review of proton exchange membranes based on protic ionic liquid/polymer blends for polymer electrolyte membrane fuel cells. J. Power Sources 2021, 484, 229197. [Google Scholar] [CrossRef]
2. Whiting, K.; Carmona, L.; Sousa, T. A review of the use of exergy to evaluate the sustainability of fossil fuels and non-fuel mineral depletion. Renew. Sustain. Energy Rev. 2017, 76, 202–211. [Google Scholar] [CrossRef]

3. Vinothkannan, M.; Kim, A.; Ramakrishnan, S.; Yu, Y.; Yoo, D. Advanced Nafion nanocomposite membrane embedded with unzipped and functionalized graphite nanofibers for high-temperature hydrogen-air fuel cell system: The impact of filler on power density, chemical durability and hydrogen permeability of membrane. Compos. Part B Eng. 2021, 215, 108828. [Google Scholar] [CrossRef]

4. Sun, C.; Zhang, H. Review of the Development of First-Generation Redox Flow Batteries: Iron-Chromium System. ChemSusChem 2022, 15, e202101798. [Google Scholar] [CrossRef] [PubMed]

5. Wu, J.; Nie, S.; Liu, H.; Gong, C.; Zhang, Q.; Xu, Z.; Liao, G. Design and development of nucleobase modified sulfonated poly(ether ether ketone) membranes for high-performance direct methanol fuel cells. J. Mater. Chem. A 2022. [Google Scholar] [CrossRef]

6. Huang, H.; Ma, Y.; Jiang, Z. Spindle-like MOFs-derived porous carbon filled sulfonated poly (ether ether ketone): A high performance proton exchange membrane for direct methanol fuel cells. J. Membr. Sci. 2021, 636, 119585. [Google Scholar] [CrossRef]

7. Simari, C.; Nicotera, I.; Aricò, A.S.; Baglio, V.; Lufrano, F. New insights into properties of methanol transport in sulfonated polysulfone composite membranes for direct methanol fuel cells. Polymers 2021, 13, 1386. [Google Scholar] [CrossRef]

8. Imaan, D.U.; Mir, F.Q.; Ahmad, B. Synthesis and characterization of a novel poly (vinyl alcohol)-based zinc oxide (PVA-ZnO) composite proton exchange membrane for DMFC. Int. J. Hydrogen Energy 2021, 46, 12230–12241. [Google Scholar] [CrossRef]

9. Lu, Y.; Liu, Y.; Li, N.; Hu, Z.; Chen, S. Sulfonated graphitic carbon nitride nanosheets as proton conductor for constructing long-range ionic channels proton exchange membrane. J. Membr. Sci. 2020, 601, 117908. [Google Scholar] [CrossRef]

10. Imaan, D.U.; Mir, F.Q.; Ahmad, B. In-situ preparation of PSSA functionalized ZWP/sulfonated PVDF composite electrolyte as proton exchange membrane for DMFC applications. Int. J. Hydrogen Energy 2022, in press. [Google Scholar] [CrossRef]

11. Ma, L.; Li, J.; Xiong, J.; Xu, G.; Liu, Z.; Cai, W. Proton conductive channel optimization in methanol resistive hybrid hyperbranched polyamide proton exchange membrane. Polymers 2017, 9, 703. [Google Scholar] [CrossRef] [PubMed]

12. Wang, H.; Zhang, J.; Ning, X.; Tian, M.; Long, Y.; Ramakrishna, S. Recent advances in designing and tailoring nanofiber composite electrolyte membranes for high-performance proton exchange membrane fuel cells. Int. J. Hydrogen Energy 2021, 46, 25225–25251. [Google Scholar] [CrossRef]
13. Sun, C.; Zhang, H.; Luo, X.; Chen, N. A comparative study of Nafion and sulfonated poly(ether ether ketone) membrane performance for iron-chromium redox flow battery. Ionics 2019, 25, 4219–4229. [Google Scholar] [CrossRef]
14. Zhang, J.; Liu, H.; Ma, Y.; Wang, H.; Chen, C.; Yan, G.; Tian, M.; Long, Y.; Ning, X.; Cheng, B. Construction of dual-interface proton channels based on γ-polyglutamic acid@cellulose whisker/PVDF nanofibers for proton exchange membranes. J. Power Sources 2022, 548, 231981. [Google Scholar] [CrossRef]
15. Hyun, J.; Doo, G.; Yuk, S.; Yuk, S.; Lee, D.; Lee, D.; Choi, S.; Kwen, J.; Kang, H.; Tenne, R.; et al. Magnetic Field-Induced Through-Plane Alignment of the Proton Highway in a Proton Exchange Membrane. ACS Appl. Energy Mater. 2020, 3, 4619–4628. [Google Scholar] [CrossRef]
16. Sun, C.; Negro, E.; Nale, A.; Pagot, G.; Vezzu, K.; Zawodzinski, T.; Meda, L.; Gambaro, C.; Noto, V. An efficient barrier toward vanadium crossover in redox flow batteries: The bilayer [Nafion/(WO3)x] hybrid inorganic-organic membrane. Electrochim. Acta 2021, 378, 138133. [Google Scholar] [CrossRef]
17. Zhang, X.; Yu, S.; Zhu, Q.; Zhao, L. Enhanced anhydrous proton conductivity of SPEEK/IL composite membrane embedded with amino functionalized mesoporous silica. Int. J. Hydrogen Energy 2019, 44, 6148–6159. [Google Scholar] [CrossRef]
18. Ahmadian-Alam, L.; Mahdavi, H. A novel polysulfone-based ternary nanocomposite membrane consisting of metal-organic framework and silica nanoparticles: As proton exchange membrane for polymer electrolyte fuel cells. Renew. Energy 2018, 126, 630–639. [Google Scholar] [CrossRef]
19. Pal, S.; Mondal, R.; Chatterjee, U. Sulfonated polyvinylidene fluoride and functional copolymer based blend proton exchange membrane for fuel cell application and studies on methanol crossover. Renew. Energy 2021, 170, 974–984. [Google Scholar] [CrossRef]
20. Patnaik, P.; Mondal, R.; Sarkar, S.; Choudhury, A.; Chatterjee, U. Proton exchange membrane from the blend of poly(vinylidene fluoride) and functional copolymer: Preparation, proton conductivity, methanol permeability, and stability. Int. J. Hydrogen Energy 2022, in press. [Google Scholar] [CrossRef]
21. Ma, L.; Xu, G.; Li, S.; Ma, J.; Li, J.; Cai, W. Design and optimization of a hyper-branched polyimide proton exchange membrane with ultra-high methanol-permeation resistivity for direct methanol fuel cells applications. Polymers 2018, 10, 1175. [Google Scholar] [CrossRef] [PubMed]

22. Sun, L.; Gu, Q.; Wang, H.; Yu, J.; Zhou, X. Anhydrous proton conductivity of electrospun phosphoric acid-doped PVP-PVDF nanofibers and composite membranes containing MOF fillers. RSC Adv. 2021, 11, 29527–29536. [Google Scholar] [CrossRef]

23. Wang, H.; Tang, C.; Zhuang, X.; Cheng, B.; Wang, W.; Kang, W.; Li, H. Novel structure design of composite proton exchange membranes with continuous and through-membrane proton-conducting channels. J. Power Sources 2017, 365, 92–97. [Google Scholar] [CrossRef]

24. Sood, R.; Giancola, S.; Donnadio, A.; Zatoń, M.; Donzel, N.; Rozière, J.; Jones, D.J.; Cavaliere, S. Active electrospun nanofibers as an effective reinforcement for highly conducting and durable proton exchange membranes. J. Membr. Sci. 2021, 622, 119037. [Google Scholar] [CrossRef]

25. Cheng, G.; Li, Z.; Ren, S.; Han, D.; Xiao, M.; Wang, S.; Meng, Y. A robust composite proton exchange membrane of sulfonated poly (fluorenyl ether ketone) with an electrospun polyimide mat for direct methanol fuel cells application. Polymers 2021, 13, 523. [Google Scholar] [CrossRef]

26. Li, H.; Lee, Y.; Lai, J.; Liu, Y. Composite membranes of Nafion and poly(styrene sulfonic acid)-grafted poly(vinylidene fluoride) electrospun nanofiber mats for fuel cells. J. Membr. Sci. 2014, 466, 238–245. [Google Scholar] [CrossRef]

27. Liu, G.; Tsen, W.; Wen, S. Sulfonated silica coated polyvinylidene fluoride electrospun nanofiberbasedcomposite membranes for direct methanol fuel cells. Mater. Des. 2020, 193, 108806. [Google Scholar] [CrossRef]

28. Ballengee, J.B.; Pintauro, P.N. Preparation of nanofiber composite proton-exchange membranes from dual fiber electrospun mats. J. Membr. Sci. 2013, 442, 187–195. [Google Scholar] [CrossRef]

29. Zhao, G.; Xu, X.; Zhao, H.; Shi, L.; Zhuang, X.; Cheng, B.; Yin, Y. Zeolitic imidazolate framework decorated on 3D nanofiber network towards superior proton conduction for proton exchange membrane. J. Membr. Sci. 2020, 601, 117914. [Google Scholar] [CrossRef]

30. Klingele, M.; Breitwieser, M.; Zengerle, R.; Thiele, S. Direct deposition of proton exchange membranesenabling high performance hydrogen fuel cells. J. Mater. Chem. A 2015, 3, 11239–11245. [Google Scholar] [CrossRef][Green Version]

31. Breitwieser, M.; Klose, C.; Klingele, M.; Hartmann, A.; Erben, J.; Cho, H.; Kerres, J.; Zengerle, R.; Thiele, S. Simple fabrication of 12 μm thin nanocomposite fuel cell membranes by direct electrospinning and printing. J. Power Sources 2017, 337, 137–144. [Google Scholar] [CrossRef]

32. Wang, H.; Li, X.; Feng, X.; Liu, Y.; Kang, W.; Xu, X.; Zhuang, X.; Cheng, B. Novel proton-conductive nanochannel membranes with modified SiO2 nanospheres

for direct methanol fuel cells. J. Membr. Sci. 2018, 22, 3475–3484. [Google Scholar] [CrossRef]

33. Wang, H.; Sun, N.; Zhang, L.; Zhou, R.; Ning, X.; Zhuang, X.; Long, Y.; Cheng, B. Ordered proton channels constructed from deoxyribonucleic acid-functionalized graphene oxide for proton exchange membranes via electrostatic layer-by-layer deposition. Int. J. Hydrogen Energy 2020, 45, 27772–27778. [Google Scholar] [CrossRef]
34. Zhang, S.; He, G.; Gong, X.; Zhu, X.; Wu, X.; Sun, X.; Zhao, X.; Li, H. Electrospun nanofiber enhanced sulfonated poly (phthalazinone ether sulfone ketone) composite proton exchange membranes. J. Membr. Sci. 2015, 493, 58–65. [Google Scholar] [CrossRef]
35. Dong, C.; Hao, Z.; Wang, Q.; Zhu, B.; Cong, C.; Meng, X.; Zhou, Q. Facile synthesis of metal oxide nanofibers and construction of continuous proton-conducting pathways in SPEEK composite membranes. Int. J. Hydrogen Energy 2017, 42, 25388–25400. [Google Scholar] [CrossRef]
36. Bisht, S.; Balaguru, S.; Ramachandran, S.K.; Gangasalam, A.; Kweon, J. Proton exchange composite membranes comprising SiO2, sulfonated SiO2, and metal–organic frameworks loaded in SPEEK polymer for fuel cell applications. J. Appl. Polym. Sci. 2021, 138, 50530. [Google Scholar] [CrossRef]
37. Wang, H.; Wang, X.; Fan, T.; Zhou, R.; Li, J.; Long, Y.; Zhuang, X.; Cheng, B. Fabrication of electrospun sulfonated poly(ether sulfone) nanofibers with amino modified SiO2 nanosphere for optimization of nanochannels in proton exchange membrane. Solid State Ion. 2020, 349, 115300. [Google Scholar] [CrossRef]

CHAPTER-12

ANALYSIS OF VIBRATIONAL EMISSIONS FROM CN AND C2 IN NYLON AND ZNO/NYLON POLYMER COMPOSITES THROUGH LASER-INDUCED BREAKDOWN SPECTROSCOPY (LIBS)

1. INTRODUCTION: AN OVERVIEW

In the realm of materials science, understanding the molecular composition and vibrational characteristics of polymers is essential for optimizing their performance in various applications. This study delves into the vibrational emissions of CN and C2 in two distinct polymer matrices: nylon and zinc oxide (ZnO)/nylon composites. Leveraging the precision and sensitivity of Laser-Induced Breakdown Spectroscopy (LIBS), this research aims to unravel the intricate vibrational signatures of carbon-containing species within these materials. The investigation not only contributes to the fundamental knowledge of polymer behaviour but also explores the influence of ZnO nanoparticles on the vibrational properties of the polymer matrix.

Nylon is a polyamide polymer that is frequently used to make a variety of different types of attire products, such as lithium batteries [1], 3-D printed nylon memory composites [2], asleep bags, rope, seat belts in vehicles, parachuting fabric, tubing pipe, sheets, and dental floss [3]. Unlike other organic or semi-synthetic structures, nylon fibers are completely synthetic. Laser-induced breakdown spectroscopy (LIBS) has been developed as a rapid, reliable, non-destructive, real-time, and in situ detection analytical technique for the spectroscopic study of different organic, inorganic and explosive textiles [4,5,6,7]. LIBS can also be coupled with further spectroscopic techniques, for instance Raman spectroscopy and photoluminescence spectroscopy, to enhance the capacities of LIBS for organic composite analysis [8]. In LIBS, a high-power laser beam is focused through the convex lens on the target surface, which causes laser–matter interaction. This interaction causes high-density micro-plasma containing electrons, ions, and neutral ingredients. The light emitted from this hot plasma shows the spectral identifications of the atomic as well as molecular optical emissions that appear in the sample, which can be used for chemical analysis. For the last couple of years, LIBS has been concerned with polymer composite analyses using spectroscopic information such as CN and C2 molecular structures recorded in an open-air environment [9,10]. The molecular optical emission spectra due to the fundamental transition give information about all the polymer composites present in the sample. In atomic emission spectra, a single line transition is produced at a distinct wavelength, whereas in molecular optical emission, a group of spectral lines appear that corresponds to the vibrational modes due to electronic transitions which are distinguished as vibrational emissions of the molecules [11,12]. Although the molecular optical emission band has already been analyzed, however, molecular optical vibrational emission is still to be explored using LIBS. Moreover, the signal-to-noise ratio (SNR) of CN and C2 molecular structures is typically the maximum in the spectra of LIBS, revealing suitable evidence for the difference between organic-polymer materials [13].

Formerly reported analyses on the polymers generally focused on molecular emissions and chemometric studies. Numerous research articles have described the study of polymers in different circumstances by applying LIBS-assisted chemometric analysis. Dong et al. [14,15] utilized a laser-ablation molecular isotopic spectrometry (LA-MIS) technique to study the CN and C2 molecular optical emission using organic substances. Boueri et al. [16] reported the artificial neural networks (ANNs) coupled with LIBS to distinguish various polymers. Mousavi et al. [17,18] examined the impact of molecular structures and chemical concentrations on C2 and CN spectral emissions in numerous organic substances, including polymers, PAHs, aliphatic carboxylic acids, aromatic carboxylic acids, and amides. Farooq et al. [19] reported the molecular spectral lines of CH, C2, CN, and CO molecules along with the LIBS spectra of the elements Ca, Al, Si, Mg, P, Br, and N in polymer samples such as polystyrene and polycarbonate, respectively. Lasheras et al. [20] studied the various polymers using the LIBS technique through normalized coordinates. Costa et al. [21] identified and classified various e-waste polymers by PCA, SIMCA, and KNN based on the LIBS spectra. The proposed SIMCA and KNN techniques can be utilized for the reprocessing of e-waste polymers through different industrial segments. Gregoire et al. [22] investigated the space-time resolved optical spectra of four various kinds of substances, including polyamide with CN bands, polyethylene containing C2 bands, and polystyrene comprising two C2 bands, in addition to polyoxymethylene, comprising neither CN nor C2 bands in laser-produced polymer plasma. Aquino et al. [23] utilized partial least-squares-discriminant analysis (PLS-DA) to detect and distinguish various polymers. More recently, Zhu et al. [24] studied the C2 and CN spectral emissions characteristics from various coals and their pyrolysis char target samples. Unfortunately, the above-mentioned literature review shows that the majority of the work was presented to the classification analysis of various polymers. Hence, a qualitative analysis including optical emission spectra and plasma diagnostics as a function of the laser power density of the synthetic polymers plays a major role in characterizing the organic polymers. Taking into account the above-mentioned characterization, the present study establishes the potential of the LIBS technique to perform qualitative analyses together with the plasma characterization of the pure and ZnO/nylon polymer samples.

ZnO has various incomparable properties such as catalytic, optical, and electronic, due to direct-wide bandgap semiconductor having band gap energy (Eg) ~3.4 eV along with an enormous excitonic binding energy of ~60 meV, signifying that it is an efficient component for luminescent and lasing devices with steady room temperature [25]. The bandgap of ZnO semiconductors is analogous to the bandgap of TiO2. ZnO ingredients are reported for the photo degradation of volatile organic compounds, antibiotics, pesticides, dyes, and surfactants. The different polymers doped with ZnO have radiative recombination processes that may be developed for broadband light and wavelength-tunable emission [25] as these processes are affected by various conditions such as the organic and inorganic properties of the elements along with their comparative chemical composition. Nylon 6/6 is useful as a potential substantial in several areas for diverse applications such as in lithium-ion batteries, antibacterial agents, and cement to enhance the properties [26,27,28]. Nylon 6/6 has extraordinary thermal stability, hydrophilicity and excessive tensile power [29]. Nylon 6/6 doped with ZnO (ZnO/nylon) can be employed as a photocatalyst for the photo degradation of alizarin red dye [30].

In the present work, the LIBS spectra of organic materials such as pure nylon and nylon doped with ZnO (ZnO/nylon) have been studied including the vibrational emission of the C2 and CN molecular optical bands. The obtained molecular optical emission is established on the optimized experimental parameters such as the integration time, laser pulse energy, and laser pulse irradiance. Laser irradiance performs a significant role in the vibrational transitions observed in the molecular optical emissions. Furthermore, the intensity variation of vibrational transitions and trend of plasma parameters has been investigated by varying laser irradiance for pure and ZnO/nylon polymers. The analysis of vibrational emission of CN (hetero-nuclear diatomic) and C2 (homo-nuclear diatomic) molecules in polymers has revealed that LIBS is a fast and consistent diagnostic technique for the study of polymers regarding their molecular structure.

2. MATERIALS AND METHODS

Two organic polymers were utilized in this study, namely: nylon, and ZnO/nylon. The molecular formula along with the 2D and 3D molecular structure of the nylon organic polymer is presented in **Figure 1**. Nylon (6/6) and ZnO was obtained from Sigma-Aldrich (USA). The composite ZnO/Nylon polymer film was prepared using solution method technique.

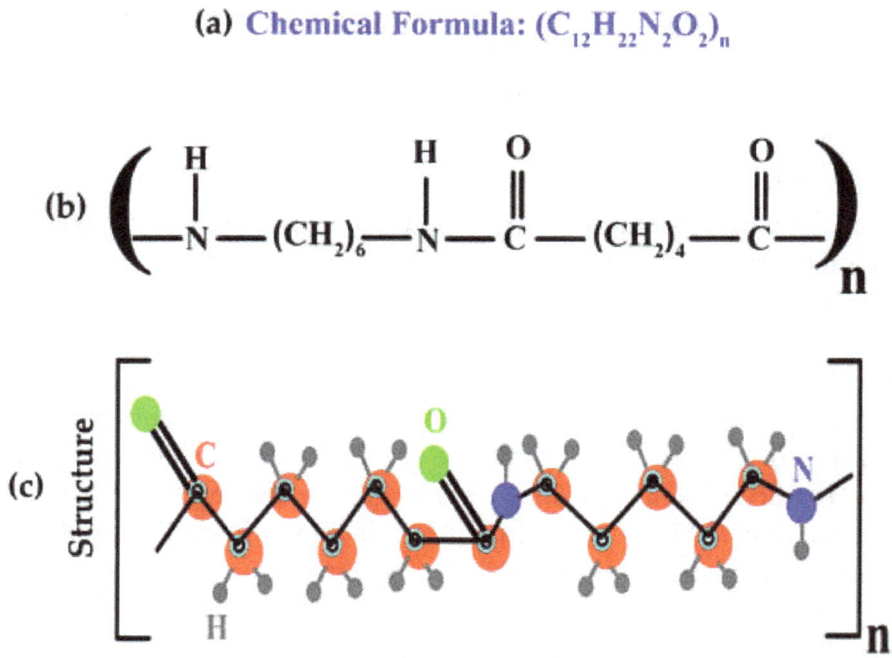

Figure 1. Investigated nylon; (a) chemical formula, (b) 2D and (c) 3D molecular structure.

Experimental Details

The details of the experimental system utilized to accomplish the optical emission of any sample have been described somewhere else [**31,32,33**]. In brief, a second harmonic (2ω), Q-switched Nd:YAG laser (Quantel (Brilliant-B), France), was used; it had a 532 nm wavelength and operated at a repetition frequency of 10 Hz with a 5 ns pulse width, and was capable of delivering the pulse energy of about 200 mJ. The LIBS experiments were performed at room temperature in an airy ambiance. The laser beam was converged onto the sample surface, which was positioned in the air at atmospheric pressure using a convex (quartz) lens with a focal length of 10 cm. The samples were kept on a rotating circular stage with a rotation speed of ~10 rpm to enable a clean surface for each fresh laser shot. The optical emission spectra were recorded using an optical fiber attached to a charged coupled devices (CCDs) array spectrometer with an optical wavelength range from 200 nm to 720 nm. An average of ten laser shots were employed to cleanse the surface of the target sample. The spectra were accomplished at an average of 20 laser pulse shots at different spots on the target surface. The averaged optical emission spectrum was then utilized to acquire the molecular optical emission of the polymer samples, which takes into account the target sample in homogeneity and lowers the statistical inaccuracies.

3. RESULTS AND DISCUSSION

Emission Studies of the CN Violet and C_2 Swan System

LIBS has been a widely utilized analytical tool for the analysis of synthetic–organic materials such as teflon, epoxy, polystyrene, and nylon. Overall, all organic materials show very analogous molecular or atomic emission spectra. However, various features such as molecular bonding dissociation and vibrational emission can be studied by varying laser irradiance, integration time, and laser power density. Figure 2a shows a schematic transition diagram for the spectra of nylon in the CN spectral region from 385 nm to 388 nm along with (0,0), (1,1), (2,2), (3,3), and (4,4) vibrational transitions at 3.2 eV excitation energy. The molecular optical emission of the CN violet system of a diatomic molecule relates to radiative transitions from $B^2\Sigma+$ to $X^2\Sigma+$ of the electronic state for $\Delta v = 0$ at 2.4 eV excitation energy. The transitions from the lowest vibrational states correspond to high probability and strong relative intensity as compared to the highest vibrational states. Figure 2b shows potential energy curves against internuclear distance for the two electronic states of the C2 Swan band system also known as vibronic (400–700 nm), revealing the transitions between d3 Πg and a^3 Πu (from the Plank equation; $\Delta E = hc/\lambda$). Emitted wavelength λ of the C^2 molecular optical emission lies between the spectral region from 400 nm to 700 nm. C2 Swan system originated due to a variation in wave-function-symmetry as of the allowed electronic dipole transition between two 3Π states from upper to lower state (g → u). The Swan band spectrum of C2 corresponds $\Delta v = +1$, $\Delta v = +2$, $\Delta v = 0$, and $\Delta v = -1$ between excited electronic states to the ground electronic states with v'= 0–10, and v'' = 0–9 vibrational transitions. Where, v'' and v' are the vibrational states' quantum parameters [34] for the lower and upper vibrational stages of the electronic levels, respectively.

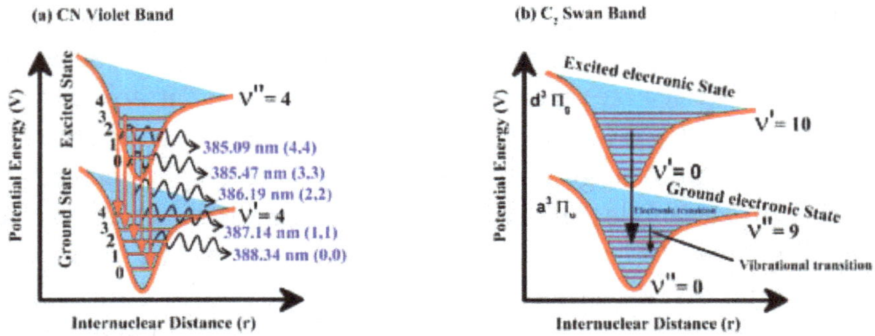

Figure 2. (a) Schematic diagram for the spectra of nylon in the CN spectral region (b) Potential energy against internuclear distance curves for the two electronic states of the C_2 Swan band system, revealing the various transitions.

Figure 3 shows time-resolved LIBS emission spectra of pure nylon in the wavelength spectral region from 350 nm to 645 nm. Nylon spectrum clearly demonstrates the presence of molecular optical emission band of CN Violet at 388.34 (0,0), 387.14 (1,1), 386.19 (2,2), 385.47 (3,3), and 385.09 (4,4) nm due to B2Σ+–X2Σ+ and C2 Swan band structure at 467.3 (5,4), 468.3 (4,3), 469.8 (3,2), 471.6 (2,1), 473.7 (1,0), 509.6 (2,2), 512.9 (1,1), 516.5 (0,0), 550.2 (3,4), 554.1 (2,3), 558.5 (1,2), 563.5 (0,1), 600.5 (3,5), 606.0 (2,4), and 612.0 (1,3) nm due to d3 Πg–a3 Πu transitions, respectively. Transitions band heads between two electronic states are identified using the difference in vibrational quantum numbers from the upper to lower electronic state such as Δν = ν′–ν″. Pure nylon spectra are observed at different laser irradiances such as 10, 20, and 30 GW/cm2 and 1–10 ms integration time. Time integrated spectra show continuous growth of the C2 and CN molecular bands emission intensities. The strongest transitions of the CN and C2 molecular optical emission bands are observed at 388.34 nm (Δν = 0), and 516.5 nm (Δν = 0) wavelength, respectively. Interestingly, the sequence Δν = 0 for the C2 structure around 600 nm is shown prominently only at 30 GW/cm2 irradiance, showing emission dependency at laser energy accumulation on the target material.

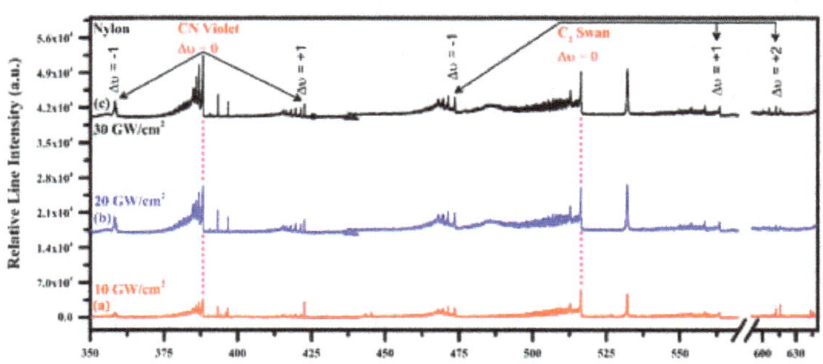

Figure 3. Time-integrated spectrum of the pure organic nylon from 350 nm to 645 nm at different laser irradiance (GW/cm²) showing CN and C² vibrational optical emissions.

For the comparative molecular optical emission analyses, organic nylon doped with ZnO was also utilized in this study. Figure 4 validates the similar molecular optical emission structures of the CN and C2 band at 10, 20, and 30 GW/cm2 laser irradiance. However, in this optical region from 300 nm to 645 nm, the expected emission spectral lines of zinc are also observed with an excellent signal-to-noise ratio (SNR) at 328.2, 330.3, 334.5, and 481.1 nm due to 4d 3D1 → 4p 3P0, 4d 3D1 → 4p 3P1, 4d 3D3 → 4p 3P2, and 5s 3S1 → 4p 3P2 radiative transitions, respectively. Nevertheless, it is worthwhile to mention that various structures of Ca emission lines at 393.37, 396.85, 442.54, 443.50, 445.48, 527.03, 558.20, 558.88, 559.01, 559.45, 559.85, 612.22, 616.13, and 616.22 nm are identified in the synthetic nylon. Carbon (C) and hydrogen (Hα) emissions at 247.86 nm (2p3s 1P1 → 2p2 1S0) and 656.28 nm (3d 2D5/2 → 2p 2P3/2) are also observed. Therefore, carbon (C) is the major source of molecular band formation in nylon.

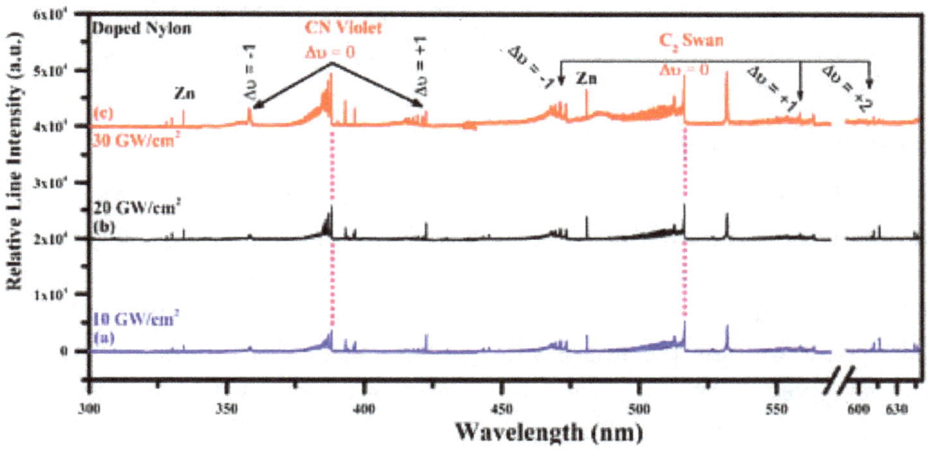

Figure 4. Time-integrated spectrum of the organic doped nylon from 300 nm to 645 nm showing CN and C2 vibrational optical emissions together with well-isolated zinc optical spectral lines.

4. PLASMA CHARACTERIZATION

To describe the laser-induced plasma established on the optical emission spectra, specific conditions have to be fulfilled, i.e., the plasma must be optically thin and should follow the local thermo-dynamical equilibrium (LTE) order. As optically thick plasma would entail self-absorption and saturation in the emitted line profiles causes an irregular peak in the spectrum, it influences the computations for electron plasma density and excitation temperature. We have utilized the line intensity proportion procedure to confirm the condition of an optically thin laser plasma [**35,36**]:

$$\frac{I_{ki}}{I_{nm}} = \left(\frac{\lambda_{nm}}{\lambda_{ki}}\right)\left(\frac{A_{ki}}{A_{nm}}\right)\left(\frac{g_k}{g_n}\right)exp\left(-\frac{E_k - E_n}{k_B T}\right) \quad (1)$$

where Iki and Inm are the detected spectral line intensities at optical wavelength λnm and λki, respectively. Aki and Anm are the transition probabilities, *gk* and *gn* are the statistical weights of the upper levels, kB is the Boltzmann constant and T is the excitation temperature. We have employed the optical emission lines with nearly the equivalent or very adjacent energies to the upper levels to minimize the temperature influence. The experimentally observed ratio of the spectral line intensities and the ratio determined from the spectroscopic factors are in excellent agreement within a 5% error. For illustration, the experimentally calculated intensity ratio using Zn I at transitions 328.23 nm and 330.29 nm is 0.45, whereas the intensity ratio calculated from the spectroscopic parameters is 0.47. The error relative to the theoretical ratio is found within 5%, which is acceptable. Thus, the optical plasma can be assumed as optically thin.

4.1. Plasma Excitation Temperature

The excitation temperature was determined by applying the Boltzmann plot technique, having established that the optical plasma is optically thin. The spectroscopic atomic factors of the optically thin spectral lines of Ca I and Zn I applied to construct the Boltzmann plots were acquired from the NIST database [**37**]. The spectroscopic parameters of the persistent lines including wavelength (nm), upper-level energy (eV), electron configuration transition, statistical weight, and transition probability (s^{-1}) are presented in **Table 1.** To draw the Boltzmann plots, these lines were carefully chosen as they lie in a thin optical region where the efficiency of the detector device stays nearly persistent, thus reducing the inaccuracies associated with the measurements of the line intensity. The excitation temperatures have been determined using the Boltzmann plot procedure in which numerous spectral emission lines along with their relative line intensities are considered [**38,39**]:

$$\ln\left(\frac{I_{ki}\lambda_{ki}}{A_{ki}g_k}\right) = -\frac{E_k}{k_B T_e} + \ln\left(\frac{FN}{U(T)}\right) \quad (2)$$

where, $F = \frac{hcd}{4\pi}$. is the characteristic plasma length, Iki is the spectral line intensity due to the k → i transition, λki is the spectral wavelength, h is the Planks constant, c is the speed of light, $Aki$is the transition probability, gk is the upper-level statistical weight, Ek is the upper-level transition energy. Similarly, kB, Te, N, and (T) are the Boltzmann constant, excitation temperature, total number density, and the temperature-dependent partition function, respectively. A plot corresponding to the logarithmic term in this equation versus the upper-level energies (Ek) yields a straight line (y = mx +c) and its slope (m) is like $1/kTe$. The Boltzmann plots corresponding to the Ca I, and Zn I lines for pure nylon and doped nylon, respectively, are shown in Figure 5, demonstrating good linearity; the linear correlation coefficients are: R_2Ca>0.950.95, and R_2Zn>0.97, respectively. The excitation temperatures have been determined from the slopes of the lines corresponding to Ca I, and Zn I. The excitation temperature at different irradiances (10–30 GW/cm^2) was calculated as; 0.65 eV to 0.76 eV for pure nylon and 0.64 eV to 0.77 eV for doped nylon. The error bars show the uncertainty (~5%) in the measurement of electron temperature, which is attributed to the error in the integrated line intensity, transition probability, and the line profile fitting method.

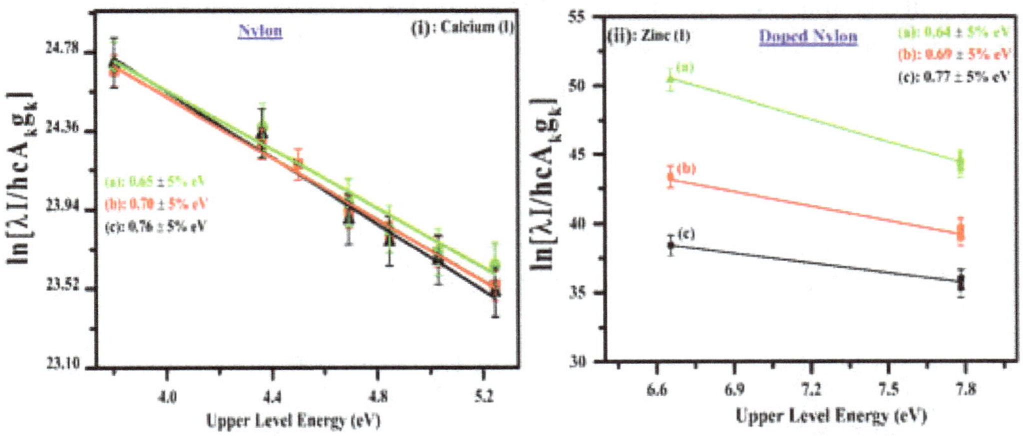

Figure 5. (i) Temperature Boltzmann plots for the pure organic nylon drawn using five neutral calcium (Ca I) emission lines. (ii) Boltzmann plots for the co-doped nylon made using zinc optically thin emission lines.

Table 1. Spectroscopic atomic parameters of the emission lines for the calculation of excitation temperature [37].

Wavelength λ (nm)	Transition Upper to Lower	Upper-Level Energy Ek (eV)	Transition Probability and Statistical Weight Akgk (108 s-1)
Ca I			
487.81	4s4f 1F3 → 3d4s 1D2	5.25	1.32
527.03	3d4p 3P2→ 3d4s 3D3	4.88	2.50
559.85	3d4p 3D1 → 3d4s 3D1	4.74	1.29
585.75	4p2 1D2 → 4s4p 1P1	5.05	3.30
612.22	4s5s 3S1 → 4s4p 3P1	3.91	0.86
649.97	4p 3F2 → 4s 3D2	4.43	0.41
671.76	4s5p 1P1 → 3d4s 1D2	4.55	0.36

Wavelength λ (nm)	Transition Upper to Lower	Upper-Level Energy Ek (eV)	Transition Probability and Statistical Weight Akgk (108 s-1)
Zn I			
328.23	4d 3D1 → 4p 3P0	7.78	2.70
330.29	4d 3D1 → 4p 3P1	7.78	6.00
334.50	4d 3D3 → 4p 3P2	7.78	11.9
481.05	5s 3S1 → 4p 3P2	6.65	0.70

4.2. Electron Number Density

The full width at a half area (FWHA) is calculated using a hydrogen Hα spectral emission line profile at 656.28 nm as demonstrated in Figure 6, displaying the experimentally analyzed line profile (blue color) corresponding to FWHA yielding as (0.99 ± 0.05) nm. The simplest formula for the scheming of electron number density using the Hα line is given as [40,41]:

$$\omega_{FWHA} = 5.49\ \text{Å} \times \left(\frac{N_e}{10^{17}\text{cm}^{-3}}\right)^{0.67965} \quad (3)$$

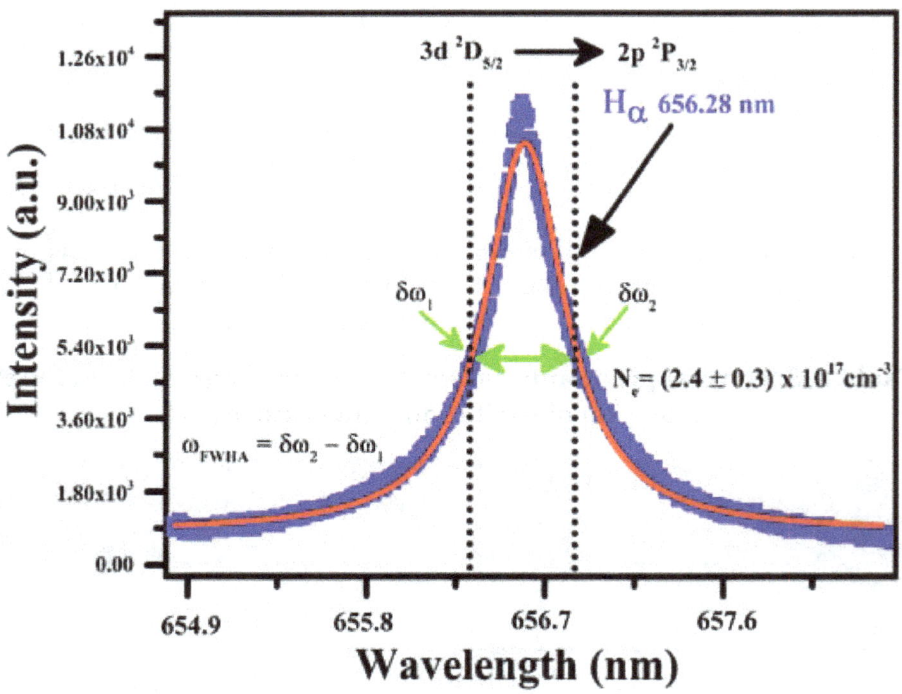

Figure 6. Stark broadened line profile of hydrogen Hα spectral line profile at 656.28 nm through the Voigt fitting.

Here, Ne is the plasma number density, and ωFWHM is the full width at the half area. The parameter ωFWHM is calculated through the relation, $\omega FWHA = \delta\omega 2 - \delta\omega 1$.

The calculated electron number density using Equation (3) is (2.4 ± 0.3)×1017(2.4 ± 0.3)×1017 cm^{-3}. The error bars show the uncertainty (~5%) in electron density, which is principally owing to the error in the electron impact parameter, FWHM, and in the deconvolution of the line width to the instrumental width.

4.3. Local Thermodynamical Equilibrium (LTE) Condition

The ionic and the excitation temperatures have come to be identical to the electron temperature for the optical plasma to be in local thermo-dynamical equilibrium (LTE). McWhirter advised a basis for a trivial limit of the electron number density to confirm the plasma to be in LTE. The above criteria was validated by calculating the smaller limit of the electron plasma density using the following relation [**42,43**]:

$$N(cm\text{-}3) \geq 1.6\times1012(T(K)\text{----}\surd(\Delta Enm\ (eV)) \quad (4)$$

where, $\Delta E(eV)$ is the highest transition energy between the upper to a lower ground level (n → m), and T is the excitation temperature in kelvin (K). In this work, the calculated electron number density using this relationship is on the order of ~1014 cm-31014 cm-3. This value of the plasma number density is lower at three orders of magnitude than that determined from Equation (3). Therefore, it can be concluded that plasma is satisfying the LTE state.

The validity of the LTE condition has also been confirmed by estimating the diffusion length using the Cristoforetti et al. criterion for an inhomogeneous plasma to be in LTE. According to this condition, the characteristic variation length "d" is significantly greater than the diffusion length ($10\lambda < d$). The diffusion length was estimated employing the following relation [**44,45**].

$$\lambda \approx \sqrt{D_{diff} \times \tau_{rel}} = 1.4 \times 10^{12} \left[\frac{(k_B T)^{\frac{3}{4}}}{N_e}\right] \cdot \left(\frac{\Delta E}{M_A\, f_{12}\, (\overline{G})}\right)^{\frac{1}{2}} \cdot \exp\left[\frac{\Delta E}{2k_B T}\right]. \quad (5)$$

$$d \approx T(x)\left({dT(x)}/{dx}\right)^{-1} \quad (6)$$

where ΔE and kBT are determined in electron volt (eV), Ne is the electron density measured in (cm-3), MA denotes the atomic mass of a specie, f12 represents the oscillator strength taken from the NIST database [37], G^{-} is the gaunt factor carefully chosen by Cristoforetti et al. [45], d is the plasma diameter (typically a few mm) and also known as characteristic variation length. In the present work, we have calculated the diffusion length by using the emission line of neutral zinc (Zn I). The calculated diffusion length, $\lambda \cong$ 1.87×10-3 mm mm is much below the characteristic variational length of the optical plasma ($10\lambda<d$), which confirms that the plasma follows the LTE condition.

4.4. Effect of Laser Irradiance on the Plasma Parameters

The effect of electron plasma number density and temperature as a function of laser irradiance was investigated, as shown in Figure 7a,b. The dependency of plasma parameters on the laser irradiance was studied by varying it from 10 to 30 GW/cm2. The electron number density for both the samples (pure and doped nylon polymer) is determined through the FWHA of the Hα line profile at 656.28 nm and plotted as a function of laser irradiance as shown in Figure 7a. The figure shows that the plasma electron number density increases exponentially from 2.4 × 1017 cm−3 to 2.8 × 1017 cm−3 and 2.49 × 1017 cm−3 to 2.9 × 1017 cm−3 for pure nylon and co-doped nylon, respectively, at the same laser irradiance from 10–30 GW/cm2. Similarly, excitation temperature was determined using the same samples and 532 nm wavelength at various laser irradiances ranging from 10 to 30 GW/cm2. Figure 7b demonstrates the variance in electron temperature as a function of laser irradiance from 0.65 eV to 0.76 eV and 0.64 eV to 0.77 eV for pure nylon and co-doped nylon, respectively. The electron temperature is showing almost linear growth concerning laser irradiance due to the sufficient laser absorption onto the target, revealing the inverse bremsstrahlung (IB) effect.

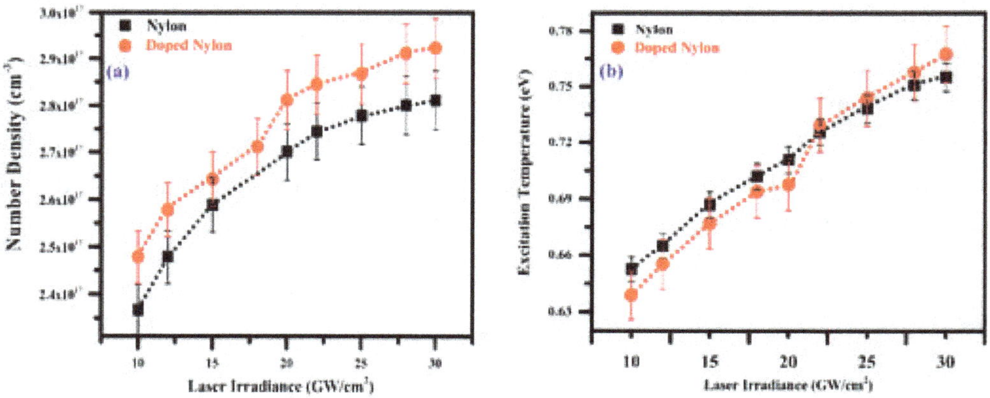

Figure 7. (a) Variation in the electron number density and (b) Variation in excitation temperature of the pure and doped nylon plasma as a function of laser irradiance.

4.5. Effect of Laser Irradiance on CN and C_2 Vibrational Emission

In this study, the intensities of the laser-produced optical plasma of the pure nylon and ZnO/nylon polymer were normalized by dividing the intensity of vibrational lines of CN and C_2 using the intensity of the neutral carbon (C I) emission spectral line at 247.86 nm under time delay of 2 μs between the laser pulse to the preliminary of the acquisition system, 1 ms integration time and 532 nm laser wavelength. A similar normalization approach has also been investigated in previously reported literature [46,47]. Figure 8 shows the normalized intensity of CN at 388.34 nm and 516.42 nm due to (0,0) vibrational transition as a function of laser irradiance from 10–30 GW/cm^2 for pure nylon and doped nylon. As shown in the figure, the normalized intensities of the vibrational transitions are growing concerning laser irradiance. The maximum intensity of the CN and C^2 molecular band for (0,0) vibrational transition is observed at 30 GW/cm^2 irradiance for both the samples (nylon and doped nylon). At below 10 GW/cm^2 and above 30 GW/cm2, we observed that the signal vanishes for all the modes due to the low absorption of the energy onto the target surface and dissociation of the molecular band structures, respectively. The structural emission at CN: 388.34 nm and C2: 516.42 nm are the dominant molecular profiles throughout the whole scale of investigated irradiance from 10–30 GW/cm2. In an innovative sense, polymers can be distinguished from the other non-polymer materials corresponding to their vibrational structure. In the nylon sample, the vibrational emission of the CN Violet and C2 Swan band due to the (0 → 0) vibrational transition is observed as strongly intense than the lower energy transitions, for instance (1 → 1), (2 → 2), and (3 → 3). Therefore, this factor is significantly very critical when exploring the LIBS spectra of polymers having organic molecular structures.

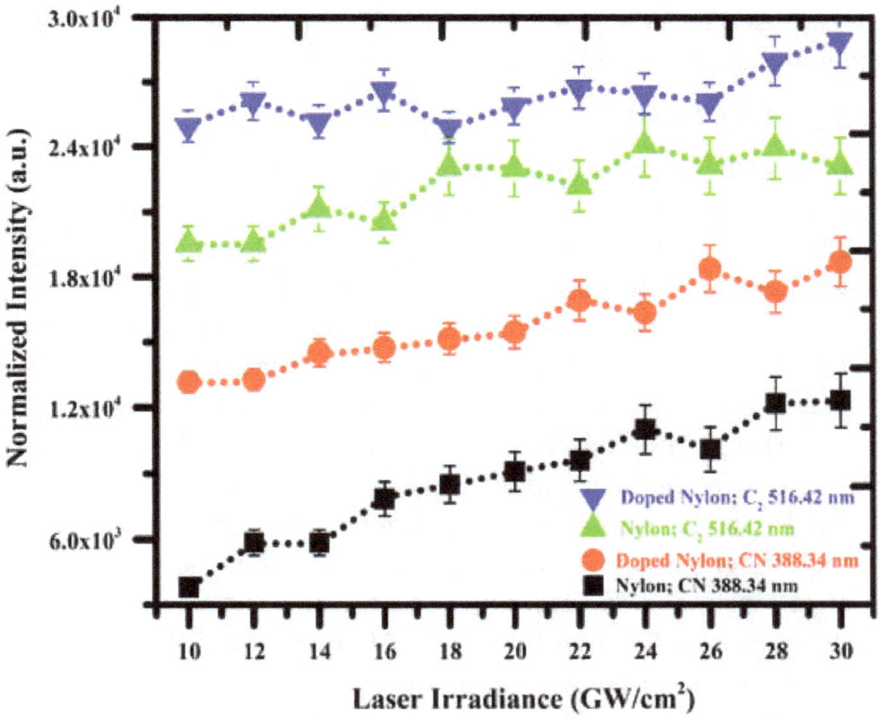

Figure 8. Normalized intensity of CN and C_2 molecular band at 388.34 nm, and 516.42 nm, respectively, as a function of laser irradiance for pure and doped nylon.

5. CONCLUSIONS

In the present work, molecular emissions, CN (B2Σ+ to X2Σ+) violet and the C2 (d3 Πg–a3 Πu) swan band system of nylon polymer plasma, at various irradiances from 10–30 GW/cm2, were noticed for the vibrational transitions Δν = –1, 0, +1 and Δν = –1, 0, +1, +2, respectively, using a laser-induced breakdown spectroscopy technique. The excitation temperature, plasma electron density, and line intensity as a function of laser irradiance were observed. Depending on the laser pulse irradiance, integration time, excitation temperature, and plasma number density, the characteristics of the CN and C2 emission spectra were found to change significantly. These analyses determine that the emission line intensities from CN and C2 bands are very sensitive to laser irradiance, integration time and delay time among laser pulse shot and spectral acquisition. At low laser irradiances such as 10 GW/cm2, the emission bands due to CN and C2 predominate while at higher irradiance the multiple neutral and ionized lines of various elements have been detected together with CN and C2 molecular optical emissions. The excitation temperature is noticed to grow with rising laser irradiance and saturates at higher irradiance levels. The saturation of excitation temperature at higher laser irradiance is caused by exhaustion of excited-level population density of CN and C2 molecules and by plasma shielding.

6. CONCISE SUMMARY

The laser-induced breakdown spectroscopy (LIBS) technique was performed on polymers to study the neutral and ionic emission lines along with the CN violet system (B2Σ+ to X2Σ+) and the C2 Swan system (d3 Πg–a3 Πu). For the laser-based emission analyses, the plasma was produced by focusing the laser beam of a Q-switched Nd: YAG laser (2ω) at an optical wavelength of 532 nm, 5 ns pulse width, and a repetition frequency of 10 Hz. The integration time of the detection system was fixed at 1–10 ms while the target sample was positioned in air ambiance. Two organic polymers were investigated in this work: nylon and nylon doped with ZnO. The molecular optical emission study of nylon and doped nylon polymer sample reveals CN and C_2 molecular structures present in the polymer. The vibrational emission analysis of CN and C2 bands gives information about the molecular structure of polymers and dynamics influencing the excitation structures of the molecules. Besides, it was further investigated that the intensity of the molecular optical emission structure strongly depends on the electron number density (cm-3), excitation temperature (eV), and laser irradiance (W/cm2). These results suggest that LIBS is a reliable diagnostic technique for the study of polymers regarding their molecular structure, identification, and compositional analysis.

REFERENCES

1. Qin, S.; Wang, Y.; Wu, X.; Zhang, X.; Zhu, Y.; Yu, N.; Zhang, Y.; Wu, Y. Nylon-Based Composite Gel Membrane Fabricated via Sequential Layer-by-Layer Electrospinning for Rechargeable Lithium Batteries with High Performance. Polymers 2020, 12, 1572. [Google Scholar] [CrossRef] [PubMed]

2. Kabir, S.; Lee, S. Study of shape memory and tensile property of 3D printed sinusoidal sample/nylon composite focused on various thicknesses and shape memory cycles. Polymers 2020, 12, 1600. [Google Scholar] [CrossRef] [PubMed]

3. Dasgupta, S.; Hammond, W.B.; Goddard, W.A. Crystal structures and properties of nylon polymers from theory. J. Am. Chem. Soc. 1996, 118, 12291–12301. [Google Scholar] [CrossRef]

4. Rai, S.; Rai, A.K. Characterization of organic materials by LIBS for exploration of correlation between molecular and elemental LIBS signals. AIP Adv. 2011, 1, 042103. [Google Scholar] [CrossRef]

5. Moros, J.; Laserna, J. Laser-induced breakdown spectroscopy (LIBS) of organic compounds: A review. Appl. Spectrosc. 2019, 73, 963–1011. [Google Scholar] [CrossRef]

6. De Lucia, F.C.; Harmon, R.S.; McNesby, K.L.; Winkel, R.J.; Miziolek, A.W. Laser-induced breakdown spectroscopy analysis of energetic materials. Appl. Opt. 2003, 42, 6148–6152. [Google Scholar] [CrossRef]

7. Baudelet, M.; Boueri, M.; Yu, J.; Mao, X.; Mao, S.S.; Russo, R. Laser ablation of organic materials for discrimination of bacteria in an inorganic background. In Ultrafast Phenomena in Semiconductors and Nanostructure Materials XIII; SPIE: Bellingham, WA, USA, 2009; Volume 7214, pp. 97–106. [Google Scholar]

8. Moros, J.; Lorenzo, J.A.; Lucena, P.; Miguel Tobaria, L.; Laserna, J.J. Simultaneous Raman Spectroscopy- Laser-Induced Breakdown Spectroscopy for instant standoff analysis of explosives using a mobile integrated sensor platform. Anal. Chem. 2010, 82, 1389–1400. [Google Scholar] [CrossRef]

9. Carbone, E.; D'Isa, F.; Hecimovic, A.; Fantz, U. Analysis of the C2 Swan bands as a thermometric probe in CO2 microwave plasmas. Plasma Sources Sci. Technol. 2020, 29, 055003. [Google Scholar] [CrossRef]

10. Gottfried, J.L.; De Lucia Jr, F.C.; Munson, C.A.; Miziolek, A.W. Double-pulse standoff laser-induced breakdown spectroscopy for versatile hazardous materials detection. Spectrochim. Acta Part B Acta Spectrosc. 2007, 62, 1405–1411. [Google Scholar] [CrossRef]

11. Jihua, G.; Ali, A.; Dagdigian, P.J. State-to-state collisional interelectronic and intraelectronic energy transfer involving CN A 2Π v= 3 and X 2Σ+ v= 7 rotational levels. J. Chem. Phys. 1986, 85, 7098–7105. [Google Scholar] [CrossRef]

12. Sauder, D.G.; Patel-Misra, D.; Dagdigian, P.J. The vibronic state distribution of the NCO $X\sim \sim 2\ \Pi$) product from the CN+O2 reaction. J. Chem. Phys. 1991, 95, 1696–1707. [Google Scholar] [CrossRef]
13. Fernández-Bravo, Á.; Delgado, T.; Lucena, P.; Laserna, J.J. Vibrational emission analysis of the CN molecules in laser-induced breakdown spectroscopy of organic compounds. Spectrochim. Acta Part B Acta Spectrosc. 2013, 89, 77–83. [Google Scholar] [CrossRef]
14. Dong, M.; Chan, G.C.Y.; Mao, X.; Gonzalez, J.J.; Lu, J.; Russo, R.E. Elucidation of C2 and CN formation mechanisms in laser-induced plasmas through correlation analysis of carbon isotopic ratio. Spectrochim. Acta Part B Acta Spectrosc. 2014, 100, 62–69. [Google Scholar] [CrossRef]
15. Dong, M.; Mao, X.; Gonzalez, J.J.; Lu, J.; Russo, R.E. Carbon isotope separation and molecular formation in laser-induced plasmas by laser ablation molecular isotopic spectrometry. Anal. Chem. 2013, 85, 2899–2906. [Google Scholar] [CrossRef]
16. Boueri, M.; Motto-Ros, V.; Lei, W.Q.; Ma, Q.L.; Zheng, L.J.; Zeng, H.P.; Yu, J. Identificationofpolymermaterialsusinglaser-inducedbreakdownspectroscopy combined with artificial neural networks. Appl. Spectrosc. 2011, 65, 307–314. [Google Scholar] [CrossRef]
17. Mousavi, S.J.; Farsani, M.H.; Darbani, S.M.R.; Asadorian, N.; Soltanolkotabi, M.; Majd, A.E. Identification of atomic lines and molecular bands of benzene and carbon disulfide liquids by using LIBS. Appl. Opt. 2015, 54, 1713–1720. [Google Scholar] [CrossRef]
18. Mousavi, S.J.; Hemati Farsani, M.; Darbani, S.M.R.; Mousaviazar, A.; Soltanolkotabi, M.; Eslami Majd, A. CN and C2 vibrational spectra analysis in molecular LIBS of organic materials. Appl. Phys. B 2016, 122, 106. [Google Scholar] [CrossRef]
19. Farooq, W.A.; Al-Johani, A.S.; Alsalhi, M.S.; Tawfik, W.; Qindeel, R. Analysis of polystyrene and polycarbonate used in manufacturing of water and food containers using laser induced breakdown spectroscopy. J. Mol. Struct. 2020, 1201, 127152. [Google Scholar] [CrossRef]
20. Lasheras, R.J.; Bello-Gálvez, C.; Anzano, J. Identification of polymers by libs using methods of correlation and normalized coordinates. Polym. Test. 2010, 29, 1057–1064. [Google Scholar] [CrossRef]
21. Costa, V.C.; Aquino, F.W.B.; Paranhos, C.M.; Pereira-Filho, E.R. Identification and classification of polymer e-waste using laser-induced breakdown spectroscopy (LIBS) and chemometric tools. Polym. Test. 2017, 59, 390–395. [Google Scholar] [CrossRef]

22. Grégoire, S.; Motto-Ros, V.; Ma, Q.L.; Lei, W.Q.; Wang, X.C.; Pelascini, F.; Surma, F.; Detalle, V.; Yu, J. Correlation between native bonds in a polymeric material and molecular emissions from the laser-induced plasma observed with space and time resolved imaging. Spectrochim. Acta Part B Acta Spectrosc. 2012, 74, 31–37. [Google Scholar] [CrossRef]

23. Aquino, F.W.; Pereira-Filho, E.R. Analysis of the polymeric fractions of scrap from mobile phones using laser-induced breakdown spectroscopy: Chemometric applications for better data interpretation. Talanta 2015, 134, 65–73. [Google Scholar] [CrossRef] [PubMed]

24. Zhu, W.; Li, X.; Sun, R.; Cao, Z.; Yuan, M.; Sun, L.; Yu, X.; Wu, J. Investigation of the CN and C2 emission characteristics and microstructural evolution of coal to char using laser-induced breakdown spectroscopy and Raman spectroscopy. Energy 2022, 240, 122827. [Google Scholar] [CrossRef]

25. Aga, A.; Mu, M. Doping of Polymers with ZnO Nanostructures for Optoelectronic and Sensor Applications. In Nanowires Science and Technology; Lupu, N., Ed.; IntechOpen: London, UK, 2010. [Google Scholar]

26. Yanilmaz, M.; Zhu, J.; Lu, Y.; Ge, Y.; Zhang, X. High-strength, thermally stable nylon 6, 6 composite nanofiber separators for lithium-ion batteries. J. Mater. Sci. 2017, 52, 5232–5241. [Google Scholar] [CrossRef]

27. Xu, Z.; Mahalingam, S.; Rohn, J.L.; Ren, G.; Edirisinghe, M. Physio-chemical and antibacterial characteristics of pressure spun nylon nanofibres embedded with functional silver nanoparticles. Mater. Sci. Eng. C 2015, 56, 195–204. [Google Scholar] [CrossRef]

28. Nguyen, T.N.; Moon, J.; Kim, J.J. Microstructure and mechanical properties of hardened cement paste including Nylon 66 nanofibers. Constr. Build. Mater. 2020, 232, 117134. [Google Scholar] [CrossRef]

29. Abd Halim, N.S.; Wirzal, M.D.H.; Bilad, M.R.; Md Nordin, N.A.H.; Adi Putra, Z.; Sambudi, N.S.; Mohd Yusoff, A.R. Improving performance of electrospun nylon 6, 6 nanofiber membrane for produced water filtration via solvent vapor treatment. Polymers 2019, 11, 2117. [Google Scholar] [CrossRef]

30. Hu, S.; Yuan, D.; Liu, Y.; Zhao, L.; Guo, H.; Niu, Q.; Zong, W.; Liu, R. The toxic effects of alizarin red S on catalase at the molecular level. RSC Adv. 2019, 9, 33368–33377. [Google Scholar] [CrossRef]

31. Alrebdi, T.A.; Fayyaz, A.; Asghar, H.; Zaman, A.; Asghar, M.; Alkallas, F.H.; Hussain, A.; Iqbal, J.; Khan, W. Quantification of Aluminum Gallium Arsenide (AlGaAs) Wafer Plasma Using Calibration-Free Laser-Induced Breakdown Spectroscopy (CF-LIBS). Molecules 2022, 27, 3754. [Google Scholar] [CrossRef]

32. Iqbal, J.; Asghar, H.; Shah, S.K.H.; Naeem, M.; Abbasi, S.A.; Ali, R. Elemental analysis of sage (herb) using calibration-free laser-induced breakdown spectroscopy. Appl. Opt. 2020, 59, 4927–4932. [Google Scholar] [CrossRef]

33. Fayyaz, A.; Liaqat, U.; Adeel Umar, Z.; Ahmed, R.; Aslam Baig, M. Elemental Analysis of Cement by Calibration-Free Laser Induced Breakdown Spectroscopy (CF-LIBS) and Comparison with Laser Ablation–Time-of-Flight–Mass Spectrometry (LA-TOF-MS), Energy Dispersive X-Ray Spectrometry (EDX), X-Ray Fluorescence Spectroscopy (XRF), and Proton Induced X-Ray Emission Spectrometry (PIXE). Anal. Lett. 2019, 52, 1951–1965. [Google Scholar]

34. Macrae, R.M. Puzzles in bonding and spectroscopy: The case of dicarbon. Sci. Prog. 2016, 99, 1–58. [Google Scholar] [CrossRef] [PubMed]

35. Zhang, S.; Wang, X.; He, M.; Jiang, Y.; Zhang, B.; Hang, W.; Huang, B. Laser-induced plasma temperature. Spectrochim. Acta Part B Acta Spectrosc. 2014, 97, 13–33. [Google Scholar] [CrossRef]

36. Unnikrishnan, V.K.; Alti, K.; Kartha, V.B.; Santhosh, C.; Gupta, G.P.; Suri, B.M. Measurements of plasma temperature and electron density in laser-induced copper plasma by time-resolved spectroscopy of neutral atom and ion emissions. Pramana 2010, 74, 983–993. [Google Scholar] [CrossRef]

37. NIST. NIST Standard Reference Database 78; NIST: Gaithersburg, MD, USA, 1995. [CrossRef]

38. Cremers, D.A.; Radziemski, L.J. Handbook of Laser-Induced Breakdown Spectroscopy; John Wiley & Sons: Hoboken, NJ, USA, 2013. [Google Scholar]

39. McWhirter, R.W.P. Plasma Diagnostic Techniques; Academic Press: New York, NY, USA, 1965; Chapter 5. [Google Scholar]

40. Gigosos, M.A.; Gonzalez, M.A.; Cardenoso, V. Computer simulated Balmer-alpha,-beta and-gamma Stark line profiles for non-equilibrium plasmas diagnostics. Spectrochim. Acta Part B: Acta Spectrosc. 2003, 58, 1489–1504. [Google Scholar] [CrossRef]

41. Praher, B.; Palleschi, V.; Viskup, R.; Heitz, J.; Pedarnig, J.D. Calibration free laser-induced breakdown spectroscopy of oxide materials. Spectrochim. Acta Part B Acta Spectrosc. 2010, 65, 671–679. [Google Scholar] [CrossRef]

42. Fujimoto, T.; McWhirter, R.W.P. Validity criteria for local thermodynamic equilibrium in plasma spectroscopy. Phys. Rev. A 1990, 42, 6588. [Google Scholar] [CrossRef]

43. Hahn, D.W.; Omenetto, N. Laser-induced breakdown spectroscopy (LIBS), part I: Review of basic diagnostics and plasma–particle interactions: Still-challenging issues within the analytical plasma community. Appl. Spectrosc. 2010, 64, 335A–366A. [Google Scholar] [CrossRef] [PubMed]

44. Hey, J.D. Criteria for local thermal equilibrium in non-hydrogenic plasmas. J. Quant. Spectrosc. Radiat. Transf. 1976, 16, 69–75. [Google Scholar] [CrossRef]

45. Cristoforetti, G.; De Giacomo, A.; Dell'Aglio, M.; Legnaioli, S.; Tognoni, E.; Palleschi, V.; Omenetto, N. Local thermodynamic equilibrium in laser-induced breakdown spectroscopy: Beyond the McWhirter criterion. Spectrochim. Acta Part B Acta Spectrosc. 2010, 65, 86–95. [Google Scholar] [CrossRef]

46. Iqhrammullah, M.; Suyanto, H.; Pardede, M.; Karnadi, I.; Kurniawan, K.H.; Chiari, W.; Abdulmadjid, S.N. Cellulose acetate-polyurethane film adsorbent with analyte enrichment for in-situ detection and analysis of aqueous Pb using Laser-Induced Breakdown Spectroscopy (LIBS). Environ. Nanotechnol. Monit. Manag. 2021, 16, 100516. [Google Scholar] [CrossRef]

47. Nisah, K.; Ramli, M.; Iqhrammullah, M.; Mitaphonna, R.; Hartadi, B.S.; Abdulmadjid, S.N.; Sani, N.D.M.; Idroes, R.; Safitri, E. Controlling the diffusion of micro-volume Pb solution on hydrophobic polyurethane membrane for quantitative analysis using laser-induced breakdown spectroscopy (LIBS). Arab. J. Chem. 2022, 15, 103812. [Google Scholar] [CrossRef]

CHAPTER-13

SYNTHESIS AND PHOTOCATALYTIC DEGRADATION CHARACTERISTICS OF CATALYSTS BASED ON CARBON NANOFIBERS

1. INTRODUCTION: AN OVERVIEW

In recent years, there has been a growing interest in the development of advanced catalysts for environmental remediation, particularly in the context of photocatalytic degradation. Carbon nanofibers (CNFs) have emerged as promising candidates due to their unique structural and electronic properties. This study focuses on the preparation and investigation of the photo-degradation properties of catalysts based on carbon nanofibers. The synthesis process involves a meticulous preparation method to tailor the catalytic properties of CNFs, aiming to enhance their efficiency in degrading various pollutants under light irradiation. Understanding and optimizing the photo-degradation capabilities of these carbon-nanofiber-based catalysts hold significant implications for addressing contemporary environmental challenges.

Dyes are commonly used in modern industries, such as textile, food, paper, printing, leather, and cosmetics. They will cause serious pollution problems and bring risks to human health, such as carcinogenesis and kidney dysfunction, if discharged directly into the natural environment without treatment [**1,2,3**]. Therefore, how to remove organic dyes from wastewater effectively has become a hot research issue in recent years [**4**].

Currently, the treatment of printing and dyeing wastewater can be divided into adsorption technologies [**5,6,7**], advanced oxidation technologies, such as Fenton oxidation [**8,9,10,11,12,13**] and ozone oxidation [**14,15**], biological technologies, such as bacterial [**16,17**], fungal [**18,19**], and algal [**20**] treatment, membrane separation technologies, such as ultrafiltration, nanofiltration, reverse osmosis and electrodialysis [**21,22**], electro-Fenton, anodicoxidation [**23**], electrocoagulation electrochemical treatment techniques, ion exchange [**24,25,26**] and some multiple processes [**27,28,29,30,31**]. Fenton oxidation is reported to be one of the most widely studied and applied advanced oxidation technologies. It can completely degrade the refractory toxic and hazardous organic compounds into water and carbon dioxide with a short treatment cycle.

The conventional Fenton technique uses ferrous salts, but some iron-containing oxides, such as Fe2O3, can be used instead of ferrous salts. It was shown previously that composites containing Fe2O3 can effectively degrade a large number of organic compounds in wastewater [32,33,34,35,36,37,38]. There are kinds of methods that could prepare Fe2O3, mainly the co-precipitation method, solution-gel method, micro-emulsion method, solvothermal method, hydrothermal method, etc. Among them, the hydrothermal method can produce nano Fe2O3 with high purity, controllable morphology and particle size under mild synthesis conditions. Electrospinning is a simple technique for the effective production of nanofibers. The device is shown in Figure 1. Since carbon nanofibers (CNFs) have a high specific surface area, few structural defects, low density and high conductivity, they can be used as a catalytic template to prevent the agglomeration of nanoparticles which can solve the problem that nano Fe2O3 is prone to agglomeration. In this study, CNFs were obtained after the carbonization of polymethyl methacrylate/polyacrylonitrile (PAN/PMMA) nanofibers and used as template materials. In addition, Fe^2O^3/CNFs composites were prepared by the hydrothermal method with high purity, controlled morphological particle size, and uniform dispersibility. Then, the Fe^2O^3/CNFs composites were used in the Fenton reaction to degrade the MO solution.

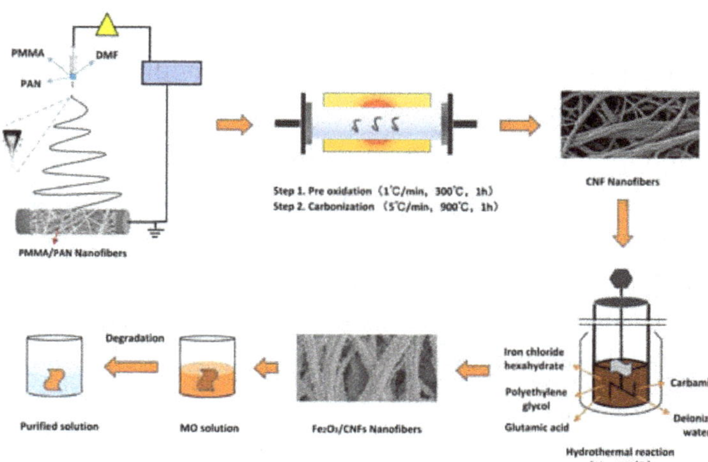

Figure 1. Schematic illustration of fabrication of Fe2O3/CNFs composite nanocatalysts and their degradation of MO.

2. EXPERIMENTAL METHODS

2.1. Materials

Polyacrylonitrile (PAN) (Mw 150,000) was purchased from Shanghai Maclean. Polymethyl methacrylate (PMMA) (high flow type) was purchased from Shanghai Maclean. N, N-dimethylformamide (DMF) (AR) was purchased from Shanghai SiXin Bio. Ferric chloride hexahydrate (FeCl3·6H2O) (EP) was purchased from 3A chemicals. Urea (AR) was purchased from Genbio. Glycine (AR) was purchased from Genbio. Polyethylene glycol 1000 monomethyl ether was purchased from Aladdin. Ethanol (AR) was purchased from Shanghai Chemical Reagent Procurement and Supply Station. Methyl orange (MO) (AR) was purchased from Shanghai Aladdin Technology & Biochemistry. All reagents and solvents were used without further purification.

2.2. Preparation of Carbon Nanofiber Membrane

PAN and PMMA were added to DMF for dissolution (the mass ratio of PAN and PMMA was 4:6, and the concentration of the spinning solution was 20%), and the spinning solution was prepared by stirring at room temperature for 24 h. The spinning solution was ultrasonicated for 2 h and spun by an electrospinning machine. The spinning time was 24 h. Then, the above nanofiber film was put into a tube furnace and pre-oxidized in the air for 1 h. After the pre-oxidation, it was continued to be carbonized in nitrogen for 1 h. After the tube furnace was naturally cooled to room temperature, carbon nanofibers were obtained.

2.3. Preparation of Fe_2O_3/CNFs Composites

The Fe2O3/CNFs were prepared by the hydrothermal method. Firstly, 0.16 g FeCl3·6H2O, 0.2 g polyethylene glycol, 0.484 g urea and 0.03 g glycine were added to 25 mL deionized water. The mixed solution was stirred until it was completely dissolved and then poured into the hydrothermal reactor. Then, 20 mg carbon nanofiber membranes were added to the hydrothermal reactor to make them uniformly dispersed in the solution. The reaction was carried out at a certain temperature for a while to obtain Fe2O3/CNFs composites.

To explore the influence of hydrothermal reaction time and temperature on iron oxide particles, the following experiments were set up:

- The hydrothermal reaction time was 1 h, and the reaction temperatures were 130 °C, 140 °C, 150 °C, 160 °C, 170 °C and 180 °C, respectively.
- The hydrothermal reaction temperature was 180 °C, and the reaction time was 1 h, 2 h, 3 h, 4 h, 5 h and 6 h, respectively.

Under visible light conditions, Fe2O3/CNFs were used as a catalyst for the degradation of MO.

2.4. Characterization

The surface structures of the composites were observed by scanning electron microscopy (SEM, s-4800) (Hitachi Manufacturing, Tokyo, Japan). The crystal structure of the composites was characterized by X-ray diffraction (XRD) (X' Pert-Pro MRD, Panaco, The Netherlands). The functional groups of the composites were determined by infrared spectroscopy (IR) (Nikoli Instruments Manufacturing, Waltham, Massachusetts, USA). The chemical composition of the composites was examined by energy dispersive spectroscopy (EDS) (Hitachi Manufacturing, Tokyo, Japan) and by X-ray photoelectron spectroscopy (XPS) using Axis Ultra HAS equipment.

2.5. Degradation Experiments

A total of 100 mg/L MO solution was prepared, and the pH of the solution was 7.4. Then the degradation experiments were carried out as follows:

- **Effect of reaction temperature:** Fe_2O_3/CNFs composite is 0.8 g/L, the concentration of hydrogen peroxide is 0.194 mol/L, and the reaction time is 0~120 min. The reaction temperatures are 50 °C, 60 °C, 70 °C and 80 °C, respectively.
- **Effect of the amount of Fe_2O_3/CNFs composite:** The initial concentration of hydrogen peroxide is 0.194 mol/L, the reaction temperature is 80 °C, and the reaction time is 0~120 min. The amounts of the Fe_2O_3/CNFs composite are 0.4 g/L, 0.6 g/L, 0.8 g/L and 1.0 g/L, respectively.
- **Effect of initial concentration of hydrogen peroxide:** Fe_2O_3/CNFs composite is 0.8 g/L, the reaction temperature is 80 °C and the reaction time is 0~240 min. The initial concentrations of hydrogen peroxide are 0.097 mol/L, 0.146 mol/L, 0.194 mol/L, and 0.243 mol/L, respectively.

2.6. Degradation Performance of Fe_2O_3/CNFs

MO solutions with the concentration of 4 mg/L, 10 mg/L and 40 mg/L (pH ≈ 7.4) were prepared. The UV-Vis spectrums were measured by Cary 5000 UV-vis-NIR spectrophotometer between the wavelengths of 190–600 nm so as to obtain the maximum absorption wavelength. MO solutions with the concentrations of 4 mg/L, 6 mg/L, 8 mg/L, 10 mg/L, 20 mg/L, 30 mg/L, 40 mg/L, 50 mg/L and 60 mg/L were prepared. The absorbances of MO solutions with different concentrations at the maximum absorption wavelength were measured, and the absorbance concentration standard curve was obtained.

After the degradation experiment, the absorbance of MO was measured at the maximum absorption wavelength. Then the solution concentration was obtained according to the standard curve of MO. The decolorization rate of MO was calculated.

Determination of the decolorization rate of MO: The absorbance of MO solutions with different concentrations at the maximum absorption wavelength was measured and the absorbance–concentration standard curve was obtained. The concentration of MO before degradation was denoted as C_0. After the degradation experiment, the absorbance was measured at the maximum absorption wavelength. Then the solution concentration was obtained according to the standard curve, denoted as C_1. The decolorization rate was calculated as follows:

$$\frac{C_0 - C_1}{C_0} \times 100\% \quad (1)$$

3. RESULTS AND DISCUSSION

3.1. Characterization of Fe_2O_3/CNFs

Figure 2 and Figure 3 are the SEM images of carbon nanofibers and Fe_2O_3/CNFs prepared under different hydrothermal reaction conditions. From the figures, it can be seen that when the reaction temperature rises to 140 °C, large amounts of nanoparticles were successfully grown on carbon nanofiber. When the temperature is lower than 150 °C, the generated Fe_2O_3 nanoparticles are uneven, indicating that the hydrothermal reaction temperature affects the fabrication of nanoparticles. Meanwhile, the Fe_2O_3 particle size gradually increased with the increase in reaction time. It was known that the specific surface area increased as the number of nanoparticles increased, and the particle size became smaller. Furthermore, the catalytic efficiency improved as the specific surface area increased. Therefore, the hydrothermal reaction temperature and time were determined to be 180 °C and 1 h. Under this reaction condition, the Fe2O3 nanoparticle size is around 90 nm and grows evenly on carbon nanofibers.

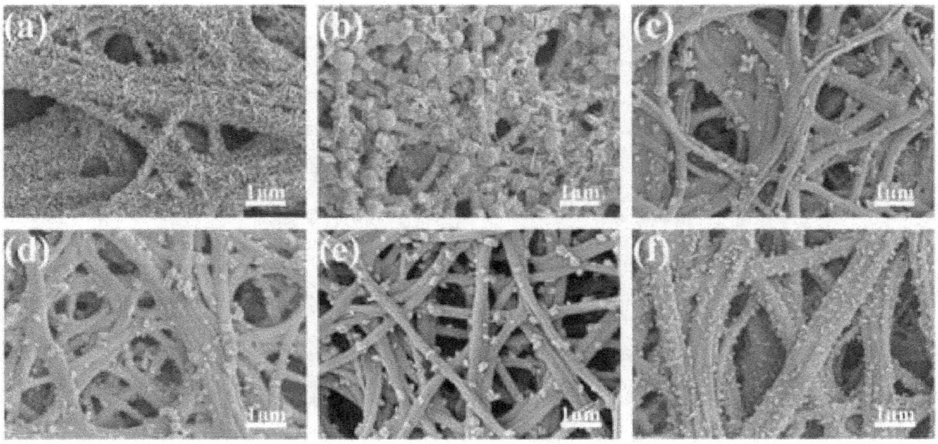

Figure 2. SEM images of Fe_2O_3/CNFs (effect of temperature): (a) 130 °C; (b) 140 °C; (c) 150 °C; (d) 160 °C; (e) 170 °C; (f) 180 °C.

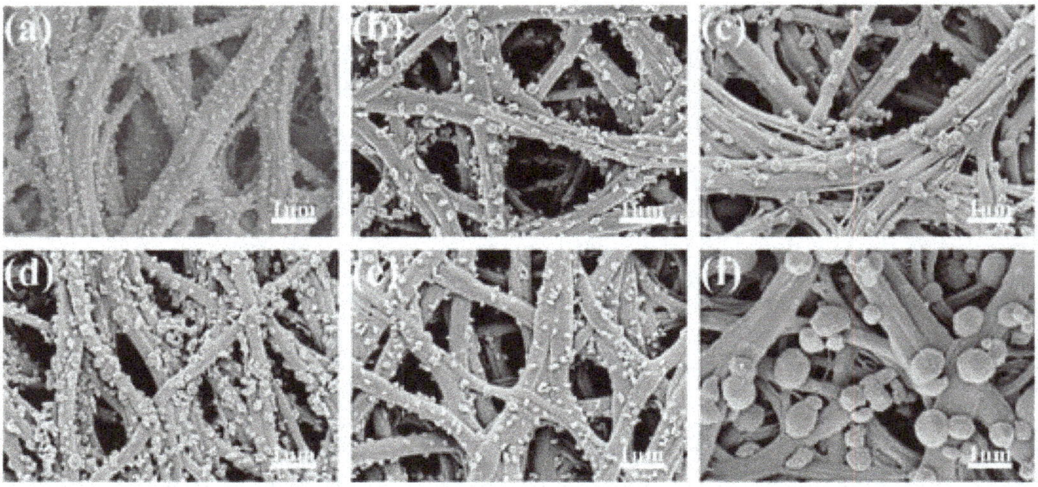

Figure 3. SEM images of Fe_2O_3/CNFs (effect of time): (a) 1 h; (b) 2 h; (c) 3 h; (d) 4 h; (e) 5 h; (f) 6 h.

Figure 4 shows the EDS element analysis chart of the Fe2O3/CNFs composites. From the figure, it can be seen that the Fe elements and O elements are uniformly distributed on the carbon fibers, indicating that the Fe2O3/CNFs composites were successfully prepared by the hydrothermal method.

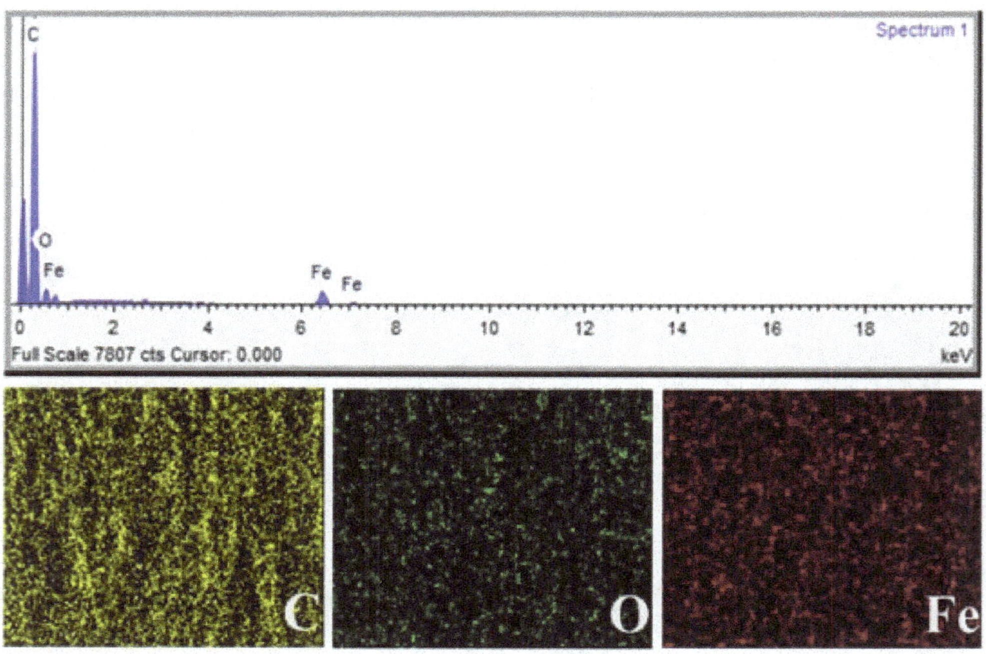

Figure 4. EDS image of Fe2O3/CNFs.

Figure 5 shows the XRD patterns of Fe_2O_3/CNFs at different reaction temperatures and times. When the hydrothermal reaction temperature raised to

140 °C, characteristic peaks of Fe2O3 appeared at 2θ = 24.2°, 33.2°, 35.6°, 40.9°, 49.5°, 54.1°, 62.4°, and 64.0°, which are consistent with the standard card mapping card JCPDS NO.33-0664, indicating the formation of Fe2O3 particles [39].

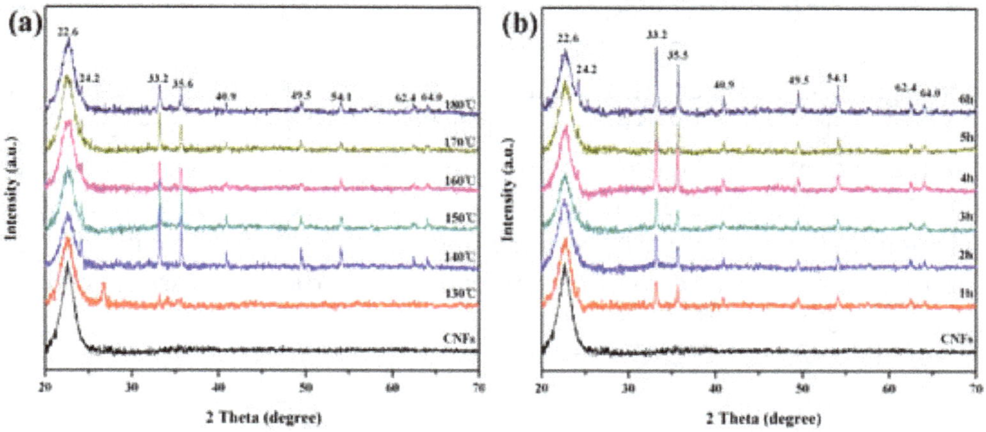

Figure 5. XRD patterns of Fe_2O_3/CNFs: (a) effect of reaction temperature; (b) effect of reaction time.

Figure 6 shows the IR spectra of CNFs and Fe_2O_3/CNFs composites. In the infrared spectrum of CNFs, the characteristic peaks at 3438 cm^{-1} are caused by -OH bond stretching vibration [40]. The characteristic peaks at 1140 cm^{-1} and 1537 cm^{-1} are caused by C-O-C and C=O vibration, respectively [41]. Compared with CNFs, Fe_2O_3/CNFs composites showed characteristic peaks at 556 cm^{-1} and 475 cm^{-1} are caused by the vibrations of Fe-O functional groups [42].

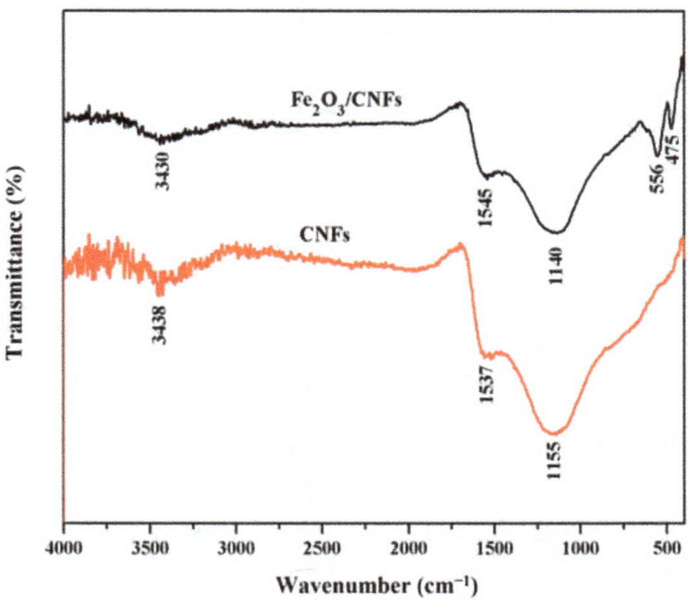

Figure 6. IR image of CNFs and Fe_2O_3/CNFs.

Figure 7 shows the XPS spectra analysis of Fe_2O_3/CNFs composites. The satellite peak of Fe2O3 at 719.2 eV is detected in Figure 7b, which shows the trivalent Fe elemental in the Fe2O3/CNFs composites. Meanwhile, there are two main peaks at 712.5 eV and 725.4 eV corresponding to Fe 2p3/2 and Fe 2p1/2, indicating the existence of Fe2+ and Fe3+ [43]. Furthermore, the energy spectrum of C 1S is shown in Figure 7c. There are three characteristic peaks. The peak at 284.5 eV is caused by the C-C bond, while the peaks at 285.8 eV and 287.9 eV represent C-O and C=O, respectively [44]. Figure 7d shows the energy spectrum of O1 s. The peak at 531.0 eV represents the O element in Fe_2O_3 and the peak at 532.7 eV represents the O element in CNFs [43,44]. XPS results show that Fe-O-C chemical bonds exist between Fe_2O_3 and CNFs. Fe_2O_3 particles grew in situ on the CNFs carrier, and the results were consistent with the XRD and IR analysis.

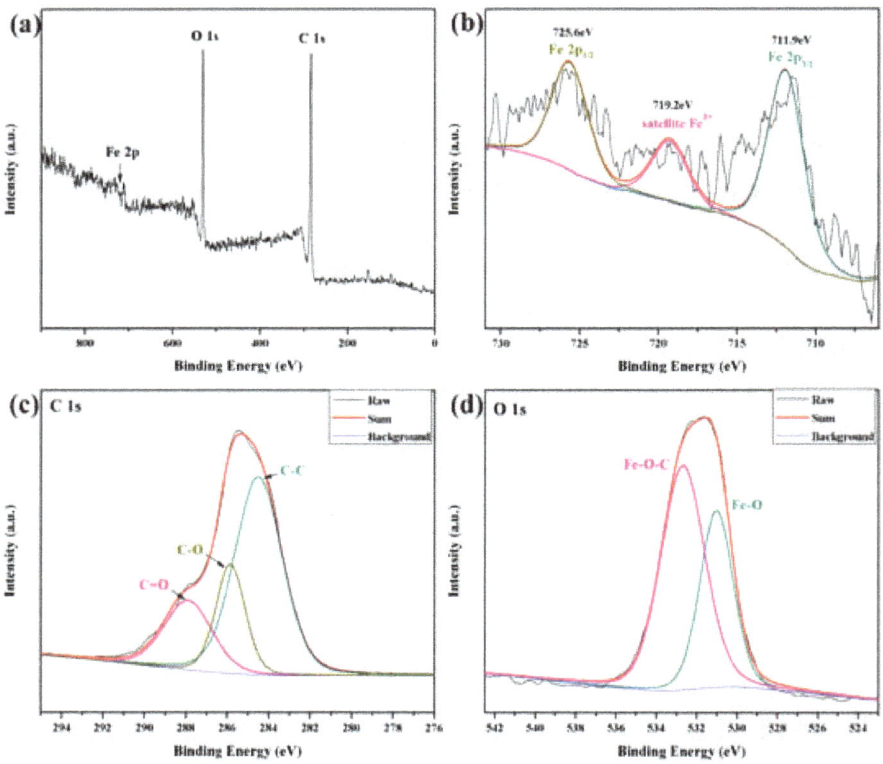

Figure 7. XPS spectra of (a) Fe2O3/CNFs; (b) Fe 2p; (c) C 1 s; (d) O 1 s.

3.2. Heterogeneous Fenton Degradation of MO by Fe_2O_3/CNFs

The MO aqueous solution has the characteristics of an acid-base indicator, and its molecular structure will change with the pH value as well colours. When pH < 3.1, it is the quinone structure and the solution is red, and when pH > 4.4, it is an azo structure and the solution is yellow. The structural changes can be expressed as follows:

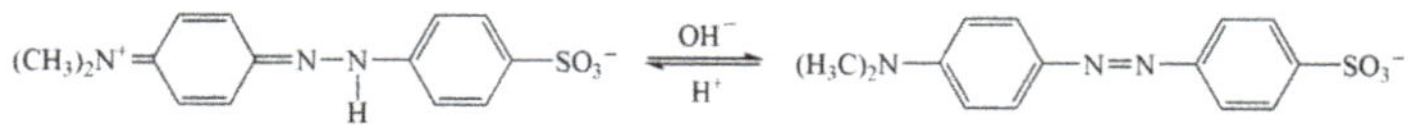

Figure 8 shows the UV spectra of MO at a different pH value. When the structure is quinone, the maximum absorption wavelength in the visible region is around 505 nm. When the structure is azo, the maximum absorption wavelength in the visible region is around 465 nm.

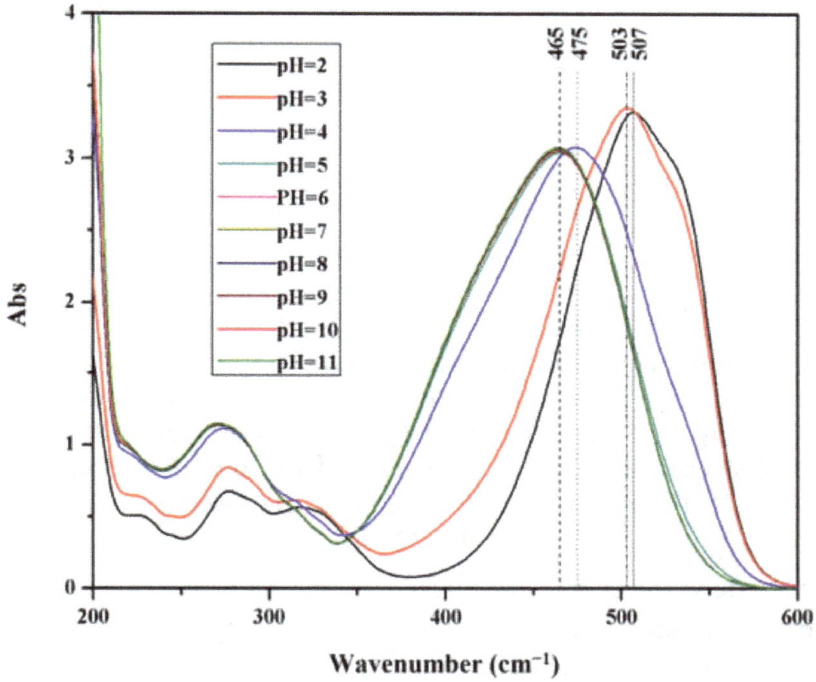

Figure 8. UV-vis spectra of MO at different pH value.

The decolorization effect of Fe2O3/CNFs composites on MO under different pH conditions is shown in Figure 9a. The initial pH of the solution has a great influence on the Fe2O3/CNFs composites catalyzing the degradation of MO by hydrogen peroxide. When pH < 3, the decolorization effect of MO is better, and the decolorization rate decreases gradually with the increase in pH. When pH is 6~8, the decolorization rate gradually increases. Furthermore, when pH > 8, the decolorization rate continues to decrease again. The best decolorization of methyl orange was achieved when pH = 2. It may be that the azo bond in the MO molecular structure changes to a quinone structure under the condition of pH < 3, and the destruction of N = N bond leads to the instability of the MO molecular structure, which makes the catalytic reaction easier to proceed. Secondly, due to the over acid condition, Fe_2O_3 is partially dissolved, more free Fe^{2+}/Fe^{3+} contacts with H^2O^2, and more ·OH is produced, thus improving the decolorization rate of MO [45].

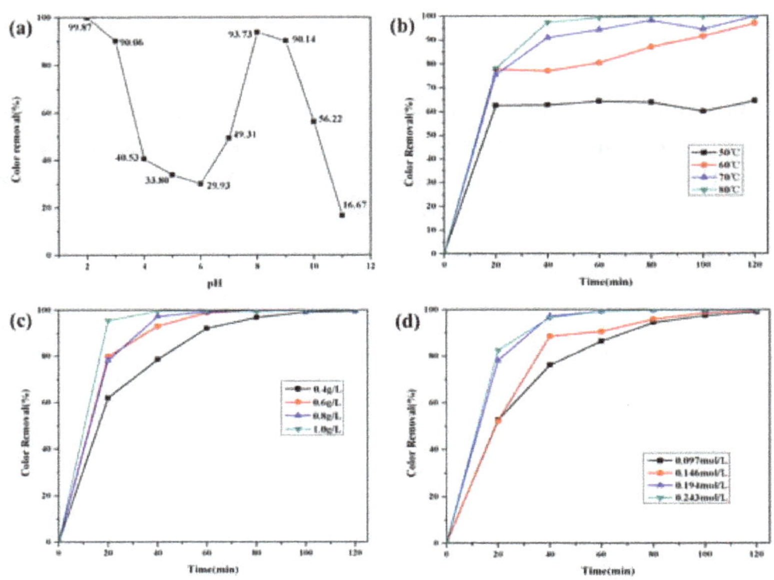

Figure 9. Decolorization effect of MO (a) with pH: ([MO]0 = 100 mg/L, [Fe2O3/ CNFs]0 = 0.6 g/L, [H_2O_2]0 = 0.155 mol/L, T = 80 °C, t = 2 h); (b) with temperature: ([MO]0 = 100 mg/L, [Fe_2O_3/CNFs]0 = 0.8 g/L, [H_2O_2]0= 0.194 mol/L, pH = 2); (c) with Fe_2O_3/CNFs content: ([MO]0 = 100 mg/L, [H_2O_2]0 = 0.194 mol/L, pH = 2, T = 80°C); (d) with H_2O_2 content: ([MO]0 = 100 mg/L, [Fe_2O_3/CNFs]0 = 0.8 g/L, pH = 2, T = 80 °C).

Figure 9b shows the effect of reaction temperature on the catalytic effect of MO. The decolorization rate of MO was improved by increasing the reaction temperature. When the temperature increased from 50 °C to 80 °C, the decolorization rate of MO increased from 66.39% to 99.87%. Moreover, the increase in reaction temperature can effectively shorten the degradation time.

The heterogeneous Fenton reaction occurs on the surface of the catalyst, and the amount of catalyst is an important factor, affecting the decolorization effect. The catalytic effect of Fe^2O^3/CNFs amount on MO is shown in Figure 9c. When the amount of catalyst increased from 0.4 g/L to 0.8 g/L, the decolorization rates of MO increased from 78.5% to 97.31%. Increasing the dosage of Fe^2O^3/CNFs has a facilitating effect on the decolorization of MO. This is because increasing the amount of catalyst can increase the number of active sites on the surface of the catalyst and accelerate the decomposition rate of H2O2 to produce ·OH. When the amount of catalyst was increased from 0.8 g/L to 1 g/L, the decolorization rate of MO increased slightly because the excessive catalyst will reduce the H2O2 adsorption per unit area [46,47].

Figure 9d shows the degradation efficiency of MO at different H^2O^2 concentrations. It can be seen from the figure that within the first 2 h when the concentration of H2O2 increases from 0.097 mol/L to 0.194 mol/L, the degradation rate of MO increases significantly. The degradation rate of MO is related to the amount of ·OH. The higher the hydrogen peroxide concentration, the more that ·OH is produced, and the degradation rate of MO increases. Continuing to increase the concentration of H2O2, the degradation rate of MO is no longer increased significantly.

It can be observed that the decolorization rate can be up to more than 99% for 100 mg/L MO solution by the Fe_2O_3/CNFs catalyst. This result was also compared with previous studies reported for the catalytic degradation of MO shown in Table 1.

Table 1. Comparison with previous studies reported for catalytic degradation of MO.

Catalysts	Initial Concentration	Decolorization Rate	Reaction Time	References
Fe2O3/CNFs	100 mg/L	More than 99%	60 min	This work
Ti_3C_2-TiO_2	40 mg/L	99%	40 min	[48]
10% Co-ZnO	100 mg/L	100%	120 min	[49]
Ag-PMOS	20 mg/L	81% & 48%	60 min	[50]
ZnO-PMOS		47% & 57%		
Ni@FP	15 mg/L	93.40%	5 min	[51]
TiO2/ZSM-5	20 mg/L	99.55%	180 min	[52]
PANI(1.5 mol)/ZnO	-	98.3%	180 min	[53]

3.3. Degradation Mechanism Analysis

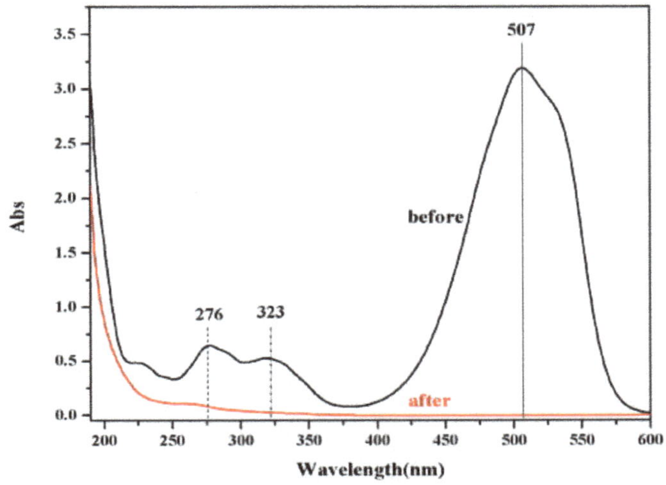

Figure 10. UV-vis spectra of MO before and after degradation.

The UV-Vis spectra of MO solution before and after degradation are shown in Figure 10, from which it can be seen that the absorption peak of MO disappeared, and no other new peaks generated. There are two possible reasons for this phenomenon: (1) the intermediate products of catalytic degradation of MO have no absorption in the range of 190–600 nm; (2) MO is directly degraded to CO_2 and H_2O without any intermediate products generated.

Fe_2O_3/CNFs composites degrade the MO solution under acidic conditions (pH = 2) when MO is a quinone structure and the hydroxyl radical HO· plays a major role in the degradation process. Combined with the Fenton reaction system, it is known that MO is not directly degraded to CO_2 and H_2O, but certain intermediate products are produced that have no absorption between 190 and 600 nm. It is speculated that there are four possible degradation pathways of MO which are shown in **Figure 11.** The intermediates produced by path (1) and path (2) will continue to be oxidized to other products in the subsequent reactions. Pathway (3) and pathway (4) are the main mechanisms of this degradation. The quinone structure of MO is decolorized by HO· oxidation, which leads to the destruction of the -N=N- group and C-N bond, thus decolorizing the MO solution. Finally, some of the intermediates are completely oxidized to CO_2 and H_2O as the reaction time increases.

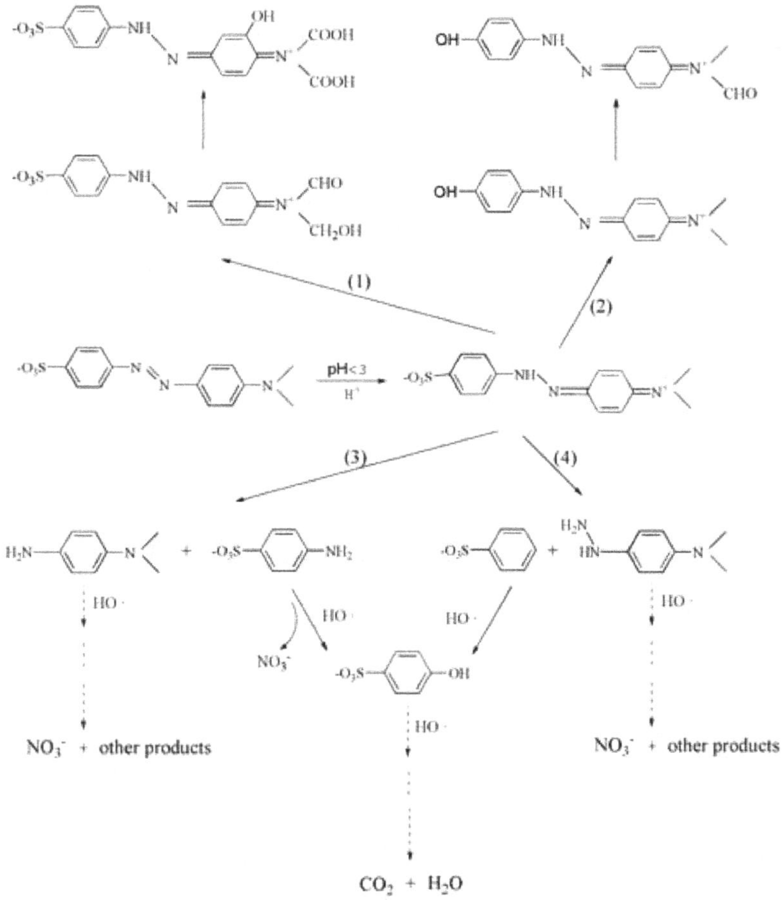

Figure 11. Speculation of possible degradation pathways of MO by Fe2O3/CNFs.

4. CONCLUSIONS

Inthisstudy,Fe_2O_3/CNFscompositeswerepreparedbyacombinationofhydrothermal and electrostatic spinning techniques. The effects of hydrothermal reaction temperature and time on the preparation of the composites were investigated. The Fe2O3/CNFs composites were successfully prepared by a hydrothermal method as proved by SEM, XRD, IR, EDS, and XPS analysis. Meanwhile, the existence of chemical interactions between Fe_2O_3 nanoparticles and carbon nanofibers was confirmed. When the hydrothermal reaction time was 180°C and the reaction time was 1 h, the nanoparticle size was about 90 nm with uniform distribution. The degradation efficiency of Fe2O3/CNFs on MO was investigated under the Fenton reaction. Under the optimal reaction conditions, the decolorization rate of MO could reach more than 99% within 60 min reaction. In addition, the degradation mechanism and pathway of the reaction were also speculated.

The prepared Fe_2O_3/CNFs composite as a heterogeneous catalyst can be separated from water easily. Furthermore, high surface area carbon nanofibers are used as the carrier of the catalysts, which can increase the degradation property. It is known that the printing and dyeing wastewater of the textile industry contains not only dyes, but also large amounts of surfactants; therefore, the degradation performance of multiple organic pollutants will be investigated in the future work.

5. CONCISE SUMMARY

In this study, an iron oxide/carbon nanofibers (Fe2O3/CNFs) composite was prepared by a combination of electrospinning and hydrothermal methods. The characterization of Fe2O3/CNFs was achieved via scanning electron microscopy (SEM), infrared spectroscopy (IR), X-ray diffraction (XRD) and X-ray photoelectron spectroscopy (XPS). It is shown that when the hydrothermal reaction time was 180 °C and the reaction time was 1 h, the Fe2O3 nanoparticle size was about 90 nm with uniform distribution. The photo-degradation performance applied to decolorize methyl orange (MO) was investigated by forming a heterogeneous Fenton catalytic system with hydrogen peroxide. The reaction conditions for the degradation of MO were optimized with the decolorization rate up to more than 99% within 1 h, which can decompose the dyes in water effectively. The degradation process of MO by Fenton oxidation was analyzed by a UV-visible NIR spectrophotometer, and the reaction mechanism was speculated as well.

REFERENCES

1. Leon, O.; Munoz-Bonilla, A.; Soto, D.; Perez, D.; Rangel, M.; Colina, M.; Fernandez-Garcia, M. Removal of Anionic and Cationic Dyes with Bioadsorbent Oxidized Chitosans. *Carbohydr. Polym.* **2018**, *194*, 375–383. **[Google Scholar]** **[CrossRef]**

2. Yagub, M.T.; Sen, T.K.; Afroze, S.; Ang, H.M. Dye and its removal from aqueous solution by adsorption: A review. *Adv. Colloid Interface Sci.* **2014**, *209*, 172–184. **[Google Scholar] [CrossRef] [PubMed]**

3. Punzi, M.; Anbalagan, A.; Borner, R.A.; Svensson, B.-M.; Jonstrup, M.; Mattiasson, B. Degradation of a textile azo dye using biological treatment followed by photo-Fenton oxidation: Evaluation of toxicity and microbial community structure. *Chem. Eng. J.* **2015**, *270*, 290–299. **[Google Scholar] [CrossRef]**

4. Li, H.; An, N.; Liu, G.; Li, J.; Liu, N.; Jia, M.; Zhang, W.; Yuan, X. Adsorption behaviors of methyl orange dye on nitrogen-doped mesoporous carbon materials. *J. Colloid Interface Sci.* **2016**, *466*, 343–351. **[Google Scholar] [CrossRef] [PubMed]**

5. Xiao, W.; Garba, Z.N.; Sun, S.; Lawan, L.; Wang, L.; Lin, M.; Yuan, Z. Preparation and evaluation of an effective activated carbon from white sugar for the adsorption of rhodamine B dye. *J. Clean. Prod.* **2020**, *253*, 119989. **[Google Scholar] [CrossRef]**

6. Mercante, L.A.; Facure, M.H.M.; Locilento, D.A.; Sanfelice, R.C.; Migliorini, F.L.; Mattoso, L.H.C.; Correa, D.S. Solution blow spun PMMA nanofibers wrapped with reduced graphene oxide as an efficient dye adsorbent. *New J. Chem.* **2017**, *41*, 9087–9094. **[Google Scholar] [CrossRef]**

7. Bu, J.; Yuan, L.; Zhang, N.; Liu, D.; Meng, Y.; Peng, X. High-efficiency adsorption of methylene blue dye from wastewater by a thiosemicarbazide functionalized graphene oxide composite. *Diam. Relat. Mater.* **2020**, *101*, 107604. **[Google Scholar] [CrossRef]**

8. Xie, X.H.; Liu, N.; Yang, F.; Zhang, Q.Y.; Zheng, X.L.; Wang, Y.Q.; Liu, J.S. Comparative study of antiestrogenic activity of two dyes after Fenton oxidation and biological degradation. *Ecotoxicol. Environ. Saf.* **2018**, *164*, 416–424. **[Google Scholar] [CrossRef]**

9. Esteves, B.M.; Rodrigues, C.S.D.; Boaventura, R.A.R.; Maldonado-Hodar, F.J.; Madeira, L.M. Coupling of acrylic dyeing wastewater treatment by heterogeneous Fenton oxidation in a continuous stirred tank reactor with biological degradation in a sequential batch reactor. *J. Environ. Manag.* **2016**, *166*, 193–203. **[Google Scholar] [CrossRef]**

10. Karthikeyan, S.; Titus, A.; Gnanamani, A.; Mandal, A.B.; Sekaran, G. Treatment of textile wastewater by homogeneous and heterogeneous Fenton oxidation processes. *Desalination* **2011**, *281*, 438–445. **[Google Scholar] [CrossRef]**

11. GilPavas, E.; Dobrosz-Gomez, L.; Gomez-Garcia, M.A. Coagulation-flocculation sequential with Fenton or Photo-Fenton processes as an alternative for the

industrial textile wastewater treatment. *J. Environ. Manag.* **2017**, *203*, 615. **[Google Scholar] [CrossRef] [PubMed]**

12. Blanco, J.; Torrades, F.; Varga, M.D.L.; Garcia-Montano, J. Fenton and biological-Fenton coupled processes for textile wastewater treatment and reuse. *Desalination* **2012**, *286*, 394–399. **[Google Scholar] [CrossRef]**
13. Solomon, D.; Kiflie, Z.; Hulle, S.V. Integration of sequencing batch reactor and homo-catalytic advanced oxidation processes for the treatment of textile wastewater. *Nanotechnol. Environ. Eng.* **2020**, *5*, 7. **[Google Scholar] [CrossRef]**
14. Asgari, G.; Faradmal, J.; Nasab, H.Z.; Ehsani, H. Catalytic ozonation of industrial textile wastewater using modified C-doped MgO eggshell membrane powder. *Adv. Powder Technol.* **2019**, *30*, 1297–1311. **[Google Scholar] [CrossRef]**
15. Sathya, U.; Keerthi Nithya, M.; Balasubramanian, N. Evaluation of advanced oxidation processes (AOPs) integrated membrane bioreactor (MBR) for the real textile wastewater treatment. *J. Environ. Manag.* **2019**, *246*, 768–775. **[Google Scholar] [CrossRef] [PubMed]**
16. Shi, C.M.; Tao, F.R.; Cui, Y.Z. Evaluation of nitriloacetic acid modified cellulose film on adsorption of methylene blue. *Int. J. Biol. Macromol.* **2018**, *114*, 400–407. **[Google Scholar] [CrossRef]**
17. Lade, H.; Kadam, A.; Paul, D.; Govindwar, S. Biodegradation and detoxification of textile azo dyes by bacterial consortium under sequential microaerophilic/aerobic processes. *EXCLI J.* **2015**, *14*, 158–174. **[Google Scholar]**
18. Guo, G.; Hao, J.X.; Tian, F.; Liu, C.; Ding, K.Q.; Xu, J.; Zhou, W.; Guan, Z.B. Decolorization and detoxification of azo dye by halo-alkaliphilic bacterial consortium: Systematic investigations of performance, pathway and metagenome. *Ecotoxicol. Environ. Saf.* **2020**, *204*, 111073. **[Google Scholar] [CrossRef]**
19. He, X.L.; Song, C.; Li, Y.Y.; Wang, N.; Xu, L.; Han, X.; Wei, D.S. Efficient degradation of Azo dyes by a newly isolated fungus Trichoderma tomentosum under non-sterile conditions. *Ecotoxicol. Environ. Saf.* **2018**, *150*, 232–239. **[Google Scholar] [CrossRef]**
20. Kumar, R.; Negi, S.; Sharma, P.; Prasher, I.B.; Chaudhary, S.; Dhau, J.S.; Umar, A. Wastewater cleanup using Phlebia acerina fungi: An insight into mycoremediation. *J. Environ. Manag.* **2018**, *228*, 130–139. **[Google Scholar] [CrossRef]**
21. Elgarahy, A.M.; Elwakeel, K.Z.; Elshoubaky, G.A.; Mohammad, S.H. Microwave-accelerated sorption of cationic dyes onto green marine algal biomass. *Environ. Sci. Pollut. Res.* **2019**, *26*, 22704–22722. **[Google Scholar] [CrossRef] [PubMed]**

22. Erkanli, M.; Yilmaz, L.; Culfaz-Emecen, P.Z.; Yetis, U. Brackish water recovery from reactive dyeing wastewater via ultrafiltration. *J. Clean. Prod.* **2017**, *165*, 1204–1214. **[Google Scholar] [CrossRef]**

23. Othmani, A.; Kesraoui, A.; Akrout, H.; Elaissaoui, I.; Seffen, M. Coupling anodic oxidation, biosorption and alternating current as alternative for wastewater purification. *Chemosphere* **2020**, *249*, 126480. **[Google Scholar] [CrossRef] [PubMed]**

24. Marin, N.M.; Pascu, L.F.; Demba, A.; Nita-Lazar, M.; Badea, I.A.; Aboul-Enein, H.Y. Removal of the Acid Orange 10 by ion exchange and microbiological methods. *Int. J. Environ. Sci. Technol.* **2019**, *16*, 6357–6366. **[Google Scholar] [CrossRef]**

25. Waly, A.I.; Khedr, M.A.; Ali, H.M.; Riad, B.Y.; Ahmed, I.M. Synthesis and Characterization of Ion Exchanger based on Waste Cotton for Dye Removal from Wastewater. *Egypt. J. Chem.* **2018**, *62*, 451–468. **[Google Scholar] [CrossRef]**

26. Bayramoglu, G.; Kunduzcu, G.; Arica, M.Y. Preparation and characterization of strong cation exchange terpolymer resin as effective adsorbent for removal of disperse dyes. *Polym. Eng. Sci.* **2020**, *60*, 192–201. **[Google Scholar] [CrossRef]**

27. Kaykioglu, G.; Coban, A.; Debik, E.; Kayacan, B.B.; Koyuncu, I. Mass transport coefficients of different nanofiltration membranes for biologically pre-treated textile wastewaters. *Desalination* **2011**, *269*, 254–259. **[Google Scholar] [CrossRef]**

28. Aouni, A.; Fersi, C.; Ali, M.B.S.; Dhahbi, M. Treatment of textile wastewater by a hybrid electrocoagulation/nanofiltration process. *J. Hazard. Mater.* **2009**, *168*, 868–874. **[Google Scholar] [CrossRef]**

29. Sahinkaya, E.; Uzal, N.; Yetis, U.; Dilek, F.B. Biological treatment and nanofiltration of denim textile wastewater for reuse. *J. Hazard. Mater.* **2008**, *153*, 1142–1148. **[Google Scholar] [CrossRef]**

30. Lin, L.; Wan, H.; Mia, R.; Jiang, H.Y.; Liu, H.H.; Mahmud, S. Bioreduction and Stabilization of Antibacterial Nanosilver Using Radix Lithospermi Phytonutrients for Azo-contaminated Wastewater Treatment: Synthesis, Optimization and Characterization. *J. Clust. Sci.* **2022**. **[Google Scholar] [CrossRef]**

31. Zhang, G.B.; Wan, H.; Mia, R.; Huang, Q.L.; Liu, H.H.; Mahmud, S. Fabrication and stabilization of nanosilver using Houttugniae for antibacterial and catalytic application. *Int. J. Environ. Anal. Chem.* **2022**, 1–21. **[Google Scholar] [CrossRef]**

32. Fouda, A.; Hassan, S.E.-D.; Saied, E.; Azab, M.S. An eco-friendly approach to textile and tannery wastewater treatment using maghemite nanoparticles

(γ-Fe2O3-NPs) fabricated by Penicillium expansum strain (K-w). *J. Environ. Chem. Eng.* **2021**, *9*, 104693. **[Google Scholar] [CrossRef]**

33. Feng, M.L.; Yu, S.C.; Wu, P.C.; Wang, Z.W.; Liu, S.H.; Fu, J.W. Rapid, high-efficient and selective removal of cationic dyes from wastewater using hollow polydopamine microcapsules: Isotherm, kinetics, thermodynamics and mechanism. *Appl. Surf. Sci.* **2021**, *542*, 148633. **[Google Scholar] [CrossRef]**

34. Liu, Y.Y.; Jin, W.; Zhao, Y.P.; Zhang, G.S.; Zhang, W. Enhanced catalytic degradation of methylene blue by α-Fe_2O_3/graphene oxide via heterogeneous photo-Fenton reactions. *Appl. Catal. B Environ.* **2017**, *206*, 642–652. **[Google Scholar] [CrossRef]**

35. Guo, S.; Wang, H.J.; Yang, W.; Fida, H.; You, L.M.; Zhou, K. Scalable synthesis of Ca-doped α-Fe_2O_3 with abundant oxygen vacancies for enhanced degradation of organic pollutants through peroxymonosulfate activation. *Appl. Catal. B Environ.* **2020**, *262*, 118520. **[Google Scholar] [CrossRef]**

36. Ding, M.M.; Chen, W.; Xu, H.; Shen, Z.; Lin, T.; Hu, K.; Lu, C.H.; Xie, Z.L. Novel α-Fe_2O_3/MXene nanocomposite as heterogeneous activator of peroxymonosulfate for the degradation of salicylic acid. *J. Hazard. Mater.* **2020**, *382*, 121064. **[Google Scholar] [CrossRef]**

37. Niu, L.J.; Zhang, G.M.; Xian, G.; Ren, Z.J.; Wei, T.; Li, Q.G.; Zhang, Y.; Zou, Z.G. Tetracycline degradation by persulfate activated with magnetic □-Fe_2O_3/CeO_2 catalyst: Performance, activation mechanism and degradation pathway. *Sep. Purif. Technol.* **2021**, *259*, 118156. **[Google Scholar] [CrossRef]**

38. Wang, W.L.; Zhao, W.L.; Zhang, H.C.; Dou, X.C.; Shi, H.F. 2D/2D step-scheme α-Fe_2O_3/Bi_2WO_6 photocatalyst with efficient charge transfer for enhanced photo-Fenton catalytic activity. *Chin. J. Catal.* **2021**, *42*, 97–106. **[Google Scholar] [CrossRef]**

39. Park, H.; Lee, Y.C.; Choi, B.G.; Choi, Y.S.; Yang, J.W.; Hong, W.H. Green one-pot assembly of iron-based nanomaterials for the rational design of structure. *Chem. Commun.* **2009**, *27*, 4058–4060. **[Google Scholar] [CrossRef]**

40. Liang, H.-W.; Cao, X.; Zhang, W.-J.; Lin, H.-T.; Zhou, F.; Chen, L.-F.; Yu, S.-H. Robust and Highly Efficient Free-Standing Carbonaceous Nanofiber Membranes for Water Purification. *Adv. Funct. Mater.* **2011**, *21*, 3851–3858. **[Google Scholar] [CrossRef]**

41. Wu, X.Y.; Yang, F.; Gan, J.; Zhao, W.Y.; Wu, Y. A flower-like waterborne coating with self-cleaning, self-repairing properties for superhydrophobic applications. *J. Mater. Res. Technol.* **2021**, *14*, 1820–1829. **[Google Scholar] [CrossRef]**

42. Lassoued, A.; Dkhil, B.; Gadri, A.; Ammar, S. Control of the shape and size of iron oxide (alpha-Fe_2O_3) nanoparticles synthesized through the chemical precipitation method. *Results Phys.* **2017**, *7*, 3007–3015. **[Google Scholar] [CrossRef]**

43. Zhang, X.D.; Yang, Y.; Song, L.; Wang, Y.X.; He, C.; Wang, Z.; Cui, L.F. High and stable catalytic activity of Ag/Fe_2O_3 catalysts derived from MOFs for CO oxidation. *Mol. Catal.* **2018**, *447*, 80–89. **[Google Scholar] [CrossRef]**

44. Qiao, J.; Zhang, X.; Xu, D.M.; Kong, L.X.; Lv, L.F.; Yang, F.; Wang, F.L.; Liu, W.; Liu, J.R. Design and synthesis of TiO_2/Co/carbon nanofibers with tunable and efficient electromagnetic absorption. *Chem. Eng. J.* **2020**, *380*, 122591. **[Google Scholar] [CrossRef]**

45. Zhu, H.Y.; Jiang, R.; Xiao, L.; Zeng, G.M. Preparation, characterization, adsorption kinetics and thermodynamics of novel magnetic chitosan enwrapping nanosized gamma-Fe_2O_3 and multi-walled carbon nanotubes with enhanced adsorption properties for methyl orange. *Bioresour. Technol.* **2010**, *101*, 5063–5069. **[Google Scholar] [CrossRef]**

46. Pereira, M.C.; Oliveira, C.A.; Murad, E. Iron oxide catalysts: Fenton and Fenton-like reactions a review. *Clay Miner.* **2012**, *47*, 285–302. **[Google Scholar] [CrossRef]**

47. Jia, L.D.; Zhang, Q.R. Heterogeneous Fenton catalytic oxidation for water treatment. *Prog. Chem.* **2020**, *32*, 978–988. **[Google Scholar]**

48. Hieu, V.Q.; Phung, T.K.; Nguyen, T.-Q.; Khan, A.; Doan, V.D.; Tran, V.A.; Le, V.T. Photocatalytic degradation of methyl orange dye by Ti_3C_2–TiO_2 heterojunction under solar light. *Chemosphere* **2021**, *276*, 130154. **[Google Scholar] [CrossRef]**

49. Adeel, M.; Saeed, M.; Khan, I.; Muneer, M.; Akram, N. Synthesis and Characterization of Co-ZnO and Evaluation of Its Photocatalytic Activity for Photodegradation of Methyl Orange. *ACS Omega* **2021**, *6*, 1426–1435. **[Google Scholar] [CrossRef]**

50. Shahzad, K.; Najam, T.; Bashir, M.S.; Nazir, M.A.; Rehman, A.U.; Bashir, M.A.; Shah, S.S.A. Fabrication of Periodic Mesoporous Organo Silicate (PMOS) composites of Ag and ZnO: Photo-catalytic degradation of methylene blue and methyl orange. *Inorg. Chem. Commun.* **2021**, *123*, 108357. **[Google Scholar] [CrossRef]**

51. Zeng, Q.Q.; Liu, Y.; Shen, L.G.; Lin, H.J.; Yu, W.M.; Xu, Y.C.; Li, R.J.; Huang, L.L. Facile preparation of recyclable magnetic Ni@filter paper composite materials for efficient photocatalytic degradation of methyl orange. *J. Colloid Interface Sci.* **2021**, *582*, 291–300. **[Google Scholar] [CrossRef] [PubMed]**

52. Znad, H.; Abbas, K.; Hena, S.; Awual, M.R. Synthesis a novel multilamellar mesoporous TiO_2/ZSM-5 for photo-catalytic degradation of methyl orange dye in aqueous media. *J. Environ. Chem. Eng.* **2018**, *6*, 218–227. **[Google Scholar]** **[CrossRef]**

53. Saravanan, R.; Sacari, E.; Gracia, F.; Khan, M.M.; Mosquera, E.; Gupta, V.K. Conducting PANI stimulated ZnO system for visible light photocatalytic degradation of coloured dyes. *J. Mol. Liq.* **2016**, *221*, 1029–1033. **[Google Scholar]** **[CrossRef]**

CHAPTER-14

DEVELOPING AN OPTICAL ALGORITHM TO ASCERTAIN THE RELATIVE THICKNESS OF EACH MONOCHROME COMPONENT WITHIN MULTILAYER TRANSPARENT MIXED FILMS

1. INTRODUCTION: AN OVERVIEW

In the realm of optical film analysis, particularly within the context of multilayer transparent mixed films, there is a growing need for precise methods to determine the relative thickness of individual monochrome components. This study introduces an innovative optical algorithm designed to address this challenge. By leveraging advanced optical techniques, the algorithm aims to provide accurate measurements of the thickness of each monochrome layer within the complex structure of multilayer transparent films. The significance of this research lies not only in its potential applications in film characterization but also in advancing our understanding of optical algorithms for intricate material analysis. This pursuit holds promise for diverse fields, ranging from materials science to optical engineering, offering valuable insights into the composition and properties of multilayer transparent mixed films.

The quantity or relative quantity distribution of each monochrome item is an important variable to configure a colour-mixed material's properties in terms of structure and uniformity in the textile industry, such as the colour blending fibers, yarn, and fabric. A potential environmental textile brand featured blending fibers after fiber colouring, with 50% reduction of water than traditional process [**1**,**2**,**3**]. Meanwhile, it is proposed to be applied in the field of photometric measurement, among which the fibrogram is a typical application, a way for measuring the length of white cotton fiber by the parallel fiber beard linear density along the fiber axis [**4,5**].

In 1932, Hertel [**4,5**] proposed a modified form of the derived Lambert-Beer (L-B) law to measure the linear density of cotton beards for bias reduction. However, experimental coefficients were different for various materials and obtained difficultly, calling for theoretical study [**4,5,6,7,8,9**]. In 1970s, the high volume instrument (HVI) was invented by Spinlab Corporation based on Hertle's study for cotton fiber length distribution, and has been an international standard method on white cotton fiber length until now [**9**]. In 2016, Wu et al. [**10**] derived a modification of two-flux Kubelka-Munk (K-M) theory for relative thickness or surface density of a turbid medium. The theory contains the up- and down-wards absorption and scattering light overcoming the L-B law's shortcoming of including down-wards absorption only [**10,11,12**]. Its results proved to be much better than L-B law, particularly for scattered wool fibers [**10,11,12**]. Although the derived transmittance K-M theory has been widely used in predicting the relative thickness of white materials [**10,11,12**], the mixed-coloured specimen of each monochrome fibers has not yet been investigated which is typical important for quality control of colour blending yarn industry.

In 2021, Chen et al. [**13,14**] of our group proposed an optical algorithm for the thickness of each colour material in a mixed multilayer transparent specimen, combined L-B law and transmission images. An estimating equation group was developed to describe the relationship between the physical thickness of each colour material and the optical depth of multilayer transparent specimen under different monochrome light from linear regression methods. The binary system of first order equations was employed to predict each coloured wool fiber material's relative physical thickness in the mixed coloured-fiber specimen. Although its results turn to be pretty good in smooth films, the L-B law's shortcoming of containing down-wards absorption only limits its usage scope in scattering and thick films [**13**] and fiber materials [**14**], particularly the fiber beards with different colours [**14**]. Hence, relative quantity distribution of each monochrome component needs further study.

In this paper, combined with the previous derived transmission K-M theory and conservation law of light flux of scattered light, a new scattering optical algorithm is proposed for the relative thickness of each colour material in a multilayer transparent specimen. This algorithm has an advantage of comprising up- and down-ward scattering and absorption lights, overcoming L-B law's shortcomings of including down-ward absorption light only. In this optical algorithm, the linear regression method was applied in obtaining linear equation between the physical thickness and optical scattered photometric of multilayer transparent monochrome specimens. According to the conservation law of light flux of scattered light, ab estimation equation system was expressed to predict relative thickness of each coloured material in the multilayer specimen, and a better affinity is achieved according to the comparison between experimental and predicted relative thickness which is compared with results of previous algorithm from L-B law.

2. THEORY

2.1. Lambert-Beer Law

L-B law **[8,9,10,11,12,13,14,15]** provides a light absorptivity relationship between the attenuation of light and the physical thickness of material when a light transmits through a material. This relationship only considers the down-wards light absorption, expressed as in Equations (1) and (2),

$$I = I0e^{-Kx} \qquad (1)$$

$$A = xK = \ln(\frac{I_0}{I}) = -\ln(T) \qquad (2)$$

where K is the absorption coefficient, x is the thickness of the material, T is the transmittance, and A is the absorbance.

2.2. Derived Kubelka-Munk Theory

Considering the up- and down-wards light scattering and absorption, Wu et al. **[12]**, the author, published a derived K-M theory indicating a scattering relationship between the transmittance of light and the physical thickness of material when light transmits through a material. According to the equation, the thickness (*x*) of specimen is proportional to the algorithm of transmittance as elaborated in Equation (3),

$$P = Sx = \frac{2r_\infty}{1 - r_\infty^2}\ln(\frac{1 - r_\infty^2 + \sqrt{(1 - r_\infty^2)^2 + 4T^2r_\infty^2}}{2T}) \qquad (3)$$

where, S denotes the coefficient of scatter defined by the corresponding thickness of layer; $r\infty$ is the light reflectivity of the material with infinite thickness; P is defined as scattered photometric here representing the ability of material's light scattered. Here, S could be obtained experimentally.

According to the conservation law of light flux of scattered light, the scattered coefficients of color-mixed material is equal to the sum of the results of scattered coefficient of each composition multiplied by its corresponding concentration ratio wi, $S_{mix} = \sum ni=1$, where $w_i = x_i/x_{mix}$ and n is number of the monochrome materials.

Hence, the scattered photometrics of mixed material is proposed in this paper to be equal to the sum of the scattered photometrics of its corresponding monochrome materials, expressed as $P_{mix} = \Sigma ni{=}1$. Two or three of these equations under different lights form the mixed-film estimation equation system for 2-mixed or 3-mixed colour multilayer films, respectively.

2.3. Proposed Estimation Procedure

In this study, our own built imaging scanner was applied to obtain the RGB transmission images at a greyscale of 0–255 with a resolution of 1000 dpi, where dpi means the number of points within per inch. These acquired R, G, and B values represent the transmitted light of red (R), green (G), and blue (B) monochromatic light, respectively.

Figure 1 shows that when monochromatic light enters the fiber aggregate, it is assumed that both light reflection and light scattering inside the fiber aggregate are considered as scattering, while scattering and absorption in air are ignored. According to the above conservation law of scattered light flux, for A and B two-colour mixed colour fiber, the scattered light amount of the mixed colour fiber is equal to the total scattered light of component A and the total scattered light of component B.

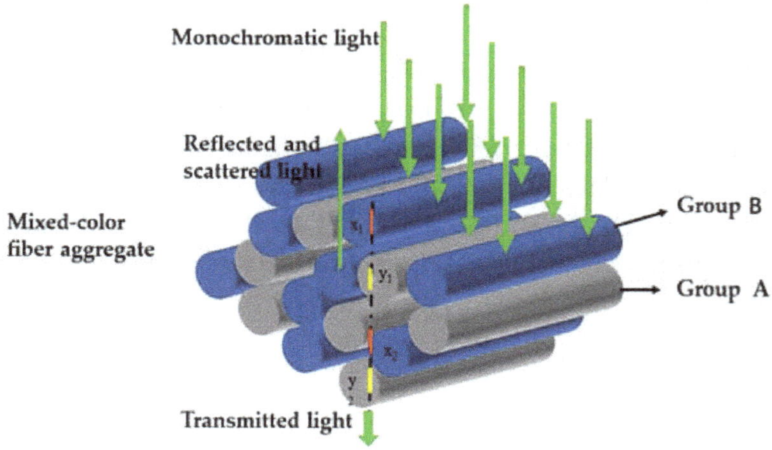

Figure 1. Monochromatic light incident analysis of two-colour hybrid fiber aggregates.

Figure 2 shows the flow chart to achieve the physical thickness of each colour film in the multilayer specimen, in which ith (i = 1, 2, 3) film represents different monochromatic film and kth (k = R, G, B) light denotes the light channel of colour images. This procedure has two steps: (1) Colour-mixed estimation system and (2) Application. For colour-mixed estimation system, the monochrome films were piled up to multilayer films one by one to scan their transmitted RGB digital images. Next, calculate their corresponding transmittance using R/R0, G/G0 and B/B0, where R0, G0 an B0 represents the amount of incident light under each channel. These transmittance and corresponding infinite reflectance were applied in Equations

(1) and (2) to get Ai, and Pi, respectively. Details of reflectance measurement are described in Section 3.2 Optical parameter. After that, train the estimating equations referred to photometric with linear regression method of ith monochromatic films and each light. Furthermore, these estimating equations were added up to form the mixed-film estimation equation under each light. Two or three of these equations under different lights form the mixed-film estimation system for 2-mixed or 3-mixed colour multilayer films, respectively. In step 2: Application, the designed mixed films were arranged according to their corresponding designed order and number of a group and accumulated to multi-groups. Next, the RGB images of these groups were scanned to get their transmittance using R/R0, G/G0 and B/B0. These transmittance and corresponding infinite reflectance were applied to obtain the absorbed and scattered photometric, $Amix$, and $Pmix$,, using Equations (1) and (2), respectively. Afterwards, these results are substituted into the mixed-film estimation equation system above.

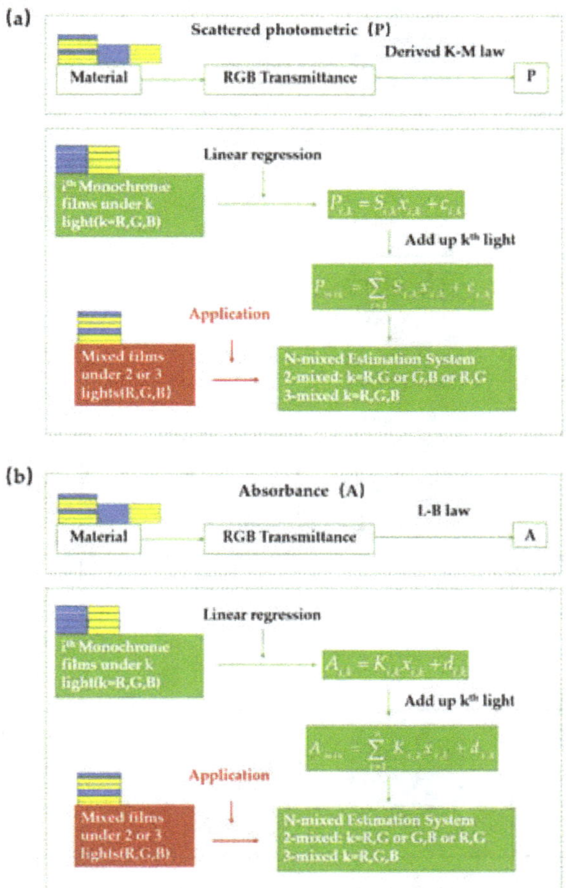

Figure 2. Flow chart to achieve relative thickness of each colour film in the multilayer films, (a) Scattered photometric (P), (b) Absorbance (A).

3. EXPERIMENT

3.1. Material

In this experiment, seven commercial transparent and uniform films with different colours are chosen as the experimental materials, numbered 1# to 7#, whose information are listed in **Table 1**. Film 1# to 5# are made of poly-ethylene terephthalate (PET) and film 6# to 7# are polypropylene (PP) films. All these coloured films have characteristics of transparent and smoothy, except 6# and 7# with rough surfaces. All images in Table 1 were captured from films with 20 layers except 5# containing 40 layers for higher transparency. These samples were employed to build the estimation equation systems for colour separation.

Table 1. Information of Monochrome Films

Number	Color	Image	Thickness/mm	Material	Surface
1#	Blue		0.02	PET	S
2#	Yellow		0.02	PET	S
3#	Pink		0.35	PET	S
4#	Cyan		0.35	PET	S
5#	White		0.01	PET	S
6#	Dark blue		0.5	PP	R
7#	Dark Yellow		0.5	PP	R

S and R means smooth and rough respectively.

To test the proposed method, 8 sets of mixed multilayer films, numbered a# to h# were designed in accordance to their corresponding order and ratio given in Table 2. For example, a group of a# mixed films turns to be 211 arranged from bottom to top, where 2 and 1 stands for a layer of 2# and 1# film respectively. Different numbers of groups are selected for the limit linear test range. Films a#–f# and film g# are of PET and PP respectively, while h# is a mixture of PP and PET with rough and smooth surfaces. a#–h# samples in Table 2 are divided into 5 sets by the compositions of each mixed material, numbered A# to E#. An estimation linear equation system of each set could be composed for thickness or quantity of monochrome material.

Table 2. Information of Mixed Multilayer Films.

		Image	Order	Ratio	Group	Material	Surface
A#	a#		2#1#	1:2	4	PET	S
	b#		1#2#	3:1	3	PET	S
	c#			1:1	3	PET	S
	d#			1:2	4	PET	S
B#	e#		4#3#	1:1	3	PET	S
C#	f#		4#5#3#	1:1:1	2	PET	S
D#	g#		6#7#	1:1	3	PP	R

		Image	Order	Ratio	Group	Material	Surface
E#	h#		7#4#	1:1	7	PP\PET	R\S
	i#			1:2	4	PP\PET	R\S

S and R means smooth and rough respectively.

3.2. Optical Parameter $r\infty$

Reflectance of infinite layers $r\infty$, is an essential optical parameter for derived K-M theory. To measure this parameter, specimens need to be piled up to enough thickness, so that no light can transmit. Samples of monochrome and mixed films are stacked to 20 layers and 10 groups of layers respectively, except 5# with 40 layers for higher transparency. Table 3 denotes the reflectance of infinite layers from reflective images of these multilayers obtained with built scanning image equipment.

Table 3. r∞ of Monochrome and Mixed Films under Monochrome Light.

Number	r∞ of Monochromatic Specimen/%			Number	r∞ of Mixed Specimen/%		
	R	G	B		R	G	B
1#	2.10	50.85	70.02	a#	27.03	55.84	42.43
2#	74.35	71.79	22.13	b#	26.50	55.89	46.89
3#	22.49	10.12	12.01	c#	35.33	59.82	40.22
4#	29.37	40.58	23.73	d#	44.64	62.54	35.07
5#	85.19	87.98	93.58	e#	30.15	26.93	20.40
6#	15.80	31.11	53.62	f#	32.93	16.85	14.82
7#	76.49	66.95	34.00	g#	17.40	32.83	37.44
				h#	73.47	71.35	33.41

3.3. Monochrome Estimation Equation

To get the linear equation between the scattered photometric and thickness, specimens with same colour were accumulated to multilayers for RGB images of 1# to 7# samples using a scanner, as shown in Figure 1. Scan images of each coloured film with multilayers at an area of 9 mm × 10 mm, ranging from 0–5 layer.

Afterwards, their scattered and absorbance were computed with Equation (2) and Equation (1), respectively. Figure 3 indicates the transmitted intensity, absorbed and scattered photometric of 0–5 layers with sample 1# to 7# under R, G and B lights. As physical thicknesses are multiples of its layer numbers, the latter was used as reference values here. It can be seen that the scattered photometric has a better linearity than the absorbed photometric with the layer of films for most films ranging from 0 to 5 layers, such as 7# in R and G channels, and different coloured film shows different linearities in R, G and B channels. Hence, optimal channels could be selected according to the transmittance under R, G, and B channels of the 2-mixed mixture and its components.

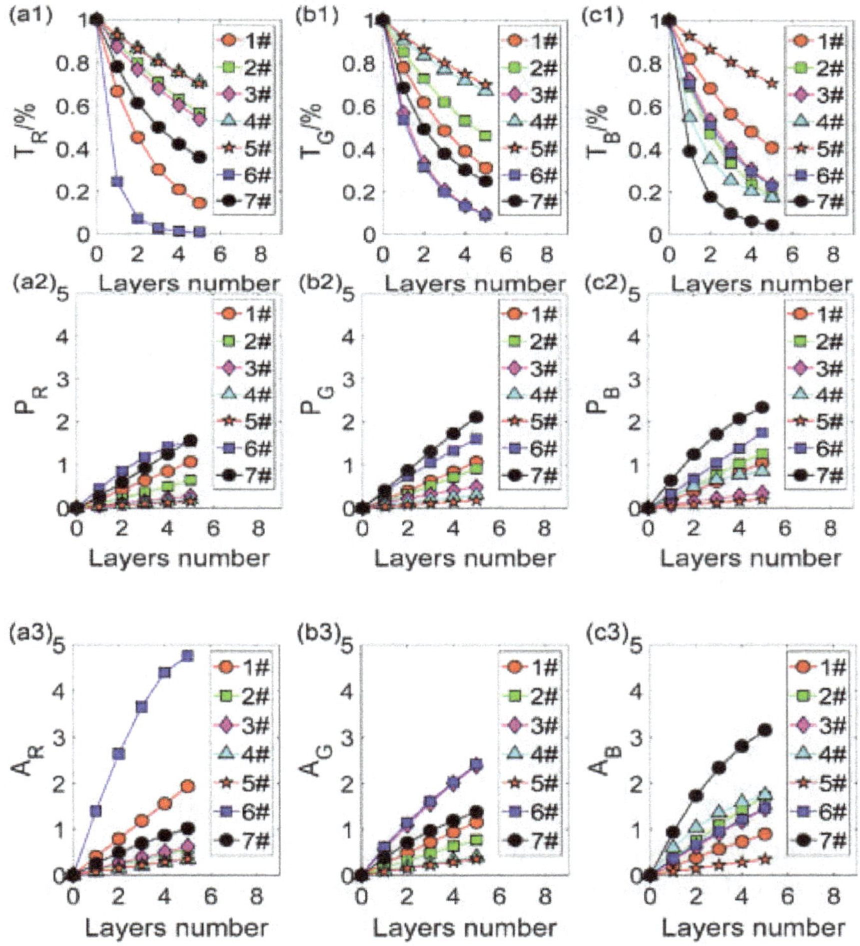

Figure 3. Relationship between transmittance (a^1–c^1), scattered photometric (a^2–c^2) and absorbance (a^3–c^3) of 0–5 layers with 1#–7# films under R, G and B lights.

In addition, linear regression method was employed for linear equation between the layer number and absorbance or scattered photometric. Data of 0–4 layers for 1#–5# and 0–3 layers for 6#–7# were regressed for better linearity of scattered photometric P = SX + C, and absorbed photometric A = KX + D, where S denotes the coefficient of scatter defined by the corresponding thickness of layer; and K is the coefficient of absorption, and C and D are constants related to noises. In this section, the experiments results indicated good linear relationship between photometric and layer numbers of monochrome films under each monochrome light, whose r2 were all above 0.98. These estimated scattering and absorption linear equations could be used in the construction of colour-mixed equations complied with the conversation of scattered and absorbed light in the following section.

3.4. Thickness of Each Component in Mixed Samples

To examine the accuracy of proposed method, specimens a# and b# were created numbered A# set as listed in Table 2. Following the conversation of scattered and absorbed light in the following section, estimation equation system was constructed by summing up linear equations from Figure 3 for each component, 1# and 2#, in mixed samples of A# about scattered photometric and absorbance under particular (G or B) light, respectively. Its results comprised 2-mixed estimation equation systems of A#. Figure 3 declared a higher degree of linearity between predicted and measured scattered photometric compared with that of the absorbance from L-B law. This may lead to more accurate predicted layer numbers for scattered method. Afterwards, compute layer number of coloured films 1# and 2# with 2-mixed estimation systems of Figure 2. Their results and the sum of each component with both methods are illustrated in Figure 4.

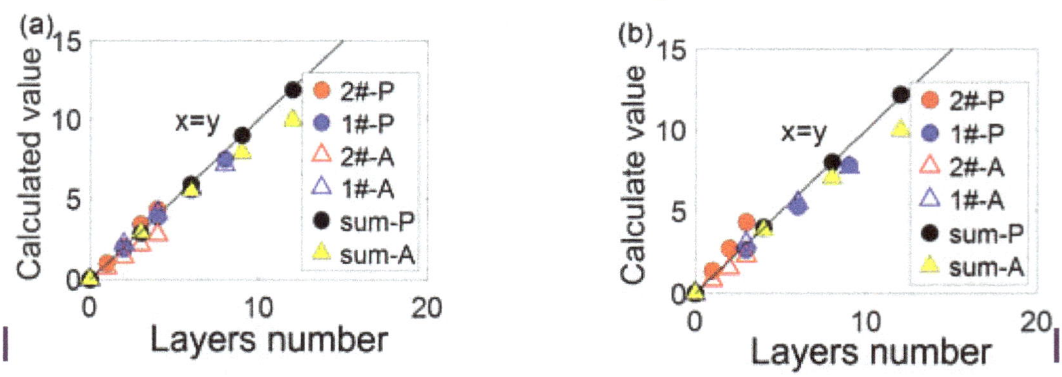

Figure 4. Comparison between physical and calculated monochrome film layers number, in mixed samples of (a) Sample a# and (b) Sample b# using scattered photometric and absorbance.

For accuracy analysis, their corresponding layer number deviation ratios σ of each composition were calculated according to Equation (4),

$$\sigma=|x-x0|x0\times 100\% \quad (4)$$

where x denotes the calculated value; $x0$ is the reference value.

The mean and maximum value of these deviation ratios turns to be 2.05%, 6.21% and 4.14%, 20.40% for derived K-M and L-B methods, respectively. Particularly, the sum of each component measured with the derived K-M has presents less difference from the true data. Hence, the scattered method exhibits a better consequence in a# and b# PET materials compared with L-B law.

3.5. Relative Thickness of Each Component in Mixed Material

To avoid effect of boundary reflectance ignorance, random error of reflectance of infinite layers and noises of equipment [**8,16**], relative optical thickness was proposed for testing the quantity distribution of each monochrome material. PET, PP and PP\PET mixed materials were applied for relative optical thicknesses of monochromatic film in multilayers as **Table 2** described. For estimation system construction, add up the linear equations of each component from 0–4 layers of coloured films in **Figure 3**, following procedures in **Figure 2** to a predicted mixed equation under a particular light (R, G and B). Afterward, two or three of these predicted equations under different monochromatic light were employed to construct the 2- or 3-mixed estimation equation system. Finally, the relative thickness for each composition of the mixed multilayers was calculated with Equation (5), shown in **Figure 5**.

$$x_r = \frac{x - x_{\min}}{x_{\max} - x_{\min}} \times 100\% \qquad (5)$$

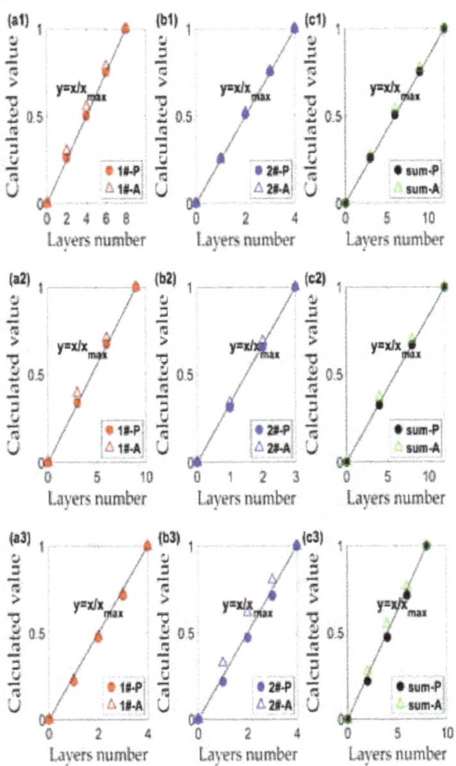

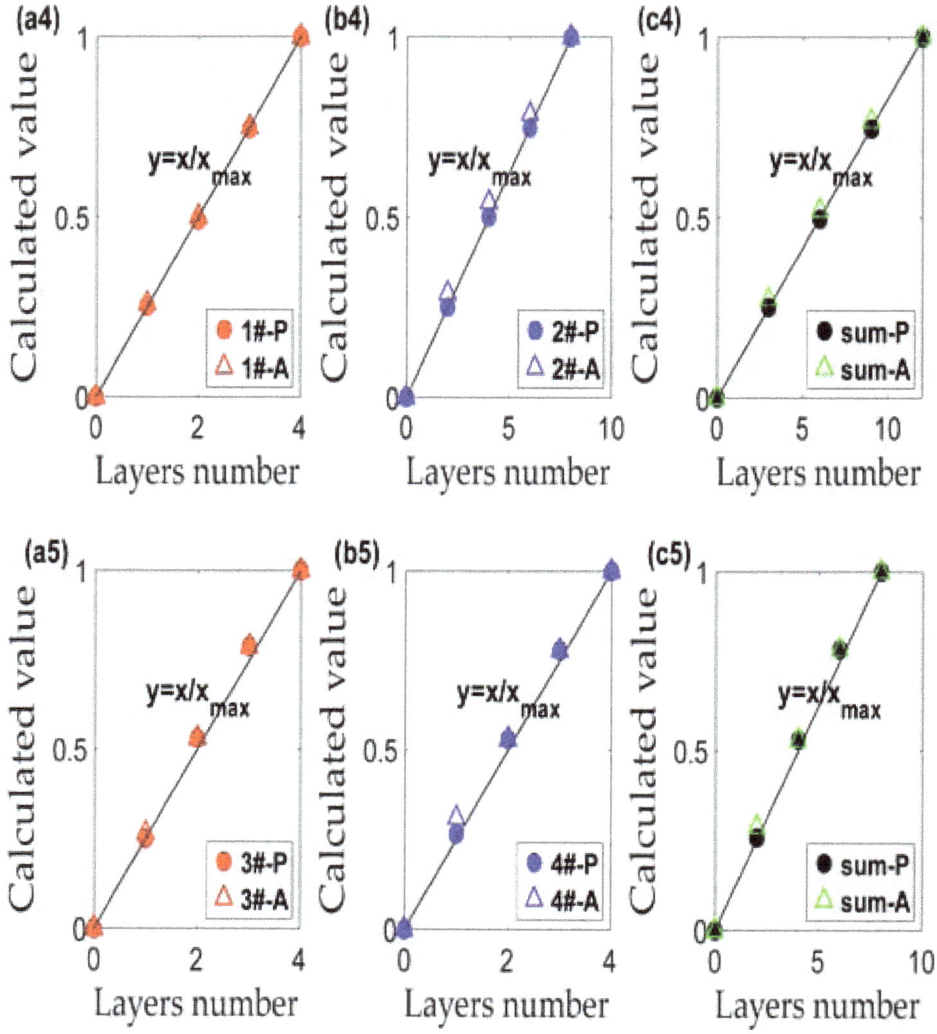

Relative thickness deviation ratios of Figure 5 were computed with Equation (4) as shown in Table 4, whose mean and maximum values from scattered photometric and absorbance were 3.56%, 14.03% and 6.77%, 42.24%, for 2-mixed PET material (1#2# and 2#1#), 2.28%, 6.86% and 1.7%, 6.74% for 3-mixed PET material (4#5#3#), 1.94%, 6.13% and 4.08%, 12.05% for 2-mixed PP material (6#7#), and 2.89%, 14.38% and 17.07%, 78.31% for PET/PP material (7#4#), respectively. This indicates a better application of the modification of K-M theory than L-B law both in average and maximum error rate. The reason for this is that the L-B model only considers the unidirectional absorption of light by materials but does not consider the reflection and scattering of light. When the light passes through the material, in addition to absorption and reflection, a large amount of scattered light will be generated inside and on the surface. Therefore, the modified K-M theory is proposed to consider not only light absorption and light transmission, but also light scattering. Its advantages are particularly obvious in PP\PET mixed sample with smooth and rough surfaces.

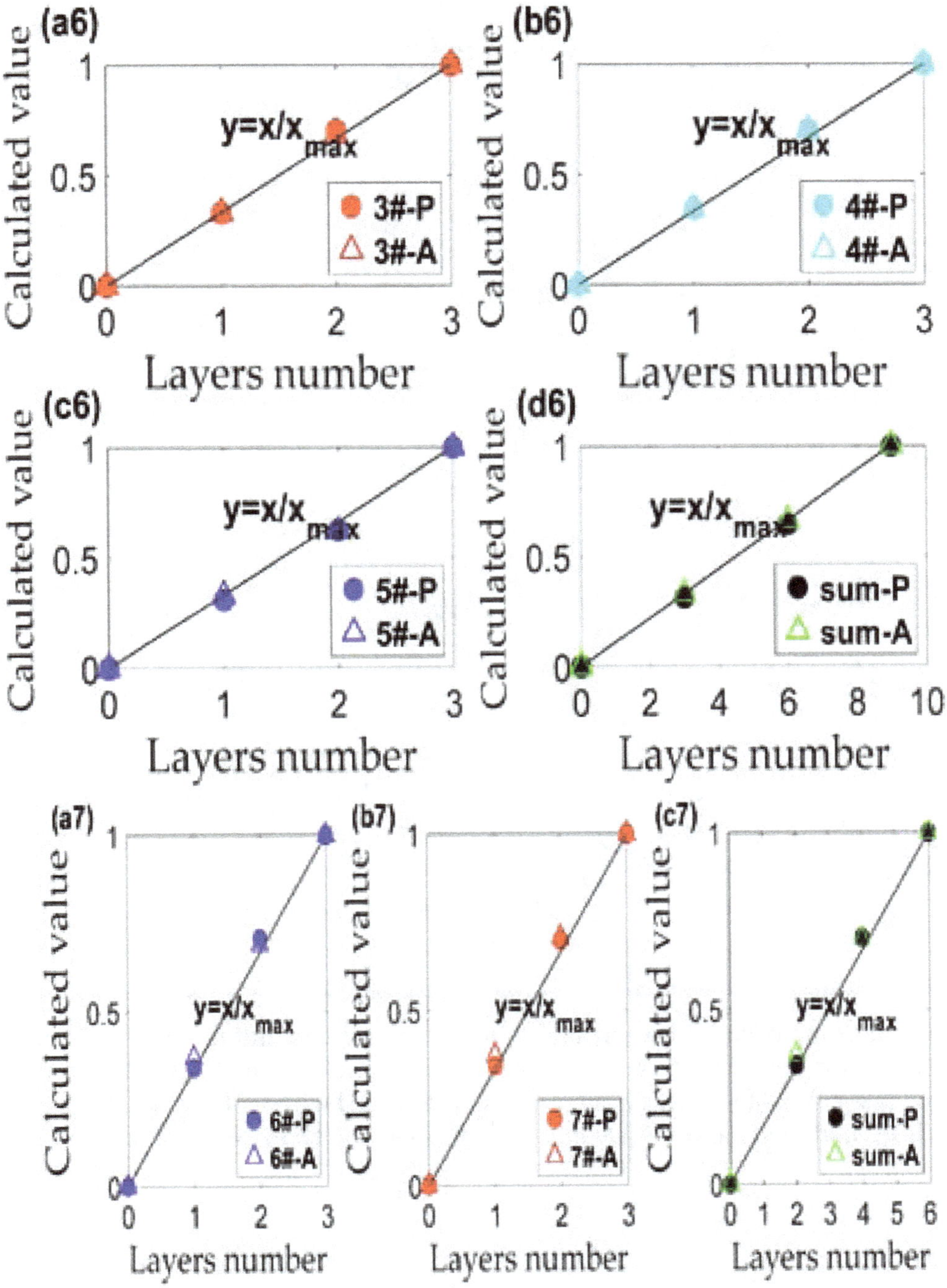

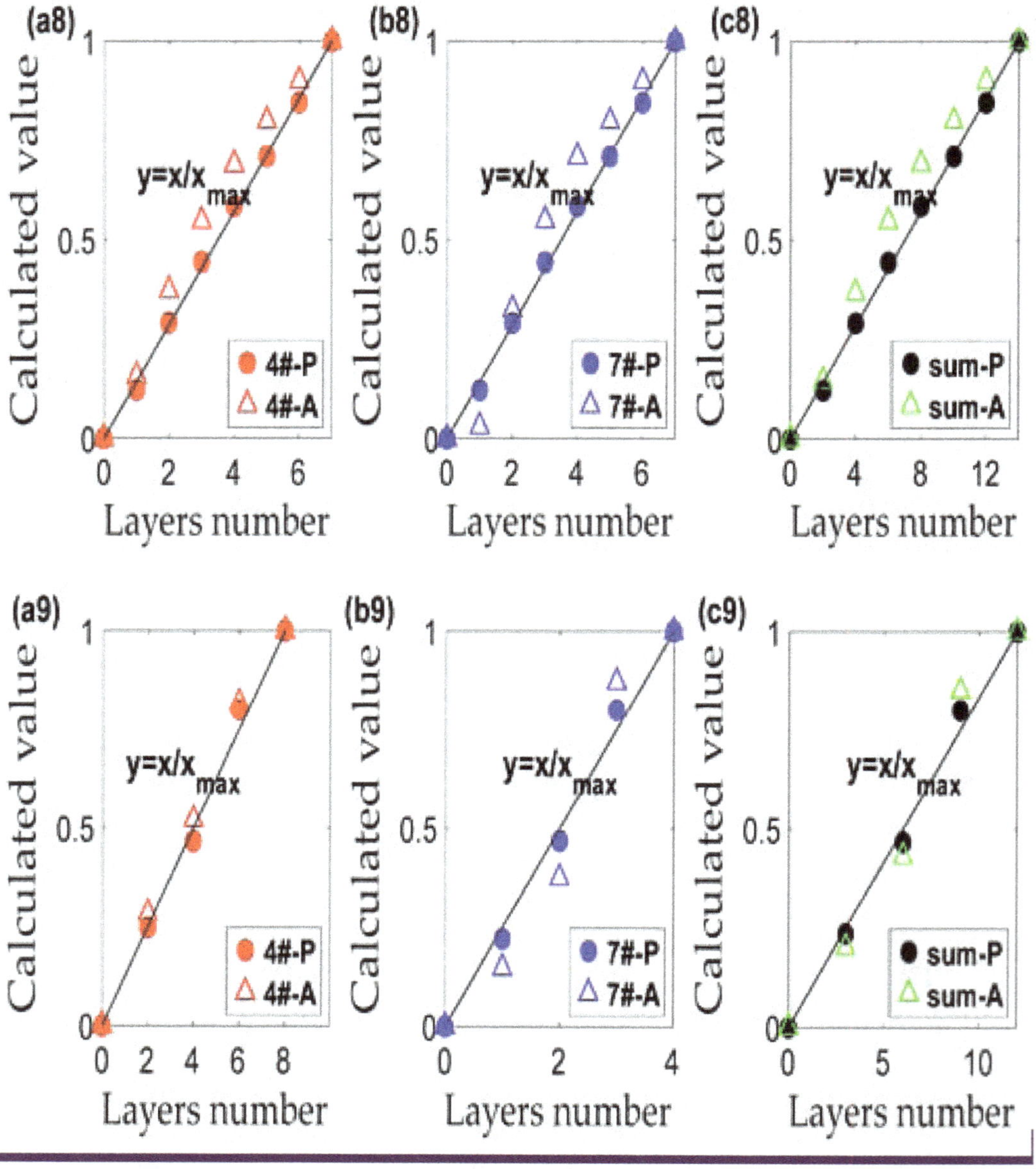

Figure 5. Comparison between layers number and calculated relative thickness (calculated value) of each component and their sum for Sample a# (a1–c1), b# (a2–c2), c# (a3–c3), d# (a4–c4), e# (a5–c5), and f# (a6–d6) with PET material, g# (a7–c7) with PP materials and h# (a8–c8), and i# (a9–c9) with PP\PET mixed material.

Table 4. The relative thickness deviation ratios of each component.

Order	Ratio	Scattered Photometric		Absorbance	
		Mean Deviation Rate%	Max Deviation Rate%	Mean Deviation Rate%	Max Deviation Rate%
2#1#	1:2	1.53	5.09	4.26	20.4
1#2#	3:1	2.56	6.21	4.029	19.57
	1:1	6.81	14.03	12.45	42.24
	1:2	3.36	8.76	6.35	23.48
4#3#	1:1	2.72	6.35	5.12	24.63
4#5#3#	1:1:1	2.28	6.86	1.7	6.74
6#7#	1:1	1.94	6.13	4.08	12.05
7#4#	1:1	2.89	14.38	17.07	78.31
	1:2	4.09	12.01	10.69	42.72

4. APPLICATION

To make sure the algorithm's applicability in fiber assemblies, cotton, wool and polymer coloured fiber assemblies were used to compare with the common L-B law. Primary fibers, white and black coloured cotton, grey and yellow coloured wool fibers, and pink coloured polymer fibers were piled up parallelly with different weight for optical weight with transmission images. Linear regression method was applied to relationship between scattered photometric or absorbance and their weight in R, G, and B channels, respectively. These linear equations for each primary fibers under same light were summed up to construct the estimation equations system, as shown in Figure 2. Relative quantity of each primary fiber could be obtained in gray-yellow blended wool and black-white cotton with ratio of 1:1 and 1:2 respectively as procedure above compared with actual weights from a balance (accurate to 0.001 g), as described in Figure 6. The mean and maximum relative weight deviation ratios of scattered photometric and absorbance are 3.57%, 15.29% and 4.44%, 15.04%, for 2-mixed wool fibers, 1.86%, 5.5% and 27.25%, 176.79% for cotton samples, and 1.46%, 5.31% and 6.65%, 16.63% for cotton and polymer mixed fibers, respectively. Hence, this new proposed method is better in fiber assemblies. The large derivation of black cotton using L-B law in Figure 6(a2) may deduced from random errors of fiber assemblies and ignorance of reflectance. Hence, based on this theory, a new method with digital image technology could be invented for primary fiber quantity distribution from blending fiber beards, which is essential data for the fibrogram for primary fiber length measurement.

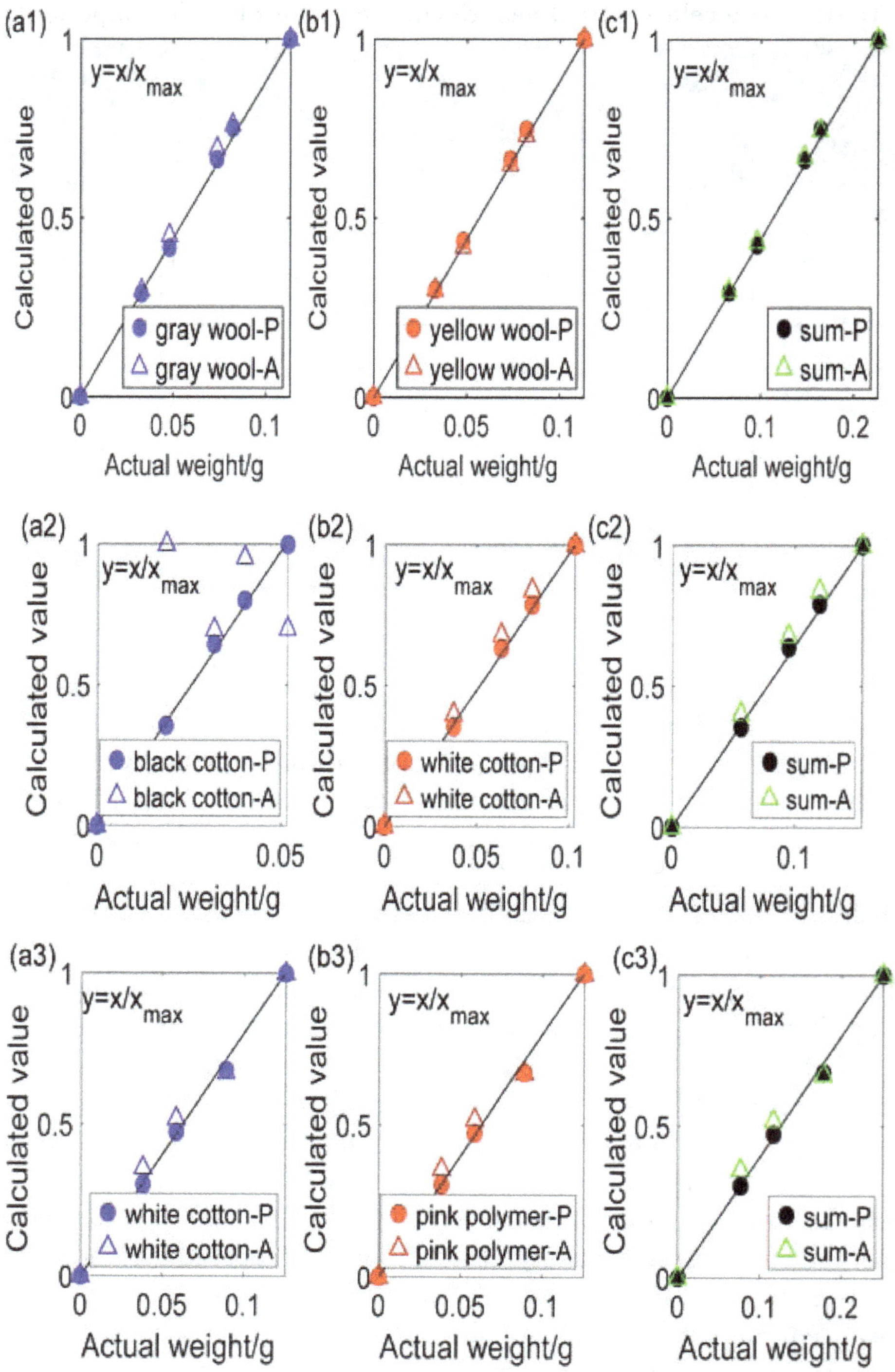

Figure 6. **Comparison between actual weight and calculated relative thickness, named calculated value in Figure of each component and their sum for color mixed wool (a1–c1), cotton (a2–c2) and cotton and polymer mixed (a3–c3) fibers.**

5. CONCLUSIONS

In this study, a scattered optical algorithm was proposed for relative quantity distribution of each monochrome component in colour mixed material based on derived K-M theory and the colour transmission image. The linear regression method and conservation of scattered light were applied to obtain the estimating equation system on a defined optical variable, scattered photometric P, from transmission images of monochrome item with different weight or thickness. The obtained results were relative quantities to avoid ignorance factors of theoretical surface reflectance, random error, and measured derivation of reflectance of infinite layers. A series of experiments were performed with colour-mixed specimens with smooth PET, rough PP, PP\PET mixed films, cotton, and wool fiber assemblies. Results show that this algorithm performs better than the commonly used L-B theory, especially in smooth PET\rough PP mixed materials and fiber assemblies. Therefore, this optical algorithm shows a potential application in assessing the primary fiber length of blending fibers, as well as testing the evenness of scattering film and fiber assembly, especially hollow fibers and other fiber materials with shape modifications for functional application [**17,18,19**], as well as to support the fibrogram of fiber beards for fiber length testing. Based on this theory, a new method with digital image technology could be invented for primary fiber length measurement from blending fiber beards, having the characteristics speed, high accuracy, and low cost.

6. CONCISE SUMMARY

A modification of the two-flux Kubelka-Munk (K-M) model was proposed to describe the energy conservation of scattered light in colored mixed material with a defined scattered photometric, which is applied for the relative quantity distribution of each coloured monochrome component in mixed material. A series of systematical experiments demonstrated a higher consistency with the reference quantity distribution than the common Lambert-Beer (L-B) law. Its application in the fibrogram of each component for measuring the cotton fiber's length was demonstrated to be good, extending its applicability to white and dark coloured blended fibers, the length of which is harder to measure using L-B law.

REFERENCES

1. Yang, R.H.; Pan, B.; Wang, L.J.; Lin, J.W. Blending effects and performance of ring-, rotor-, and air-jet-spun color-blended viscose yarns. *Cellulose* **2021**, *28*, 1769–1780. [**Google Scholar**] [**CrossRef**]
2. Lam, N.Y.K.; Zhang, M.; Guo, H.F.; Ho, C.P.; Li, L. Effect of fiber length and blending method on the tensile properties of ring spun chitosan–cotton blend yarn. *Text. Res. J.* **2017**, *87*, 244–257. [**Google Scholar**] [**CrossRef**]

3. Li, S.Y.; Fu, H. Image analysis and evaluation for internal structural properties of cellulosic yarn. *Cellulose* **2021**, *28*, 6739–6756. **[Google Scholar] [CrossRef]**
4. Hertel, K.L.; Zervigon, M.G. An optical method for the length analysis of cotton fibres. *Text. Res. J.* **1936**, *6*, 331–339. **[Google Scholar] [CrossRef]**
5. Hertel, K.L. A method of fibre-length analysis using the fibrograph. *Text. Res. J.* **1940**, *10*, 510–520. **[Google Scholar] [CrossRef]**
6. Wu, H.Y.; Wang, F.M. Image measuring method for fiber length measurements. *Ind. Text.* **2013**, *64*, 321–325. **[Google Scholar]**
7. Jin, J.Y.; Xu, B.G.; Wang, F.M. Measurement of short fiber contents in raw cotton using dual-beard images. *Text. Res. J.* **2018**, *88*, 14–26. **[Google Scholar] [CrossRef]**
8. Gordon, H.R. Can the Lambert-Beer law be applied to the diffuse attenuation coefficient of ocean water? *Limnol. Oceanogr.* **1989**, *34*, 1389–1409. **[Google Scholar] [CrossRef]**
9. Naylor, G.R.; Delhom, C.D.; Cui, X.; Gourlot, J.P.; Rodgers, J. Understanding the influence of fiber length on the high volume instrument measurement of cotton fiber strength. *Text. Res. J.* **2014**, *84*, 979–988. **[Google Scholar] [CrossRef]**
10. Wu, M.Q.; Wang, F.M. Optical algorithm for calculating the quantity distribution of fiber assembly. *Appl. Opt.* **2016**, *55*, 7157–7162. **[Google Scholar] [CrossRef] [PubMed]**
11. Wu, M.Q.; Jin, J.Y.; Zhang, J.; Wang, F.M. Calculation method of a random beard fibrogram based on the derived Kubelka–Munk theory. *Text. Res. J.* **2019**, *89*, 2281–2293. **[Google Scholar] [CrossRef]**
12. Lang, C.H.; Wu, M.Q.; Pan, X.X.; Jin, J.Y.; Wang, F.M.; Xu, B.G.; Qiu, Y.P. Algorithm for measuring fiber length distributions of raw cotton and combed wool using dual-beard image method. *Text. Res. J.* **2020**, *90*, 2149–2160. **[Google Scholar] [CrossRef]**
13. Chen, L.J.; Shen, H.; Wang, F.M. Quantifying the thickness of each color material in multilayer transparent specimen based on transmission image. *Text. Res. J.* **2020**, *90*, 2522–2532. **[Google Scholar] [CrossRef]**
14. Chen, L.J.; Shen, H.; Heng, C.; Wang, F.M. Algorithm for predicting the length of each color fiber in mixed-wool fiber assemblies based on the transmission image. *Text. Res. J.* **2020**, *90*, 357–366. **[Google Scholar] [CrossRef]**
15. Strong, F.C. Theoretical basis of Bouguer-Beer law of radiation absorption. *Anal. Chem.* **1952**, *24*, 338–342. **[Google Scholar] [CrossRef]**

16. Molenaar, R.; Jaap, J.; Zijp, J.R. Determination of Kubelka–Munk scattering and absorption coefficients by diffuse illumination. *Appl. Opt.* **1999**, *38*, 2068–2077. **[Google Scholar] [CrossRef] [PubMed]**

17. Dong, T.; Li, Q.; Tian, N.; Zhao, H.G.; Zhang, Y.M.; Han, G.T. Concus Finn Capillary driven fast viscous oil-spills removal by superhydrophobic cruciate polyester fibers. *J. Hazard. Mater.* **2021**, *417*, 126133. **[Google Scholar] [CrossRef] [PubMed]**

18. Tian, N.; Wu, S.H.; Han, G.T.; Zhang, Y.M.; Li, Q.; Dong, T. Biomass-derived oriented neurovascular network-like superhydrophobic aerogel as robust and recyclable oil droplets captor for versatile oil/water separation. *J. Hazard. Mater.* **2022**, *424*, 127393. **[Google Scholar] [CrossRef] [PubMed]**

19. Guo, Z.J.; Lu, Z.; Li, Y.; Liu, W. Highly Performed Fiber-Based Supercapacitor in a Conjugation of Mesoporous MXene. *Adv. Mater. Interfaces* **2022**, *9*, 2101977. **[Google Scholar] [CrossRef]**

CHAPTER-15

A NOVEL APPROACH HAS BEEN DEVELOPED FOR THE SEM-BASED ANALYSIS OF HELICAL-FIBER NONWOVENS

1. INTRODUCTION: AN OVERVIEW

In the realm of material science and textile engineering, the characterization of fibrous structures plays a pivotal role in understanding and optimizing their properties. Helical-fiber nonwovens, in particular, pose unique challenges in terms of their intricate geometry and arrangement. This necessitates advanced analytical techniques for a comprehensive understanding of their structural characteristics. In response to this need, a cutting-edge methodology has been devised for the Scanning Electron Microscope (SEM)-based characterization of helical-fiber nonwovens. This innovative approach promises to unveil previously inaccessible insights into the morphology and composition of these materials, thereby opening new avenues for enhanced design and applications in various industrial domains.

Nature's creatures have been endowed with a tremendous number of excellent biological materials with fascinating structures, which exhibit numerous optimized functions [**1,2,3**]. Helical structure is one of them. Helical micro/nanofibers, which combine a helical structure with micro/nanofibers, have attracted increasing interest due to their unique structure and characteristics [**4,5,6,7**]. Electrospinning is a versatile technique to produce fibers with a diameter ranging from nanometers to several micrometers effectively. Electrospun micro/nanoproducts have shown their potential in the sorption and reuse of crude oil, tissue engineering, artificial muscles, and other smart systems [**8,9,10**]. The traditional setup for electrospinning uses a single needle, the need of complex nanostructure-based advanced functional nanomaterials has promoted the appearance of several kinds of multifluid electrospinning processes, such as tri-axial electrospinning [**11**], tri-fluid electrospinning [**12**], coaxial electrospinning [**13**] with a side-by-side core, and co-electrospinning. Micro/nanofibrous nonwovens that contain helical fibers (called "helical-fiber nonwovens"), which can be produced via electrospinning, have the advantage of large superficial area, large porosity, and good mechanical properties [**12,14,15,16,17**]. However, one major challenge in the research is the lack of quantitative analysis of helical fiber morphology. Image processing technique, which is a powerful tool to acquire information from images, has been utilized to characterize and analyze nonwovens since the 1990s [**18,19,20**]. Recently, Moll et al. measured fiber orientation of fiber injection molded nonwovens by image analysis [**21**].

In this work, two polymers, cellulose acetate (CA) and thermoplastic polyurethane (TPU), which have different molecular flexibility, were chosen to produce nonwovens with helical nanofibers. The objective of this work was to provide an efficient software package for characterizing fiber morphology in helical-fiber nonwovens, including fiber diameter, fiber orientation, as well as helical diameter and curvature. The presence of the characterization for the helical-fiber nonwovens can lead to modelling of the nonwovens' behaviours. A new algorithm was produced to quantify helical diameter and curvature of helical fibers. This image analysis was achieved by application of MATLAB.

In the following sections, the image acquisition, implementation details, and analysis procedures are proposed.

2. EXPERIMENTS

2.1. Materials and Solution Preparation

Thermoplastic polyurethane (TPU; Desmopan DP 2590A) was purchased from the Bayer Materials Science, Leverkusen, Germany. Cellulose acetate powders (CA; 39.8 wt% acetyl, molecular of mass of Mn ~ 30,000). N,N-dimethylformamide (DMF; 0.944 g/mL) and N,N-dimethylacetamide (DMAc; 0.937 g/mL at 25 °C) were purchased from Sigma-Aldrich, St. Louis, MO, USA. Acetone was purchased from Shanghai Lingfeng Chemical Reagent Co. Ltd., Shanghai, China. Lithium chloride anhydrous (LiCl, Mw = 42.39 g/mol) was provided by the Shanghai Chemical Reagents, Shanghai, China. All of these materials were directly used without any further purification.

TPU was dissolved in DMF with different concentrations of 12 wt% and 14 wt%. CA solution of 17 wt% was prepared by dissolving CA powder in a mixture of DMAc/acetone (1:2) with 2 wt% LiCl. CA solution of 15 wt% was prepared by dissolving CA powders in DMAc. All of these solutions were stirred for 8 h at room temperature.

2.2. Co-Electrospinning

The co-electrospinning system used in this study is shown in **Figure 1**. The system is an off-centered core-shell spinneret in which the inner needle is eccentrically situated inside the outer one. This study focused on the characterization of helical nanofibers in nonwovens, according to our previous study [**22**]—we used the optimized processing conditions to obtain the target nonwoven sample. The helical-fiber nonwoven sample was electrospun by CA (15 wt%)/TPU (12 wt%), while the nonwoven sample with straight fibers was electrospun by CA (17 wt%)/TPU (14 wt%). A voltage of 20 kV was applied to the co-electrospinning. The distance from the spinneret to the collector was 15 cm. In this research, all the experiments were performed at ambient temperature of about 25 °C and relative humidity of 40–60%.

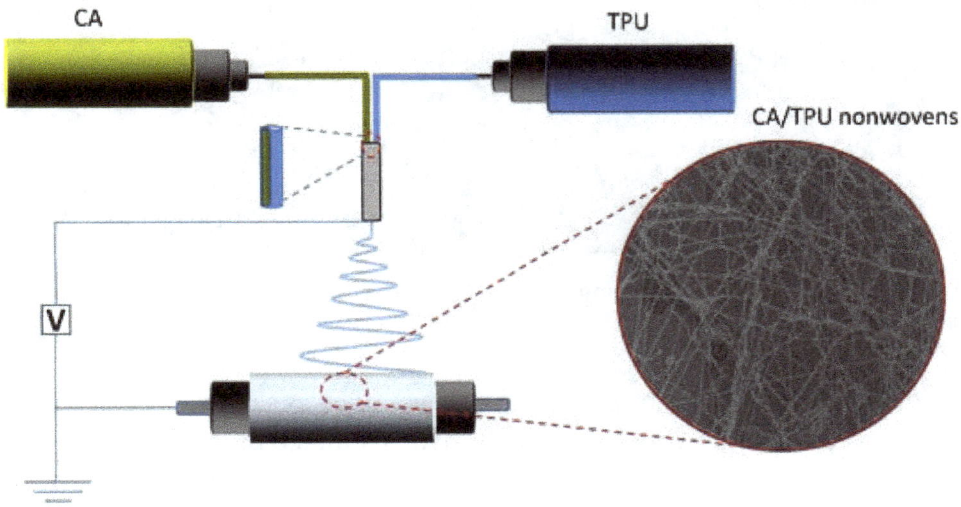

Figure 1. The schematic of co-electrospinning system and the SEM image of CA/TPU nonwovens.

3. IMAGE PROCESSING-METHODOLOGY

3.1. Image Acquisition

Figure 2 shows the image of the obtained helical-fiber nonwoven sample. The field emission scanning electron microscope (FE-SEM, SU8010, HITACHI, Tokyo, Japan, 5 kV, 10 mA, 8 mm working distance) was used to create the digital image of the nonwoven sample. In digital images, the geometric information of fibers is measured by pixels. The format of the image acquired by SEM was classified as the grayscale image, where each pixel was represented by a value from 0 to 255. A zoom factor (i.e., the real length of one pixel), which is determined from the SEM image (**Figure 2**), was used to transform the pixel to geometrical real size. In this case, each pixel had a dimension of 0.0246 μm in length and width.

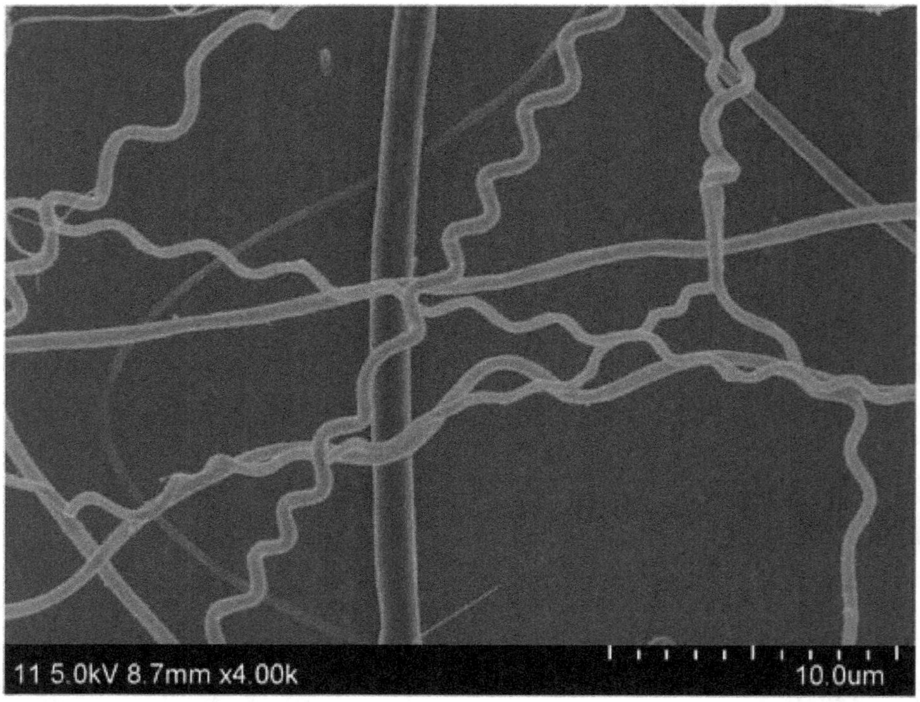

Figure 2. The SEM image of the helical-fiber nonwoven sample.

3.2. Image Vectorization

The digital image acquired by SEM was read and saved as a matrix of discrete pixels. An image vectorization progress, which involves image binarization, connected-component labeling, and boundary extraction, was implemented to transform the digital image to a vector graph and to extract the fiber geometry consequently.

The $m \times n$ pixel matrix can be expressed by a discrete function $(i,)$ in the Cartesian coordinate system:

$$f(i,j) = \begin{bmatrix} f(1,1) & f(1,2) & \cdots & f(1,m) \\ f(2,1) & f(2,2) & & f(2,m) \\ & \vdots & \ddots & \vdots \\ f(n,1) & f(n,2) & \cdots & f(n,m) \end{bmatrix}, \ i = 1,\ 2,\ \ldots, n,\ j = 1,\ 2, \ldots, m. \quad (1)$$

To binarize the digital image, a global threshold was determined according to Otsu's [23] method and turned out to be 109 (0.4275 × 255) in this case. All the pixel values in $(i,)$ were set to two values, 1 (white) or 0 (black). The pixel values above 109 were set as 1, and the other ones were set as 0. The final binary image is shown in Figure 3a.

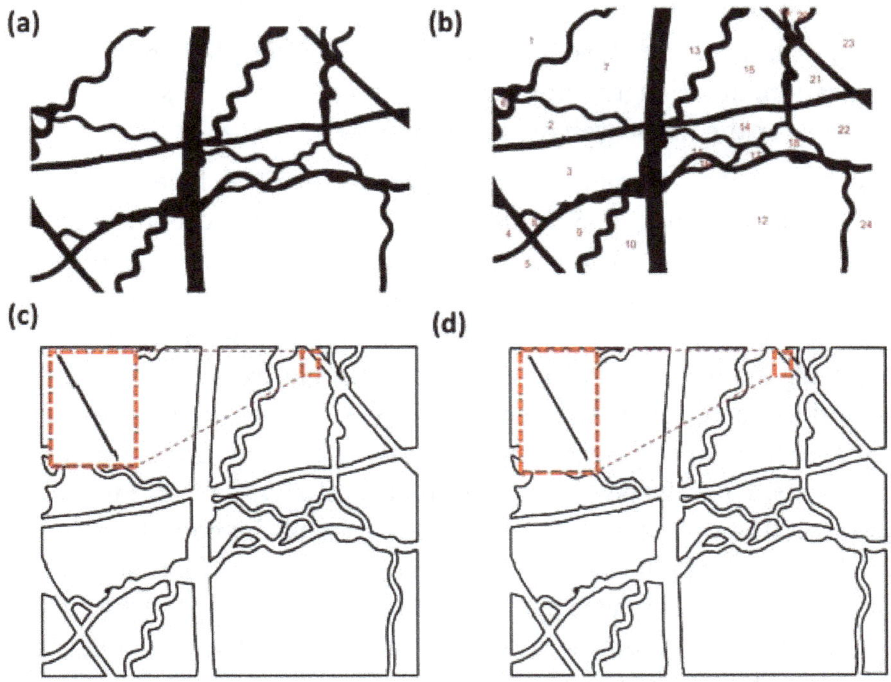

Figure 3. Vectorization of the helical-fiber nonwoven sample: (a) binarization image; (b) labelled binarization image; (c) boundaries extraction without refinement; (d) boundaries extraction after refinement.

We used boundary of pores to obtain the geometric information of fibers. Consequently, the binary image (**Figure 3**a) was segmented into a series of discrete pore domains. To label the pore domains, a connected-component labeling algorithm [**24**] was adopted. The pore domain, whose pixel value is 1 (white), was labeled and recorded by searching the eight neighboring domains until no white pixel was found. The total labeled pore domains form a label matrix. Regionprops function was applied to sort the label matrix into different label sets to isolate individual pore domains. **Figure 3**b shows the labeled binary image. It can be seen that the fibers are distinguished from the pores.

Followed by the labeling procedure, boundaries of pores were extracted and their coordinates registered using the boundary extraction algorithm. The vector graph, which is constituted of the profiles of the boundaries, is shown in **Figure 3**c. It can be seen that the boundaries present as jagged polylines containing huge numbers of points (i.e., pixel positions). We used the method proposed by Prasad [**25,26**] to make dominant point detection and refine the boundaries based on the dominant points. **Figure 3**d shows the pore boundaries after refinement. It can be seen that the boundaries are smoothed under the condition where their curvature properties are retained.

3.3. Fiber Diameter, Length, and Orientation

The fiber geometric characteristics, including fiber morphology (fiber diameter, fiber orientation, fiber length) and helix geometry (helix radius, helix pitch), were characterized for the nonwoven model generation.

The average fiber diameter was analyzed from the labeled binary image (**Figure** 3b) and is given by $D=S/L$ where S is the total area occupied by the fibers, and L is the total length of the fibers. S can be determined by the number of the black pixels and the unit dimension of each pixel, and L is the pixel number of the boundary between black and white pixels. Fiber length, fiber orientation, and helix geometry were analyzed from the vector graphs. **Figure** 4a shows an individual pore domain extracted from the vector graph. A bounding rectangle of the pore domain was determined by searching the nearest and the farthest points in the x and y directions. Then the boundary was divided into five curves (**Figure** 4b) based on the intersecting points between the boundary and the bounding rectangle. As mentioned above, pores in the nonwoven sample are formed by interlacing fibers. Therefore, the pore boundary in **Figure** 4a is simultaneously the boundary of the constituent fibers. The curves possess the information of fiber length, fiber orientation, and helix geometry of the helical fiber. In the following, Curve 1 was chosen as a fiber segment to characterize the fiber geometry.

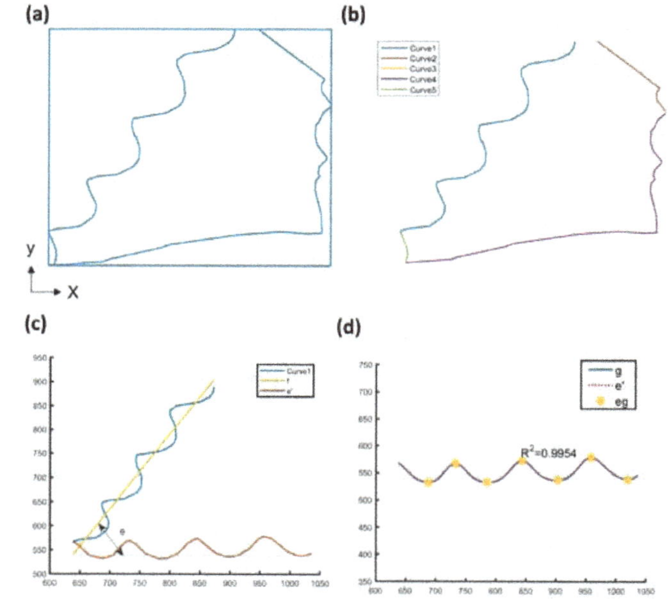

Figure 4. The characterization of fibers geometries by analyzing the helical-fibers of nonwoven sample: (a) edge detection; (b) dominant points edge detection; (c) the original Curve 1 and the rotated curve e'; (d) the Fourier fit curve and extreme points eg.

Curve 1 can be expressed as e = {P_1 P_2 ... PN}, where Pi (xi, yi) is the *ith* point in the Cartesian coordinate system, and N is the number of pixel positions in the curve. l is given by

$$l = \sum_i^N \left[(x_i - x_{i-1})^2 + (y_i - y_{i-1})^2\right]^{\frac{1}{2}} \quad (2)$$

The inclination of the best-fit line of Curve 1 serves as the orientation of the fiber segment in the x-y plane. Curve 1 and its fit line (f), which is derived using the least-square error method, is shown in Figure 4c. The angle θ, which represents the inclination angle of M relative to the x-direction, can be derived from the slope *a* of M.

$$a = \frac{\sum_{i=1}^N x_i y_i - \frac{1}{N}\sum_{i=1}^N x \sum_{i=1}^N y}{\sum_{i=1}^N x_i^2 - \frac{1}{N}\left(\sum_{i=1}^N x_i\right)^2} \quad (3)$$

$$\theta = \arctan a \quad (4)$$

where N is the number of points in the curve, and (*xi,yi*) is the coordinate of the *ith* point.

3.4. Helix Diameter and Curvature

The curves shown in the vector graph represent the projection of helical fibers in the x-y plane. To characterize helix geometry of the helical fiber, the Fourier series with sine-cosine form was used to describe the curve due to the period characteristic of a helix. To start with, to eliminate the influence of fiber orientation on curve fitting, Curve 1 (e) rotates -θ around P1 to the x-direction and becomes e' = {P1 P2' ... PN'}. **Figure** 4c shows the curves e' and the Fourier fit curve is given by the expression

$$(x)=a0+\Sigma nc1[an\cos(n\omega x)+bn\sin(n\omega x)] \quad (5)$$

The number of Fourier coefficients employed in the present calculation is taken as nc = 8, leading to a 0.9954 R2 value. This correlation coefficient (close to 1) provides the reliability of the curve fit. The parameters a0 (the distance of the axial line of e' away from the x-axial), w (the minimum frequency of the trigonometric functions), Fourier coefficients an and bn are as follows:

a_0 = 551.3 ω = 0.01437

a_1 = 0.5819 b_1 = 5.045

a_2 = 0.6606- b_2 = 0.7522-

a_3 = 1.465- b_3 = 2.75-

a_4 = 0.4519- b_4 = 18.24-

$a_5 = 0.3821-\ b_5 = 2.303$

$a_6 = 0.166-\ b_6 = 0.02012-$

$a_7 = 0.8761-\ b_7 = 0.1932-$

$a_8 = 2.499-\ b_8 = 1.339$

As the parameter x varies from P1 to PN', the Fourier fit curve g is shown in **Figure 4**d.

Helix geometry can be characterized by the pitch, radius, and curvature in terms of pitch and radius. The helix pitch (p) is determined by the x-distance of the adjacent extreme points of g (**Figure 4**d), while the helix diameter (d_h) is determined by the y-distance of the extreme points of g. The corresponding helix curvature is expressed by

$$\kappa = \frac{\frac{d_h}{2}}{\left(\left(\frac{d_h}{2}\right)^2 + \left(\frac{p}{2*\pi}\right)^2\right)} \quad (6)$$

4. CHARACTERIZATION RESULTS

According to the developed soft package, the characterization based on the images of two types of nonwoven samples were carried out. Figure 2 and Figure 5 show the helical-fiber nonwoven sample and the nonwoven sample with straight fibers, respectively. Figure 6 shows the comparison of characterization of fiber orientation, helix diameter, and helix curvature for the two types of nonwoven samples.

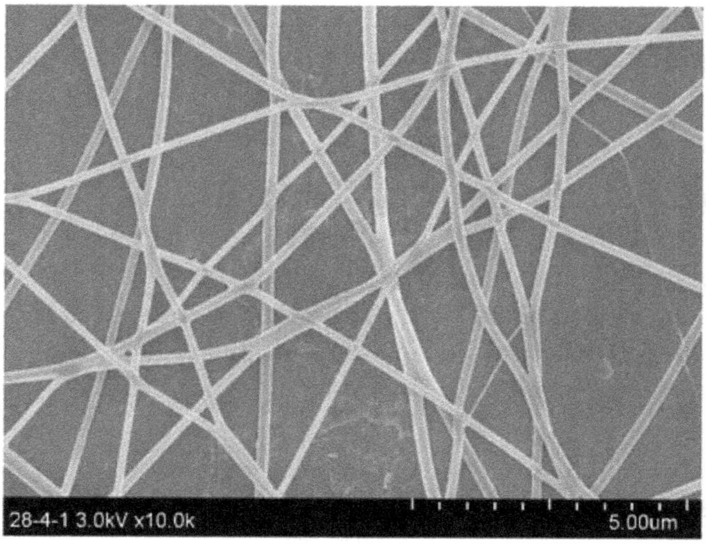

Figure 5. The SEM image of the nonwoven sample with straight fibers.

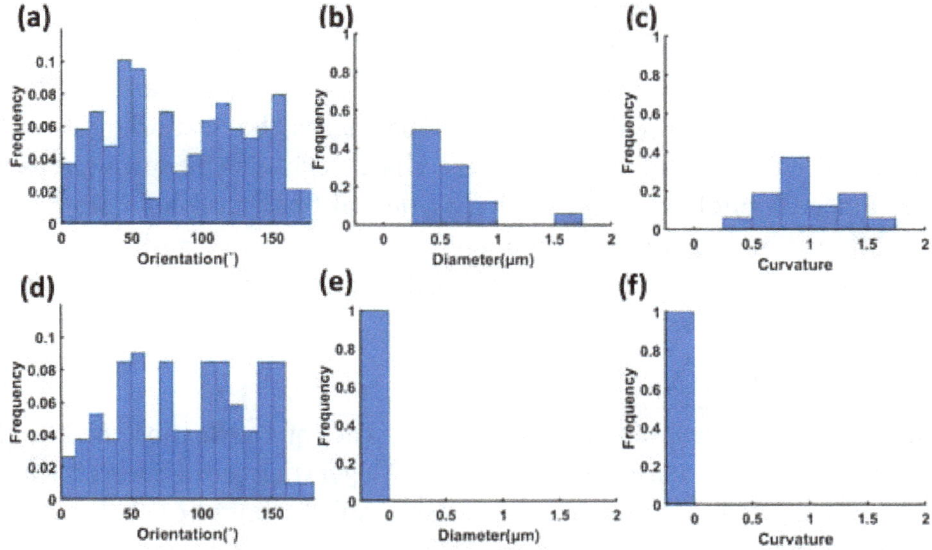

Figure 6. Comparison of characterization results for (a–c) helical-fiber nonwoven sample and (d–f) nonwoven sample with straight fiber, where (a,d) are fiber orientation distributions; (b,e) are helix diameter distributions; (c,f) are curvatures of helix distributions.

To verify the characterization of fiber diameter, which is 0.6 μm for the helical-fiber nonwoven sample and 0.22 μm for the sample with straight fibers, DiameterJ plugin [27], which is an open source for nanofiber diameter measurement tool, was used to analyze the two samples. The results (0.57 μm for the helical-fiber sample and 0.19 μm for the sample with straight fibers) turn out to be in accordance with the results using our method.

Figure 6a,d shows that the orientation distributions of fibers for the two samples have similar characteristic and present the anisotropy of the nonwovens. **Figure** 6b,e shows the helix diameter distributions of fibers for the two cases. The fiber helix diameters of the helical-fiber nonwoven sample are in the range of 0.25–1.5 μm, with the average diameter of 0.6 μm. As expected, the helix diameter frequency of the sample with straight fibers is 0.

Figure 6c,f shows the curvatures of helixes of the fibers for the two cases. For the helical-fiber nonwoven sample, the curvature of the helical fibers ranges from 0.25–1.75 μm^{-1}, with most of the curvatures concentrated at 0.5–1.5 μm^{-1}. As expected, the frequency of fibers processing curvature for the nonwoven sample with straight fibers is 0.

5. CONCLUSIONS

The morphology of fibers in non-woven fabrics is essential for fabric performance. However, the lack of tools for characterization of the helix microstructure observed in SEM micrographs makes it difficult to quantify the morphological parameters of helical-fiber nonwovens. This study proposes a method for the characterization of helical fibers in nonwovens. By developing a new method to extract the fiber morphology from scanning electron microscope images, the fiber orientation, helix diameter, and curvature of the helix were characterized. The software package provides a user-friendly code for analyzing helical fiber morphologies of digital images obtained by scanning electron microscope images. This software package can encourage other researchers to investigate micromechanical behaviour using such numerical methods.

6. CONCISE SUMMARY

The lack of tools particularly designed for the quantification of the fiber morphology in nonwovens, especially the multi-level structured fibers, is the main reason for the limited research studies on the establishment of realistic nonwoven structure. In this study, two polymers, cellulose acetate (CA) and thermoplastic polyurethane (TPU), which have different molecular flexibility, were chosen to produce nonwovens with helical nanofibers. Focusing on the nonwovens with helical fibers, a soft package was developed to characterize fiber morphologies, including fiber orientation, helix diameter, and curvature of helix. The novelty of this study is the proposal of a method for the characterization of nanofibrous nonwovens with special fiber shape (helical fibers) which can be used for curve fibers. The characterization results for the helical-fiber nonwoven sample and the nonwoven sample with straight fibers were compared and analyzed.

REFERENCES

1. Wu, J.; Wang, N.; Zhao, Y.; Jiang, L. Electrospinning of multilevel structured functional micro-/nanofibers and their applications. *J. Mater. Chem. A* **2013**, *1*, 7290–7305. **[Google Scholar] [CrossRef]**
2. Jiang, L.; Zhao, Y.; Zhai, J. A lotus-leaf-like superhydrophobic surface: A porous microsphere/nanofiber composite film prepared by electrohydrodynamics. *Angew. Chem.* **2004**, *116*, 4438–4441. **[Google Scholar] [CrossRef]**
3. Ceylan, H.; Urel, M.; Erkal, T.S.; Tekinay, A.B.; Dana, A.; Guler, M.O. Mussel inspired dynamic cross-linking of self-healing peptide nanofiber network. *Adv. Funct. Mater.* **2013**, *23*, 2081–2090. **[Google Scholar] [CrossRef]**
4. Lin, T.; Wang, H.X.; Wang, X.G. Self-crimping bicomponent nanoribers efectrospun from polyacrylonitrile and elastomeric polyurethane. *Adv. Mater.* **2005**, *17*, 2699–2703. **[Google Scholar] [CrossRef]**

5. Chen, S.; Hou, H.; Hu, P.; Wendorff, J.H.; Greiner, A.; Agarwal, S. Effect of different bicomponent electrospinning techniques on the formation of polymeric nanosprings. *Macromol. Mater. Eng.* **2009**, *294*, 781–786. **[Google Scholar] [CrossRef]**

6. Chen, S.; Hou, H.; Hu, P.; Wendorff, J.H.; Greiner, A.; Agarwal, S. Polymeric nanosprings by bicomponent electrospinning. *Macromol. Mater. Eng.* **2009**, *294*, 265–271. **[Google Scholar] [CrossRef]**

7. Wu, H.H.; Zheng, Y.S.; Zeng, Y.C. Fabrication of Helical Nanofibers via Co-Electrospinning. *Ind. Eng. Chem. Res.* **2015**, *54*, 987–993. **[Google Scholar] [CrossRef]**

8. Iacob, A.-T.; Drăgan, M.; Ionescu, O.-M.; Profire, L.; Ficai, A.; Andronescu, E.; Confederat, L.G.; Lupașcu, D. An Overview of Biopolymeric Electrospun Nanofibers Based on Polysaccharides for Wound Healing Management. *Pharmaceutics* **2020**, *12*, 983. **[Google Scholar] [CrossRef]**

9. Smith, S.A.; Park, J.H.; Williams, B.P.; Joo, Y.L. Polymer/ceramic co-continuous nanofiber membranes via room-curable organopolysilazane for improved lithiumion battery performance. *J. Mater. Sci.* **2017**, *52*, 3657–3669. **[Google Scholar] [CrossRef]**

10. Gao, J.; Li, B.; Wang, L.; Huang, X.; Xue, H. Flexible membranes with a hierarchical nanofiber/microsphere structure for oil adsorption and oil/water separation. *J. Ind. Eng. Chem.* **2018**, *68*, 416–424. **[Google Scholar] [CrossRef]**

11. Yang, Y.; Li, W.; Yu, D.-G.; Wang, G.; Williams, G.R.; Zhang, Z. Tunable drug release from nanofibers coated with blank cellulose acetate layers fabricated using tri-axial electrospinning. *Carbohyd. Polym.* **2019**, *203*, 228–237. **[Google Scholar] [CrossRef] [PubMed]**

12. Zhao, T.N.; Zheng, Y.S.; Zhang, X.M.; Teng, D.F.; Xu, Y.Q.; Zeng, Y.C. Design of helical groove/hollow nanofibers via tri-fluid electrospinning. *Mater. Des.* **2021**, *205*, 109705. **[Google Scholar] [CrossRef]**

13. Yoon, J.; Yang, H.-S.; Lee, B.-S.; Yu, W.-R. Recent Progress in Coaxial Electrospinning: New Parameters, Various Structures, and Wide Applications. *Adv. Mater.* **2018**, *30*, 1704765. **[Google Scholar] [CrossRef] [PubMed]**

14. Chen, P.; Xu, Y.; He, S.; Sun, X.; Pan, S.; Deng, J.; Chen, D.; Peng, H. Hierarchically arranged helical fibre actuators driven by solvents and vapours. *Nat. Nanotechnol.* **2015**, *10*, 1077–1083. **[Google Scholar] [CrossRef]**

15. Fleischer, S.; Feiner, R.; Shapira, A.; Ji, J.; Sui, X.; Wagner, H.D.; Dvira, T. Spring-like fibers for cardiac tissue engineering—ScienceDirect. *Biomaterials* **2013**, *34*, 8599–8606. **[Google Scholar] [CrossRef]**

16. Zhao, Y.Y.; Miao, X.R.; Lin, J.Y.; Li, X.H.; Bian, F.G.; Wang, J.; Zhang, X.Z.; Yue, B.H. Coiled Plant Tendril Bioinspired Fabrication of Helical Porous Microfibers for Crude Oil Cleanup. *Glob. Chall.* **2017**, *1*, 6. **[Google Scholar]** **[CrossRef]** **[PubMed]**

17. Sim, H.J.; Jang, Y.; Kim, H.; Choi, J.G.; Park, J.W.; Lee, D.Y.; Kim, S.J. Self-helical fiber for glucose-responsive artificial muscle. *ACS Appl. Mater. Interfaces* **2020**, *12*, 20228–20233. **[Google Scholar]** **[CrossRef]** **[PubMed]**

18. Pourdeyhimi, B.; Ramanathan, R.; Dent, R. Measuring fiber orientation in nonwovens .1. Simulation. *Text. Res. J.* **1996**, *66*, 713–722. **[Google Scholar]** **[CrossRef]**

19. Pourdeyhimi, B.; Kim, H.S. Measuring fiber orientation in nonwovens: The Hough transform. *Text. Res. J.* **2002**, *72*, 803–809. **[Google Scholar]** **[CrossRef]**

20. Hou, J.; Xu, B.; Gao, H.; Wang, R. Measuring fiber orientations in nonwoven web images using corner detection by Bézier fitting curves. *Text. Res. J.* **2017**, *88*, 2120–2131. **[Google Scholar]** **[CrossRef]**

21. Moll, P.; Wang, S.F.; Coutandin, S.; Fleischer, J. Fiber orientation measurement of fiber injection molded nonwovens by image analysis. *Text. Res. J.* **2021**, *91*, 664–680. **[Google Scholar]** **[CrossRef]**

22. Zhang, X.; Chen, J.; Zeng, Y. Morphology development of helical structure in bicomponent fibers during spinning process. *Polymer* **2020**, *201*, 122609. **[Google Scholar]** **[CrossRef]**

23. Otsu, N. A threshold selection method from gray-level histograms. *IEEE Trans. Syst. Man Cybern.* **1979**, *9*, 62–66. **[Google Scholar]** **[CrossRef]**

24. Wilson, J.N.; Ritter, G.X. *Handbook of Computer Vision Algorithms in Image Algebra*; CRC press: Boca Raton, FL, USA, 2000. **[Google Scholar]**

25. Prasad, D.K.; Quek, C.; Leung, M.K.; Cho, S.Y. A Parameter Independent Line Fitting Method. In Proceedings of the 1st Asian Conference on Pattern Recognition (ACPR), Beijing, China, 28–28 November 2011; pp. 441–445. **[Google Scholar]**

26. Prasad, D.K.; Leung, M.K.H.; Quek, C.; Cho, S.-Y. A novel framework for making dominant point detection methods non-parametric. *Image Vis. Comput.* **2012**, *30*, 843–859. **[Google Scholar]** **[CrossRef]**

27. Hotaling, N.A.; Bharti, K.; Kriel, H.; Simon, C.G. DiameterJ: A validated open source nanofiber diameter measurement tool. *Biomaterials* **2015**, *61*, 327–338. **[Google Scholar]** **[CrossRef]**

CHAPTER-16

REVIEWING CALCIUM ALGINATE FIBER: FABRICATION, CHARACTERISTICS, AND UTILIZATIONS

1. INTRODUCTION: AN OVERVIEW

In recent years, Calcium Alginate Fiber has emerged as a subject of significant interest in the field of materials science and biotechnology. This review explores the diverse aspects of Calcium Alginate Fiber, encompassing its fabrication methods, inherent properties, and wide-ranging applications. As a biopolymer derived from seaweed, calcium alginate exhibits unique characteristics that make it well-suited for various industrial and biomedical applications. Understanding the intricacies of its fabrication, the distinct properties it possesses, and the potential applications it holds is pivotal for harnessing the full spectrum of its capabilities.

In recent years, alginate has become one of the most preferred materials as an abundant natural biopolymer [1]. Alginate is a linear polymer polysaccharide composed of β-D-mannuronic acid (M block) and α-L-guluronic acid (G block) jointed by 1,4-linkages, which is extracted from either brown algae or some genera of bacteria. The molecular chain is arranged in an irregular blockwise pattern of varying proportions of G-G, M-G and M-M blocks [2,3], shown in Figure 1. The percentage of M and G blocks and their distribution has an impact on the physicochemical properties of alginate, such as alginate of rich M units displays a flexible structure and better biocompatibility, while alginate of enriched G units exhibits a rigid molecular structure [4]. Alginate is known to be rich in carboxyl and hydroxyl groups distributed along the backbone, making it open to chemical functionalization and cross-linking treatment [5]. Typically, alginate can be cross-linked to form a hydrogel in the presence of divalent or trivalent metal cations, such as Fe^{3+}, Al^{3+}, Cr^{3+}, Cu^{2+}, Ba^{2+}, Sr^{2+}, Ca^{2+}, et al. [6]. The gelation mechanism is the coordination between the carboxyl groups of alginate and the metal ions. Taking Ca^{2+} as an example, each calcium ion forms coordination bonds with two G units of the alginate molecular chain, which is called the egg-box structure [7,8], as shown in Figure 2.

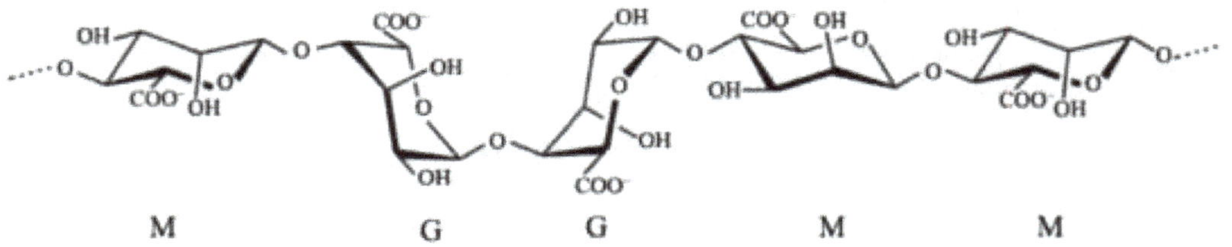

Figure 1. The structural schematic diagram of chain conformation and M/G block distribution of alginate molecular chain.

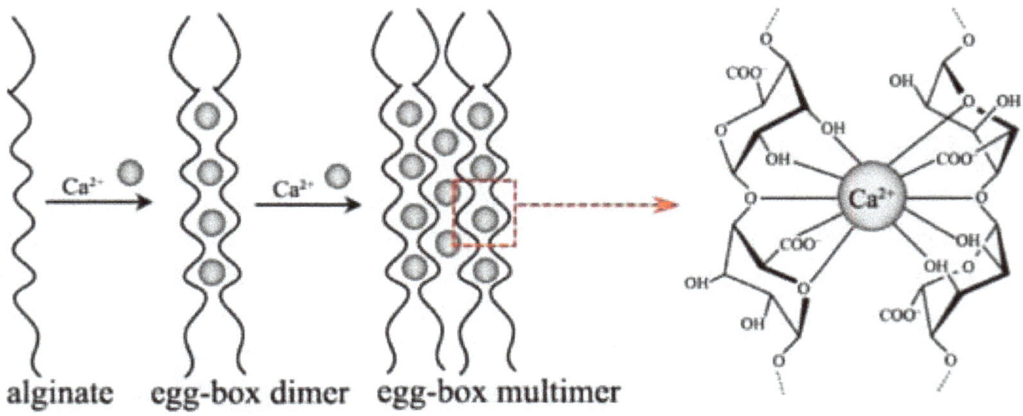

Figure 2. Egg-box junctions of calcium ion and the G block of alginate.

Based on the ability to gel with metal ions, different alginate-based materials with various morphology were developed, such as the porous scaffold **[9,10,11]**, hydrogel **[12,13]**, fiber **[14,15,16]**, nonwoven fabric **[17,18]**, membrane **[19,20,21]**, and so forth. The common way to fabricate the alginate-based materials includes ion cross-linking **[22]**, microfluidic spinning technique **[23]**, freeze-drying **[24]**, wet spinning technology **[25]** and immersive rotary/centrifugal jet spinning technique **[26,27,28]**. Recently, alginate hydrogel has been prepared by the freeze-drying technique or ionic cross-linking method. Commonly, the calcium salt solution was dropped into the homogeneous alginate solution dropwise to induce the cross-linking and form calcium alginate hydrogel. Moreover, the hydrogel can be also endowed with freeze-drying treatment to eliminate water, and then the porous scaffold is obtained. As displayed in **Figure 3**, Che et al. **[11]** used the alginate to cross-link with cellulose to fabricate the alginate-based composite hydrogel, which was then lyophilized to be the hydrogel porous scaffold. The pore diameter of the scaffold was precisely tuned by adjusting the lyophilization process parameters such as the lyophilization temperature and time. As well known, pores of the lyophilized scaffold can provide the oxygen and nutrient substance for the host cell to facilitate the growth of new tissue. Moreover, the alginate porous scaffold can be acquired without ionic cross-linking, just by the freeze-drying technique. In order to enrich the application properties of the alginate scaffold, some other polymers such as gelatin **[29,30]**, chitosan **[31,32]** and collagen **[33,34]** were also incorporated into the alginate polymer to lyophilize and form the hybrid multi-functional lyophilized scaffold.

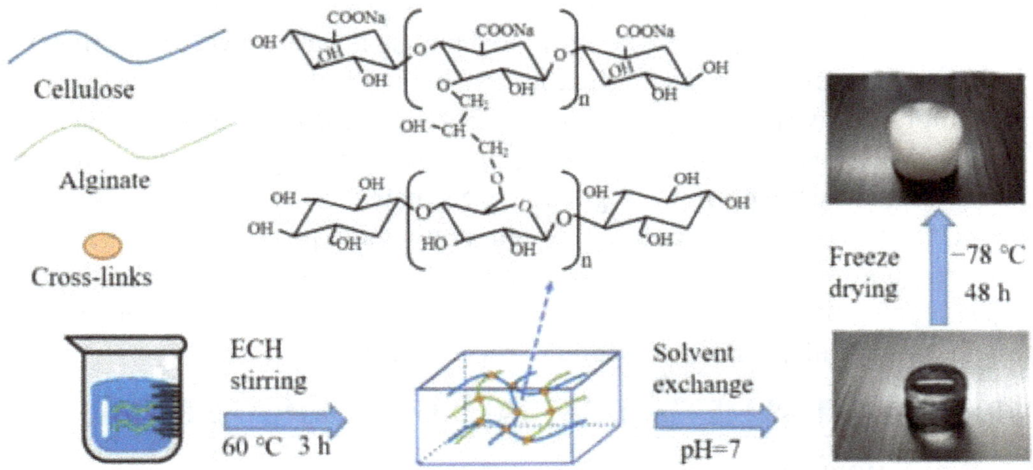

Figure 3. Fabrication of the alginate hydrogel and its porous scaffold by combining ionic cross-linking and freeze-drying techniques.

The alginate fiber is a research hotspot in the field of textile material, but the shortage of related reviews results in a lack of systematic reviews. Hence, this review focused on the fabrication and application properties of alginate fiber, which relied on the coordination between the metal ions exchanging the alkaline metal ions, usually sodium, and forming a strong bond with the G unit to precipitate fiber. The most classic way to obtain alginate fiber is either the microfluidic spinning or wet spinning technique **[35,36,37,38,39]**, or electrospinning for fiber with less diameter requirement. The specific preparation process is described in the following text. The alginate nonwoven fabric can be prepared by the needle punching technology as shown in **Figure 4,** where alginate fiber is firstly combed into the fiber web and then the needled felt is obtained by the needle-punching. The alginate nonwoven fabric has multiple applications, such as wound dressing, tissue scaffolds, or facial masks due to its excellent hygroscopic property and water retention capacity **[40]**.

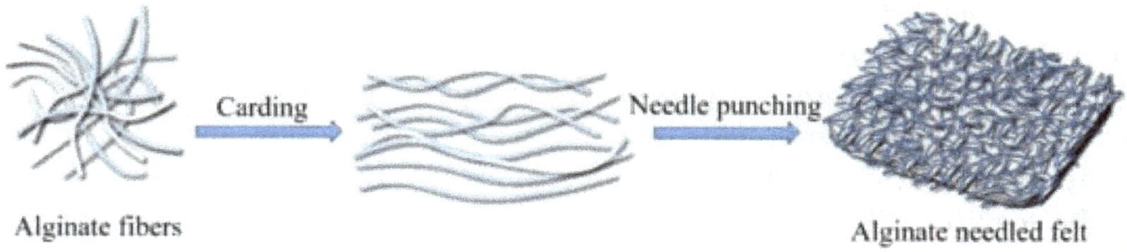

Figure 4. Preparation of the alginate nonwoven fabric by the acupuncture technique.

There are several reviews on the alginate materials, which present the biomedical application properties of alginate hydrogel or the chemical modification of alginate polymer, but the review on alginate fibers is rare. However, the fiber has a broad application in the field of textiles and biomedical materials. It is urgent to give comprehensive knowledge of alginate fiber including the preparation method, the application characteristics and future progress, assisting researchers to better understand and develop novel alginate-based materials. Hence, the fabrication, physicochemical performance and different application properties of the alginate fiber are going to be introduced in detail in this review. The physicochemical properties included its mechanical performance, moisture absorption performance and biocompatibility, that correspond to the application.

2. PREPARATION METHOD OF CALCIUM ALGINATE FIBER

2.1. Wet Spinning

The wet spinning technique is a pioneering approach for preparing the alginate fiber and its schematic diagram is exhibited in Figure 5. The homogeneous alginate spinning solution extruded from the spinneret is introduced into a coagulation bath consisting of calcium salt solution to induce ion cross-linking and then form the primary fiber. The multiple draft rollers are installed in the coagulation bath to endow the primary fiber with a proper drawing ratio, which is beneficial for the improvement of fibrous mechanical performance. Moreover, to further enrich the application properties of alginate fiber, some researchers combined Ca^{2}+ with other metal ions such as Zn^{2+}, Ba^{2+}, Cu^{2+}, Al^{3+} and so forth to form the multi-metal ions coagulation bath [41]. There are some differences in terms of the chelation interaction of various metal ions with alginate molecules, resulting in various formation rates of fiber.

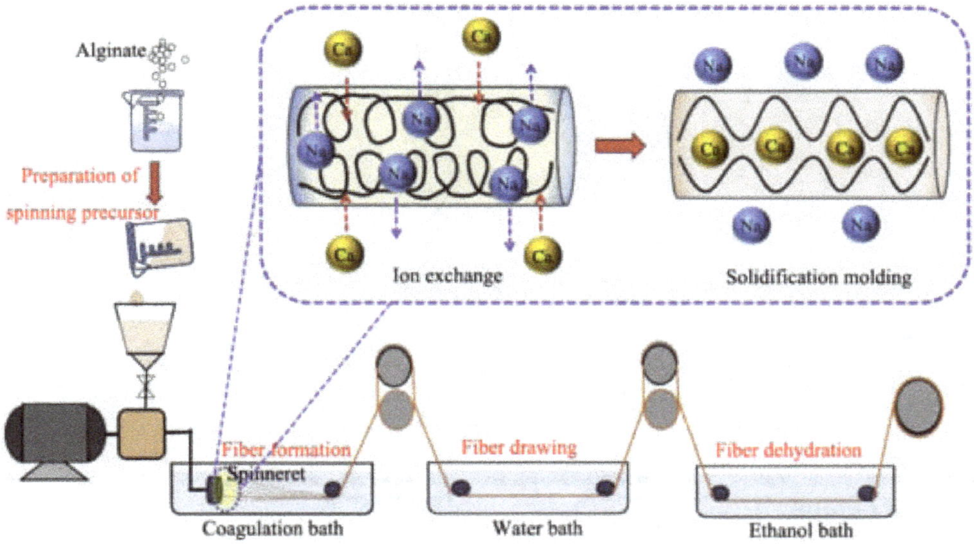

Figure 5. Schematic diagram of wet spinning process of calcium alginate fiber.

The wet spinning device is large and is not versatile enough for the fabrication of alginate fiber with a special shape and function. Therefore, a modified version, mini wet spinning device was proposed and developed by some researchers [33,34] as displayed in Figure 6. During the fabrication, the alginate spinning dope in the syringe is extruded into the CaCl2 solution to finish the ion exchange and then the primary fibers will be obtained without further stretching. In addition, to enhance the mechanical performance and extend the application properties of alginate fiber, some other functional polymers or cross-linkers were added to the spinning dose to produce the hybrid fiber [40,42,43]. For instance, the catechol [44], quaternary ammonium chitosan (QAC) polysaccharide polyelectrolyte complex (PEC) [45] and the hydrolysis compound of vinyl triethoxy silane (VTES) [46] was incorporated in the alginate solution, respectively, to fabricate the composite fiber by the homemade wet spinning device. The introduction of those polymers can form the intermolecular cross-linking and thus improve the fibrous mechanical properties as well as endowing special functions such as the photo-thermal and antibacterial performance. Compared with the traditional wet spinning device, its modified version can realize the variety in fibrous function and structure. However, the lack of stretching dramatically decreased the mechanical properties of the alginate fiber.

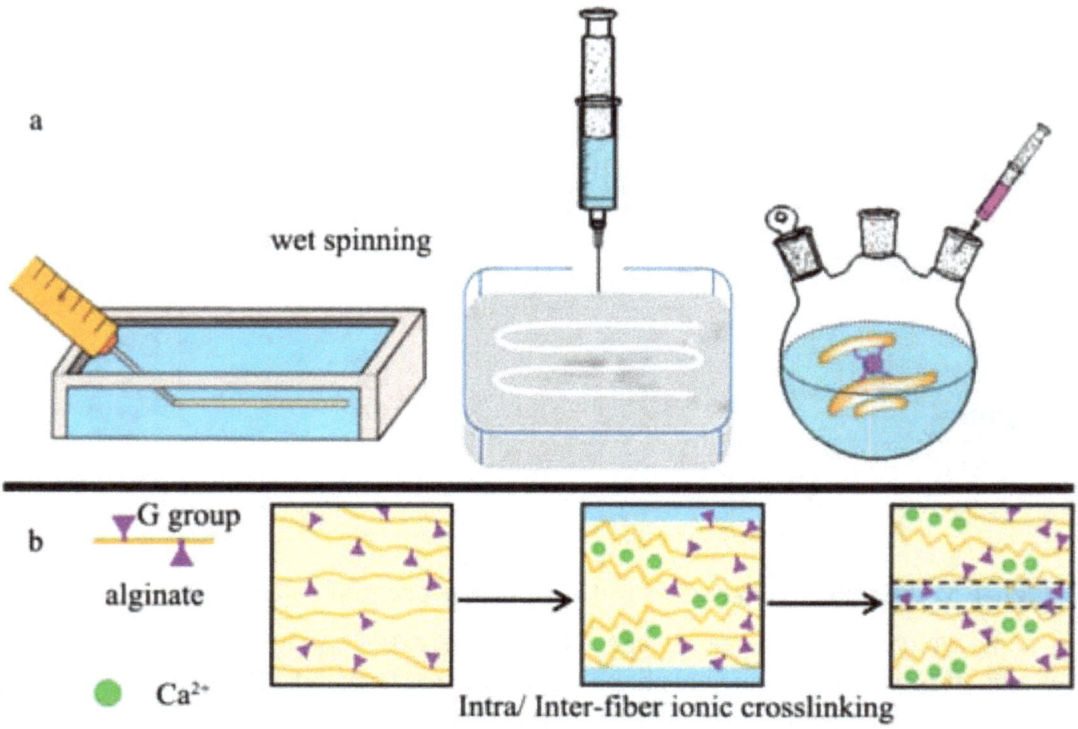

Figure 6. A mini modified version wet spinning device, (a) schematic diagram of the fiber spinning and (b) the formation mechanism of fiber.

2.2. Microfluidic Spinning

Based on the development of wet-spun alginate fiber, some researchers developed the microfluidic spinning device, which is an efficient and facile strategy to fabricate alginate fiber. The preparation of alginate fiber by the microfluidic spinning technique (MST) has received much attention recently, due to its eco-friendly, simple and effective implementation. That also makes MST a popular method to function as a microbioreactor, including the bio-chip. The schematic diagram of fabricating alginate fiber by MST is illustrated in Figure 7a,b, in which the core and sheath flows are the alginate and CaCl2 aqueous solution, respectively. The micro-channel can be endowed with various cross sections (Figure 7a), resulting in the generated fibers with various structures containing tubular, porous, flat, hybrid and hollow (Figure 7c) [47,48]. Calcium cation is the most common cation used to cross-link alginate to form fiber, but other metal cations including Ba2+, Al3+, Cu2+ and Zn2+, et al. (Figure 7d) can also coordinate with G units of alginate to form a fiber. Significantly, some cells with signaling molecules such as extracellular matrix or growth factors can be added to the sample solution to fabricate the multi-functional alginate fiber application in the tissue engineering scaffold including skin, liver, heart and microvessel (Figure 7e) [49,50]. The micro-channel of MST is mainly composed of the core and sheath micro-channel (Figure 8a,b), which can be made of the pulled glass tube and the stainless steel tube. Furthermore, the channel can also be produced by using the micro-electromechanical systems technology to engrave on the surface of the polydimethylsiloxane (PDMS) platform, which then creates the PDMS microfluidic spinning device with various topological constructions (Figure 8c). The method can endow the channel with a variety of parallel-grooved substrata with various pitches and depths, which is beneficial for producing the orientated-microgroove structure of alginate fiber (Figure 8d) [51]. There was a report on the tuning core-sheath flow rate to achieve the helical alginate microfiber as well as the incorporation of Fe3O4 magnetic substance via the MST [36].

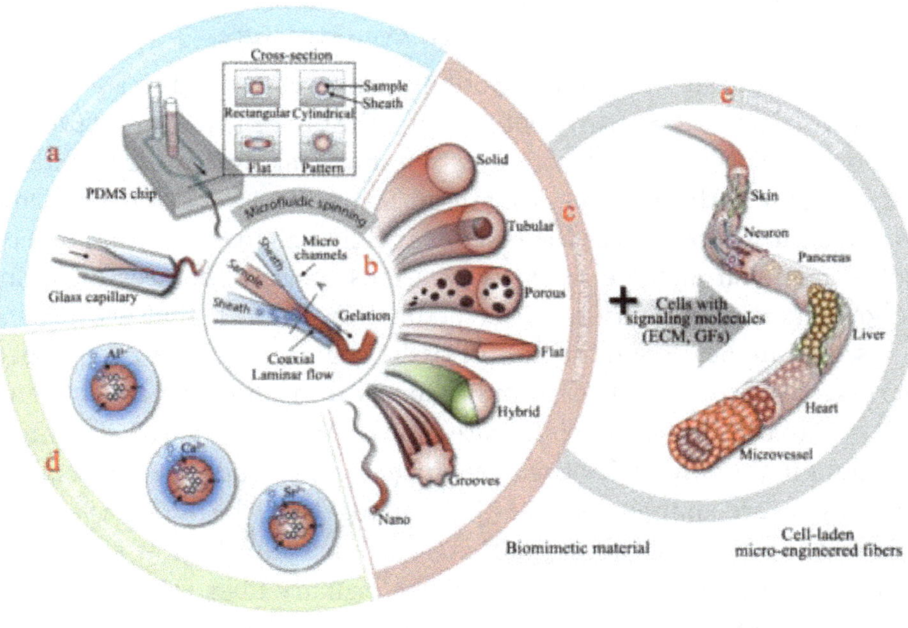

Figure 7. (a) The micro-channel device with various cross sections, (b) Overview of the microfluidic platforms composed of coaxial core and sheath fluids, (c) Anisotropic structure of alginate fiber fabricated by the MST, (d) Various metal cations cross-linked with alginate polymer to fabricate fiber and (e) The cell load micro-engineered fibers as the desired biomimetic material.

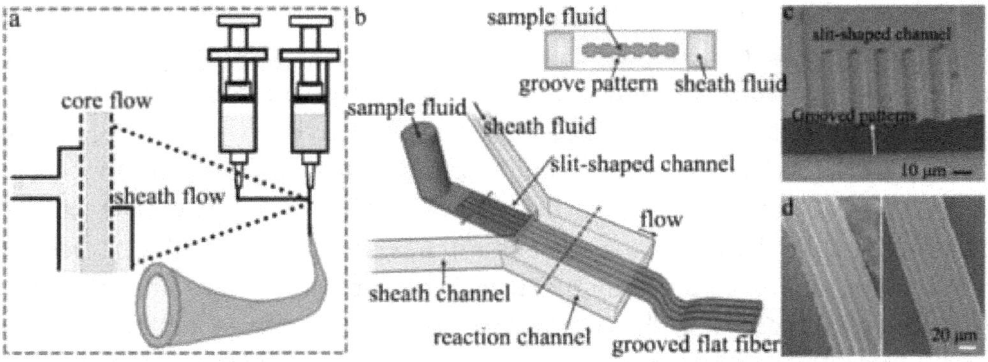

Figure 8. Schematic diagram of (a) microfluidic spinning system and (b) generating flat fibers with micro-grooves, SEM images of (c) the slit-shaped channel and (d) alginate fiber with grooved structure.

The microfluidic spinning device not only can be used to fabricate the alginate microfiber with various structures but also offer the fiber with different functions by the addition of a bioactive small molecule substance, making it an ideal candidate for controlled release as the biomimetic material. Some other works have incorporated the bioactive substance such as the protein or cell in alginate solution to fabricate the multi-functional alginate fiber with various micro-structures including the flat, hollow and grooved anisotropic structure via the MST, that was used as the micro-vascular application in the field of tissue engineering [**47,52,53**]. Especially, the micro-grooved flat fiber has not only been used to guide the morphogenesis of various types of cells but also to integrate topographic control over cell alignment with the design of scaffolds for tissue engineering purposes, such as scaffolds for reconnecting severed muscle tissue [**54**]. Taken together, the MST is a versatile and effective method to prepare the multi-structure and function of alginate fibers. However, the mechanical performance of the primary fiber was poor because of the lack of stretching, and it was just suitable for small batch production.

3. PHYSICOCHEMICAL PROPERTIES OF CALCIUM ALGINATE FIBER

3.1. Mechanical Performance

Calcium alginate fiber and its nonwoven fabric are ideal materials in the field of textiles due to the abundant marine resource and being environmentally friendly. Among them, the mechanical performance of fiber is the focal spot, which directly affects its range of application. The well-known setback of alginate fiber fabricated by different techniques is the formation rate is too fast to control, resulting in uneven structure in the fibers. The prevalence of structural defects of fiber can easily cause stress concentration and lead to fiber failure. A slower gelation and formation rate is an effective method to optimize the fiber structure and improve mechanical performance. Based on this, some researchers have resorted to a calcium salt composite solution with different solubility by choosing the compound coagulation bath system to retard the formation rate of fiber [55]. That was the original calcium chloride (CaCl2) solution combined with the indissolvable calcium salt solution to reduce free calcium cation concentration. Some other scholars chose other metal cations with better gelation ability to tune the ion exchange rate with alginate polymer, which retarded the Ca–Na cross-linking rate and achieved the homogeneous structure of metal-alginate fiber, following the order of the metal ion exchange rate: Pb2+ > Cu2+ > Cd2+ > Ba2+ > Sr2+ > Ca2+ > Co2+ ≈ Ni2+ ≈ Zn2+ > Mn2+ [56]. These works aimed to decrease calcium ion content for prolonging the formation rate of fiber, but the longer formation time resulted in forming a hierarchical structure in the fiber, as shown in Figure 9. The breaking strength of alginate fiber without the hierarchical structure is larger than that of the hierarchical structure. That was not beneficial for the improvement of the mechanical performance of the fiber.

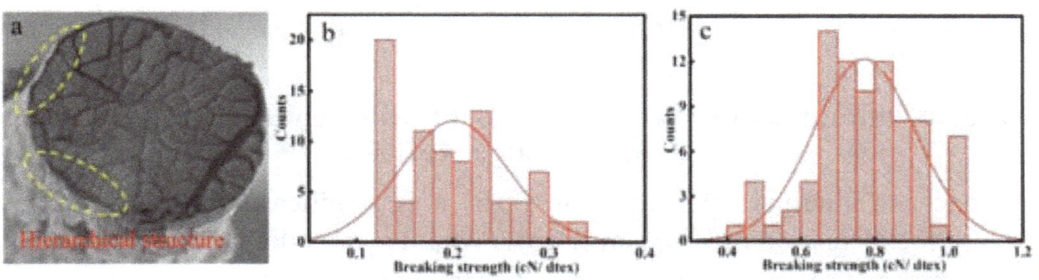

Figure 9. (a) SEM images of the hierarchical structure in the alginate fiber, Breaking strength of alginate fiber with the hierarchical structure (b) and without the hierarchical structure (c).

While the above works succeed in retarding the formation rate of fiber by reducing the Ca–Na ion exchange rate, and decreasing the uneven structure in the fiber, the prolonged formation time caused another problem, delamination between the core and sheath layers of fiber. Some other researchers focused on the incorporation of reinforced materials in the spinning solution, which helped to strengthen the cross-linking interaction between the fiber molecular chain. The popular doping additive materials include the hydrophilic polymers such as collagen [45,57], chitosan [58,59], cellulose nanocrystal [40,60] and its derivative [55] or the nanoparticles including silica [61,62], hydroxyapatite [63,64] and graphene [42,65]. These strategies introduce extra hydrogen bond, ionic bond or covalent bond in the alginate fiber. For example, doping the hydroxypropyl methylcellulose (HPMC), the inorganic nanoparticles (VSNP NPs) or polyacrylamide, respectively, in the solution can improve the breaking strength and stretching performance of the fiber, as shown in **Figure 10** [43,46,66]. Though these strategies yielded an improved result that the breaking strength of calcium alginate fiber was about 2.0 cN/dtex, a 30% increase, researchers are still not satisfied with the processing requirement, when compared with the common cotton or synthetic fiber.

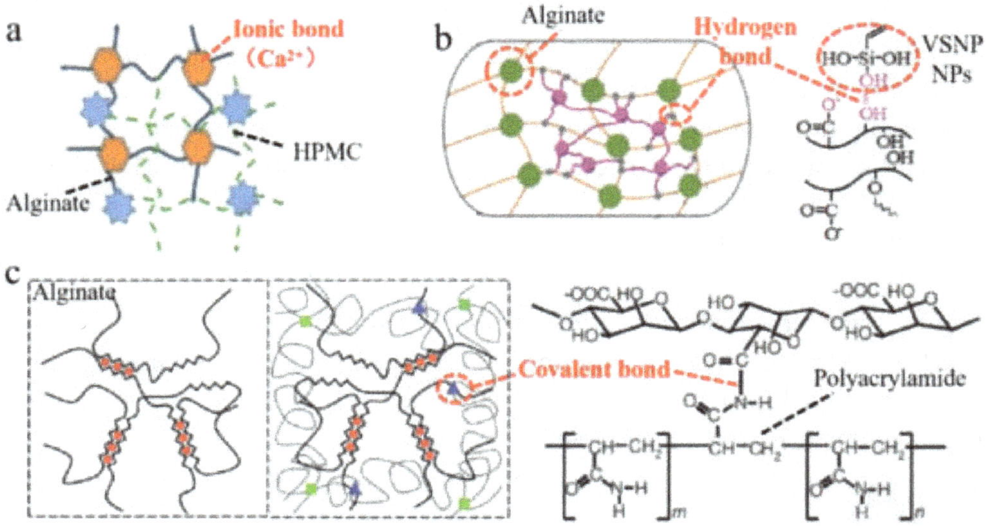

Figure 10. Introduction of (a) ionic bond, (b) hydrogen bond and (c) covalent bond in the alginate fiber.

Hence, to eliminate the defect in fiber, it is significant to investigate and disclose the formation mechanism of fiber. What is more, the impact of defects on fibrous mechanical performance is also needed to be further studied. Controlling the formation rate and improving the uniformity of fiber is the key to enhancing fibrous mechanical properties.

3.2. Moisture Absorption and Biological Compatibility Performance

Alginate is one of the most preferred biomaterials as an abundant natural biopolymer, which was approved by the United States Food and Drugs Administration [**3,67**]. Calcium alginate fiber is nontoxic and has high hygroscopicity and biocompatibility, making it a perfect candidate for biomaterials such as wound dressing and tissue engineering scaffolds [**68,69**]. Significantly, alginate fiber can imitate the physicochemical environment of tissue and its degradable product in vivo can be efficiently cleared by the renal, which is beneficial for tissue repair and regeneration. As an ideal biomaterial, the hygroscopic properties and biocompatibility of alginate fiber have become the research focus.

The alginate fiber has excellent moisture absorption capacity, which can absorb 10 times more than its own weight of water [70,71]. Its molecular chain displays abundant hydrophilic groups including -OH and -COOH groups and can combine a great deal of polarized water molecules. This high absorbency can keep the wound moisture and reduce local pain by providing a cooling effect. It does not adhere to the wound bed and the new granulation tissue will not be affected by washing away the alginate fiber, which can form a self-adherence process in the peri-wound area with a good cover of the infected area. Therefore, the remarkable hygroscopicity of alginate fiber is consistent with the modern theory of moist wound healing [28], holding the potential as a candidate for wound dressing. Moreover, biocompatibility is the other critical factor for its utilization as a biomaterial. It was demonstrated that this fiber can degrade into small molecules including the polysaccharide or its derivative and then be naturally excreted from the body through metabolism [52], during which no hazardous substances remain in the human. Some researchers [15,62] used alginate fibrous dressing to cure the simulated scratch wound, and the result is shown in Figure 11. The simulated scratch wound formed by culturing the fibroblasts and keratinocytes in a culture plate and then created with 20 μL micropipette tips, was completely covered and healed after 48 h, respectively. This result proved that the dressing facilitates the proliferation and migration of fibroblasts and keratinocytes, implying no significant cytotoxicity of the alginate fiber dressing [62]. Furthermore, other works have also implanted the alginate fibrous scaffold into a rat model of S. aureus bacterial infected wound, and the histological analysis indicated that the alginate fibrous scaffold acted a positive effect on the acceleration of wound healing [52]. The healing process involved the overlapping phases of inflammation, cell migration and proliferation, neo-vascularization and extracellular matrix production, all of which certified the biocompatibility of alginate fiber. Taken together, the preferable hygroscopic and biocompatible properties of alginate fiber make it become a hot biomaterial.

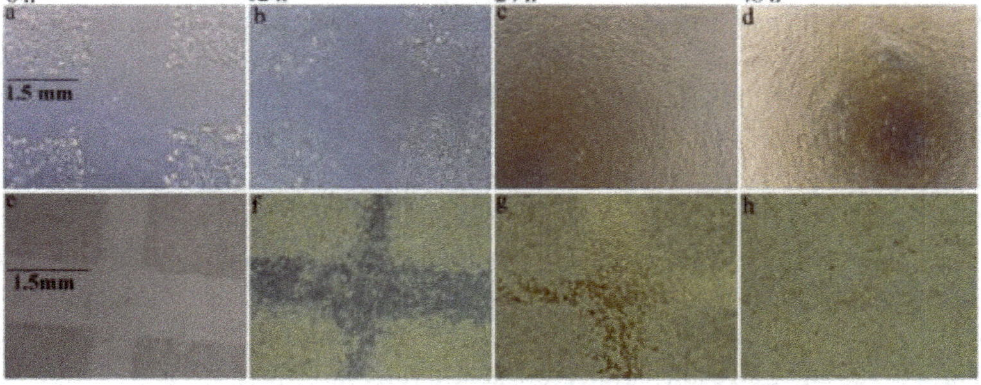

Figure 11. The migration of fibroblasts (a–d) and keratinocytes (e–h) covered on the simulated scratch wound [15,58].

3.3. Flame Retardant Property

Alginate can cross-link with most of the divalent or trivalent metal ions such as Ca2+ by the supramolecular interaction sites to achieve outstanding inherent flame retardancy, because of the inert metal ions during the heating. Calcium alginate fiber as a common alginate-based material exhibited excellent intrinsic flame retardant properties due to the presence of calcium cation and abundant oxygen atoms in its molecule, as displayed in Figure 12. The fiber forms calcium salt and releases the noncombustible carbon dioxide (CO2) gas and gaseous water vapor in burning, which cover the fiber surface and then impede the penetration of oxygen (O2) and the heat, which will effectively retard the further combustion. Furthermore, the heat release rate, the total heat release, the smoke production rate and the total smoke production of the calcium alginate fiber are far less than other materials in the whole burning [72], making it an ideal candidate for the fireproof textile and furniture construction material.

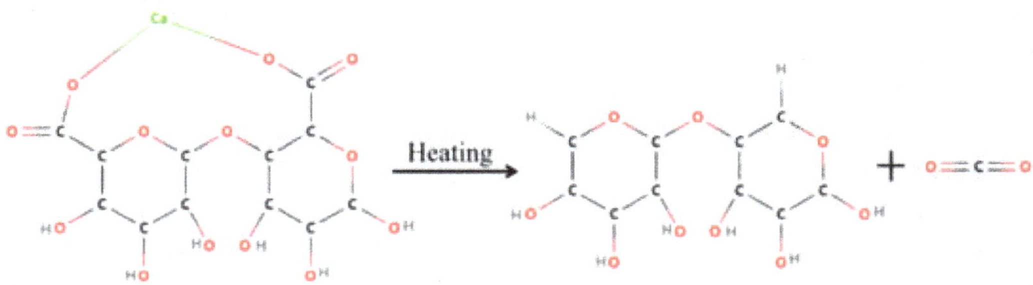

Figure 12. Schematic diagram of thermal decomposition of alginate fiber.

Xia et al. [**73**] and Zhu et al. [**74**] have demonstrated the wet spun alginate fiber to have outstanding flame retardancy with a limiting oxygen index (LOI) as high as 48%, while that of cotton and viscose fiber is about 18% and 20%, respectively. Moreover, the time to ignition (TTI) of alginate is 142 s, which is much longer than any other fibers. Thus, alginate as a kind of functional and high-value material was used to construct fire-resistant textile or as a component and modifier of fireproof coating to reduce the flammability of textile. Driven by this encouraging result, a number of works have investigated the pyrolysis behaviour of alginate fiber by the measurements of synchronous thermal analysis, then studied its flame retardant properties by carrying out the vertical flame test (VFT), UL 94 and cone calorimetry test (CCT), and finally disclosed and established the fireproof mechanism by the combustion product analysis of Thermogravimetric Infrared Spectroscopy (TG-IR) and Pyrolysis Gas Chromatography–Mass Spectrometry (PY-GC–MS). As exhibited in **Figure 13**, all the results indicate that the alginate fiber pyrolyzes according to the decarboxylation or esterification pathway in the burning, in which 2,3-butanedione and furfural are the main combustion compounds [**3,73,74**]. There were fewer volatile combustible matter, heat and smoke in burning, illustrating the excellent flame retardancy and environmental friendless of alginate fiber compared with other nonflammable fiber.

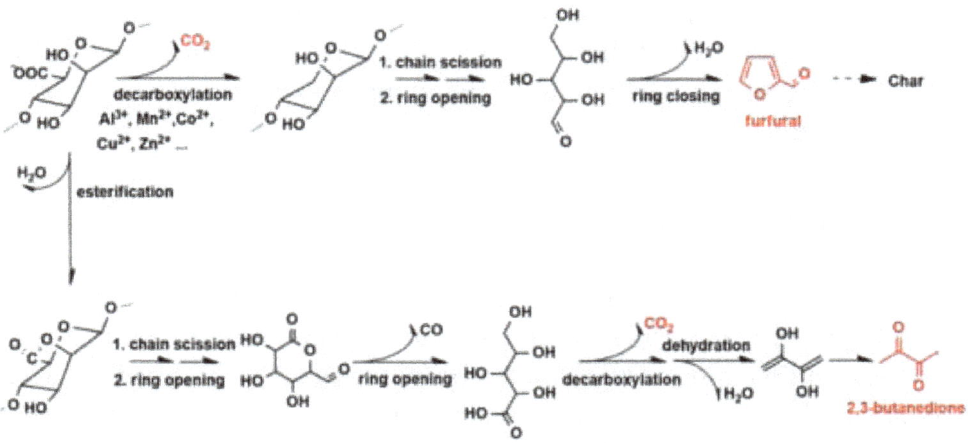

Figure 13. The proposed pyrolysis pathways of alginate fiber.

4. THE APPLICATION OF CALCIUM ALGINATE FIBER

4.1. Wound Dressing

The treatment of acute and chronic wounds with inflammation caused by burns, scalds or the complications of chronic diseases, is an urgent medical need [**75,76,77**]. Traditional wound dressing such as cotton gauze, petrolatum gauze, bandages and so forth, has displayed a serious barrier to wound regeneration. The fibrous dressing absorbs the wound exudates and then helps them to evaporate to keep the wound drying, which can prevent the contamination with bacteria and pathogens [**78,79**]. The dressing protects the wound from external disturbance, but the dry wound is not helpful in the proliferation and migration of skin cells. In addition, the new granulation tissue of the dry wound can be torn easily in the dressing replaces, which may trigger secondary damage. The dry dressing cannot provide and maintain the needed temperature of the wound, making it retard the wound healing. In contrast, the novel modern dressings such as alginate dressing (e.g., hydrogel dressing, the porous scaffold, the fibrous or fabric dressing) can provide and keep a moist wound environment, and then accelerate the migration and re-epithelialization of epithelial cells of the wound, which was a benefit for the wound healing.

Alginate wound dressing can appear in the form of a hydrogel, fiber, nonwoven fabric, freeze-dried scaffold and foam, all of which are prepared by Ca–Na ionic cross-linking interaction. Calcium alginate fiber or the corresponding nonwoven fabric employed as a kind of typical wound dressing, can absorb the wound exudates and then formed gel to keep the wound wet. That provided a safe, sealed and physiological microenvironment, and minimized the bacterial infection possibility around the wound. Furthermore, the hypoxia microenvironment stimulates angiogenesis and facilitates cells to secret the growth factor, which promotes the new granulation tissue formation and accelerated wound healing [**80**]. However, the only gelation function of the original alginate fiber dressing cannot satisfy the actual clinical needs. A variety of multi-functional and bioactive wound dressings are an urgent clinical need.

For enriching the function of alginate fiber dressing, many works have combined alginate with some special therapeutic effects of drugs [**81**], various growth factors [**82**], metal ions [**83**] and polymers [**84,85**]. For instance, Tang et al. [**86**] prepared the functional alginate dressing by incorporating quaternized chitosan (hydroxypropyltrimethyl ammonium chloride chitosan) and magnesium (Mg) in the spinning solution to cure diabetic foot ulcers. The modified chitosan and Mg metal particles can effectively eradicate methicillin-resistant *Staphylococcus aureus* and methicillin-resistant *Staphylococcus epidermidis*, displaying an outstanding antibacterial ability. The results of in vivo microbiological and histological analysis illustrated that alginate dressing containing the functional addition facilitated the migration of human dermal fibroblasts and human umbilical vein endothelial cells, which stimulated angiogenesis and accelerated wound healing. In addition, Vieira et al. [**84**] immobilized the papain on alginate fiber wound dressing to endow the dressing with excellent wound healing ability, which was also capable of promoting the debridement of devitalized or necrotic tissues. The added papain can stimulate the production of cytokines to promote the local cell multiplication and narrow the wound edge, which was a benefit for reducing the scar and assisting wound healing. In conclusion, the incorporation of multi-functional additives in alginate fiber wound dressing has yielded good progress and become the future development trend.

4.2. Tissue Engineering Scaffold

The other major application of calcium alginate-based fiber is tissue engineering scaffold, as a cost-effective material for cell immobilization and encapsulation. Compared to the wound dressing, the tissue engineering scaffold has additional requirements such as serving as an extracellular medium matrix to support cell growth, migration, differentiation, and eventually cell normal function. Multiple pieces of literature reported alginate hydrogel serving as the tissue engineering scaffold, in a bulky form [**87**], or as an ink to be printed into a designed form [**88**], while investigation on calcium alginate fiber is also underway extensively. The fiber provides a few features that hydrogel does not have. As the assembly of fibers, the space between fibers allows a fast transport of nutrients and oxygen to the regeneration site and quick release of waste [**89**]. Additionally, the manipulation of fiber orientation into an aligned form mimics the physiological environment, especially for the purpose to cue multiple cells to arrange into an aligned form, including muscle cell [**90**] or neuron [**91**], where the pattern is crucial for the activation of normal physiological function.

There are a few challenges that remained. Though the microfluidic spinning technique described above was used by several reports as the bioactive scaffold, the general problem of micrometer-diameter fiber is that the space between them is too large for cells, which may require extra time for the cell to fill the space. People turn to electrospinning to reduce the diameter of the fiber to match the cell size. However, there are more challenges, particularly for alginate in electrospinning. Alginate is an ionic polymer that hydrolyzes in water to increase its conductivity, making the solution extremely difficult to electrospin because of the risk of a power surge. Moreover, its lack of shear strength in the sodium salt form and being too dense in the calcium salt form makes it difficult to electrospin. Additionally, there is a very limited choice of solvents because of the hydrophilic nature of the alginate molecule, leaving the only choice to be water with a high boiling point. Many researchers overcome this problem by adding other water-soluble polymers to help overcome mentioned problems. Commonly used polymers include PCL [**92**], chitosan [**93**], collagen, gelatin, and PEO [**94**] to either make electrospinning feasible or bring additional properties needed in tissue engineering. Researchers have used these fibers for the regeneration of bone, muscle, neuron, and skin.

As a tissue engineering fiber, the alginate was used to carry different drugs to facilitate the regeneration speed, including small chemical drugs, nucleic acid [**95**] and protein [**96**]. Additionally, the cations in the calcium alginate fiber can be substituted with other bivalent charge cations that have physiological functions, such as Zn^{2+}, Mg^{2+} and Cu^{2+}. Those bivalent charge cations can replace calcium without destroying its gelation property. For example, calcium can be replaced with copper cations [**97**]. A small amount of copper ions was proven to have a profound effect in enhancing tissue sealing and repair, because of the using angiogenesis ability of copper to help form new blood vessels, and its photo-thermal ability to help the fiber to heat up. Additionally, the calcium in the fiber can be the nucleation site for mineralization, which is the key step to introducing hydroxyapatite to the scaffold and using it to enhance the osteogenesis ability [**94,98**].

4.3. Smart Material

With the development of smart materials, some researchers used the active groups of alginate molecular chains such as -OH and -COOH groups to cross-link with polymers to obtain special intelligent response performance. The covalent compounds such as polyacrylamide or poly (vinyl alcohol) were incorporated into the alginate molecular chain to introduce the covalent bond for fabricating the highly stretchable fiber strain sensor utilization as the ionic skin, wearable and implantable sensors for innovative electronics [99,100]. The concomitant ionic bond and covalent bond endowed the alginate fiber with remarkable mechanical performance especially its high stretchability based on the energy dissipation and deformation hysteresis [101,102]. The fiber is full of metal cations, thus the fiber can have a desirable electrical conductivity, making it a potential platform for smart wearable and implantable innovative electronics. In addition, other researchers used alginate fiber to fabricate the soft actuators and robots [103], thermosensitivity [104] and pH-sensitivity sensor [105] by the addition of some functional temperature/pH sensitive materials as exhibited in Figure 14, which were demonstrated to achieve a preferable intelligent response effect to the surrounding environment. Taken together, it is believed that the alginate fiber displays the fabrication possibilities for novel components of soft actuators and micropumps, as well as smart wearable devices for various sensors.

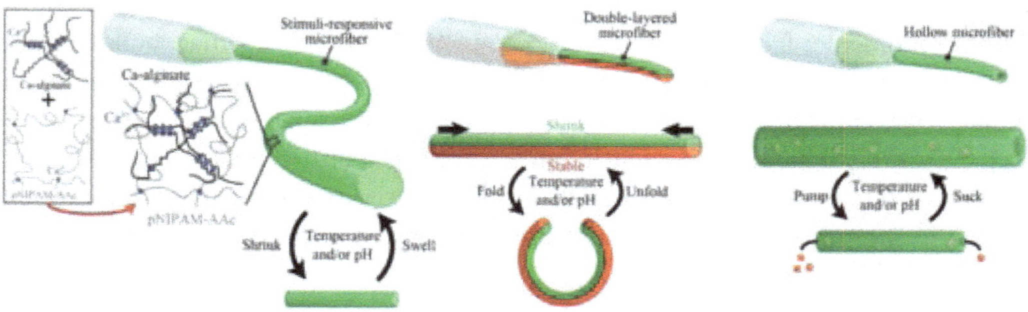

Figure 14. Schematic diagram of a pH/temperature-responsive alginate microfiber with single or double layered structure fabricated by the microfluidic system.

4.4. Fire-Resistant Material

With a tightening low-carbon policy, bio-based materials have received more attention than petroleum-based materials. The commercial fire-resistant materials receive flameproof treatment by using halogen or phosphorus-based flame retardants. However, the fabric treated by the fire-retardant finishing will release some toxic volatile gaseous compounds in the thermal degradation process, which is harmful to human health and violates the eco-friendly concept. What is more, the flame retardancy of that fabric is not durable, decreasing with the increase in usage times. Therefore, the bio-based flame-resistant material has become a research hotspot due to its sustainability, environmental friendliness and recyclability, such as protein, chitosan, DNA, starch, phytic acid and so forth. Among them, the abundant marine material has been widely used to prepare film [**106,107**], hydrogel [**108**], fiber [**109,110,111**], fabric [**74,112**] and aerogel [**113,114,115**] for fire prevention. Alginate fiber as an intrinsic fireproof material has become an ideal flame-resistant textile material application in the field of industrial and academic.

The environmentally sustainable flame retardants for textiles and inherently flame-retardant fiber materials thereby have become mainstream. There were some reports that illustrate that the toxic gaseous compound including CO, CO2 and smoke production rates of the alginate fiber in burning were very little, which was in accordance with a low carbon concept [116,117]. Recently, the alginate fibers were mixed with viscose fibers to produce the nonflammable composite nonwoven felt by the acupuncture technology, serving as the doll filling materials [107]. The alginate fibers and cotton fibers were twisted into yarns at various mass ratios to fabricate the eco-friendly flame-retardant textile fabric by the knitting or weaving technique, which can be applied in the fire-fighter uniform. Those works have achieved the great flame retarding effect and the introduction of alginate fiber was demonstrated to improve the flame retardancy of fabric. Although the alginate fiber displayed an excellent flame retardant performance, it experienced a minor afterglow phenomenon after removal from the fire, which was a drawback for its fireproof abilities. To address this hurdle, some researchers combined the alginate fiber with some synthetic fibers such as polyester to obtain high flame resistance materials [118]. The mixed polyester fiber can form the molten drop in burning and then reduce the afterglow of alginate fiber. As shown in Figure 15, SEM images of the alginate composite fibrous mat before and after the burning, illustrated the droplet covered on the fiber to inhibit the afterglow. Facile blending has become an effective approach to obtaining a self-extinguishing composite alginate fibrous mat or fabric. However, the smoldering process of the original alginate fiber has not been eliminated, and the incorporation of flammable petroleum-based fiber decreased the fire-resistant property and produced some toxic gaseous products in burning, which limited its application as a flame retardant protective material.

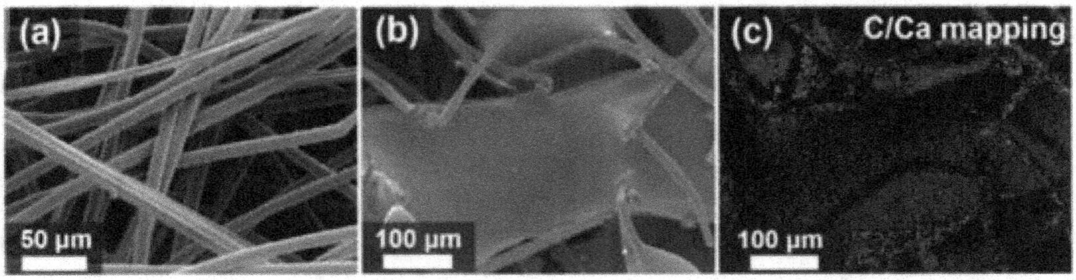

Figure 15. SEM images of polyester/alginate mixed fiber (a) before and (b) after the flammability). test, and (c) EDX element mapping image (purple for carbon mainly from melt polyesters and yellow for calcium in fibrous charred alginate

Hence, it is urgent to eliminate the smoldering combustion behaviour of alginate fiber and the simple mixture of other synthetic fibers is not a good strategy. The molecular chain of alginate fiber displays a good deal of active groups, which can be easily modified and grafted with the corresponding groups to optimize the flame retardant property of alginate fiber in the future.

5. CONCLUSIONS

Compared to the numerous research and review articles on alginate hydrogel, a review on fabrication, physicochemical performance and application properties of alginate fiber is very rare, resulting in the lack of corresponding knowledge. Nevertheless, alginate fiber, as a family of emerging advanced functional materials, has demonstrated great utility and potential, particularly in the wound healing, tissue engineering and flame retardancy fields. The most attractive features of alginate fiber for these applications are attributed to its excellent moisture absorption performance, biocompatibility and intrinsic flame retardant property. Therefore, this work mainly summarized the fabrication of alginate fiber and compared their respective characteristics, and then analyzed the corresponding application properties.

As a wound dressing or tissue engineering scaffold, alginate fiber has a track record of clinical safety and was safely implanted in various hosts including the islet transplantation for treatment of type 1 diabetes, chondrocyte transplantation for treatment of urinary incontinence and vesicoureteral reflux, and the infected wound transplantation for healing of the inflammatory. Although, alginate fiber is widely used in the field of biomaterials and is likely to evolve considerably. It already plays a fairly passive role during treatment due to a shortage of multi-functions. Future dressing or tissue scaffold will likely play a much more active role, which requires the material itself to cure and accelerate the wound healing or repairing impaired tissue in the host. One or more bioactive agents such as proteins, DNA, antibiotics and other polymers that reduce inflammation and promote bone growth, can be incorporated into the alginate fiber as a facile and efficient method to endow it with a multi-functional design, which has been proven to be the future trend.

In addition, the alginate fiber as a burgeoning bio-based flame retardant material has proven to be the future trend, due to the rare limited significant toxicity combustion product and the intrinsic eco-friendly behaviour. However, the decrease in its minor afterglow profile and the enhancement of mechanical properties is an urgent need before its widespread application in the fireproof construction and building field. As one looks to the future, the elimination of fibrous structure irregularity is the future development. For addressing the hurdle of the afterglow behaviour of alginate fiber in burning, the grafted modification of the alginate molecular chain is probably a new and efficient strategy. In conclusion, alginate fiber as a novel eco-friendly bio-based material was proven to have a broad application prospect in the biomaterial and fireproof construction fields.

6. CONCISE SUMMARY

As a natural linear polysaccharide, alginate can be gelled into calcium alginate fiber and exploited for functional material applications. Owing to its high hygroscopicity, biocompatibility, nontoxicity and non-flammability, calcium alginate fiber has found a variety of potential applications. This article gives a comprehensive overview of research on calcium alginate fiber, starting from the fabrication technique of wet spinning and microfluidic spinning, followed by a detailed description of the moisture absorption ability, biocompatibility and intrinsic fire-resistant performance of calcium alginate fiber, and briefly introduces its corresponding applications in biomaterials, fire-retardant and other advanced materials that have been extensively studied over the past decade. This review assists in better design and preparation of the alginate bio-based fiber and puts forward new perspectives for further study on alginate fiber, which can benefit the future development of the booming eco-friendly marine biomass polysaccharide fiber.

REFERENCES

1. Su, H.C.; Bansal, N.; Bhandari, B. Alginate gel particles—A review of production techniques and physical properties. *Crit. Rev. Food Sci. Nutr.* **2017**, *57*, 1133–1152. **[Google Scholar]**
2. Sikorski, P.; Mo, F.; Skjåk-Bræk, G.; Stokke, B.T. Evidence for Egg-Box-Compatible Interactions in Calcium-Alginate Gels from Fiber X-ray Diffraction. *Biomacromolecules* **2007**, *8*, 2098–2103. **[Google Scholar] [CrossRef]**
3. Xu, Y.J.; Qu, L.Y.; Liu, Y.; Zhu, P. An overview of alginates as flame-retardant materials: Pyrolysis behaviors, flame retardancy, and applications. *Carbohydr. Polym.* **2021**, *260*, 117827. **[Google Scholar] [CrossRef]**
4. Zhang, X.; Wang, L.; Weng, L.; Deng, B. Strontium ion substituted alginate-based hydrogel fibers and its coordination binding model. *J. Appl. Polym. Sci.* **2020**, *137*, 48571. **[Google Scholar] [CrossRef]**
5. Benslima, A.; Sellimi, S.; Hamdi, M.; Nasri, R.; Jridi, M.; Cot, D.; Li, S.; Nasri, M.; Zouari, N. The brown seaweed *Cystoseira schiffneri* as a source of sodium alginate: Chemical and structural characterization, and antioxidant activities. *Food Biosci.* **2020**, *40*, 100873. **[Google Scholar] [CrossRef]**
6. Bierhalz, A.C.; da Silva, M.A.; Braga, M.E.; Sousa, H.J.; Kieckbusch, T.G. Effect of calcium and/or barium crosslinking on the physical and antimicrobial properties of natamycin-loaded alginate films. *LWT Food Sci. Technol.* **2014**, *57*, 494–501. **[Google Scholar] [CrossRef][Green Version]**

7. Brus, J.; Urbanova, M.; Czernek, J.; Pavelkova, M.; Kubova, K.; Vyslouzil, J.; Abbrent, S.; Konefal, R.; Horský, J.; Vetchy, D.; et al. Structure and Dynamics of Alginate Gels Cross-Linked by Polyvalent Ions Probed via Solid State NMR Spectroscopy. *Biomacromolecules* **2017**, *18*, 2478–2488. **[Google Scholar] [CrossRef] [PubMed][Green Version]**

8. Fang, Y.; Al-Assaf, S.; Phillips, G.O.; Nishinari, K.; Funami, T.; Williams, P.A.; Li, L. Multiple steps and critical behaviors of the binding of calcium to alginate. *J. Phys. Chem. B* **2007**, *111*, 2456–2462. **[Google Scholar] [CrossRef]**

9. Vakilian, S.; Jamshidi-Adegani, F.; Yahmadi, A.A.; Al-Broumi, M.; Al-Hashmi, S. A Competitive Nature-derived Multilayered Scaffold based on Chitosan and Alginate, for Full-thickness Wound Healing. *Carbohydr. Polym.* **2021**, *262*, 117921. **[Google Scholar] [CrossRef]**

10. García-Gareta, E. Poly-ε-Caprolactone/Fibrin-Alginate Scaffold: A New Pro-Angiogenic Composite Biomaterial for the Treatment of Bone Defects. *Polymers* **2021**, *13*, 3399. **[Google Scholar]**

11. Zheng, Y.; Wang, L.; Bai, X.; Xiao, Y.; Che, J. Bio-inspired composite by hydroxyapatite mineralization on (bis)phosphonate-modified cellulose-alginate scaffold for bone tissue engineering. *Colloids Surf. A Physicochem. Eng. Asp.* **2021**, *635*, 127958. **[Google Scholar] [CrossRef]**

12. Kumar, A.; Bhatt, A.N.; Singh, L.; Karim, Z.; Ansari, M.S. Alginate-based hydrogels for tissue engineering. In *Polysaccharide-Based Nanocomposites for Gene Delivery and Tissue Engineering*; Woodhead Publishing: Sawston, UK, 2021. **[Google Scholar]**

13. Gunatilake, U.B.; Garcia-Rey, S.; Ojeda, E.; Basabe-Desmonts, L.; Benito-Lopez, F. TiO2 Nanotubes Alginate Hydrogel Scaffold for Rapid Sensing of Sweat Biomarkers: Lactate and Glucose. ACS Appl. Mater. Interfaces 2021, 13, 37734–37745. [Google Scholar] [CrossRef]

14. Dangi, Y.R.; Lin, X.; Choi, J.W.; Lim, C.R.; Song, M.H.; Han, M.; Bediako, J.K.; Cho, C.-W.; Yun, Y.-S. Polyethyleneimine functionalized alginate composite fiber for fast recovery of gold from acidic aqueous solutions. *Environ. Technol. Innov.* **2022**, *28*, 102605. **[Google Scholar] [CrossRef]**

15. Weng, L.; Zhang, X.; Fan, W.; Lu, Y. Development of the inorganic nanoparticles reinforced alginate-based hybrid fiber for wound care and healing. *J. Appl. Polym. Sci.* **2021**, *138*, 51228. **[Google Scholar] [CrossRef]**

16. Chen, Z.; Song, J.; Xia, Y.; Jiang, Y.; Li, Y. High strength and strain alginate fibers by a novel wheel spinning technique for knitting stretchable and biocompatible wound-care materials. *Mater. Sci. Eng. C* **2021**, *127*, 112204. **[Google Scholar] [CrossRef]**

17. Li, S.-Q.; Tang, R.-C.; Yu, C.-B. Flame retardant treatment of jute fabric with chitosan and sodium alginate. *Polym. Degrad. Stab.* **2022**, *196*, 109826. **[Google Scholar] [CrossRef]**

18. Wang, F.; Jiang, J.; Sun, F.; Sun, L.; Li, M. Flexible wearable graphene/alginate composite non-woven fabric temperature sensor with high sensitivity and anti-interference. *Cellulose* **2020**, *27*, 2369–2380. **[Google Scholar] [CrossRef]**

19. Huang, C.C. Characteristics and Preparation of Designed Alginate-Based Composite Scaffold Membranes with Decellularized Fibrous Micro-Scaffold Structures from Porcine Skin. *Polymers* **2021**, *13*, 3464. **[Google Scholar] [CrossRef]**

20. Dodero, A.; Alloisio, M.; Vicini, S.; Castellano, M. Preparation of composite alginate-based electrospun membranes loaded with ZnO nanoparticles. *Carbohydr. Polym.* **2019**, *227*, 115371. **[Google Scholar] [CrossRef]**

21. Anokhina, T.; Dmitrieva, E.; Volkov, A. Recovery of Model Pharmaceutical Compounds from Water and Organic Solutions with Alginate-Based Composite Membranes. *Membranes* **2022**, *12*, 235. **[Google Scholar] [CrossRef]**

22. Costa, M.J.; Marques, A.M.; Pastrana, L.M.; Teixeira, J.A.; Sillankorva, S.M.; Cerqueira, M.A. Physicochemical properties of alginate-based films: Effect of ionic crosslinking and mannuronic and guluronic acid ratio. *Food Hydrocoll.* **2018**, *81*, 442–448. **[Google Scholar] [CrossRef][Green Version]**

23. Cai, J.; Ye, D.; Wu, Y.; Fan, L.; Yu, H. Injectable alginate fibrous hydrogel with a three-dimensional network structure fabricated by microfluidic spinning. *Compos. Commun.* **2019**, *15*, 1–5. **[Google Scholar] [CrossRef]**

24. Jamnezhad, S.; Motififard, M.; Saber-Samandari, S.; Asefnejad, A.; Khandan, A. Development and investigation of novel alginate-hyaluronic acid bone fillers using freeze drying technique for orthopedic field. *Nanomed. Res. J.* **2020**, *5*, 306–315. **[Google Scholar]**

25. Zhang, X.; Chen, H.; Jin, X. Influence of K+ and Na+ ions on the degradation of wet-spun alginate fibers for tissue engineering. J. Appl. Polym. Sci. 2017, 134, 44396. [Google Scholar] [CrossRef]

26. Gonzalez, G.M.; Macqueen, L.A.; Lind, J.U.; Fitzgibbons, S.A.; Chantre, C.O.; Huggler, I.; Golecki, H.M.; Goss, J.A.; Parker, K.K. Production of Synthetic, Para-Aramid and Biopolymer Nanofibers by Immersion Rotary Jet-Spinning. *Macromol. Mater. Eng.* **2017**, *302*, 1600365. **[Google Scholar] [CrossRef]**

27. Bielanin, J. The Development and Evaluation of Alginate Nanofibers as a Neuroprotective Nano-scaffold for Amyotrophic Lateral Sclerosis (ALS).

Chemistry & Biochemistry Undergraduate Honors Theses. Bachelor's Thesis, University of Arkansas, Fayetteville, AR, USA, 2020. **[Google Scholar]**

28. Parker, A.C.; Hannah, L.; Prashanth, R.; Tara, S.; Jeffrey, L.; Kartik, B.; Ryan, T.Z. Anti-microbial alginate for wound healing applications. *Am. Chem. Soc.* **2018**, *256*, 1155. **[Google Scholar]**

29. Homem, N.C.; Tavares, T.D.; Miranda, C.; Antunes, J.C.; Felgueiras, H.P. Functionalization of Crosslinked Sodium Alginate/Gelatin Wet-Spun Porous Fibers with Nisin Z for the Inhibition of *Staphylococcus aureus*-Induced Infections. *Int. J. Mol. Sci.* **2021**, *22*, 1930. **[Google Scholar] [CrossRef] [PubMed]**

30. Ghanbari, M.; Salavati-Niasari, M.; Mohandes, F.; Dolatyar, B.; Zeynali, B. In vitro study of alginate–gelatin scaffolds incorporated with silica NPs as injectable, biodegradable hydrogels. *RSC Adv.* **2021**, *11*, 16688–16697. **[Google Scholar] [CrossRef]**

31. Pska, B.; Np, A.; Skaa, B. Fabrication and characterization of Chrysin—A plant polyphenol loaded alginate -chitosan composite for wound healing application. *Colloids Surf. B Biointerfaces* **2021**, *206*, 111922. **[Google Scholar]**

32. Zhang, M.; Wang, G.; Wang, D.; Zheng, Y.; Lee, S. Ag@MOF-loaded chitosan nanoparticle and polyvinyl alcohol/sodium alginate/chitosan bilayer dressing for wound healing applications. *Int. J. Biol. Macromol.* **2021**, *175*, 481–494. **[Google Scholar] [CrossRef] [PubMed]**

33. Yan, M.; Shi, J.; Tang, S.; Liu, L.; Zhu, H.; Zhou, G.; Zeng, J.; Zhang, H.; Yu, Y.; Guo, J. Strengthening and toughening sodium alginate fibers using a dynamically cross-linked network of inorganic nanoparticles and sodium alginate through the hydrogen bonding strategy. *New J. Chem.* **2021**, *45*, 10362–10372. **[Google Scholar] [CrossRef]**

34. Kim, K.; Choi, J.H.; Shin, M. Mechanical Stabilization of Alginate Hydrogel Fiber and 3D Constructs by Mussel-Inspired Catechol Modification. *Polymers* **2021**, *13*, 892. **[Google Scholar] [CrossRef]**

35. Zhang, Z.; Li, Z.; Li, Y.; Wang, Y.; Yao, M.; Zhang, K.; Chen, Z.; Yue, H.; Shi, J.; Guan, F. Sodium alginate/collagen hydrogel loaded with human umbilical cord mesenchymal stem cells promotes wound healing and skin remodeling. *Cell Tissue Res.* **2021**, *383*, 809–821. **[Google Scholar] [CrossRef]**

36. Ma, S.; Zhou, J.; Huang, T.; Zhang, Z.; Xing, Q.; Zhou, X.; Zhang, K.; Yao, M.; Cheng, T.; Wang, X. Sodium alginate/collagen/stromal cell-derived factor-1 neural scaffold loaded with BMSCs promotes neurological function recovery after traumatic brain injury. *Acta Biomater.* **2021**, *131*, 185–197. **[Google Scholar] [CrossRef]**

37. Lim, J.; Choi, G.; Joo, K.I.; Cha, H.J.; Kim, J. Embolization of Vascular Malformations via In Situ Photocrosslinking of Mechanically Reinforced Alginate Microfibers using an Optical-Fiber-Integrated Microfluidic Device. *Adv. Mater.* **2021**, *33*, 2006759. **[Google Scholar] [CrossRef]**

38. Liu, R.; Kong, B.; Chen, Y.; Liu, X.; Mi, S. Formation of helical alginate microfibers using different G/M ratios of sodium alginate based on microfluidics. *Sens. Actuators B Chem.* **2020**, *304*, 127069. **[Google Scholar] [CrossRef]**

39. Sa, V.; Kornev, K.G. A method for wet spinning of alginate fibers with a high concentration of single-walled carbon nanotubes. *Carbon* **2011**, *49*, 1859–1868. **[Google Scholar] [CrossRef]**

40. He, Y.; Zhang, N.; Gong, Q.; Qiu, H.; Wei, W.; Yu, L.; Gao, J. Alginate/graphene oxide fibers with enhanced mechanical strength prepared by wet spinning. *Carbohydr. Polym.* **2012**, *88*, 1100–1108. **[Google Scholar] [CrossRef]**

41. Zhang, X.; Weng, L.; Liu, Q.; Li, D.; Deng, B. Facile fabrication and characterization on alginate microfibres with grooved structure via microfluidic spinning. *R. Soc. Open Sci.* **2019**, *6*, 181928. **[Google Scholar] [CrossRef][Green Version]**

42. Ci, M.; Liu, J.; Shang, S.; Jiang, Z.; Sui, S. The Effect of HPMC and CNC on the Structure and Properties of Alginate Fibers. *Fibers Polym.* **2020**, *21*, 2179–2185. **[Google Scholar] [CrossRef]**

43. Wang, Q.; Zhang, L.; Liu, Y.; Zhang, G.; Zhu, P. Characterization and functional assessment of alginate fibers prepared by metal-calcium ion complex coagulation bath. *Carbohydr. Polym.* **2020**, *232*, 115693. **[Google Scholar] [CrossRef]**

44. Fu, X.; Liang, Y.; Wu, R.; Shen, J.; Chen, Z.; Chen, Y.; Wang, Y.; Xia, Y. Conductive core-sheath calcium alginate/graphene composite fibers with polymeric ionic liquids as an intermediate. *Carbohydr. Polym.* **2019**, *206*, 328–335. **[Google Scholar] [CrossRef]**

45. Ci, M.; Liu, J.; Liu, L.; Shang, S.; Sui, S. Preparation and Characterization of Hydroxypropyl Methylcellulose Modified Alginate Fiber. *J. Phys. Conf. Ser.* **2021**, *1790*, 012073. **[Google Scholar] [CrossRef]**

46. Zhao, T.; Li, X.; Gong, Y.; Guo, Y.; Shi, Q. Study on polysaccharide polyelectrolyte complex and fabrication of alginate/chitosan derivative composite fibers. *Int. J. Biol. Macromol.* **2021**, *184*, 181–187. **[Google Scholar] [CrossRef]**

47. Jun, Y.; Kang, E.; Chae, S.; Lee, S.H. Microfluidic spinning of micro- and nano-scale fibers for tissue engineering. *Lab A Chip* **2014**, *14*, 2145–2160. **[Google Scholar] [CrossRef]**

48. Xie, R.; Xu, P.; Liu, Y.; Li, L.; Luo, G.; Ding, M.; Liang, Q. Necklace-Like Microfibers with Variable Knots and Perfusable Channels Fabricated by an

Oil-Free Microfluidic Spinning Process. *Adv. Mater.* **2018**, *30*, e1705082. **[Google Scholar] [CrossRef]**

49. Hancock, M.J.; Piraino, F.; Camci-Unal, G.; Rasponi, M.; Khademhosseini, A. Anisotropic material synthesis by capillary flow in a fluid stripe. *Biomaterials* **2011**, *32*, 6493–6504. **[Google Scholar] [CrossRef][Green Version]**
50. Annabi, N.; Selimović, Š.; Cox JP, A.; Ribas, J.; Bakooshli, M.A.; Heintze, D.; Weiss, A.S.; Cropek, D.; Khademhosseini, A. Hydrogel-coated microfluidic channels for cardiomyocyte culture. *Lab A Chip* **2013**, *13*, 3569–3577. **[Google Scholar] [CrossRef][Green Version]**
51. Shi, X.; Ostrovidov, S.; Zhao, Y.; Liang, X.; Kasuya, M.; Kurihara, K.; Nakajima, K.; Bae, H.; Wu, H.; Khademhosseini, A. Microfluidic Spinning of Cell-Responsive Grooved Microfibers. *Adv. Funct. Mater.* **2015**, *25*, 2250–2259. **[Google Scholar] [CrossRef]**
52. Ahn, S.Y.; Mun, C.H.; Lee, S.H. Microfluidic spinning of fibrous alginate carrier having highly enhanced drug loading capability and delayed release profile. *RSC Adv.* **2015**, *5*, 15172–15181. **[Google Scholar] [CrossRef]**
53. Chung, B.G.; Lee, K.H.; Khademhosseini, A.; Lee, S.H. Microfluidic fabrication of microengineered hydrogels and their application in tissue engineering. *Lab A Chip* **2011**, *12*, 45–59. **[Google Scholar] [CrossRef]**
54. Kang, E.; Choi, Y.Y.; Choi, Y.J.; Sang, H.L. Microfludics spinning of flat fiber with micro grooves for cell-aligning scaffolds. In Proceedings of the 16th International Conference on Miniaturized Systems for Chemistry and Life Sciences, MicroTAS, Okinawa, Japan, 28 October–1 November 2012. **[Google Scholar]**
55. Kuo, C.K.; Ma, P.X. Ionically crosslinked alginate hydrogels as scaolds for tissue engineering: Part 1. structure, gelation rate and mechanical properties. *Biomaterials* **2001**, *22*, 511–521. **[Google Scholar] [CrossRef]**
56. Mørch, Ý.A.; Donati, I.; Strand, B.L.; Skjåk-Bræk, G. Effect of Ca^{2+}, Ba^{2+}, and Sr^{2+} on Alginate Microbeads. *Biomacromolecules* **2006**, *7*, 1471–1480. **[Google Scholar] [CrossRef]**
57. Sharabi, M.; Benayahu, D.; Benayahu, Y.; Isaacs, J.; Haj-Ali, R. Laminated collagen-fiber bio-composites for soft-tissue bio-mimetics. *Compos. Sci. Technol.* **2015**, *117*, 268–276. **[Google Scholar] [CrossRef]**
58. Elkady, M.; Salama, E.; Amer, W.A.; Ebeid, E.; Ayad, M.M.; Shokry, H. Novel eco-friendly electrospun nanomagnetic zinc oxide hybridized PVA/alginate/chitosan nanofibers for enhanced phenol decontamination. *Environ. Sci. Pollut. Res.* **2020**, *27*, 43077–43092. **[Google Scholar] [CrossRef]**

59. Zhang, X.; Huang, C.; Zhao, Y.; Jin, X. Ampicillin-incorporated alginate-chitosan fibers from microfluidic spinning and for vitro release. *J. Biomater. Sci.* **2017**, *28*, 1408–1425. **[Google Scholar]** **[CrossRef]**

60. Kim, H.J.; Jeong, J.H.; Choi, Y.H.; Eom, Y. Review on cellulose nanocrystal-reinforced polymer nanocomposites: Processing, properties, and rheology. *Korea Aust. Rheol. J.* **2021**, *33*, 165–185. **[Google Scholar]** **[CrossRef]**

61. Shu, H.J.; Chun-Xuan, W.U.; Yang, K.; Liu, T.W.; Chen, L.I.; Cao, C.L. Preparation of rapid expansion alginate/silica fiber composite scaffold and application of rapid hemostatic function. *J. Mater. Eng.* **2019**, *47*, 124–129. **[Google Scholar]**

62. Zhang, X.; Huang, C.; Zhao, Y.; Jin, X. Preparation and characterization of nanoparticle reinforced alginate fibers with high porosity for potential wound dressing application. *RSC Adv.* **2017**, *7*, 39349–39358. **[Google Scholar]** **[CrossRef][Green Version]**

63. Adhikari, J.; Perwez, M.S.; Das, A.; Saha, P. Development of hydroxyapatite reinforced alginate-chitosan based printable biomaterial-ink. *Nano Struct. Nano Objects* **2021**, *25*, 100630. **[Google Scholar]** **[CrossRef]**

64. D'Elía, N.; Silva, R.R.; Sartuqui, J.; Ercoli, D.; Mestres, G. Development and characterisation of bilayered periosteum-inspired composite membranes based on sodium alginate-hydroxyapatite nanoparticles. *J. Colloid Interface Sci.* **2020**, *572*, 408–420. **[Google Scholar]** **[CrossRef]** **[PubMed]**

65. Zhou, Q.; Li, H.; Li, D.; Wang, B.; Wang, G. A graphene assembled porous fiber-based Janus membrane for highly effective solar steam generation. *J. Colloid Interface Sci.* **2021**, *592*, 77–86. **[Google Scholar]** **[CrossRef]** **[PubMed]**

66. Sun, J.Y.; Zhao, X.; Illeperuma, W.; Chaudhuri, O.; Oh, K.H.; Mooney, D.J.; Vlassak, J.J.; Suo, Z. Highly stretchable and tough hydrogels. *Nature* **2012**, *489*, 133–136. **[Google Scholar]** **[CrossRef]**

67. Yang, J.S.; Xie, Y.J.; He, W. Research progress on chemical modification of alginate: A review. *Carbohydr. Polym.* **2011**, *84*, 33–39. **[Google Scholar]** **[CrossRef]**

68. Lee, K.Y.; Mooney, D.J. Alginate: Properties and biomedical applications. *Prog. Polym. Sci.* **2012**, *37*, 106–126. **[Google Scholar]** **[CrossRef][Green Version]**

69. Krishnan, R.; Ko, D.; Rd, F.C.; Liu, W.; Smink, A.M.; De, H.B.; De, V.P.; Lakey, J.R. Immunological Challenges Facing Translation of Alginate Encapsulated Porcine Islet Xenotransplantation to Human Clinical Trials. *Methods Mol. Biol.* **2017**, *1479*, 305–333. **[Google Scholar]**

70. Zhe, S.; Xia, Y.; Quan, F.; Li, H.; Zhu, L. Adsorption properties of alginate fiber for basic fuchsin. *China Synth. Fiber Ind.* **2012**, *4*, 4–7. **[Google Scholar]**

71. Kong, Q.S.; Jiang, L.P.; Gao, J.X.; Quan, J.I.; Xia, Y.Z.; Jian, Y.U. Study on the Adsorption Property of Alginate Fiber to Zn~(2+). *Synth. Fiber China* **2008**, *40*, 1130–1136. **[Google Scholar]**

72. Zhang, X.; Yang, J.; Weng, L.; Fan, W.; Xu, Y. Enhanced mechanical performance of cellulose nanocrystal doped eco-friendly calcium-alginate based bio-composite fiber with superior flame retardancy. *Text. Res. J.* **2022**, *92*, 1820–1829. **[Google Scholar] [CrossRef]**

73. Zhang, J.; Quan, J.; Wang, F.; Tan, L.; Xia, Y. Effects of divalent metal ions on the flame retardancy and pyrolysis products of alginate fibres. *Polym. Degrad. Stab.* **2012**, *97*, 1034–1040. **[Google Scholar] [CrossRef]**

74. Liu, Y.; Tao, Y.; Wang, B.; Li, P.; Xu, Y.; Jiang, Z.; Dong, C.; Zhu, P. Fully bio-based fire-safety viscose/alginate blended nonwoven fabrics: Thermal degradation behavior, flammability, and smoke suppression. *Cellulose* **2020**, *27*, 6037–6053. **[Google Scholar] [CrossRef]**

75. Suzuki, Y.; Tanihara, M.; Nishimura, Y.; Suzuki, K.; Yamawaki, Y.; Kudo, H.; Kakimaru, Y.; Shimizu, Y. In vivo evaluation of a novel alginate dressing. *J. Biomed. Mater. Res.* **1999**, *48*, 522–527. **[Google Scholar] [CrossRef]**

76. Mirean, V. Current Trends in Advanced Alginate-Based Wound Dressings for Chronic Wounds. *J. Pers. Med.* **2021**, *11*, 890. **[Google Scholar]**

77. Ahmad, F.; Mushtaq, B.; Butt, F.A.; Rasheed, A.; Ahmad, S. Preparation and characterization of wool fiber reinforced nonwoven alginate hydrogel for wound dressing. *Cellulose* **2021**, *28*, 7941–7951. **[Google Scholar] [CrossRef]**

78. Hopper, G.P.; Deakin, A.H.; Crane, E.O.; Clarke, J.V. Enhancing patient recovery following lower limb arthroplasty with a modern wound dressing: A prospective, comparative audit. *J. Wound Care* **2013**, *21*, 200. **[Google Scholar] [CrossRef]**

79. Erwin, E.; Etriwati, E.; Zamzami, R.S.; Hosea, C. Moist wound dressing and its application in distant skin flap in cats. *Vet. World* **2021**, *14*, 734. **[Google Scholar] [CrossRef]**

80. Zhong, J.; Wang, H.; Yang, K.; Wang, H.; Duan, C.; Ni, N.; An, L.; Luo, Y.; Zhao, P.; Gou, Y.; et al. Reversibly immortalized keratinocytes (iKera) facilitate re-epithelization and skin wound healing: Potential applications in cell-based skin tissue engineering. *Bioact. Mater.* **2022**, *9*, 523–540. **[Google Scholar] [CrossRef]**

81. Asadi, L.; Mokhtari, J.; Abbasi, M. An alginate–PHMB–AgNPs based wound dressing polyamide nanocomposite with improved antibacterial and hemostatic properties. *J. Mater. Sci. Mater. Med.* **2021**, *32*, 7. **[Google Scholar] [CrossRef]**

82. Das, D.; Zhang, S.; No, I. Synthesis and characterizations of alginate-□-tricalcium phosphate microparticle hybrid film with flexibility and high mechanical property as a biomaterial. *Biomed. Mater.* **2018**, *13*, 025008. **[Google Scholar] [CrossRef]**

83. Liang, L.; Hou, T.; Ouyang, Q.; Xie, L.; Li, C. Antimicrobial sodium alginate dressing immobilized with polydopamine-silver composite nanospheres. *Compos. Part B Eng.* **2020**, *188*, 107877. **[Google Scholar] [CrossRef]**

84. Filho, R.; Vasconcelos, N.F.; Andrade, F.K.; Rosa, M.; Vieira, R.S. Papain immobilized on alginate membrane for wound dressing application. *Colloids Surf. B Biointerfaces* **2020**, *194*, 111222. **[Google Scholar] [CrossRef] [PubMed]**

85. Cesar, P.; Natarelli, C.; Oliveira, J.; Andrade, P.A.; Marcussi, S. Development and characterization of a poly (vinyl alcohol) and sodium alginate blend foam for wound dressing loaded with propolis and alltrans retinoic acid. *J. Appl. Polym. Sci.* **2021**, *138*, 50480. **[Google Scholar] [CrossRef]**

86. Wang, M.; Yang, Y.; Yuan, K.; Yang, S.; Tang, T. Dual-functional hybrid quaternized chitosan/Mg/alginate dressing with antibacterial and angiogenic potential for diabetic wound healing. *J. Orthop. Transl.* **2021**, *30*, 6–15. **[Google Scholar] [CrossRef]**

87. Ashton, R.S.; Banerjee, A.; Punyani, S.; Schaffer, D.V.; Kane, R.S. Scaffolds based on degradable alginate hydrogels and poly(lactide-co-glycolide) microspheres for stem cell culture. *Biomaterials* **2007**, *28*, 5518–5525. **[Google Scholar] [CrossRef]**

88. Roopavath, U.; Soni, R.; Mahanta, U.; Deshpande, A.S.; Rath, S.N. 3D printable SiO_2 nanoparticle ink for patient specific bone regeneration. *RSC Adv.* **2019**, *9*, 23832–23842. **[Google Scholar] [CrossRef][Green Version]**

89. Cao, H.; Liu, T.; Chew, S.Y. The application of nanofibrous scaffolds in neural tissue engineering. *Adv. Drug Deliv. Rev.* **2009**, *61*, 1055–1064. **[Google Scholar] [CrossRef]**

90. Yeo, M.; Kim, G. Nano/microscale topographically designed alginate/PCL scaffolds for inducing myoblast alignment and myogenic differentiation. *Carbohydr. Polym.* **2019**, *223*, 115041. **[Google Scholar] [CrossRef]**

91. Tachizawa, S.; Takahashi, H.; Kim, Y.J.; Odawara, A.; Pauty, J.; Ikeuchi, Y.; Suzuki, I.; Kikuchi, A.; Matsunaga, Y.T. Bundle Gel Fibers with Tunable Microenvironment for In Vitro Neuron Cell Guiding. *ACS Appl. Mater. Interfaces* **2017**, *9*, 43250–43257. **[Google Scholar] [CrossRef]**

92. Yeo, M.G.; Kim, G.H. Fabrication of cell-laden electrospun hybrid scaffolds of alginate-based bioink and PCL microstructures for tissue regeneration. *Chem. Eng. J.* **2015**, *275*, 27–35. **[Google Scholar] [CrossRef]**

93. Hu, W.W.; Yu, H.N. Coelectrospinning of chitosan/alginate fibers by dual-jet system for modulating material surfaces. *Carbohydr. Polym.* **2013**, *95*, 716–727. **[Google Scholar] [CrossRef]**

94. Chae, T.; Yang, H.; Moon, H.; Troczynski, T.; Ko, F.K. Biomimetically Mineralized Alginate Nanocomposite Fibers for Bone Tissue Engineering: Mechanical Properties and in Vitro Cellular Interactions. *ACS Appl. Bio Mater.* **2020**, *3*, 6746–6755. **[Google Scholar] [CrossRef] [PubMed]**

95. Hu, W.W.; Ting, J.C. Gene immobilization on alginate/polycaprolactone fibers through electrophoretic deposition to promote in situ transfection efficiency and biocompatibility. *Int. J. Biol. Macromol.* **2019**, *121*, 1337–1345. **[Google Scholar] [CrossRef] [PubMed]**

96. Mohammadi, S.; Ramakrishna, S.; Laurent, S.; Shokrgozar, M.A.; Semnani, D.; Sadeghi, D.; Bonakdar, S.; Akbari, M. Fabrication of Nanofibrous PVA/Alginate-Sulfate Substrates for Growth Factor Delivery. *J. Biomed. Mater. Res. Part A* **2019**, *107*, 403–413. **[Google Scholar] [CrossRef] [PubMed]**

97. Ghosh, D.; Godeshala, S.; Nitiyanandan, R.; Islam, M.S.; Rege, K. Correction to Copper-Eluting Fibers for Enhanced Tissue Sealing and Repair. *ACS Appl. Mater. Interfaces* **2020**, *12*, 53568–53569. **[Google Scholar] [CrossRef]**

98. Chae, T.; Yang, H.; Leung, V.; Ko, F.; Troczynski, T. Novel biomimetic hydroxyapatite/alginate nanocomposite fibrous scaffolds for bone tissue regeneration. *J. Mater. Sci. Mater. Med.* **2013**, *24*, 1885–1894. **[Google Scholar] [CrossRef]**

99. Chen, H.; Gao, Y.; Ren, X.; Gao, G. Alginate Fiber Toughened Gels Similar to Skin Intelligence as Ionic Sensors. *Carbohydr. Polym.* **2020**, *235*, 116018. **[Google Scholar] [CrossRef]**

100. Kim, D.; Ahn, S.; Yoon, J. Highly Stretchable Strain Sensors Comprising Double Network Hydrogels Fabricated by Microfluidic Devices. *Adv. Mater. Technol.* **2019**, *4*, 1800739. **[Google Scholar] [CrossRef]**

101. Gong, J.P.; Katsuyama, Y.; Kurokawa, T.; Osada, Y. Double-Network Hydrogels with Extremely High Mechanical Strength. *Adv. Mater.* **2003**, *15*, 1155–1158. **[Google Scholar] [CrossRef]**

102. Huang, T.; Xu, H.G.; Jiao, K.X.; Zhu, L.P.; Wang, H.L. A Novel Hydrogel with High Mechanical Strength: A Macromolecular Microsphere Composite Hydrogel. *Adv. Mater.* **2010**, *19*, 1622–1626. **[Google Scholar] [CrossRef]**

103. Nakajima, S.; Kawano, R.; Onoe, H. Stimuli-responsive hydrogel microfibers with controlled anisotropic shrinkage and cross-sectional geometries. *Soft Matter.* **2017**, *13*, 3710–3719. **[Google Scholar] [CrossRef]**

104. Zou, X.; Shang, S.; Liu, J.; Ci, M.; Zhu, P. Facile Fabrication of Temperature Triggered Thermochromic Core-sheath Alginate Microfibers from Microfluidic Spinning. *Fibers Polym.* **2021**, *22*, 1535–1542. **[Google Scholar] [CrossRef]**

105. Cui, L.; Hu, J.J.; Wang, W.; Yan, C.; Tu, C. Smart pH response flexible sensor based on calcium alginate fibers incorporated with natural dye for wound healing monitoring. *Cellulose* **2020**, *27*, 6367–6381. **[Google Scholar] [CrossRef]**

106. Liu, Y.; Wang, J.S.; Zhu, P.; Zhao, J.C.; Zhang, C.J.; Guo, Y.; Cui, L. Thermal degradation properties of biobased iron alginate film. *J. Anal. Appl. Pyrolysis* **2016**, *119*, 87–96. **[Google Scholar] [CrossRef][Green Version]**

107. Zhang, C.J.; Zhao, J.C.; Guo, Y.; Zhu, P.; Wang, D.Y. Bio-based barium alginate film: Preparation, flame retardancy and thermal degradation behavior. *Carbohydr. Polym.* **2016**, *139*, 106–114. **[Google Scholar] [CrossRef] [PubMed]**

108. Hou, J.; Li, C.; Guan, Y.; Zhang, Y.; Zhu, X.X. Enzymatically crosslinked alginate hydrogels with improved adhesion properties. *Polym. Chem.* **2015**, *6*, 2204–2213. **[Google Scholar] [CrossRef]**

109. Ma, X.; Li, R.; Zhao, X.; Ji, Q.; Xing, Y.; Sunarso, J.; Xia, Y. Biopolymer composite fibres composed of calcium alginate reinforced with nanocrystalline cellulose. *Compos. Part A Appl. Sci. Manuf.* **2017**, *96*, 155–163. **[Google Scholar] [CrossRef]**

110. Tian, G.; Quan, J.; Xu, D.; Tan, L.; Xia, Y. The effect of zinc ion content on flame retardance and thermal degradation of alginate fibers. *Fibers Polym.* **2013**, *14*, 767–771. **[Google Scholar] [CrossRef]**

111. Zhang, X.; Xia, Y.; Yan, X.; Shi, M. Efficient suppression of flammability in flame retardant viscose fiber through incorporating with alginate fiber. *Mater. Lett.* **2018**, *215*, 106–109. **[Google Scholar] [CrossRef]**

112. Zhang, C.J.; Liu, Y.; Cui, L.; Yan, C.; Zhu, P. Bio-based calcium alginate nonwoven fabrics: Flame retardant and thermal degradation properties. *J. Anal. Appl. Pyrolysis* **2016**, *122*, 13–23. **[Google Scholar] [CrossRef]**

113. Chen, H.B.; Shen, P.; Chen, M.; Zhao, H.B.; Schiraldi, D.A. Highly Efficient Flame Retardant Polyurethane Foam with Alginate/Clay Aerogel Coating. *ACS Appl. Mater. Interfaces* **2016**, *8*, 32557–32564. **[Google Scholar] [CrossRef] [PubMed]**

114. Wang, Y.; Li, Z.; Li, Y.; Wang, J.; Liu, X.; Song, T.; Yang, X.; Hao, J. Spray drying assisted Layer-by-Layer assembly of alginate, 3-aminopropyltriethoxysilane, and magnesium hydroxide flame retardant and its catalytic graphitization in ethylene-vinyl acetate resin. *ACS Appl. Mater. Interfaces* **2018**, *10*, 10490–10500. **[Google Scholar] [CrossRef]**

115. Shang, K.; Liao, W.; Wang, J.; Wang, Y.T.; Wang, Y.Z.; Schiraldi, D.A. Nonflammable Alginate Nanocomposite Aerogels Prepared by a Simple Freeze-

Drying and Post-Cross-Linking Method. *ACS Appl. Mater. Interfaces* **2016**, *8*, 643–650. **[Google Scholar] [CrossRef]**

116. Liu, Y.; Zhao, X.-R.; Peng, Y.-L.; Wang, D.; Yang, L.; Peng, H.; Zhu, P.; Wang, D.-Y. Effect of reactive time on flame retardancy and thermal degradation behavior of bio-based zinc alginate film. *Polym. Degrad. Stab.* **2016**, *127*, 20–31. **[Google Scholar] [CrossRef]**
117. Li, X.-L.; Chen, M.-J.; Chen, H.-B. Facile fabrication of mechanically-strong and flame retardant alginate/clay aerogels. *Compos. Part B Eng.* **2019**, *164*, 18–25. **[Google Scholar] [CrossRef]**
118. Li, P.; Wang, Q.Z.; Wang, B.; Liu, Y.Y.; Zhu, P. Blending alginate fibers with polyester fibers for flame-retardant filling materials: Thermal decomposition behaviors and fire performance. *Polym. Degrad. Stab.* **2021**, *183*, 109470. **[Google Scholar] [CrossRef]**

CHAPTER-17

DEVELOPING AN ULTRA-SENSITIVE AND FLEXIBLE STRAIN SENSOR UTILIZING DOPAMINE-MODIFIED ELECTROSPUN STYRENE-ETHYLENE-BUTYLENE-STYRENE BLOCK COPOLYMER YARNS AND MULTI-WALLED CARBON NANOTUBES

1. INTRODUCTION: AN OVERVIEW

The contemporary landscape of sensor technology is continually advancing to meet the demands of diverse applications, from healthcare to structural monitoring. In this context, the development of highly sensitive and flexible strain sensors has garnered significant attention. This study introduces a novel approach, focusing on the utilization of dopamine-modified electrospun Styrene-Ethylene-Butylene-Styrene (SEBS) block copolymer yarns combined with multi-walled carbon nanotubes (MWCNTs). The synergy between the unique properties of dopamine-modified SEBS and the conductive nature of MWCNTs presents an innovative solution for achieving both high sensitivity and flexibility in strain sensing applications.

Strain sensors have the ability to convert physical deformations into measurable electrical signals. As wearable electronic devices have become commonplace in daily life, great advances in strain sensors occurred in various fields including healthcare, robotics, virtual reality and other sectors [**1**]. However, conventional strain sensors fabricated by using metals and semiconductor materials, despite their toughness, have low sensitivity and work over small strain ranges, so do not meet the requirements of high strain and high flexibility [**2**]. To resolve these deficiencies, an elastic polymer has been used as a substrate to provide higher mechanical deformation for strain sensors. Recently, thermoplastic polyurethane (TPU) [**3,4**], styrene-butadiene-styrene copolymer (SBS) [**5,6**], polydimethylsiloxane (PDMS) [**7,8**], Ecoflex [**9**], styrene-ethylene-butene-styrene copolymer (SEBS) [**10,11**] amongst others have been extensively employed in the fabrication of flexible strain sensors. Recently, resistance-based strain sensors have gradually attracted more attention because of their simple manufacturing process, low energy consumption in operation and relatively simple reading system [**12,13**]. At the same time, conductive nanomaterials are an important constituent in the preparation of resistance-based strain sensors. Carbon-based nanomaterials including carbon nanotubes (CNTs) [**14,15**], carbon blacks (CBs), graphene [**2,7,16**], and graphene oxide (GO) [**17**] are good candidates, because of their excellent electrical and robust mechanical properties [**18,19**].

At present, various methods have been applied to prepare strain sensors, but most sensors cannot simultaneously possess both high workable strains and high sensitivity [**20**]. Chen et al. [**16**] chose CBs as the conductive material and rubber as the substrate material to fabricate conductive composites, and then modified the composite material with PDMS to fabricate a flexible strain sensor. The gauge factor (GF, which is used for the characterization of sensitivity) reached 242.6 under a 71.4% strain. Wang et al. [**21**] reported a stretchable strain sensor fabricated by coating single-wall carbon nanotubes (SWCNTs) into an elastic cotton/polyurethane (PU) core-spun yarn. It had a maximum workable strain up to 200%, but the GF was only 0.06. Lee et al. [**22**] fabricated a strain sensor by incorporating a cracked transparent epitaxial layer of indium tin oxide (ITO) on a transparent PET substrate, which had a maximum GF of 4000 while its sensing strain was only 2%. Consequently, the fabrication of a strain sensor with both a high strain capability and a high GF presented a challenge, which is described in this study.

SEBS is a kind of thermoplastic elastomer with high elongations and elastic recovery properties [**23,24,25,26**]. However, due to its low surface energy and poor compatibility with inorganic materials and weak adhesion, SEBS requires modification before use. Dopamine (DA) has strong adhesion because it can form a polydopamine (PDA) layer via self-polymerization. PDA is similar to mucin, which is the most abundant macromolecule in mucus and secreted by mussels, exhibiting strong adhesion [**27**].

In this work, carboxylic multi-walled carbon nanotubes (MWCNTs), as the conductive fillers, were coated onto dopamine-modified SEBS yarn. Thus, a simple and practical method for fabricating PDA@SEBS/MWCNTs yarn-based strain sensor is reported. Firstly, SEBS yarn was prepared by electrospinning, rolling and twisting. Secondly, the surface of SEBS yarn was modified by dopamine. Thirdly, MWCNTs were coated onto dopamine-modified SEBS yarn by brushing and ultrasonication. Fourthly, the chemical composition, mechanical properties, sensing performance, durability and washing fastness were studied. Finally, the sensing performance of SEBS@PDA/MWCNTs yarn was demonstrated by monitoring various human joint motions (such as fingers, elbow, knuckle and wrist bending).

2. EXPERIMENTAL PROCEDURES

2.1. Preparation of Yarn-Based Strain Sensors

The procedure for preparing the yarn-based strain sensor shown in Figure 1 includes three steps: (1) electrospinning the SEBS yarn; (2) DA modification; (3) MWCNTs coating.

2.1.1. Preparation of Electrospun SEBS Yarn

SEBS (weight ratio of EP/PS = 67/33, Kraton, Houston, TX, USA) and Tetrahydrofuran (THF, Sinopharm, Beijing, China) were mixed with a SEBS concentration of 25 wt.% for 12 h at room temperature. The spinning solution was loaded into a 20 mL syringe with a 22G needle and fixed on the syringe pump. The whole electrospinning process was conducted at room temperature, while the injection speed was 3 mL/h, the voltage was 10 kV, and the receiving distance was 15 cm (Figure 1a). After 5 min, the SEBS fibrous membrane was dried at 70 °C for 1 h. Then the electrospun SEBS fibrous membrane was rolled and twisted (Figure 1b), forming the fabricated electrospun SEBS yarn.

2.1.2. Dopamine Modification of the Electrospun SEBS Yarn

DA solution with a concentration of 10 mmol/L was prepared by dissolving dopamine hydrochloride (DA·HCl, Shanghai Macklin, Shanghai, China) powder into a tris(hydroxymethyl)methyl aminomethane (Tris, Beijing Solarbio, Beijing, China)-hydrochloric acid (HCl, Sinopharm, Beijing, China) solution (pH = 8.5). Then sodium periodate (SP, Shanghai Macklin, Shanghai, China) was added into the DA solution with a DA to SP mole ratio of 3:2. Finally, SEBS yarn was immersed in the DA solution for 6 h (**Figure 1**c), then washed with distilled water and dried in the oven at 70 °C. The mechanism of dopamine polymerization can be seen in **Figure 1**e. The DA-modified SEBS yarn obtained was denoted as PDA@SEBS.

2.1.3. Preparation of MWCNTs Coated Composite Yarn

MWCNTs (Shenzhen Tuling, Shenzhen, China) were added to absolute alcohol (EtOH, Sinopharm, Beijing, China) to prepare MWCNTs/EtOH suspensions with different concentrations. As shown in Figure 1d, the MWCNTs/EtOH suspension was coated on the yarn using a brush. Subsequently, the MWCNTs coated yarn was soaked into suspension for ultrasonic treatment at 240 W and 50 Hz for a range of times before drying at 70 °C. The composite yarn was denoted as PDA@SEBS/MWCNTs.

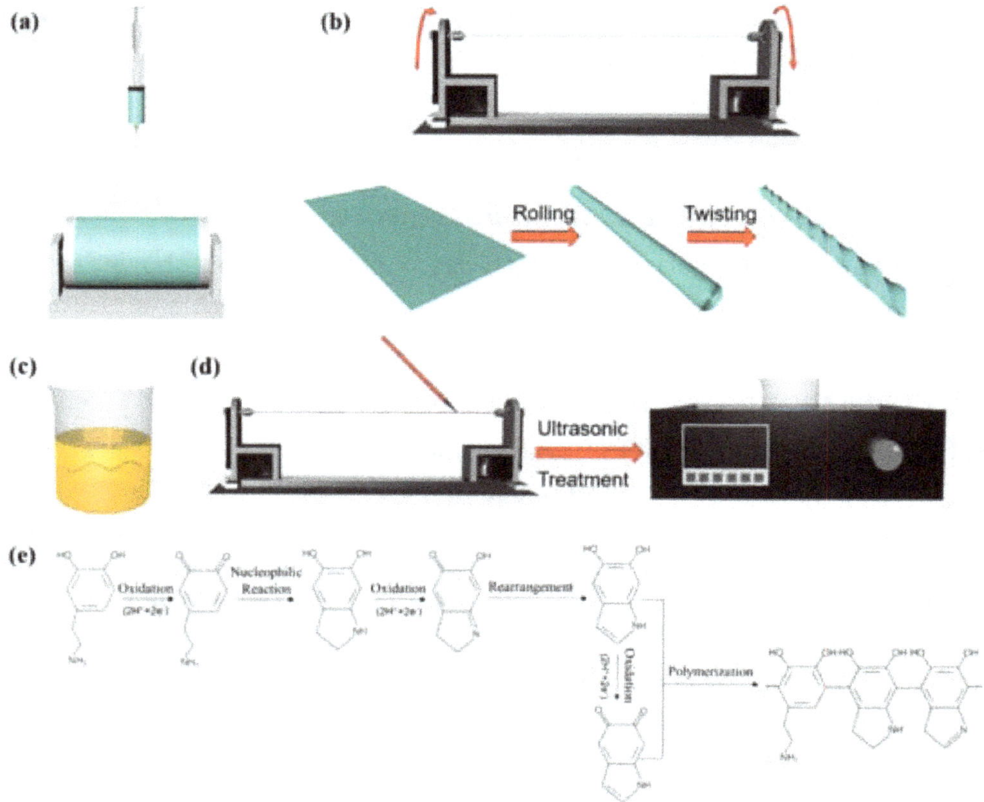

Figure 1. Illustrations of fabrication of the yarn-based strain sensors: (a) Electrospinning the SEBS fibrous membrane. (b) The electrospun SEBS yarn twisting process. (c) The DA modification of the SEBS yarn. (d) MWCNTs coated on PDA@SEBS yarn. (e) The mechanism of dopamine polymerization.

2.2. Characterization

The morphology of the yarns was observed by using scanning electron microscopy (SEM, TESCAN VEGA3, TESCAN, Brno, Czech Republic) with different magnifications and at an accelerated voltage of 10 kV and an electron beam intensity of 10 A/cm^2.

An energy-dispersive X-ray spectroscope (EDX, EDAX), equipped with SEM, and X-ray photoelectron spectroscopy (XPS, Axis Supra$^+$, Kratos, Kawasaki, Japan) were used to analyze the elemental composition of the yarn surface.

Thermogravimetric analysis was conducted by a DSC/TG synchronous thermal analyzer (STA449 F3 Jupiter, NETZSCH, Bavaria, Germany) in a nitrogen atmosphere.

Mechanical properties were measured by a universal tensile testing machine (Instron 5965, Illinois Tool Works Inc., Glenview, IL, USA) at a feed rate of 100 mm/min, while the distance between gauge points was 20 mm.

The electrical properties of the composite yarn were characterized by using a digital multimeter (KEYSIGHTB2901A, Keysight Technology, Santa Rosa, CA, USA).

3. RESULTS AND DISCUSSION

Figure 2 shows SEM images of the SEBS yarn and the PDA@SEBS yarn and an EDS spectrum of the PDA@SEBS yarn. As can be seen from Figure 2a,b, the surfaces of the SEBS and PDA@SEBS yarns were smooth. However, the EDS spectrum and mapping images in Figure 2c–g show there were N and O elements on the surface of the PDA@SEBS yarn, which indicated that SEBS yarn was modified by DA successfully.

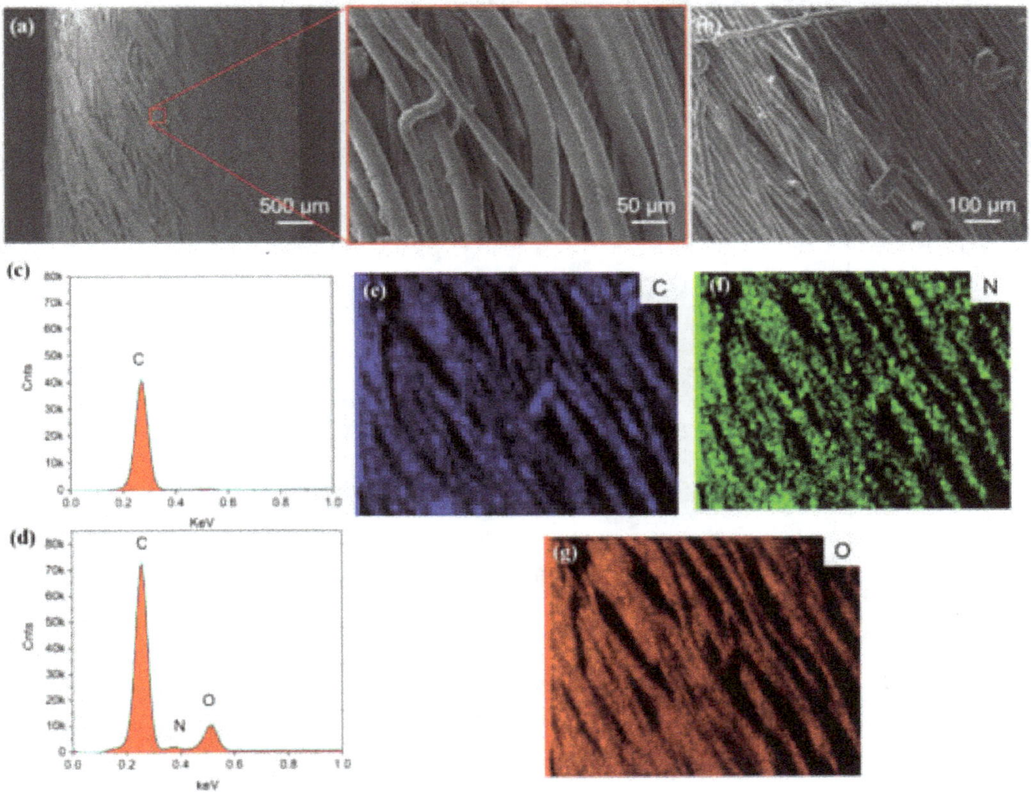

Figure 2. SEM images of (a) pure SEBS yarn and (b) PDA@SEBS yarn; EDX spectroscopy of (c) SEBS and (d) PDA@SEBS yarn; EDX mapping of SEBS@PDA yarn for (e) C, (f) N, and (g) O.

TGA was applied to quantify the content of MWCNTs coated on the surface of the untwisted PDA@SEBS yarn (PDA@u-SEBS) and twisted PDA@SEBS yarn (PDA@t-SEBS). As shown in **Figure 3**, the weight percentage of the materials decreased with the increase in temperature. When the temperature was up to 800 °C, the weight percentages of PDA@SEBS, PDA@u-SEBS/MWCNTs and PDA@t-SEBS/MWCNTs were 3.89%, 11.2% and 13.94%, respectively. The content of MWCNTs can be calculated by using Equation (1):

$$w_{MWCNTs} = \frac{w_a - w_b}{1 - w_b} \times 100\% \qquad (1)$$

where Wa is the residual mass of PDA@SEBS/MWCNTs yarn and Wb is the residual mass of PDA@SEBS yarn. By calculation, the contents of MWCNTs coated onto the PDA@u-SEBS and PDA@t-SEBS were 7.61% and 10.46%, respectively. This was because after twisting, the material surface was rougher, which caused a higher interfacial bond strength [28]. Thus, the twisted SEBS yarn was chosen as the matrix material of the strain sensor.

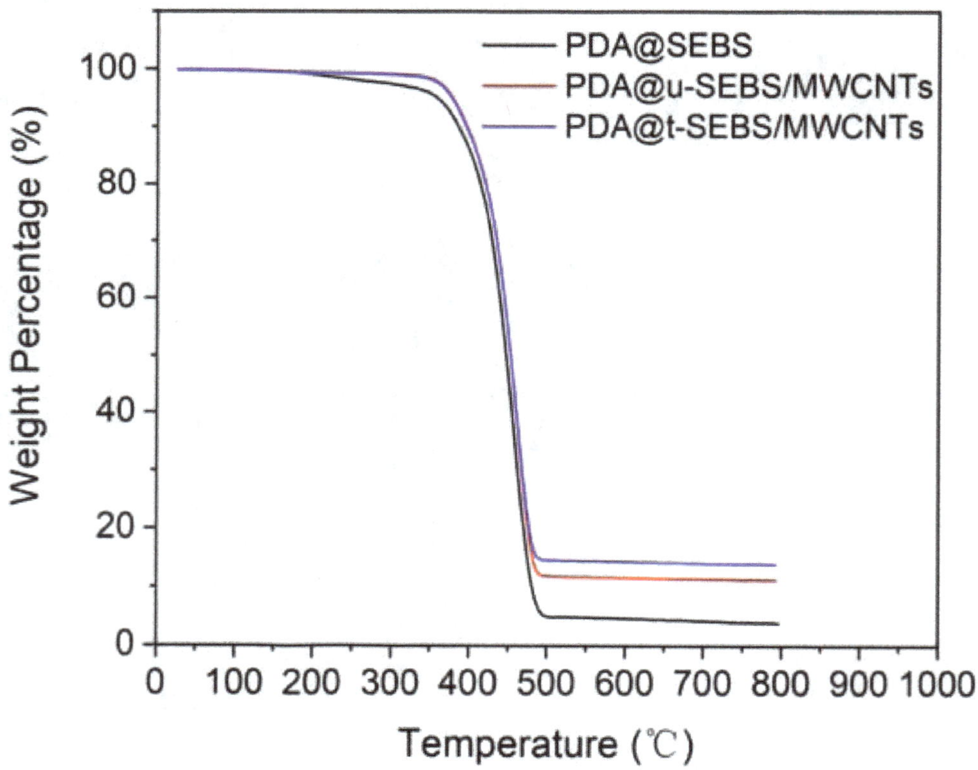

Figure 3. TGA curves of PDA@SEBS, PDA@u-SEBS/MWCNTs and PDA@t-SEBS/MWCNTs.

To fabricate the composite yarn so that it possessed good conductivity, the preparation process of PDA@SEBS/MWCNTs yarn was explored by a single factor experiment.

The conductivity of composited yarns coated with MWCNTs/EtOH suspension of different concentrations and 0.5-hour ultrasonic treatment can be seen in **Figure 4**a. With the increase in the concentration of suspension, the content of MWCNTs coated on the yarn increased (**Figure 4**b–h), which led to the conductivity of the composite yarn increasing. When the concentration of the MWCNTs/EtOH suspension was 12 g/L, the conductivity was 0.036 S/cm, while the conductivity was 0.037 S/cm when the concentration reached 14 g/L. Interestingly, between a concentration of 12 g/L and 14 g/L, there was almost no change in the conductivity of the composite yarns. This can be explained by employing the percolation theory **[29,30]**. The conductivity of the material is related to the critical concentration of conductive filler in polymer composites. The critical concentration of the conductive filler is called the percolation threshold. In the percolation threshold region, the continuous conductive network is formed only through the arrangement of fillers in conducive substrate. Above the percolation threshold, the conductivity enhanced slightly and stabilized thereafter. Accordingly, an MWCNTs/EtOH suspension concentration of 12 g/L was selected.

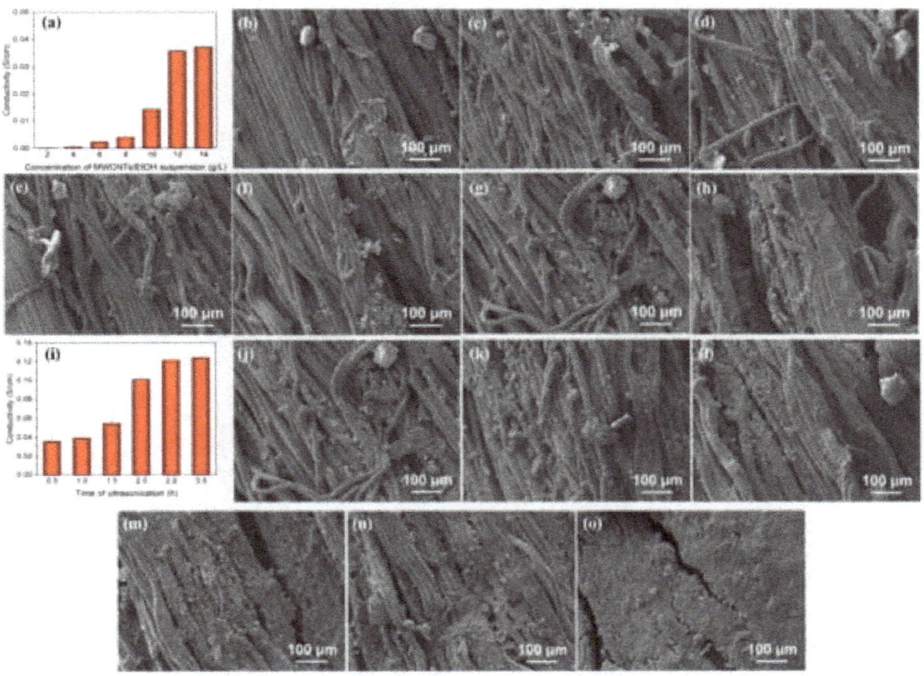

Figure 4. (a) Conductivity of PDA@SEBS/MWCNTs yarns; (b–h) SEM images of PDA@SEBS/MWCNTs yarn coated with MWCNTs/EtOH suspension concentrations of (b) 2 g/L, (c) 4 g/L, (d) 6 g/L, (e) 8 g/L, (f) 10 g/L, (g) 12 g/L and (h) 14 g/L; (i) Conductivity of PDA@SEBS/MWCNTs yarn; (j–o) SEM images of PDA@SEBS/MWCNTs sonicated for (j) 0.5 h, (k) 1 h, (l) 1.5 h, (m) 2 h, (n) 2.5 h, and (o) 3 h.

Figure 4i shows the conductivity of a composite yarn coated with MWCNT/EtOH suspension with a concentration of 10 g/L and ultrasonic treatment at varied times. With the increase in ultrasonic treatment time, more MWCNTs were deposited (**Figure 4**j–o), which was due to the cavitation effect caused by ultrasound, resulting in cavitation bubbles in the dispersion. When bubbles collapsed, powerful energy was provided to conductive particles to impact the matrix materials causing a firm adhesion to the matrix materials. The longer the time, the more conductive particles adhere [**31**]. However, when the ultrasonic treatment time exceeded 2.5 h, the conductivity of the composite yarn increased only slightly. Based on this consideration, the composite yarn coated with 12 g/L MWNCTs/EtOH suspension and sonicated for 2.5 h, was chosen for the strain sensor.

Figure 5 shows the XPS spectra of SEBS, PDA@SEBS and PDA@SEBS/MWCNTs yarn. It can be seen from Figure 5a that there were spectral peaks at 284.8 eV and 532 eV, representing the binding energy of C1s and O1s, respectively, while O1s peak on the curve of pure SEBS was caused by air pollution. Compared with pure SEBS, a new peak at 399 eV for N1s appeared in the spectra of PDA@SEBS and PDA@SEBS/MWCNTs, indicating that the DA successfully modified SEBS yarn.

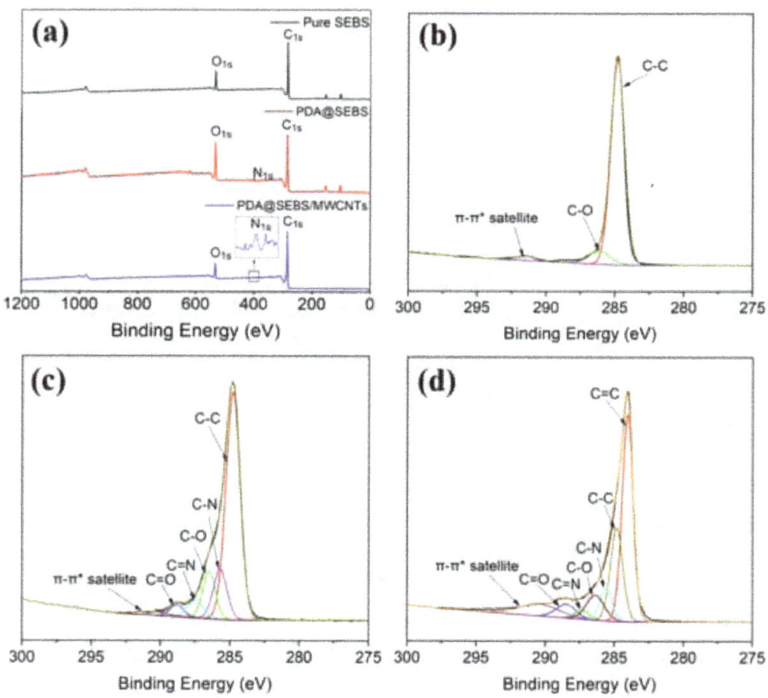

Figure 5. (a) XPS wide scan spectra of pure SEBS, PDA@SEBS and PDA@SEBS/MWCNTs. C1s core-level spectra of (b) pure SEBS, (c) PDA@SEBS, (d) PDA@SEBS/MWCNTs.

In order to further analyze the chemical composition of the materials, the C1s was fitted. As shown in Figure 5b, the C1s peak can be decomposed into C-C (284.8 eV), π-π* satellite peak (291.5 eV) and C-O (286 eV), while the appearance of C-O was due to air pollution. After DA modification, C-N, C=N, and C-O peaks appeared at 285.5 eV, 287.5 eV and 289 eV, respectively (Figure 5c). After deposition of MWCNTs, a new peak, C=C, appeared at 284 eV (Figure 5d). It proved that PDA and MWCNTs were coated onto the SEBS yarn successfully.

The mechanical properties are of great significance for the practical application of materials [32]. Figure 6a shows the stress-strain curves of SEBS, PDA@SEBS, and PDA@SEBS/MWCNTs yarns, the corresponding tensile strength and elongation at break are presented in Table 1. The tensile strength of SEBS yarn was 17.1 MPa. With the successive deposition of PDA and MWCNTs, the tensile strength increased, while PDA@SEBS had a tensile strength of 18.17 MPa and the value for PDA@SEBS/MWCNTs was 20.03 MPa. It was primarily because PDA act as a glue-like adhesive to firmly bond the yarn and MWCNTs. The strong interface interactions between MWCNTs and yarn act effectively to transfer stress from the yarn to the MWCNTs, so improving tensile strength. However, the elongation at break of pure SEBS (□at break = 1270.96%), compares favourably with those of PDA@SEBS and PDA@SEBS/MWCNTs yarns (1158.4% and 1095.8%, respectively). This was probably due to local stress concentrations caused by the aggregation of PDA and MWCNTs which would initiate cracks and lead to early fracture [33].

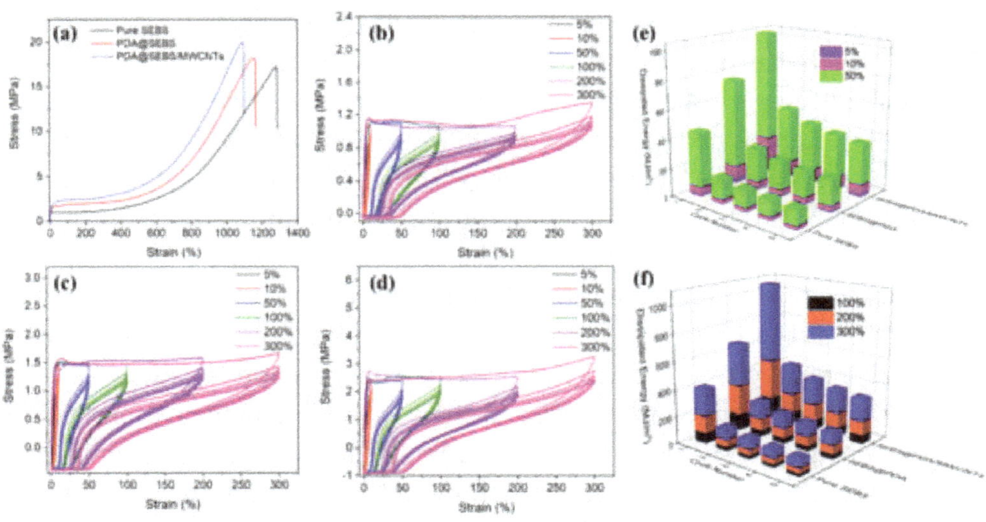

Figure 6. (a) Stress vs. strain curves of SEBS, PDA@SEBS, and PDA@SEBS/MWCNTs yarns at a feed rate of 100 mm/min. Stress vs. strain curves of the initial five tensile cycles for (b) SEBS yarn, (c) PDA@SEBS yarn, and (d) PDA@SEBS/MWCNTs yarn all at different strains. Dissipated energy of SEBS, SEBS@PDA and SEBS@PDA/MWCNTs yarns at the strain of (e) 5%, 10% and 50%, (f) 100%, 200% and 300%.

Table 1. Tensile strength (σ) and elongation at break ($\varepsilon_{at\ break}$) for PDA@SEBS and PDA@SEBS/MWCNTs yarns.

Yarn	σ (MPa)	$\varepsilon_{at\ break}$ (%)
SEBS	17.1 ± 0.45	1270.96 ± 100.15
PDA@SEBS	18.17 ± 0.42	1158.4 ± 97.75
PDA@SEBS/MWCNTs	20.03 ± 0.51	1095.8 ± 91.43

To investigate the relation between viscoelastic behavior and material repeatability, cyclic tensile tests were conducted at a tensile speed of 100 mm/min under applied strains of 5%, 10%, 50%, 100%, 200% and 300%. As can be determined from **Figure 6**b–d, the stress-strain for loading and unloading cycles are not coincident. There is markedly more hysteresis in the PDA@SEBS/MWCNTs yarn curves when compared with those of SEBS yarn and PDA@SEBS yarn, this is unsurprising since the extensibility of the PDA@SEBS/MWCNTs yarn was far greater than that of the two other materials. The PDA@SEBS yarn exhibited only slightly more hysteresis than the SEBS yarn. Internal friction in a material subjected to a tensile loading/unloading cycle will cause strain to lag behind stress as the polymeric chains need to achieve equilibrium as mechanical energy is converted to heat energy. So, hysteresis loops of different sizes and shapes were formed, and the area of the hysteresis loops represented the dissipated work characterized as mechanical hysteresis [**34**]. Useful future research can be undertaken to study the influence of hysteresis on the sensitivity of PDA@SEBS/MWCNTs devices and if the material suffers from 'set'. **Figure 6**e,f show the dissipated energy of SEBS, PDA@SEBS, PDA@SEBS/MWCNTs yarns at different strains in 5 stretching-releasing cycles. With the increase in strain and the deposition of PDA and MWCNTs, the dissipated energy also increased. In addition, the dissipated energy showed the highest value in the first cycle, and then gradually decreased in the subsequent cycles until it tended to a constant value. This phenomenon is called the Mullins Effect [**35**]. This phenomenon was due to the irreversible deformation caused by the slip between PDA, MWCNTs and SEBS fibers and between SEBS chain segments, resulting in mechanical hysteresis. Therefore, in the first cycle, greater stress was required to reach the required elongation, while in subsequent and particularly the later cycles, only a lower force was required to reach the required elongation [**2,34**].

To determine the strain sensing performance, samples of 40 mm length were held in the grips, where the gauge lengths were 20 mm and copper sheets were used as electrodes. The change of resistance is characterized by relative resistance, which is given by Equation (2).

$$\frac{\Delta R}{R_0} = \left(\frac{R - R_0}{R_0}\right) \times 100\% \tag{2}$$

where R is the real-time resistance, R0 is the initial resistance of the strain sensor. It can be seen from Figure 7a that the relative resistance increased exponentially with the increase in strain. This may be because, with an increase in strain, the MWCNTs were separated from one another, which led to the change in the number of conductive paths and the change of conductive tunnel distances between conductive fillers [30]. Further, it shows that PDA@SEBS/MWCNTs yarn had a sensing strain range of 0~215%. To study the strain sensing mechanism, the change of relative resistance of a composite yarn with a change in strain was fitted by Tunnel Effect [2,36]. The change of resistance of the composite yarn can be calculated by using Equation (3).

$$R = \left(\frac{L}{D}\right)\left(\frac{8\pi hs}{3\gamma a^2 e^2}\right) exp(\gamma s) \tag{3}$$

where D is the number of conductive paths and L is the number of particles forming a single path, h and e are Planck constant and quantum of electricity, respectively, s is the minimum distance between MWCNTs, and a2 is the cross-section area. γ is a parameter representing the height of the barrier (φ) and a function of the electron mass (m), which can be calculated by Equation (4).

$$\gamma = \frac{4\pi}{h}\sqrt{2\varphi m} \tag{4}$$

The dependance of the number of conductive paths (*N*) on the strain (ε) under stress is shown in Equation (5).

$$N = \frac{N_0}{exp(A_1\varepsilon + B\varepsilon^2 + C\varepsilon^3 + D\varepsilon^4)} \tag{5}$$

where N0 isthenumberofconductivepathswithoutstretch; b, A1, B, C and D are constants. Assuming that the initial distance (s0) change to s when suffering stretch, s can be obtained by Equation (6).

$$s = s0(1 + b\varepsilon) \tag{6}$$

The relationship between the resistance change (Δ*R*) and strain is given by Equation (7).

$$\frac{\Delta R}{R_0} = \frac{R - R_0}{R_0} = \left(\frac{Ns}{N_0 s_0}\right) exp[\gamma(s - s_0)] - 1 \tag{7}$$

By substituting Equations (5) and (6) into Equation (7), the relationship between relative resistance and strain is represented by Equation (8).

$$\frac{\Delta R}{R_0} = \left(1 + \frac{b\varepsilon}{100}\right) exp\left(\frac{A\varepsilon}{100} + B\left(\frac{\varepsilon}{100}\right)^2 + C\left(\frac{\varepsilon}{100}\right)^3 + D\left(\frac{\varepsilon}{100}\right)^4\right) \times 100 \tag{8}$$

where the units of relative resistance and strain are both %, $A = A_1 + \gamma s$.

The relative resistance of composite fibers was calculated using Equation (8). As shown in Figure 7a, R2 is 0.99752, meaning the experimental results were very consistent with the theoretical results, in which the fitting constants b, A, B, C and D are 949.77242, -2.49077, 5.99487, -4.22125 and 1.01419, respectively.

The GF was given by Equation (9).

$$GF = d(\Delta R/R_0)/d\varepsilon \tag{9}$$

where ΔR = R - R0, R is the real-time resistance, and R0 is the initial resistance of the strain sensor. Figure 7b shows the GF of PDA@SEBS/MWCNTs yarn. When the strain reached a maximum of 215%, the GF was 1.13 × 105. As shown in Figure 7c, compared with maximum GFs and sensing strains recently reported [2,3,6,9,10,14,15,16,21,22,37], the strain sensor in this study offered excellent, comprehensive sensing performance.

To measure the response time, composite yarn was stretched at a stretching speed of 500 mm/min and strained from 0 to 50%. The time to achieve this strain was 1.2 s. However, as shown in Figure 7d, the change of relative resistance needed 1.23 s, so the response time of the PDA@SEBS/MWCNTs yarn was 30 ms.

To investigate the reliability and stability of the PDA@SEBS/MWCNTs yarn, the dynamic strain sensing behaviour at different strain ranges and stretching speeds was determined as shown in Figure 7e,f. As can be observed from Figure 7e the PDA@SEBS/MWCNTs yarn exhibited excellent repeatability and stability in the strain range of 0~5%, 0~10%, 0~50% and 0~100% at a stretching speed of 10 mm/min. Figure 7f shows the synchronous change of resistive resistance under stretching speeds of 5, 10, 50, 100 and 200 mm/min when the PDA@SEBS/MWCNTs was strained from 0 to 50%. There was no obvious difference in the change of relative resistance. These outcomes revealed the material's ability to detect different external stimuli.

Figure 7g shows the current (I) and voltage (V) curves of composite yarn at different strains. The I-V curves at various strains show a linear relationship, indicating compliance with Ohm's Law. Concurrently, the long-term sensing behaviour of the composite yarn was evaluated. In Figure 7h, the relative resistance of the composite yarn changed when strained from 0 to 50% and at a frequency of 0.25 Hz for 15,000 cycles. In the initial cycles, the relative resistance increased before stabilizing. This can be attributed to the continuous destruction and reconstruction of MWCNTs pathways, and then the conductive pathways stabilized at the commencement of strain cycles [38,39,40,41,42,43]. The change of relative resistance exhibited a good response in 15,000 tensile cycles, indicating PDA@SEBS/MWCNTs yarn had good resilience.

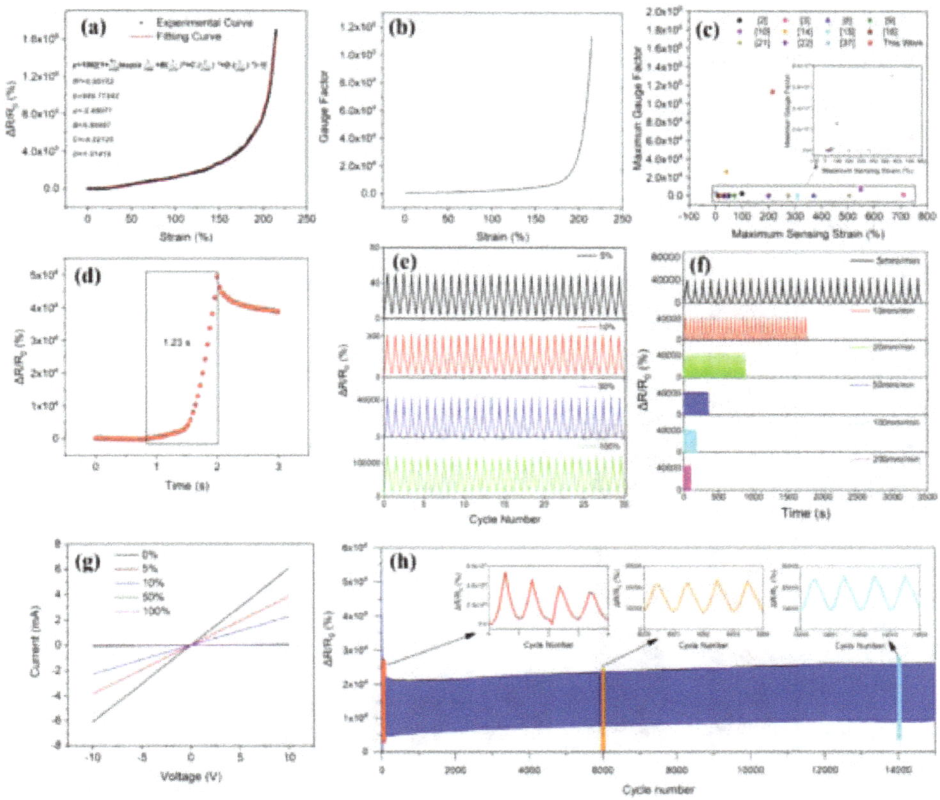

Figure 7. (a) The relative resistance change–strain curve of PDA@SEBS/MWCNTs yarn at a stretching speed of 10 mm/min. (b) The gauge factor–strain curve of PDA@SEBS/MWCNTs yarn at a stretching speed of 10 mm/min. (c) Comparison of the maximum GF and the maximum sensing strain reported in the literature in recent years and those in this work. (d) The response time of PDA@SEBS/MWCNTs yarn. (e) The relative resistance change of PDA@SEBS/MWCNTs yarn under different strains of 5%, 10%, 50%, and 100% at a stretching speed 10 mm/min. (f) The relative resistance change of PDA@SEBS/MWCNTs yarn under different stretching speeds of 5, 10, 50, 100 and 200 mm/min at a strain of 50%. (g) Current and voltage of PDA@SEBS/MWCNTs yarn under strains of 0%, 5%, 10%, 50%, and 100%. (h) 15,000 cycles in the strain range of 0 to 50% at a frequency of 0.25 Hz.

To characterize the washing fastness, PDA@SEBS/MWCNTs yarn was soaked in distilled water and ultrasonicated in an ultrasonic cleaner for 20, 40, 60, 80, 100 and 120 min, respectively. Then the conductivity of the composite yarn after different washing times was determined. As shown in **Figure 8**, the original conductivity of the composite yarn was 0.122 S/cm. Thereafter, the conductivity decreased with increases in the time of ultrasonication. After 40-min ultrasonication, the conductivity decreased to 0.16 S/cm. However, when the ultrasonic time exceeded 40 min, the conductivity changed little. After 120 min, the conductivity was 0.01 S/cm, a change rate of 91.85%. This was primarily due to the PDA layer firmly adhering MWCNTs to the SEBS matrix, so consequently the PDA@SEBS/MWCNTs had excellent water washing fastness.

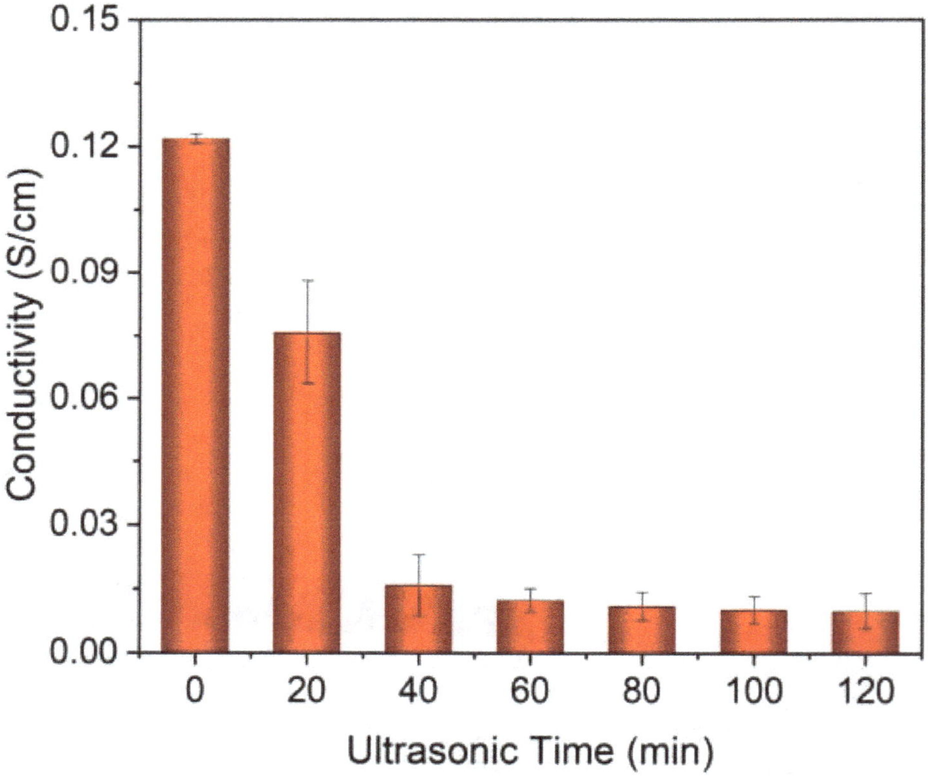

Figure 8. The conductivity of PDA@SEBS/MWCNTs yarn after different washing time.

As mentioned previously, PDA@SEBS/MWCNTs yarn has high flexibility, extensibility, and durability. This means that the composite yarn had potential applications in flexible wearable strain sensors. The composite yarn developed was employed to monitor human motions. As shown in **Figure 9**a, finger motion under different bending angles was detected when the composite yarn was attached to a consenting volunteer's finger. Moreover, the composite yarn showed the ability to detect other joints (wrist, elbow and knee) motions under different bending angles. As can be seen from **Figure 9**b–d, in all cases the relative resistance increased with the increase in joint bending angle and decreased with the return of the joint to its original position.

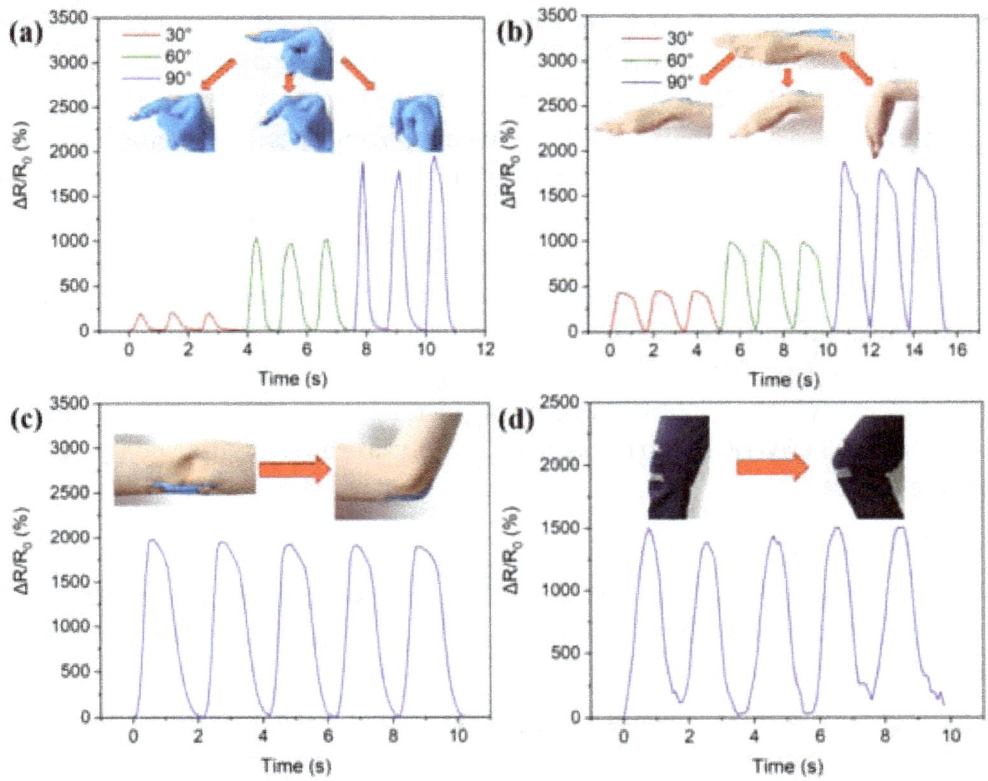

Figure 9. PDA@SEBS/MWCNTs yarn used to monitor human motions. (a) finger bending. (b) wrist bending. (c) elbow bending. (d) knee bending.

4. CONCLUSIONS

In this study, a highly stretchable and sensitive strain sensor based on electrospun SEBS yarn modified by DA and coated with MWCNTs was reported. The strain sensor exhibited a high GF (1.13×10^5) under a maximum strain of 215% while exhibiting good stability, repeatability and durability under a wide range of applied strains, at different stretching speeds and under long-term cyclic loading (15,000 cycles). Additionally, the strain sensor exhibited reliable washing fastness over an ultrasonic testing time of 120 min. Furthermore, it showed an enhanced sensing performance in monitoring joint motions. Although further development may lead to a lowering of hysteresis and avoidance of set when the material is used in dynamic applications, these results demonstrated that PDA@SEBS/MWNCTs yarn has great potential to be used in the field of wearable devices.

5. CONCISE SUMMARY

As wearable electronic devices have become commonplace in daily life, great advances in wearable strain sensors occurred in various fields including healthcare, robotics, virtual reality and other sectors. In this work, a highly stretchable and sensitive strain sensor based on electrospun styrene-ethylene-butene-styrene copolymer (SEBS) yarn modified by dopamine (DA) and coated with multi-walled carbon nanotubes (MWCNTs) was reported. Due to the process of twisting, a strain senor stretched to a strain of 1095.8% while exhibiting a tensile strength was 20.03 MPa. The strain sensor obtained a gauge factor (GF of 1.13×10^5) at a maximum strain of 215%. Concurrently, it also possessed good stability, repeatability and durability under different strain ranges, stretching speeds and 15,000 stretching-releasing cycles. Additionally, the strain sensor exhibited robust washing fastness under an ultrasonic time of 120 min at 240 W and 50 Hz. Furthermore, it had a superior sensing performance in monitoring joint motions of the human body. The high sensitivity and motion sensing performance presented here demonstrate that PDA@SEBS/MWNCTs yarn has great potential to be used as components of wearable devices.

REFERENCES

1. Rogers, J.A.; Someya, T.; Huang, Y. Materials and mechanics for stretchable electronics. *Science* **2010**, *327*, 1603–1607. **[Google Scholar] [CrossRef] [PubMed][Green Version]**

2. Wang, X.; Meng, S.; Tebyetekerwa, M.; Li, Y.; Pionteck, J.; Sun, B.; Qin, Z.; Zhu, M. Highly sensitive and stretchable piezoresistive strain sensor based on conductive poly(styrene-butadiene-styrene)/few layer graphene composite fiber. *Compos. Part A Appl. Sci. Manuf.* **2018**, *105*, 291–299. **[Google Scholar] [CrossRef]**

3. Huang, J.; Li, D.; Zhao, M.; Mensah, A.; Lv, P.; Tian, X.; Huang, F.; Ke, H.; Wei, Q. Highly Sensitive and Stretchable CNT-Bridged AgNP Strain Sensor Based on TPU Electrospun Membrane for Human Motion Detection. *Adv. Electron. Mater.* **2019**, *5*, 1900241. **[Google Scholar] [CrossRef]**

4. Wang, J.; Lou, Y.; Wang, B.; Sun, Q.; Zhou, M.; Li, X. Highly Sensitive, Breathable, and Flexible Pressure Sensor Based on Electrospun Membrane with Assistance of AgNW/TPU as Composite Dielectric Layer. *Sensors* **2020**, *20*, 2459. **[Google Scholar] [CrossRef]**

5. Li, W.Y.; Zhou, Y.F.; Wang, Y.H.; Li, Y.; Jiang, L.; Ma, J.W.; Chen, S.J. Highly Stretchable and Sensitive SBS/Graphene Composite Fiber for Strain Sensors. *Macromol. Mater. Eng.* **2020**, *305*, 1900736. **[Google Scholar] [CrossRef]**

6. Li, W.; Zhou, Y.; Wang, Y.; Jiang, L.; Ma, J.; Chen, S.; Zhou, F.L. Core–Sheath Fiber-Based Wearable Strain Sensor with High Stretchability and Sensitivity for Detecting Human Motion. *Adv. Electron. Mater.* **2020**, *7*, 2000865. **[Google Scholar] [CrossRef]**

7. Wang, L.; Chen, Y.; Lin, L.W.; Wang, H.; Huang, X.W.; Xue, H.G.; Gao, J.F. Highly stretchable, anti-corrosive and wearable strain sensors based on the PDMS/CNTs decorated elastomer nanofiber composite. *Chem. Eng. J.* **2019**, *362*, 89–98. **[Google Scholar] [CrossRef]**

8. Jin, C.C.; Liu, D.M.; Li, M.; Wang, Y. Application of highly stretchy PDMS-based sensing fibers for sensitive weavable strain sensors. *J. Mater. Sci. Mater. Electron.* **2020**, *31*, 4788–4796. **[Google Scholar] [CrossRef]**

9. Son, W.; Kim, K.B.; Lee, S.; Hyeon, G.; Hwang, K.G.; Park, W. Ecoflex-Passivated Graphene-Yarn Composite for a Highly Conductive and Stretchable Strain Sensor. *J. Nanosci. Nanotechnol.* **2019**, *19*, 6690–6695. **[Google Scholar] [CrossRef]**

10. Shen, Z.M.; Feng, J.C. Mass-produced SEBS/graphite nanoplatelet composites with a segregated structure for highly stretchable and recyclable strain sensors. *J. Mater. Chem. C* **2019**, *7*, 9423–9429. **[Google Scholar] [CrossRef]**

11. Zhou, B.Z.; Liu, Z.X.; Li, C.C.; Liu, M.S.; Jiang, L.; Zhou, Y.F.; Zhou, F.L.; Chen, S.J.; Jerrams, S.; Yu, J.Y. A Highly Stretchable and Sensitive Strain Sensor Based on Dopamine Modified Electrospun SEBS Fibers and MWCNTs with Carboxylation. *Adv. Electron. Mater.* **2021**, *7*, 2100233. **[Google Scholar] [CrossRef]**

12. Chossat, J.B.; Park, Y.L.; Wood, R.J.; Duchaine, V. A Soft Strain Sensor Based on Ionic and Metal Liquids. *IEEE Sens. J.* **2013**, *13*, 3405–3414. **[Google Scholar] [CrossRef]**

13. Gullapalli, H.; Vemuru, V.S.; Kumar, A.; Botello-Mendez, A.; Vajtai, R.; Terrones, M.; Nagarajaiah, S.; Ajayan, P.M. Flexible piezoelectric ZnO-paper nanocomposite strain sensor. *Small* **2010**, *6*, 1641–1646. **[Google Scholar] [CrossRef]**

14. Wang, Y.H.; Li, W.Y.; Li, C.C.; Zhou, B.Z.; Zhou, Y.F.; Jiang, L.; Wen, S.P.; Zhou, F.L. Fabrication of ultra-high working range strain sensor using carboxyl CNTs coated electrospun TPU assisted with dopamine. *Appl. Surf. Sci.* **2021**, *566*, 150705. **[Google Scholar] [CrossRef]**

15. Wang, Y.H.; Li, W.Y.; Zhou, Y.F.; Jiang, L.; Ma, J.W.; Chen, S.J.; Jerrams, S.; Zhou, F.L. Fabrication of high-performance wearable strain sensors by using CNTs-coated electrospun polyurethane nanofibers. *J. Mater. Sci.* **2020**, *55*, 12592–12606. **[Google Scholar] [CrossRef]**

16. Chen, Y.; Wang, L.; Wu, Z.F.; Luo, J.C.; Li, B.; Huang, X.W.; Xue, H.G.; Gao, J.F. Super-hydrophobic, durable and cost-effective carbon black/rubber composites for high performance strain sensors. *Compos. Part B Eng.* **2019**, *176*, 107358. **[Google Scholar] [CrossRef]**

17. Xu, M.; Qi, J.; Li, F.; Zhang, Y. Highly stretchable strain sensors with reduced graphene oxide sensing liquids for wearable electronics. *Nanoscale* **2018**, *10*, 5264–5271. **[Google Scholar] [CrossRef]**

18. Baptista, F.R.; Belhout, S.A.; Giordani, S.; Quinn, S.J. Recent developments in carbon nanomaterial sensors. *Chem. Soc. Rev.* **2015**, *44*, 4433–4453. **[Google Scholar] [CrossRef]**

19. Zhou, B.; Li, C.; Zhou, Y.; Liu, Z.; Gao, X.; Wang, X.; Jiang, L.; Tian, M.; Zhou, F.-L.; Jerrams, S.; et al. A flexible dual-mode pressure sensor with ultra-high sensitivity based on BTO@MWCNTs core-shell nanofibers. *Compos. Sci. Technol.* **2022**, *224*, 109478. **[Google Scholar] [CrossRef]**

20. Hu, S.; Shi, Z.; Zhao, W.; Wang, L.; Yang, G. Multifunctional piezoelectric elastomer composites for smart biomedical or wearable electronics. *Compos. B Eng.* **2019**, *160*, 595–604. **[Google Scholar] [CrossRef]**

21. Wang, Z.; Huang, Y.; Sun, J.; Huang, Y.; Hu, H.; Jiang, R.; Gai, W.; Li, G.; Zhi, C. Polyurethane/Cotton/Carbon Nanotubes Core-Spun Yarn as High Reliability Stretchable Strain Sensor for Human Motion Detection. *ACS Appl. Mater. Interfaces* **2016**, *8*, 24837–24843. **[Google Scholar] [CrossRef]**

22. Lee, T.; Choi, Y.W.; Lee, G.; Pikhitsa, P.V.; Kang, D.; Kim, S.M.; Choi, M. Transparent ITO mechanical crack-based pressure and strain sensor. *J. Mater. Chem. C* **2016**, *4*, 9947–9953. **[Google Scholar] [CrossRef]**

23. Li, X.; Huang, K.; Wang, X.; Li, H.; Shen, W.; Zhou, X.; Xu, J.; Wang, X. Effect of montmorillonite on morphology, rheology, and properties of a poly[styrene–(ethylene-co-butylene)–styrene]/poly(ε-caprolactone) nanocomposite. *J. Mater. Sci.* **2017**, *53*, 1191–1203. **[Google Scholar] [CrossRef]**

24. Kuester, S.; Barra, G.M.O.; Ferreira, J.C.; Soares, B.G.; Demarquette, N.R. Electromagnetic interference shielding and electrical properties of nanocomposites based on poly (styrene-b-ethylene-ran-butylene-b-styrene) and carbon nanotubes. *Eur. Polym. J.* **2016**, *77*, 43–53. **[Google Scholar] [CrossRef]**

25. Shin, M.; Oh, J.Y.; Byun, K.E.; Lee, Y.J.; Kim, B.; Baik, H.K.; Park, J.J.; Jeong, U. Polythiophene nanofibril bundles surface-embedded in elastomer: A route to a highly stretchable active channel layer. *Adv. Mater.* **2015**, *27*, 1255–1261. **[Google Scholar] [CrossRef]**

26. Xu, J.; Wang, S.; Wang, G.N.; Zhu, C.; Luo, S.; Jin, L.; Gu, X.; Chen, S.; Feig, V.R.; To, J.W.; et al. Highly stretchable polymer semiconductor films through the nanoconfinement effect. *Science* **2017**, *355*, 59–64. **[Google Scholar] [CrossRef]**

27. Wei, W.; Liang, H.; Parvez, K.; Zhuang, X.; Feng, X.; Mullen, K. Nitrogen-doped carbon nanosheets with size-defined mesopores as highly efficient metal-free catalyst for the oxygen reduction reaction. *Angew. Chem. Int. Ed.* **2014**, *53*, 1570–1574. **[Google Scholar] [CrossRef]**

28. Caliskan, S.; Karihaloo, B.L. Effect of surface roughness, type and size of model aggregates on the bond strength of aggregate/mortar interface. *Interface Sci.* **2004**, *12*, 361–374. **[Google Scholar] [CrossRef]**

29. Seager, C.H.; Pike, G.E. Percolation and conductivity: A computer study. II. *Phys. Rev. B* **1974**, *10*, 1435–1446. **[Google Scholar] [CrossRef]**

30. Yu, S.L.; Wang, X.P.; Xiang, H.X.; Zhu, L.P.; Tebyetekerwa, M.; Zhu, M.F. Superior piezoresistive strain sensing behaviors of carbon nanotubes in one-dimensional polymer fiber structure. *Carbon* **2018**, *140*, 1–9. **[Google Scholar] [CrossRef]**

31. Gao, J.F.; Li, W.; Shi, H.C.; Hu, M.J.; Li, R.K.Y. Preparation, morphology, and mechanical properties of carbon nanotube anchored polymer nanofiber composite. *Compos. Sci. Technol.* **2014**, *92*, 95–102. **[Google Scholar] [CrossRef]**

32. Cai,M.;He,H.;Zhang,X.;Yan,X.;Li,J.;Chen,F.;Yuan,D.;Ning,X.EfficientSynthesis of PVDF/PI Side-by-Side Bicomponent Nanofiber Membrane with Enhanced Mechanical Strength and Good Thermal Stability. *Nanomaterials* **2018**, *9*, 39. **[Google Scholar] [CrossRef][Green Version]**

33. Sang, Z.; Ke, K.; Manas-Zloczower, I. Effect of carbon nanotube morphology on properties in thermoplastic elastomer composites for strain sensors. *Compos. Part A Appl. Sci. Manuf.* **2019**, *121*, 207–212. **[Google Scholar] [CrossRef]**

34. Wang, N.; Xu, Z.Y.; Zhan, P.F.; Dai, K.; Zheng, G.Q.; Liu, C.T.; Shen, C.Y. A tunable strain sensor based on a carbon nanotubes/electrospun polyamide 6 conductive nanofibrous network embedded into poly(vinyl alcohol) with self-diagnosis capabilities. *J. Mater. Chem. C* **2017**, *5*, 4408–4418. **[Google Scholar] [CrossRef]**

35. Ogden, R.W.; Roxburgh, D.G. A pseudo–elastic model for the Mullins effect in filled rubber. *Proc. R. Soc. Lond. Ser. A* **1999**, *455*, 2861–2877. **[Google Scholar] [CrossRef]**

36. Simmons, J.G. Generalized Formula for the Electric Tunnel Effect between Similar Electrodes Separated by a Thin Insulating Film. *J. Appl. Phys.* **1963**, *34*, 1793–1803. **[Google Scholar] [CrossRef][Green Version]**

37. Li, L.; Du, Z.; Sun, B.; Li, W.; Jiang, L.; Zhou, Y.; Ma, J.; Chen, S.; Zhou, F.-L. Fabrication of electrically conductive poly(styrene-b-ethylene-ran-butylene-b-styrene)/multi-walled carbon nanotubes composite fiber and its application in ultra-stretchable strain sensor. *Eur. Polym. J.* **2022**, *169*, 111121. **[Google Scholar] [CrossRef]**

38. Liu, H.; Gao, J.; Huang, W.; Dai, K.; Zheng, G.; Liu, C.; Shen, C.; Yan, X.; Guo, J.; Guo, Z. Electrically conductive strain sensing polyurethane nanocomposites with synergistic carbon nanotubes and graphene bifillers. *Nanoscale* **2016**, *8*, 12977–12989. **[Google Scholar] [CrossRef]**

39. Li, T.; Sun, M.; Wu, S. State-of-the-Art Review of Electrospun Gelatin-Based Nanofiber Dressings for Wound Healing Applications. *Nanomaterials* **2022**, *12*, 784. **[Google Scholar] [CrossRef]**

40. Liu, J.; Li, T.; Zhang, H.; Zhao, W.; Qu, L.; Chen, S.; Wu, S. Electrospun strong, bioactive, and bioabsorbable silk fibroin/poly (L-lactic-acid) nanoyarns for constructing advanced nanotextile tissue scaffolds. *Mater. Today Bio* **2022**, *14*, 100243. **[Google Scholar] [CrossRef]**

41. Jiang, L.; Wang, Y.; Wang, X.; Ning, F.; Wen, S.; Zhou, Y.; Chen, S.; Betts, A.; Jerrams, S.; Zhou, F.-L. Electrohydrodynamic printing of a dielectric elastomer actuator and its application in tunable lenses. *Compos. Part A Appl. Sci. Manuf.* **2021**, *147*, 106461. **[Google Scholar] [CrossRef]**

42. Li, C.; Zhou, B.; Zhou, Y.; Ma, J.; Zhou, F.; Chen, S.; Jerrams, S.; Jiang, L. Carbon Nanotube Coated Fibrous Tubes for Highly Stretchable Strain Sensors Having High Linearity. *Nanomaterials* **2022**, *12*, 2458. **[Google Scholar] [CrossRef]**

43. Liu, Z.; Li, C.; Zhang, X.; Zhou, B.; Wen, S.; Zhou, Y.; Chen, S.; Jiang, L.; Jerrams, S.; Zhou, F. Biodegradable Polyurethane Fiber-Based Strain Sensor with a Broad Sensing Range and High Sensitivity for Human Motion Monitoring. *ACS Sustain. Chem. Eng.* **2022**, *10*, 8788–8798. **[Google Scholar] [CrossRef]**

CHAPTER-18

MODELING THE DYNAMIC EQUIVALENT RESISTANCE OF A KNITTED STRAIN SENSOR FOR IN-PLANE AND THREE-DIMENSIONAL SURFACE ELONGATION

1. INTRODUCTION: AN OVERVIEW

In the realm of wearable technology and smart textiles, knitted strain sensors play a pivotal role in capturing and interpreting deformation across various applications. Understanding their behaviour under different modes of strain, particularly in response to in-plane and three-dimensional surface elongation, is crucial for optimizing their performance. This study delves into the development of a dynamic equivalent resistance model for a knitted strain sensor, offering insights into its electrical response under varying types of mechanical deformation. By exploring the intricacies of in-plane and three-dimensional surface elongation, this research contributes to enhancing the reliability and applicability of knitted strain sensors in diverse fields, from healthcare and sports to industrial and robotics applications. The dynamic equivalent resistance model presented herein serves as a valuable tool for comprehending the nuanced behaviour of these sensors, fostering advancements in the design and implementation of next-generation wearable technologies.

The application of wearable strain sensors in detecting body motion posture, exercise rehabilitation, and healthcare makes our life much convenient [**1**,**2**,**3**,**4**]. Knitting fabrics, due to their multifunctional characteristics in terms of elasticity, flexibility, and deformability, are suitable to be employed in wearable garments as body monitoring sensors for movement [**5**,**6**,**7**]. Compared to woven and non-woven fabrics, knitting fabrics have unique properties [**8**,**9**,**10**,**11**,**12**]. Most textile-based sensors are based on the change in resistance [13,14,15,16], which can be used to predict fabric deformation and mechanical properties [17,18,19]. Therefore, there is a huge need to study the basic electro-mechanical properties of conductive knitted fabric, and develop the application of knitted strain sensors [20,21,22,23,24].

In the study of conductive fabric, some authors have studied the equivalent resistance model of textile sensors in their original state and under uniaxial stretching [22,25,26]; for example, Zhao et al. calculated and predicted the resistance of woven conductive fabric through the radius of the warp yarn and the resistance of one unit of conductive yarn [27]. Li et al. [28,29] explored the electro-mechanical properties of conductive yarn and conductive knitted fabric, with results given in terms of length-related resistance. It was shown that the resistance related to the length plays a dominant role in the total equivalent resistance [30]. However, the resistance change of the loop transfer was not mentioned [31]. Wang et al. examined the electro-mechanical property of conductive elastic knitted fabric based on a loops structure under biaxial extensions. Tokarska evaluated the planar anisotropy of conductive knitted fabrics and determined the resistance value on the surface of the knitted fabrics [32,33]. However, the fact is that in the process of stress deformation of conductive knitted fabric, which is affected by many factors such as environmental friction, both macroscopic fabric deformation and microscopic loop resistance changes should be considered.

The strain knitted sensor must consider the superimposed resistance problem of mutual contact between the coils in the initial state [34,35]. Therefore, the superimposed resistance of the loops squeezing against each other also plays an important role when the resistance changes [36,37,38,39]. Few studies can take into account the close contact resistance of the knitted loops in the initial state, and no appropriate theoretical model is given for exploring the equivalent resistance model of the knitted sensor under the initial state strain [40]. To explore the in-plane tensile properties, several papers have described corresponding equivalent resistance models under uniaxial and biaxial elongation and verified the validity of the corresponding model [41,42,43,44,45]. However, there are limited studies on the equivalent resistance model based on three-dimensional curved surface strain sensing.

This paper proposes two equivalent resistance models of knitted strain sensors under an in-plane deformation. They are macro–micro combined models based on length resistance and a topological equivalent resistance network based on segmented loop structure. In addition, a three-dimensional curved surface equivalent resistance model based on volume resistivity is offered. Comparing the theoretical model and experimental results, it is proven that in-plane and three-dimensional curved surface resistance models can be used to calculate the knitted strain sensor network. Key factors affecting the in-plane knitted strain sensing performance are the length resistance and the loop transfer, as shown in **Figure 1**. In three-dimensional surface sensing, the key factor influencing the sensing performance is the depth of the surface deformation.

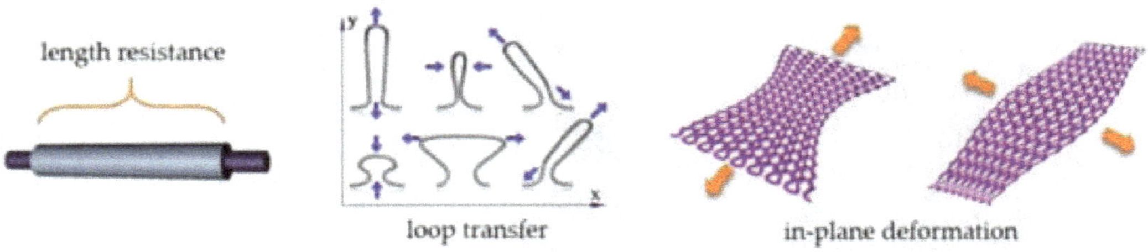

Figure 1. The concepts of length resistance, loop transfer, and in-plane deformation.

2. MATERIALS AND METHODS

2.1. Materials

The knitted strain sensor samples used in the experiments included a commercial silver-coated nylon filament yarn (40 dtex/12 f) purchased from the Tianyin Textile Technology Company (Qingdao, China). The silver-coated nylon yarn's normalized electrical resistance was 14 Ω/cm, with a resistivity of 0.1 ± 0.05 Ω·m and a tensile strength of 5.6 cN/dtex. Ordinary nylon filament yarn (75 dtex/36 f) and nylon-wrapped spandex yarn (50 dtex spandex filament covered by 20 dtex nylon staple fiber) were purchased from the Kejia Textile Fiber Products Company (Nantong, China). The ordinary nylon yarn was used as received for making a knitted base fabric, and was treated as an insulating material. A single circular knitting (SM8-TOP2 MP2, Pitch 0.907 mm, E28, Diameter 15 inch, Santoni Spa, Italy) machine was used for making conductive knitted fabric. The silver-coated yarn was knitted into the fabric as plating stitches. The specimen's off-loom course density was 130 wales/5 cm, and its wale density was 85 courses/5 cm. The size of the sensors was 5 cm × 1.5 cm (126 courses × 26 wales).

2.2. Sensing Performance Test

The load and strain of the knitted strain sensor were measured by an Instron Model 5966, in which two pairs of clamps were used to fix the fabric in the wale direction. After selecting this machine, a sensor was fixed in the bottom carrier with a bursting tool, and a ball was installed on the top to perform a curved strain sensing test on the sensor, as shown in **Figure 2**. The resistance variation of the knitted sensor was tested by a resistance acquisition program using a two-wire method to connect with the sample. This program was a real-time resistance acquisition program specially designed for knitted sensors.

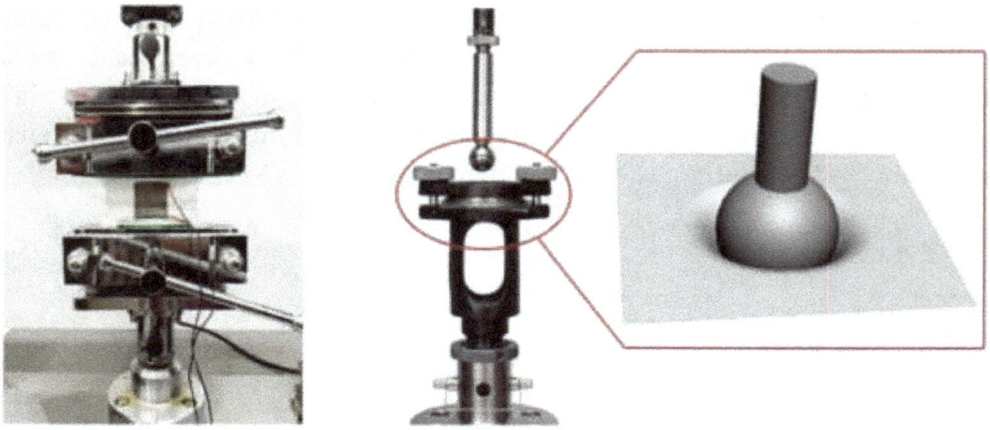

Figure 2. Experimental equipment

In accordance with the ASTM D5035 standard test method for the breaking force and elongation of textile fabrics, a specimen was stretched in the wale direction. In the experiments, the speed of the clamps was 100 mm/min, the gauge length was 100 mm, and the pre-load was 0.1 N. The relationship between strain and resistance was tested, and repeatability was tested with 2800 stretching cycles at 250 mm/min speed. The in-plane strain-resistance properties were tested by stretching the samples to produce a 100% strain 30 times under the same environment and calculating their resistance ratios, and the testing of the resistance change of the strain sensor was carried out in 2 courses × 27 wales with 50% stretch and calculating the ratios. Three-dimensional surface sensing performance was tested with a set of 400 mm/min for tensile rate and 60% strain.

2.3. Theoretical Model

In-plane elongation is generally divided into two categories, course- and wale-stretch elongation [**46**]. The longitudinal elongation of the knitted fabric is greater than the transverse direction, and the longitudinal direction of the conductive fabric is used as the sensor's sensing direction.

Alternatively, the three-dimensional curved surface deformation can be approximated as a bursting experimental principle. Through the vertical drop of the sphere, the fabric is required to produce strains in all directions in the three-dimensional space. These two strain methods include the mechanical behaviour of the sensor in various spatial directions, which is the basic premise for further research on the in-plane and three-dimensional curved surface strain sensing characteristics of knitted strain sensors.

The change of loop form is one of the main changing characteristics of a knitted strain sensor structure under both in-plane and three-dimensional stretching.

The lengths of the loop segments change with loop transmission. When the arc of a parallel and sinker loop is transferred to a leg of loop under the longitudinal tensile, the loop section is transferred to the loop column part. Its tensile resistance deformation process is assumed to be the following steps:

(1)In the unstretched state, fabric loops are tightly abutted, and the resistance of the circuit is low. When subjected to longitudinal stretching, according to the law of resistance as given in Equation (1), the overall circuit resistance increases due to the length change.

$$R = \frac{\rho L}{S} \quad (1)$$

where ρ is the proportionality constant and known as the resistivity or the specific resistance of the material of the conductor or substance; LL is the length of the substance; SS is the cross-sectional area of the substance.

(2)When the force is stretched until loops are separated from each other, the resistance is constant at the maximum value.

(3)When the fabric continues to be stressed, according to the contact resistance [23,24] theory as given in Equation (2), the contact pressure between the loops increases and the resistance decreases.

$$R_C = \frac{\rho}{2}\sqrt{\frac{\pi H}{np}} \quad (2)$$

where ρ (Ωm-1m-1), HH (Nm-2Nm-2), nn, and pp (NN) are the electrical resistivity, material hardness, number of contact points, and contact pressure between the conductive yarn, respectively.

2.4. Model 1: Macro–Micro Equivalent Resistance Models Based on Length Resistance

From a macro perspective, it was assumed that the conductive area of knitted fabric was a whole yarn, the overall length was L0L0, the cross-sectional area of the fabric was S0S0, and the resistance was R0R0. When the fabric was stretched and deformed, the overall length changed. Therefore, the length, cross-sectional area, and resistance were L1L1, S1S1 and R1R1, respectively, as shown in **Figure 3**a. Then, from a microscopic point of view, there was resistance on each course of the loop. Taking 2 courses × 2 wales as an example, the arc of a parallel and leg of loop column constituted the first red line of course resistance (R1R1), the sinker loop and the second-course arc of a parallel/leg of loop column constituted a second yellow line of resistance (R2R2), and the second-course sinker loop constituted a third blue line of resistance (R3R3), as shown in **Figure 3**b. The combination of the red and blue line segments was the yellow loop, and the resistance was Equation (3). According to the 2 courses × 2 wales circuit networks and the series-parallel resistance calculation method, the equivalent resistance equation was Equation (4).

$$R2 = R1 + R3 \quad (3)$$

$$\frac{1}{R_{(2,2)}} = 2\left(\frac{1}{R_1} + \frac{1}{R_2} + \frac{1}{R_3}\right), R_{(2,2)} = 2\left(\frac{R_1{}^2 + R_3{}^2 + 3R_1R_3}{R_1{}^2R_3 + R_3{}^2R_1}\right) \quad (4)$$

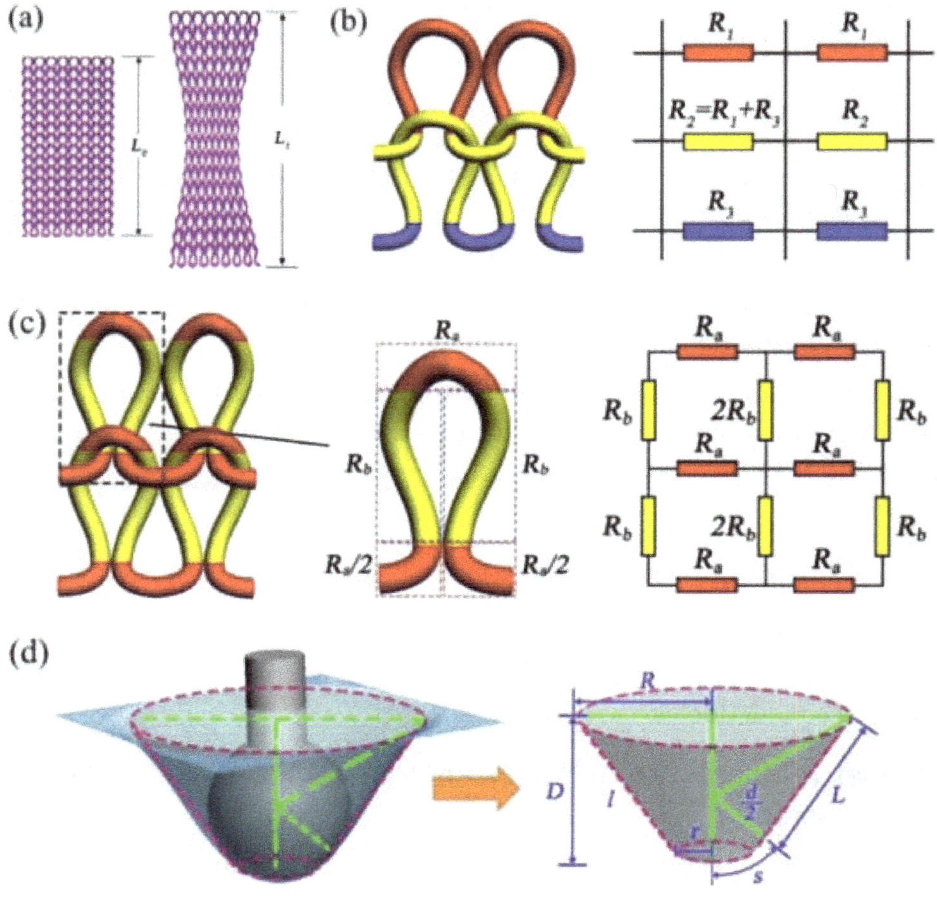

Figure 3. (a) macro structure equivalent resistance model; (b) micro structure equivalent resistance model; (c) topological structure equivalent resistance model; (d) three-dimensional surface equivalent resistance model.

2.5. Model 2: The Equivalent Resistance of the Knitted Sensor Based on the Topology Model

Figure 3c shows the loop structure of the knitted strain sensor, and the equivalent resistance is represented by the topology model. The knitting loop contained by the arc of parallel, sinker loop, and loop column. The arc of parallel had the same resistance as the sinking loop, the arc of parallel resistance was RaRa, the sinking arc was RaRa/2, and the loop column resistance was RbRb. Taking the structure of 2 courses × 2 wales as an example, in the initial state, the conductive knitted fabric was under the influence of spandex, the loops were in close contact, and the loop column and the loop column and the settlement arc and the settlement arc were close to each other. Therefore, the mutual contact resistance of the loop columns was 2Rb2Rb, and the mutual contact resistance of the sinking loop was RaRa.

2.6. Model 3: The Equivalent Resistance of the Knitted Sensor Based on the Topology Model

Depending on the characteristics of the three-dimensional curved surface strain sensor, the sensors strain surface area change was approximated as a frustum of a cone. As shown in Figure 3d, in the burst test, RR was the upper-bottom surface cone radius, rr was the bottom surface cone radius, and ll was the generatrix of a cone. Therefore, the three-dimensional surface resistance was calculated according to the volume resistivity formula, which refers to material resistance of current per unit volume, and is used to characterize material electrical properties.

$$R_v = \frac{\rho D}{S_v} \qquad (5)$$

where RvRv is the three-dimensional curved surface deformation resistance value, SvSv is the curved surface deformation area, DD is the sensor's burst depth, and ρ is volume resistivity in Ω/cm.

3. RESULTS

3.1. Calculation of the Macro–Micro Equivalent Resistance Model Based on Length Resistance

Calculated from the macrostructure at a relaxed state (according to the law of resistance), when the fabric is in its unstretched state, its length, cross-sectional area, and resistance are L0L0, S0S0, and R0R0, respectively. The strain rate is ε for the in-plane stretching. Its length, cross-sectional area, and resistance are L1L1, S1S1, and R1R1 after the in-plane stretching, as shown in Equation (6). Therefore, the theoretical model of the equivalent resistance ratio of the in-plane strain can be deduced from Equation (7). The calculation resistance ratio R′/RR′/R is assumed to be y1y1, where the relationship between y1y1 and strain ε is:

$$\begin{cases} R_0 = \frac{\rho L_0}{S_0} = \frac{\rho L_0}{\pi r^2} \\ L_1 = \varepsilon L_0 + L_0 \\ S_1 = \pi\left(\frac{r}{\varepsilon+1}\right)^2 \\ R_1 = \frac{\rho(\varepsilon+1)L_0}{\pi\left(\frac{r}{\varepsilon+1}\right)^2} = \frac{\rho L_0(\varepsilon+1)^3}{\pi r^2} \\ \frac{R_1}{R_0} = \varepsilon^3 + 3\varepsilon^2 + 3\varepsilon + 1 \end{cases} \quad (6)$$

$$y_1 = \varepsilon^3 + 3\varepsilon^2 + 3\varepsilon + 1 \quad (7)$$

Calculated from the microstructure, according to the 2 courses × 2 wales equivalent circuit and by using the same method, Equation (8) shows that in the relationship among the equivalent resistances of 2 courses × 1 wale, 2 courses × 3 wales, 3 courses × 2 wales, and 4 courses × 2 wales, the series-parallel circuit's connection of equivalent resistance of knitting sensor exists in course and wale direction. Therefore, the M course × N wale equivalent resistance of the knitted sensor can be deduced from Equation (9); it should be noted that M and N refer to the number of loop courses and wales in the sensing area, respectively, as shown in **Figure 4**.

$$\begin{cases} \frac{1}{R_{(2,1)}} = \frac{1}{R_1} + \frac{1}{R_2} + \frac{1}{R_3}, R_{(2,1)} = \frac{R_1{}^2+R_3{}^2+3R_1R_3}{R_1{}^2R_3+R_3{}^2R_1} \\ \frac{1}{R_{(2,2)}} = 2\left(\frac{1}{R_1} + \frac{1}{R_2} + \frac{1}{R_3}\right), R_{(2,2)} = 2\left(\frac{R_1{}^2+R_3{}^2+3R_1R_3}{R_1{}^2R_3+R_3{}^2R_1}\right) \\ \frac{1}{R_{(2,3)}} = 3\left(\frac{1}{R_1} + \frac{1}{R_2} + \frac{1}{R_3}\right), R_{(2,3)} = 3\left(\frac{R_1{}^2+R_3{}^2+3R_1R_3}{R_1{}^2R_3+R_3{}^2R_1}\right) \\ \frac{1}{R_{(3,2)}} = 2\left(\frac{1}{R_1} + \frac{1}{R_2} + \frac{1}{R_2} + \frac{1}{R_3}\right), R_{(3,2)} = 2\left(\frac{R_1{}^2+R_3{}^2+4R_1R_3}{R_1{}^2R_3+R_3{}^2R_1}\right) \\ \frac{1}{R_{(4,2)}} = 4\left(\frac{1}{R_1} + \frac{1}{R_2} + \frac{1}{R_2} + \frac{1}{R_2} + \frac{1}{R_3}\right), R_{(4,2)} = 4\left(\frac{R_1{}^2+R_3{}^2+5R_1R_3}{R_1{}^2R_3+R_3{}^2R_1}\right) \end{cases} \quad (8)$$

$$\frac{1}{R_{(M,N)}} = N\left(\frac{1}{R_1} + (M-1)\frac{1}{R_2} + \frac{1}{R_3}\right), R_{(M,N)} = N\left(\frac{R_1{}^2 + R_3{}^2 + (M+1)R_1R_3}{R_1{}^2R_3 + R_3{}^2R_1}\right)(M>1) \quad (9)$$

When knitted fabrics are in-plane stretched an extending of each loop and yarn segment occurs, transferring among the arc of parallel, leg and sinker loop. Then, the adjacent loops become tightly interlinked with each other. Due to the large strain caused by the range of extension, the following assumptions are considered to simplify the structure of knitted loop under a large strain. Assuming that the resistance changes of R1R1, R2R2, and R3R3 are approximately equal (for the simple calculation, the three resistances are approximately equal to R2R2), then the total resistance (RR) of the network and the resistance after stretching (R'R') can be calculated as in Equation (10). The equivalent resistance ratio R′/RR′/R is assumed to be y2y2, where the relationship between y2y2, strain ε, course, and wale is presented as Equation (11).

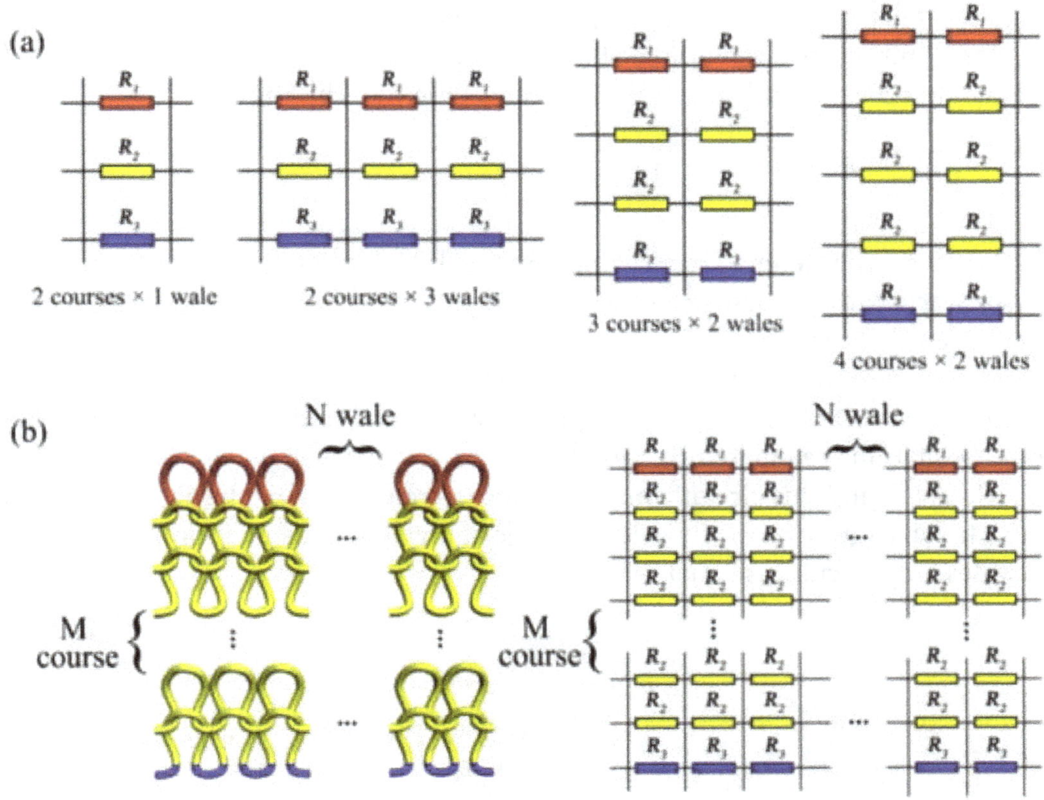

Figure 4. (a) 2 courses × 1 wale, 2 courses × 3 wales, 3 courses × 2 wales, and 4 courses × 2 wales equivalent circuits; (b) The equivalent circuit of M course × N wale.

$$\begin{cases} \frac{1}{R} = \frac{1}{N}\left(\frac{1}{R_2} + \frac{1}{R_2} + \cdots + \frac{M}{R_2}\right) \\ R = N\frac{R_2}{M+1} = N\frac{\frac{\rho L}{S}}{M+1} \\ L' = (\varepsilon+1)L,\ S' = \pi\left(\frac{r}{\varepsilon+1}\right)^2 \\ R' = \frac{\rho L'}{S'} = \frac{\rho 2(\varepsilon+1)L}{\pi\left(\frac{r}{\varepsilon+1}\right)^2} \\ \frac{R'}{R} = N\frac{2(\varepsilon+1)^3}{M+1} \end{cases} \quad (10)$$

$$y_2 = N\frac{2(\varepsilon+1)^3}{M+1} \quad (11)$$

3.2. Calculation of the Equivalent Resistance Based on the Topology Model

According to Kirchhoff's current and voltage law, the equivalent resistance of the topology model and the corresponding equations can be calculated [47]. The total current (I_1) in the circuit can be calculated by Matlab, and the equivalent resistance (R) of the knitted strain sensor can be defined as Equation (12), where V is the total voltage that the circuit loads; (I_1) is the total current in the circuit.

$$R = \frac{V}{I_1} \qquad (12)$$

Based on the geometric structure of the knitting loop, the knitting strain sensor starts with 2 courses × 2 wales and extends to M course × N wale. Kirchhoff's current and voltage law is used to calculate the total circuit I1I1 of the knitted fabric circuit. Figure 5a is a loop circuit with 2 courses × 2 wales. The knitting sensor is connected to the power supply at both ends, and its voltage is VV. The circuit equations calculated by Kirchhoff's current and voltage law are shown in Equation (13), and this circuit contains 5 loop currents. Solving the above equation with Matlab comprehensively calculates the equivalent resistance of the 2 courses × 2 wales loop circuits as Equation (14).

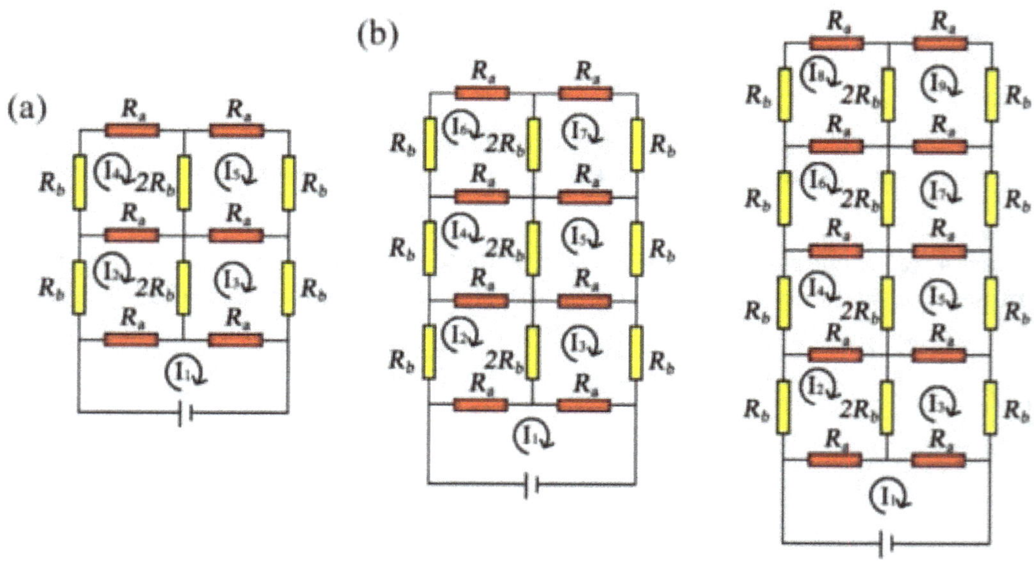

Figure 5. (a) 2 courses × 2 wales loop circuit; (b) 3 courses × 2 wales and 4 courses × 2 wales loop circuits.

The same method can be used to solve for the equivalent resistances of the 3 courses × 2 wales loop circuits and 4 courses × 2 wales loop circuits, with the circuit equation set calculated by Kirchhoff's current and voltage laws, as shown in **Figure 5b** and the Equations (15) and (16). The 3 courses × 2 wales circuits contain seven loop circuits, and the 4 courses × 2 wales circuits contain nine loop currents.

$$\begin{cases} 2I_1R_a - I_2R_a - I_3R_a = V \\ I_2(2R_a + 3R_b) - I_1R_a - I_4R_a - 2I_3R_b = 0 \\ I_3(2R_a + 3R_b) - I_1R_a - I_5R_a - 2I_2R_b = 0 \\ I_4(2R_a + 3R_b) - 2I_5R_b - I_2R_a = 0 \\ I_5(2R_a + 3R_b) - I_3R_a - 2I_4R_b = 0 \end{cases} \quad (13)$$

$$R_{(2,2)} = \frac{2R_a\left(R_a{}^2 + 3R_aR_b + R_b{}^2\right)}{3R_a{}^2 + R_b{}^2 + 4R_aR_b} \quad (14)$$

$$\begin{cases} 2I_1R_a - I_2R_a - I_3R_a = V \\ I_2(2R_a + 3R_b) - I_1R_a - I_4R_a - 2I_3R_b = 0 \\ I_3(2R_a + 3R_b) - I_1R_a - I_5R_a - 2I_2R_b = 0 \\ I_4(2R_a + 3R_b) - I_2R_a - I_6R_a - 2I_5R_b = 0 \\ I_5(2R_a + 3R_b) - I_3R_a - I_7R_a - 2I_4R_b = 0 \\ I_6(2R_a + 3R_b) - I_4R_a - 2I_7R_b = 0 \\ I_7(2R_a + 3R_b) - I_5R_a - 2I_6R_b = 0 \end{cases} \quad (15)$$

$$\begin{cases} 2I_1R_a - I_2R_a - I_3R_a = V \\ I_2(2R_a + 3R_b) - I_1R_a - I_4R_a - 2I_3R_b = 0 \\ I_3(2R_a + 3R_b) - I_1R_a - I_5R_a - 2I_2R_b = 0 \\ I_4(2R_a + 3R_b) - I_2R_a - I_6R_a - 2I_5R_b = 0 \\ I_5(2R_a + 3R_b) - I_3R_a - I_7R_a - 2I_4R_b = 0 \\ I_6(2R_a + 3R_b) - I_4R_a - I_8R_a - 2I_7R_b = 0 \\ I_7(2R_a + 3R_b) - I_5R_a - I_9R_a - 2I_6R_b = 0 \\ I_8(2R_a + 3R_b) - I_6R_a - 2I_9R_b = 0 \\ I_9(2R_a + 3R_b) - I_7R_a - 2I_8R_b = 0 \end{cases} \quad (16)$$

The equivalent resistance of the 3 courses × 2 wales and 4 courses × 2 wales loop circuits can be comprehensively calculated by solving the above equation with Matlab, as in the Equations (17) and (18). The mathematical transformation of the equivalent resistance Equations (14), (17) and (18) for 2 courses × 2 wales, 3 courses × 2 wales, and 4 courses × 2 wales results in the following formulas being obtained:

$$R_{(3,2)} = \frac{2R_a\left(R_a{}^3 + 6R_a{}^2R_b + 5R_aR_b{}^2 + R_b{}^3\right)}{4R_a{}^3 + R_b{}^3 + 10R_a{}^2R_b + 6R_aR_b{}^2} \quad (17)$$

$$R_{(4,2)} = \frac{2R_a\left(R_a{}^4 + 10R_a{}^3R_b + 15R_a{}^2R_b{}^2 + 7R_aR_b{}^3 + R_b{}^4\right)}{5R_a{}^4 + R_b{}^4 + 20R_a{}^3R_b + 21R_a{}^2R_b{}^2 + 8R_aR_b{}^3} \quad (18)$$

$$\begin{cases} R_{(2,2)} = \frac{2R_a\left[(R_a+R_b)^2+R_aR_b\right]}{(R_a+R_b)^2+2R_a(R_a+R_b)} = (R_a+R_b) - \frac{R_a{}^3+R_a{}^2R_b+3R_aR_b{}^2+R_b{}^3}{3R_a{}^2+4R_aR_b+R_b{}^2} \\ R_{(3,2)} = \frac{2}{3}(R_a+R_b) - \frac{2(R_a{}^4-4R_a{}^3R_b+R_a{}^2R_b{}^2+4R_aR_b{}^3+R_b{}^4)}{3(4R_a{}^3+10R_a{}^2R_b+6R_aR_b{}^2+R_b{}^3)} \\ R_{(4,2)} = \frac{1}{2}(R_a+R_b) - \frac{R_a{}^5-15R_a{}^4R_b-19R_a{}^3R_b{}^2+R_a{}^2R_b{}^3+5R_aR_b{}^4+R_b{}^5}{2\left(5R_a{}^4+20R_a{}^3R_b+21R_a{}^2R_b{}^2+8R_aR_b{}^3+R_b{}^4\right)} \end{cases} \quad (19)$$

It is impossible to find the rule of M course × 2 wales from this expression. Therefore, if Equation (19) is factorized, the denominator of the last term in the equation is much larger than the numerator. When there are more courses, the last term in the three expressions is approximately zero and can be ignored. Thus, the following relationship is obtained:

$$\begin{cases} R_{(2,2)} \approx (R_a+R_b) \\ R_{(3,2)} \approx \frac{2}{3}(R_a+R_b) \\ R_{(4,2)} \approx \frac{1}{2}(R_a+R_b) \\ R_{(M,2)} \approx \frac{2}{M}(R_a+R_b) \end{cases} \quad (20)$$

It is inferred that along the wale direction of the knitted sensor, the equivalent resistance decreases with the increase of courses. The circuit is a parallel circuit, and the equivalent resistance expression of M course × 2 wales is Equation (21). **Figure 6** shows the equivalent resistances of the circuits with 2 courses × 3 wales, 2 courses × 4 wales, and 2 courses × 5 wales. The equivalent resistances are calculated by the same method, and the equations obtained are shown in Equations (22)–(24).

$$R_{(M,2)} = \frac{2}{M}(R_a+R_b) \quad (21)$$

$$\begin{cases} 3I_1R_a - I_2R_a - I_3R_a - I_4R_a = V \\ I_2(2R_a+3R_b) - I_1R_a - 2I_3R_b - I_5R_a = 0 \\ I_3(2R_a+4R_b) - I_1R_a - 2I_2R_b - I_6R_a - 2I_4R_b = 0 \\ I_4(2R_a+3R_b) - 2I_3R_b - I_1R_a - I_7R_a = 0 \\ I_5(2R_a+3R_b) - I_2R_a - 2I_6R_b = 0 \\ I_6(2R_a+4R_b) - I_3R_a - 2I_5R_b - 2I_7R_b = 0 \\ I_7(2R_a+3R_b) - I_4R_a - 2I_6R_b = 0 \end{cases} \quad (22)$$

$$\begin{cases} 4I_1R_a - I_2R_a - I_3R_a - I_4R_a - I_5R_a = V \\ I_2(2R_a + 3R_b) - I_1R_a - 2I_3R_b - I_6R_a = 0 \\ I_3(2R_a + 4R_b) - I_1R_a - 2I_2R_b - I_7R_a - 2I_4R_b = 0 \\ I_4(2R_a + 4R_b) - 2I_3R_b - 2I_5R_b - I_1R_a - I_8R_a = 0 \\ I_5(2R_a + 3R_b) - I_1R_a - I_9R_a - 2I_4R_b = 0 \\ I_6(2R_a + 3R_b) - I_2R_a - 2I_7R_b = 0 \\ I_7(2R_a + 4R_b) - I_3R_a - 2I_6R_b - 2I_8R_b = 0 \\ I_8(2R_a + 4R_b) - I_4R_a - 2I_7R_b - 2I_9R_b = 0 \\ I_9(2R_a + 3R_b) - I_5R_a - 2I_8R_b = 0 \end{cases} \quad (23)$$

$$\begin{cases} 5I_1R_a - I_2R_a - I_3R_a - I_4R_a - I_5R_a - I_6R_a = V \\ I_2(2R_a + 3R_b) - I_1R_a - I_32R_b - I_7R_a = 0 \\ I_3(2R_a + 4R_b) - I_1R_a - I_8R_a - I_22R_b - I_42R_b = 0 \\ I_4(2R_a + 4R_b) - I_1R_a - I_9R_a - I_32R_b - I_52R_b = 0 \\ I_5(2R_a + 4R_b) - I_1R_a - I_{10}R_a - I_42R_b - I_62R_b = 0 \\ I_6(2R_a + 3R_b) - I_1R_a - I_{11}R_a - I_52R_b = 0 \\ I_7(2R_a + 3R_b) - I_2R_a - I_82R_b = 0 \\ I_8(2R_a + 4R_b) - I_3R_a - I_72R_b - I_92R_b = 0 \\ I_9(2R_a + 4R_b) - I_4R_a - I_82R_b - I_{10}2R_b = 0 \\ I_{10}(2R_a + 4R_b) - I_5R_a - I_92R_b - I_{11}2R_b = 0 \\ I_{11}(2R_a + 3R_b) - I_6R_a - I_{10}2R_b = 0 \end{cases} \quad (24)$$

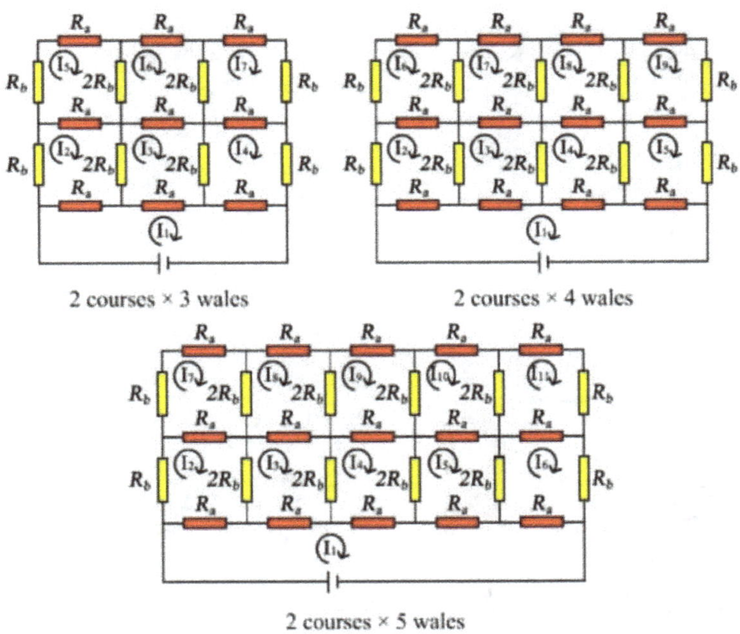

Figure 6. The equivalent resistances of the circuits with 2 courses × 3 wales, 2 courses × 4 wales, and 2 courses × 5 wales.

Through the resistance formulas of 2 courses × 3 wales, 2 courses × 4 wales, and 2 courses × 5 wales (Equation (25)), the resistance expression law of 2 courses × N wale cannot be directly obtained, so the equivalent resistance of the knitted sensor topological model can instead be obtained through the corresponding equations of M course × N wale, as in Equation (26), and as shown in **Figure 7** (the equivalent circuit of the knitted strain sensor with M course × N wale).

$$\begin{cases} R_{(2,3)} = \dfrac{R_a\left(9R_a{}^4+94R_a{}^3R_b+271R_a{}^2R_b{}^2+260R_aR_b{}^3+48R_b{}^4\right)}{9R_a{}^4+16R_b{}^4+84R_a{}^3R_b+187R_a{}^2R_b{}^2+112R_aR_b{}^3} \\ R_{(2,4)} = \dfrac{2R_a\left(6R_a{}^4+45R_a{}^3R_b+92R_a{}^2R_b{}^2+62R_aR_b{}^3+8R_b{}^4\right)}{9R_a{}^4+4R_b{}^4+60R_a{}^3R_b+95R_a{}^2R_b{}^2+40R_aR_b{}^3} \\ R_{(2,5)} = \\ \dfrac{R_a\left(45R_a{}^6+690R_a{}^5R_b+3627R_a{}^4R_b{}^2+8348R_a{}^3R_b{}^3+8592R_a{}^2R_b{}^4+3520R_aR_b{}^5+320R_b{}^6\right)}{27R_a{}^4+64R_b{}^4+396R_a{}^6R_b+1929R_a{}^4R_b{}^2+3920R_a{}^3R_b{}^3+3232R_a{}^2R_b{}^4+896R_aR_b{}^5} \end{cases} \quad (25)$$

$$\begin{cases} (N+1)I_1R_a - I_2R_a - I_3R_a - I_4R_a - \cdots - I_{N+1}R_a = V \\ I_2(2R_a+3R_b) - I_1R_a - I_32R_b - I_{N+2}R_a = 0 \\ I_3(2R_a+4R_b) - I_1R_a - I_22R_b - I_42R_b - I_{N+3}R_a = 0 \\ \vdots \\ I_{N+1}(2R_a+3R_b) - I_1R_a - I_{2N+1}R_a - I_N2R_b = 0 \\ I_{N+2}(2R_a+3R_b) - I_2R_a - I_{2N+2}R_a - I_{N+3}2R_b = 0 \\ I_{N+3}(2R_a+4R_b) - I_3R_a - I_{2N+3}R_a - I_{N+2}2R_b - I_{N+4}2R_b = 0 \\ \vdots \\ I_{2N+1}(2R_a+3R_b) - I_{N+1}R_a - I_{3N+1}R_a - I_{2N}2R_b = 0 \\ \vdots \\ I_{(M-2)N+2}(2R_a+3R_b) - I_{(M-1)N+2}R_a - I_{(M-3)N+2}R_a - I_{(M-2)N+3}2R_b = 0 \\ I_{(M-2)N+3}(2R_a+4R_b) - I_{(M-1)N+3}R_a - I_{(M-3)N+3}R_a - I_{(M-2)N+2}2R_b - I_{(M-2)N+4}2R_b = 0 \\ \vdots \\ I_{(M-1)N+1}(2R_a+3R_b) - I_{MN+1}R_a - I_{(M-2)N+1}R_a - I_{(M-1)N}2R_b = 0 \\ I_{(M-1)N+2}(2R_a+3R_b) - I_{(M-2)N+2}R_a - I_{(M-1)N+3}2R_b = 0 \\ I_{(M-1)N+3}(2R_a+4R_b) - I_{(M-2)N+3}R_a - I_{(M-1)N+2}2R_b - I_{(M-1)N+4}2R_b = 0 \\ \vdots \\ I_{MN+1}(2R_a+3R_b) - I_{(M-1)N+1}R_a - I_{MN}2R_b = 0 \end{cases} \quad (26)$$

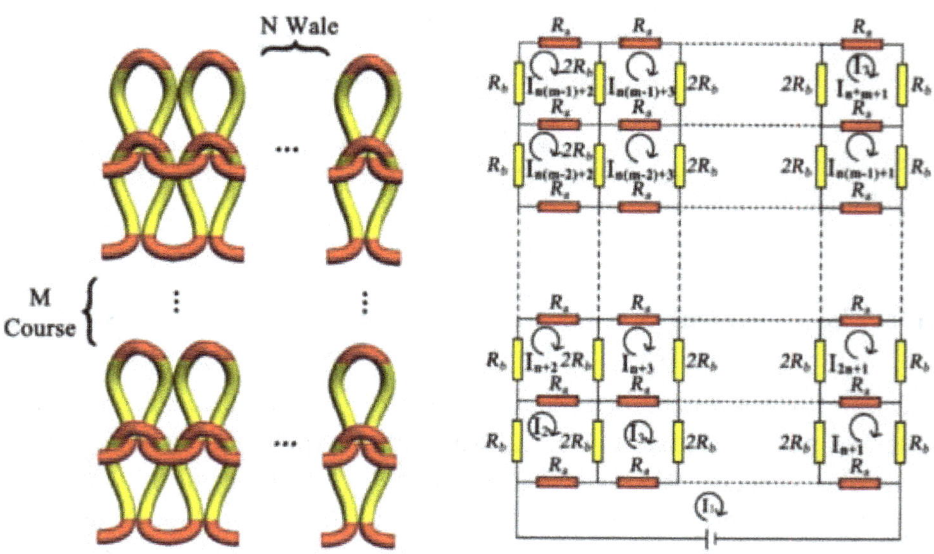

Figure 7. The equivalent circuit of the knitted strain sensor with M course × N wale.

According to the M course × N wale topology model equations on loop transfer and length resistance (take 2 courses × 2 wales as an example to calculate the equivalent resistance ratio during longitudinal strain), Ra′Ra′, Rb′Rb′, and the equivalent resistance ratio R′(2,2)/R(2,2)R′2,2/R2,2 of 2 courses × 2 wales can be calculated, as shown in Equation (27).

$$\begin{cases} R_{(2,2)} = \frac{2R_a\left(R_a{}^2+3R_aR_b+R_b{}^2\right)}{3R_a{}^2+R_b{}^2+4R_aR_b} \\ R'_a = R_a\left(1-\frac{\varepsilon}{2}\right) \\ R'_b = R_b\left(1+\varepsilon\right) \\ R'_{(2,2)} = \frac{2R_a\left(1-\frac{\varepsilon}{2}\right)\left(\left[R_a\left(1-\frac{\varepsilon}{2}\right)\right]^2+3R_a\left(1-\frac{\varepsilon}{2}\right)R_b(1+\varepsilon)+\left[R_b(1+\varepsilon)\right]^2\right)}{3\left[R_a\left(1-\frac{\varepsilon}{2}\right)\right]^2+\left[R_b(1+\varepsilon)\right]^2+4R_a\left(1-\frac{\varepsilon}{2}\right)R_b(1+\varepsilon)} \\ \frac{R'_{(2,2)}}{R_{(2,2)}} = \frac{\left(3R_a{}^2+4R_aR_b+R_b{}^2\right)\left[2R_a{}^3\left(\frac{\varepsilon}{2}-1\right)^3-R_b{}^2(1+\varepsilon)^2+3R_aR_b\left(\frac{\varepsilon}{2}-1\right)(1+\varepsilon)\right]}{2R_a\left[R_b{}^2(1+\varepsilon)^2+3R_a{}^2\left(\frac{\varepsilon}{2}-1\right)^2-4R_aR_b\left(\frac{\varepsilon}{2}-1\right)(1+\varepsilon)\right]\left(R_a{}^2+3R_aR_b+R_b{}^2\right)} \end{cases} \quad (27)$$

3.3. Calculation of the Equivalent Resistance Model Based on the Three-Dimensional Curved Surface Strain

According to the change of the sensor surface area, the relationship between the burst depth (DD) and curved surface strain is established. The volume resistivity formula is used to calculate the resistance (RvRv) before breaking and the equivalent resistance (R'vRv') after the curved surface strain. The ratio of surface strain resistance is derived as shown in Equation (28). To avoid the burst depth of the formula being zero in an initial state, the burst depth (DD) in the initial state is set to 1 mm, and bottom surface cone radius (rr) is set to one-half of a stainless steel radius. Therefore, the resistance change ratio of the curved surface strain is defined as yvyv, and the relationship between yvyv and the burst depth (DD) is expressed as Equation (29).

$$\begin{cases} R_v = \frac{\rho D}{S_v} \\ R'_v = \frac{\rho D'}{S'_v} = \frac{\rho D'}{\pi l'(R+r)+\pi r^2} \\ l' = \sqrt{(R-r)^2+D'} \\ \frac{R'_v}{R_v} = \frac{S_v D'}{\pi l' D(R+r)+\pi r^2 D} = \frac{7.5D'}{0.88705\sqrt{D'^2-2.5D'+4.84}+0.12265625} \end{cases} \quad (28)$$

$$y_v = \frac{7.5D}{0.88705\sqrt{D^2-2.5D+4.84}+0.12265625} \quad (29)$$

3.4. Fabric Cycle Test Results under In-Plane Stretching

The initial strain responsiveness was tested by stretching the samples to produce a 100% strain at a speed of 300 mm/min. Following 2800 occurrences of in-plane stretching, 50 groups of cycle data were extracted according to the cyclic stretching data for the analysis of variance, as shown in **Figure 8**. The results show that the p-value of the in-plane stretching is 0.955 > 0.05, as shown in **Table 1**, indicating that the extracted actual resistance data have no significant positive influence on all cycle data. It proved that an equivalent resistance model established for the same knitted sensor is feasible.

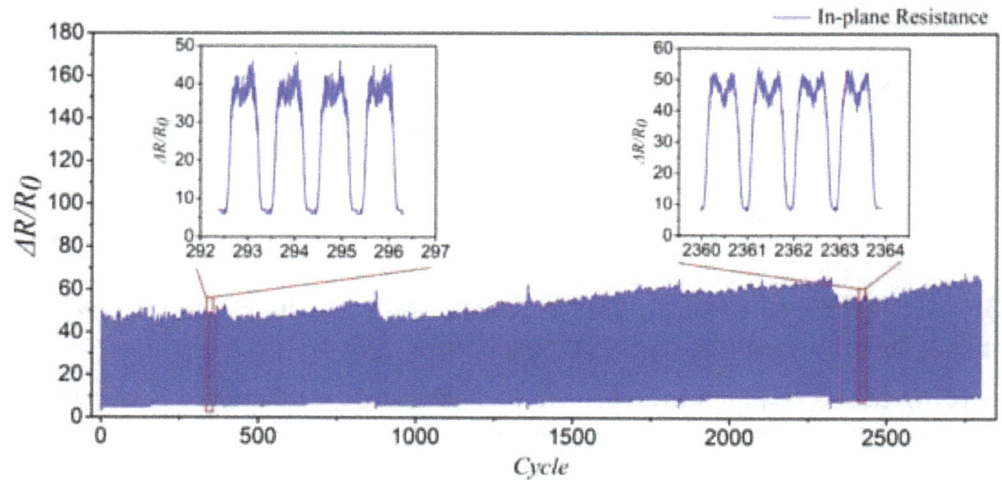

Figure 8. In-plane cycle stretching experiment.

Table 1. ANOVA for the in-plane tensile cycle stretching experiment.

	Sum of Squares	df	Mean Square	F	p Value
Regression	6081.70	50	121.63	0.688	0.95
Residual	4,326,928.20	24471	176.81		
Total	433,009.90	24521			

3.5. Comparison of the Experimental Data and the Equivalent Resistance Model Calculation Results during In-Plane Stretching

The equivalent resistance model during in-plane stretching is composed of macro–micro models; that is, the combination of the Equations (7) and (11). Therefore, the in-plane tensile equivalent resistance model Equation (30) can be derived. Figure 9 illustrates the simulation results of the equivalent resistance and the experimental data of the in-plane stretching, where the two curves YY and yy are close to each other (within acceptable accuracy), and further analyzes the fit of the two curves. It can be seen from Table 2 that the $R2$ value of the model is 0.975, meaning that the in-plane strain fitting resistance can explain 97.5% of the change of real resistance. When the F test was performed on the model, it was found that the F test (F=39060.82, p=0.00<0.05) was applied to the model, which indicated that the in-plane fitting resistance curve will have a significant positive impact on a real resistance curve. The equivalent resistance theoretical model of in-plane tensile properties can be used to predict the same knitted fabric resistance according to the given strain, and the same method can be used in future studies to predict the deformation of the fabric based on the actual resistance obtained.

$$Y = y_1 + y_2 = \varepsilon^3 + 3\varepsilon^2 + 3\varepsilon + \frac{2n(\varepsilon+1)^3}{m+1} + 1 \qquad (30)$$

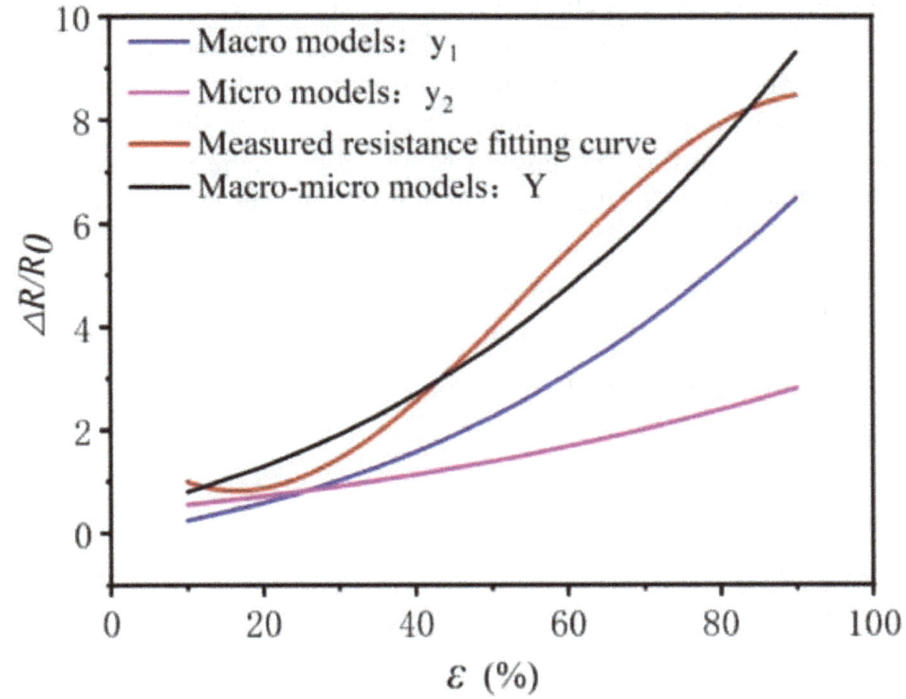

Figure 9. Comparison of the actual resistance and the in-plane tensile equivalent resistance model calculation results.

Table 2. Parameter estimates for in-plane tensile equivalent resistance model.

	Unstandardized Coefficients		Standardized Coefficients	t	p	*R2*	F
	B	Std. Error	Beta				
Constant	-0.235	0.026	-	-8.959	0	0.975	39,060.82
Fitting resistance	1.084	0.005	0.987	197.638	0		

3.6. Comparison of the In-Plane Tensile Test Results and the Equivalent Resistance Calculation Results of the Topology Model

Comparing the simulation results of the knitting sensor topology resistance model with the results of longitudinal stretching shows that the two curves are close to each other within acceptable accuracy, further verifying the fit of the two curves, as shown in Figure 10. It can be seen from Table 3 that the *R*2 value of the model is 0.99, meaning that the topology model fitting resistance can explain 99% of the change of real resistance. When the F test was performed on the model, it was found that the F test (F = 43,544.39, p = 0.00<<0.05) was applied to the model, which indicated that the topology model fitting the resistance curve will have a significant positive impact on the real resistance curve. The equivalent resistance theoretical model of the topology model can be used to predict the same knitted fabric resistance according to a given strain, and the same method can be used in future studies to predict the deformation of the fabric based on the actual resistance obtained.

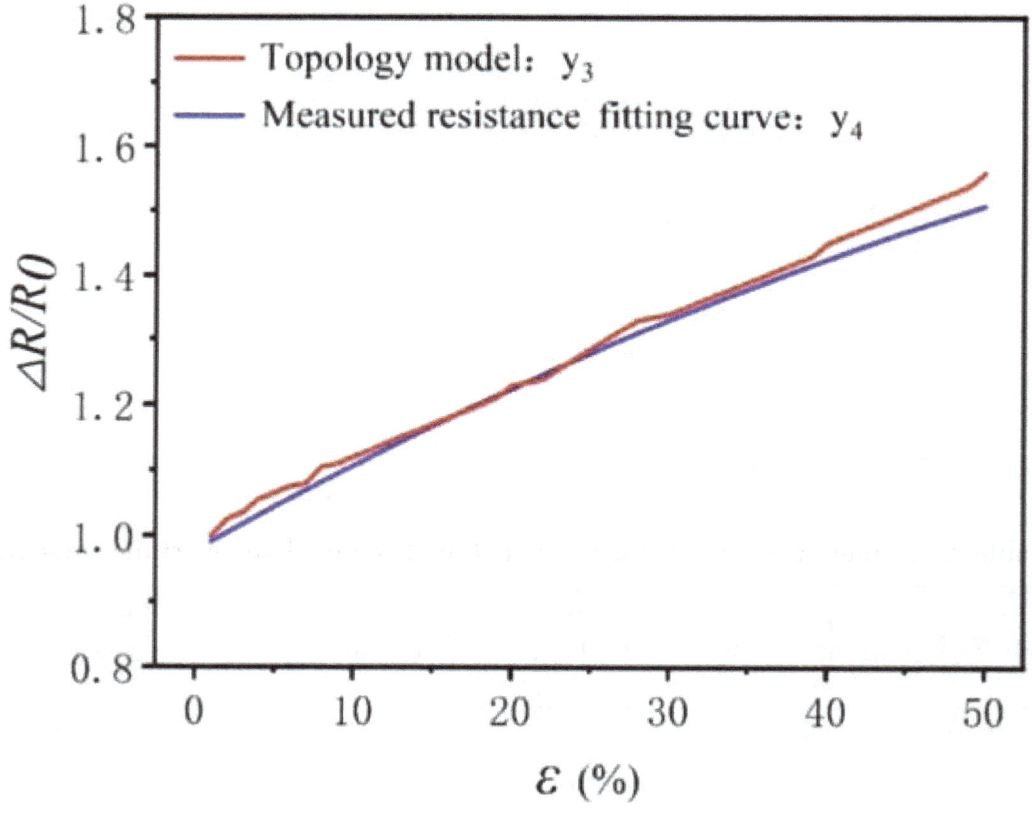

Figure 10. Comparison of actual resistance and topology equivalent resistance model calculation results.

Table 3. Parameter Estimates for topology equivalent resistance model.

	Unstandardized Coefficients		Standardized Coefficients	t	p	R2	F
	B	Std. Error	Beta				
Constant	-15.801	0.082	-	-92.94	0	0.99	43,544.393
Fitting resistance	16.965	0.081	0.99	208.67	0		

Comparing the two equivalent resistance models, the macro-micro model establishes the equivalent resistance of the knitted sensor from two perspectives. When the selected strain sensor is larger in size, the macro-micro model can be used to calculate the equivalent resistance more easily and quickly, and the calculation efficiency is higher. The topology equivalent resistance model comprehensively considers the loop transfer and length resistance. When selecting a smaller sensor application, it is more accurate to use the topology resistance model to calculate the equivalent resistance. Compared with the macro-micro models, the topological model has a higher resistance prediction accuracy, but the calculation is more complicated.

3.7. Comparison of Three-Dimensional Strain Equivalent Resistance Model and Test Results

Equation (29) demonstrates a proportional relationship between burst depth and surface strain. Therefore, the relationship between the equivalent resistance ratio and the curved surface strain can be established, as shown in Figure 9. The red curve in Figure 9 aligns with the smooth curve of the burst test data, and the black curve is the predicted result of the equivalent resistance model.

From the analysis in Figure 9, two curves are close to each other within acceptable accuracy, which further verifies the fit of two curves, as shown in Table 4. The fitted resistance is used as an independent variable and actual resistance is used as the dependent variable for linear regression analysis. The accuracy of the fit $R2$ is 0.867, which means that the fitted resistance can explain 86.7% of the actual resistance. The model passes the F test ($F = 366274.180$, $p = 0.000 \ll 0.05$), which means that the fitted resistance will affect the actual resistance. The specific analysis shows that the regression coefficient value of the fitting resistance is 0.888 ($t = 605.206$, $p = 0.000 \ll 0.01$), which means that the fitting resistance will have a significant positive influence on the actual resistance. Therefore, a three-dimensional curved surface equivalent resistance model can be used to predict the curved surface resistance change.

Table 4. Linear regression analysis of the three-dimensional equivalent resistance model and the actual resistance.

	Unstandardized Coefficients		Standardized Coefficients	t	p	R^2	F
	B	Std. Error	*Beta*				
Constant	0.218	0.005	-	41.265	0	0.867	366,274.18
Fitting resistance	0.888	0.001	0.99	605.26	0		

From the analysis in **Figure 11**, there is a power function relationship between equivalent resistance and burst depth. As the burst depth increases, the resistance change rate tends to be flat after rapid increases in resistance. This change rule better explains three-dimensional curved surface strain-resistance changes, which are consistent with the results of the curved surface resistance test. Therefore, a theoretical formula of equivalent resistance can explain the three-dimensional curved surface strain resistance change rule.

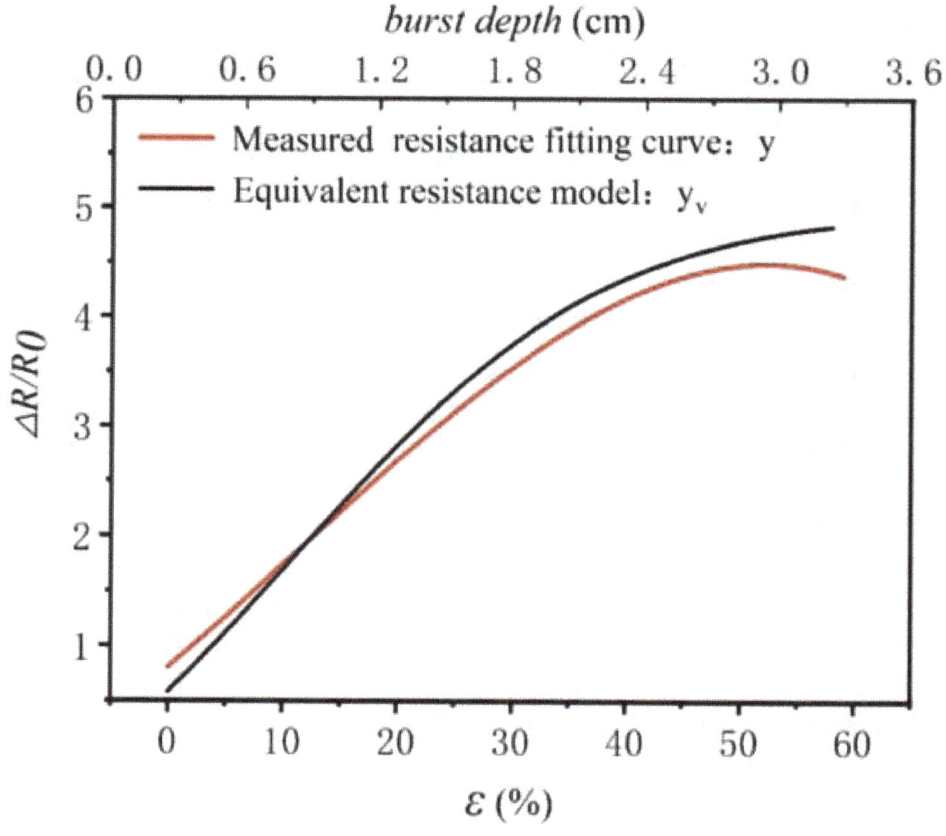

Figure 11. The theoretical model of equivalent resistance and the results of the burst test when the three-dimensional surface was strained.

4. CONCLUSIONS

In this study, two in-plane equivalent resistance models of knitted strain sensors and a three-dimensional equivalent resistance were presented. It is verified that these three resistance models can predict the resistance change of knitted strain sensors. Based on the in-plane strain, a macro–micro equivalent resistance model and topological equivalent resistance model are established. The macro–micro equivalent resistance model is a combination of two structural models, based on length resistance from a macro point of view and loop transfer from a micro point of view. The topological equivalent resistance model is based on a single loop and considers the superimposed resistance between the loops.

Comparing the in-plane tensile experimental data with the resistance models, the topological resistance model has a higher prediction and fitting accuracy, but the calculation is more complex, which is suitable for small-area knitted sensor applications. The macro–micro equivalent resistance model is simple and quick to calculate, suitable for large-area knitting sensor applications. Based on volume resistivity, a three-dimensional curved surface equivalent resistance model is established, which has a significant positive correlation effect on the change of the curved surface strain resistance. Furthermore, the results strongly suggest that the resistance change occurrence is directly related to the length resistance, loop transfer, and burst depth of the knitted sensor.

5. CONCISE SUMMARY

The dynamic equivalent resistance is a major index that determines the sensing performance of knitted strain sensors, and has the characteristics of in-plane and three-dimensional curved strain sensing. Therefore, in addition to establishing the in-plane equivalent resistance, it is necessary to establish a three-dimensional equivalent resistance model to fully explain the surface sensing performance. This project establishes two equivalent resistance models of knitted strain sensors under in-plane deformation and one equivalent resistance model of three-dimensional curved surface strain. Based on the length of resistance and the geometric topological structure, an in-plane strain macro–micro equivalent resistance model and a topological equivalent resistance model are established, respectively. In addition, a three-dimensional curved surface equivalent resistance model is created based on the volume resistance. By comparing the theoretical model with the experimental data, the results prove that the proposed in-plane and three-dimensional models can be utilized to calculate the resistance change of knitted strain sensors. Length resistance, coil transfer, and curved surface deformation depth are the main factors that affect the equivalent resistance of knitted strain sensors.

REFERENCES

1. Jakubas, A.; Łada-Tondyra, E. A Study on Application of the Ribbing Stitch as Sensor of Respiratory Rhythm in Smart Clothing Designed for Infants. *J. Text. Inst.* **2018**, *109*, 1208–1216. **[Google Scholar] [CrossRef]**
2. Jost, K.; Durkin, D.P.; Haverhals, L.M.; Brown, E.K.; Langenstein, M.; De Long, H.C.; Trulove, P.C.; Gogotsi, Y.; Dion, G. Natural Fiber Welded Electrode Yarns for Knittable Textile Supercapacitors. *Adv. Energy Mater.* **2015**, *5*, 1401286. **[Google Scholar] [CrossRef]**
3. Babu, V.J.; Anusha, M.; Sireesha, M.; Sundarrajan, S.; Abdul Haroon Rashid, S.S.A.; Kumar, A.S.; Ramakrishna, S. Intelligent Nanomaterials for Wearable and Stretchable Strain Sensor Applications: The Science behind Diverse Mechanisms, Fabrication Methods, and Real-Time Healthcare. *Polymers* **2022**, *14*, 2219. **[Google Scholar] [CrossRef] [PubMed]**
4. Chen, Q.; Shu, L.; Zheng, R.; Fu, B.; Fan, J. Electrical Resistance of Stainless Steel/Polyester Blended Knitted Fabrics for Application to Measure Sweat Quantity. *Polymers* **2021**, *13*, 1015. **[Google Scholar] [CrossRef] [PubMed]**
5. Wang, J.; Lu, C.; Zhang, K. Textile-Based Strain Sensor for Human Motion Detection. *Energy Environ. Mater.* **2020**, *3*, 80–100. **[Google Scholar] [CrossRef] [Green Version]**
6. Agcayazi, T.; Chatterjee, K.; Bozkurt, A.; Ghosh, T.K. Flexible Interconnects for Electronic Textiles. *Adv. Mater. Technol.* **2018**, *3*, 1700277. **[Google Scholar] [CrossRef]**
7. Zhou, X.; Hu, C.; Lin, X.; Han, X.; Zhao, X.; Hong, J. Polyaniline-Coated Cotton Knitted Fabric for Body Motion Monitoring. *Sens. Actuators A Phys.* **2021**, *321*, 112591. **[Google Scholar] [CrossRef]**
8. Luo, Y.; Verpoest, I. Biaxial Tension and Ultimate Deformation of Knitted Fabric Reinforcements. *Compos. Part A Appl. Sci. Manuf.* **2002**, *33*, 197–203. **[Google Scholar] [CrossRef]**
9. Isaia, C.; McNally, D.S.; McMaster, S.A.; Branson, D.T. Effect of Mechanical Preconditioning on the Electrical Properties of Knitted Conductive Textiles during Cyclic Loading. *Text. Res. J.* **2019**, *89*, 445–460. **[Google Scholar] [CrossRef]**
10. Zhang, X.; Zhong, Y. An Improved Theoretical Model of the Resistive Network for Woven Structured Electronic Textile. *J. Text. Inst.* **2020**, *111*, 235–248. **[Google Scholar] [CrossRef]**
11. Zhao, Y.; Li, L. A Simulation Model of Electrical Resistance Applied in Designing Conductive Woven Fabrics—Part II: Fast Estimated Model. *Text. Res. J.* **2018**, *88*, 1308–1318. **[Google Scholar] [CrossRef]**

12. Li, X.; Koh, K.H.; Xue, J.; So, C.H.; Xiao, N.; Tin, C.; Wai, K.; Lai, C. 1D-2D Nanohybrid-Based Textile Strain Sensor to Boost Multiscale Deformative Motion Sensing Performance. *Nano Res.* **2022**, 1–12. **[Google Scholar] [CrossRef]**

13. Xie, J.; Jia, Y.; Miao, M. High Sensitivity Knitted Fabric Bi-Directional Pressure Sensor Based on Conductive Blended Yarn. *Smart Mater. Struct.* **2019**, *28*, 035017. **[Google Scholar] [CrossRef]**

14. Wang, J.; Long, H.; Soltanian, S.; Servati, P.; Ko, F. Electromechanical Properties of Knitted Wearable Sensors: Part 1—Theory. *Text. Res. J.* **2014**, *84*, 3–15. **[Google Scholar] [CrossRef]**

15. Gioberto, G.; Dunne, L.E. Overlock-Stitched Stretch Sensors: Characterization and Effect of Fabric Property. *J. Text. Appar. Technol. Manag.* **2013**, *8*, 3. **[Google Scholar]**

16. Li, W.; Pei, Z. Strain-Sensing Fiber with a Core–Sheath Structure Based on Carbon Black/Polyurethane Composites for Smart Textiles. *Text. Res. J.* **2021**, *91*, 1907–1923. **[Google Scholar] [CrossRef]**

17. Liu, S.; Tong, J.; Yang, C.; Li, L. Smart E-Textile: Resistance Properties of Conductive Knitted Fabric—Single Pique. *Text. Res. J.* **2017**, *87*, 1669–1684. **[Google Scholar] [CrossRef]**

18. Ehrmann, A.; Heimlich, F.; Brücken, A.; Weber, M.; Haug, R. Suitability of Knitted Fabrics as Elongation Sensors Subject to Structure, Stitch Dimension and Elongation Direction. *Text. Res. J.* **2014**, *84*, 2006–2012. **[Google Scholar] [CrossRef]**

19. Liu, S.; Liu, Y.; Li, L. The Impact of Different Proportions of Knitting Elements on the Resistive Properties of Conductive Fabrics. *Text. Res. J.* **2019**, *89*, 881–890. **[Google Scholar] [CrossRef]**

20. Vallett, R.; Knittel, C.; Christe, D.; Castaneda, N.; Kara, C.D.; Mazur, K.; Liu, D.; Kontsos, A.; Kim, Y.; Dion, G. Digital Fabrication of Textiles: An Analysis of Electrical Networks in 3D Knitted Functional Fabrics. In Proceedings of the Micro- and Nanotechnology Sensors, Systems, and Applications IX, Anaheim, CA, USA, 9–13 April 2017; Volume 10194, p. 1019406. **[Google Scholar]**

21. Liang, X.; Cong, H.; Dong, Z.; Jiang, G. Size Prediction and Electrical Performance of Knitted Strain Sensors. *Polymers* **2022**, *14*, 2354. **[Google Scholar] [CrossRef]**

22. Zhang, Y.; Long, H. Resistive Network Model of the Weft-Knitted Strain Sensor with the Plating Stitch-Part 1: Resistive Network Model under Static Relaxation. *J. Eng. Fiber. Fabr.* **2020**, *15*, 1558925020944563. **[Google Scholar] [CrossRef]**

23. Zhang, Y.; Long, H. Resistive Network Model of the Weft-Knitted Strain Sensor with the Plating Stitch-Part 2: Resistive Network Model during the Elongation along Course Direction. *J. Eng. Fiber. Fabr.* **2020**, *15*, 1558925020969475. **[Google Scholar] [CrossRef]**

24. Ayodele, E.; Zaidi, S.A.R.; Scott, J.; Zhang, Z.; Hafeez, M.; McLernon, D. The Effect of Miss and Tuck Stitches on a Weft Knit Strain Sensor. *Sensors* **2021**, *21*, 358. **[Google Scholar] [CrossRef]**

25. Tajin, M.A.S.; Amanatides, C.E.; Dion, G.; Dandekar, K.R. Passive UHF RFID-Based Knitted Wearable Compression Sensor. *IEEE Internet Things J.* **2021**, *8*, 13763–13773. **[Google Scholar] [CrossRef]**

26. Wang, X.; Yang, B.; Li, Q.; Wang, F.; Tao, X. Modeling the Stress and Resistance Relaxation of Conductive Composites-Coated Fabric Strain Sensors. *Compos. Sci. Technol.* **2021**, *204*, 108645. **[Google Scholar] [CrossRef]**

27. Zhao, Y.; Tong, J.; Yang, C.; Chan, Y.F.; Li, L. A Simulation Model of Electrical Resistance Applied in Designing Conductive Woven Fabrics. *Text. Res. J.* **2016**, *86*, 1688–1700. **[Google Scholar] [CrossRef]**

28. Li, L.; Liu, S.; Ding, F.; Hua, T.; Au, W.M.; Wong, K.S. Electromechanical Analysis of Length-Related Resistance and Contact Resistance of Conductive Knitted Fabrics. *Text. Res. J.* **2012**, *82*, 2062–2070. **[Google Scholar] [CrossRef]**

29. Li, L.; Au, W.M.; Hua, T.; Wong, K.S. Design of a Conductive Fabric Network by the Sheet Resistance Method. *Text. Res. J.* **2011**, *81*, 1568–1577. **[Google Scholar] [CrossRef]**

30. Veeramuthu, L.; Venkatesan, M.; Benas, J.S.; Cho, C.J.; Lee, C.C.; Lieu, F.K.; Lin, J.H.; Lee, R.H.; Kuo, C.C. Recent Progress in Conducting Polymer Composite/Nanofiber-Based Strain and Pressure Sensors. *Polymers* **2021**, *13*, 4281. **[Google Scholar] [CrossRef]**

31. Wang, J.; Long, H.; Soltanian, S.; Servati, P.; Ko, F. Electro-Mechanical Properties of Knitted Wearable Sensors: Part 2—Parametric Study and Experimental Verification. *Text. Res. J.* **2014**, *84*, 200–213. **[Google Scholar] [CrossRef]**

32. Tokarska, M.; Orpel, M. Study of Anisotropic Electrical Resistance of Knitted Fabrics. *Text. Res. J.* **2019**, *89*, 1073–1083. **[Google Scholar] [CrossRef]**

33. Tokarska, M. Characterization of Electro-Conductive Textile Materials by Its Biaxial Anisotropy Coefficient and Resistivity. *J. Mater. Sci. Mater. Electron.* **2019**, *30*, 4093–4103. **[Google Scholar] [CrossRef][Green Version]**

34. Xie, J.; Long, H. Investigation on the Relation between Structure Parameters and Sensing Properties of Knitted Strain Sensor under Strip Biaxial Elongation. *J. Eng. Fiber. Fabr.* **2015**, *10*, 155892501501000. **[Google Scholar] [CrossRef]**

35. Yu, R.; Zhu, C.; Wan, J.; Li, Y.; Hong, X. Review of Graphene-Based Textile Strain Sensors, with Emphasis on Structure Activity Relationship. *Polymers* **2021**, *13*, 151. **[Google Scholar] [CrossRef] [PubMed]**

36. Xie, J.; Long, H. Equivalent Resistance Calculation of Knitting Sensor under Strip Biaxial Elongation. *Sens. Actuators A Phys.* **2014**, *220*, 118–125. **[Google Scholar] [CrossRef][Green Version]**

37. Li, L.; Au, W.M.; Wan, K.M.; Wan, S.H.; Chung, W.Y.; Wong, K.S. A Resistive Network Model for Conductive Knitting Stitches. *Text. Res. J.* **2010**, *80*, 935–947. **[Google Scholar] [CrossRef]**

38. Yang, K.; Song, G.L.; Zhang, L.; Li, L.W. Modelling the Electrical Property of 1 × 1 Rib Knitted Fabrics Made from Conductive Yarns. In Proceedings of the 2009 2nd International Conference on Information and Computing Science, ICIC 2009, Manchester, UK, 21–22 May 2009; Volume 4, pp. 382–385. **[Google Scholar]**

39. Gioberto, G.; Dunne, L. Theory and Characterization of a Top-Thread Coverstitched Stretch Sensor. In Proceedings of the Conference Proceedings—IEEE International Conference on Systems, Man and Cybernetics, Seoul, Korea, 14–17 October 2012; pp. 3275–3280. **[Google Scholar]**

40. Zhang, H.; Tao, X.; Wang, S.; Yu, T. Electro-Mechanical Properties of Knitted Fabric Made From Conductive Multi-Filament Yarn Under Unidirectional Extension. *Text. Res. J.* **2005**, *75*, 598–606. **[Google Scholar] [CrossRef]**

41. Zhang, H.; Tao, X.; Yu, T.; Wang, S. Conductive Knitted Fabric as Large-Strain Gauge under High Temperature. *Sens. Actuators A Phys.* **2006**, *126*, 129–140. **[Google Scholar] [CrossRef]**

42. Atalay, O.; Tuncay, A.; Husain, M.D.; Kennon, W.R. Comparative Study of the Weft-Knitted Strain Sensors. *J. Ind. Text.* **2017**, *46*, 1212–1240. **[Google Scholar] [CrossRef]**

43. Lin, Z.I.; Lou, C.W.; Pan, Y.J.; Hsieh, C.T.; Huang, C.H.; Huang, C.L.; Chen, Y.S.; Lin, J.H. Conductive Fabrics Made of Polypropylene/Multi-Walled Carbon Nanotube Coated Polyester Yarns: Mechanical Properties and Electromagnetic Interference Shielding Effectiveness. *Compos. Sci. Technol.* **2017**, *141*, 74–82. **[Google Scholar] [CrossRef]**

44. Hong, J.; Pan, Z.; Zhewang; Yao, M.; Chen, J.; Zhang, Y. A Large-Strain Weft-Knitted Sensor Fabricated by Conductive UHMWPE/PANI Composite Yarns. *Sens. Actuators A Phys.* **2016**, *238*, 307–316. **[Google Scholar] [CrossRef]**

45. Li, Q.; Tao, X. A Stretchable Knitted Interconnect for Three-Dimensional Curvilinear Surfaces. *Text. Res. J.* **2011**, *81*, 1171–1182. **[Google Scholar] [CrossRef]**

46. Yan, Y.; Jiang, S.; Zhou, J. Three-Dimensional Stretchable Knitted Design with Transformative Properties. *Text. Res. J.* **2021**, *91*, 1020–1036. [**Google Scholar**] [**CrossRef**]

47. Zhang, Y.T.; Hu, X.; Chen, H.X.; Wang, M.Y.; Chen, W.J.; Fang, X.Y.; Tan, Z.Z. Resistance Theory of General 2 × n Resistor Networks. *Adv. Theory Simul.* **2021**, *4*, 2000255. [**Google Scholar**] [**CrossRef**]

CHAPTER-19

UTILIZING FINITE ELEMENT METHOD FOR THE DESIGN AND ANALYSIS OF COMPOSITE MOTOR CASE CONNECTORS IN SOLID ROCKET PROPULSION SYSTEMS

1. INTRODUCTION: AN OVERVIEW

The development of advanced propulsion systems, particularly in the context of solid rocket motors, demands meticulous attention to structural design and analysis. This study focuses on the intricate task of designing and analyzing a critical component within these systems—the composite motor case connector. Employing the Finite Element Method (FEM), this research aims to provide a comprehensive understanding of the connector's structural behaviour under diverse operational conditions. By delving into the complexities of solid rocket technology, the investigation addresses the challenges associated with optimizing the performance and reliability of the composite motor case connector. The utilization of FEM serves as a powerful tool in this endeavor, offering detailed insights into the mechanical aspects of the connector's design. This study not only contributes to the advancement of solid rocket propulsion technology but also underscores the significance of employing sophisticated analytical methods for ensuring the structural integrity and efficiency of crucial components in such high-performance systems.

Composite materials can be classified according to the type of strengthening material into particle-reinforced and fiber-reinforced composite materials. Furthermore, fiber-reinforced composite materials can be divided into short fiber-reinforced and long fiber-reinforced composite materials [1]. A solid rocket composite shell is a kind of long fiber-reinforced composite material, which is preimpregnated with carbon fiber (or glass fiber) and wound around the core mold layer by layer, before solidification at a certain temperature. The solid rocket motor case (SRMC) is mainly composed of a tube structure, head, insulation layer, and skirt, and it has been widely applied in space vehicles, missile weapons, and other fields as a crucial part of the rocket motor. An SRMC connector is employed to connect the engine nozzle and ignition device, as shown in **Figure 1**. The design of the case connector structure significantly contributes to the development of the composite case. Moreover, its performance directly affects the blasting performance of the composite case. Due to their light weight, high strength, and high stiffness, fiber-reinforced composites can efficiently decrease the structure mass of the SRMC and increase the range of the rocket, providing great military and economic benefits [**2**]. For example, the range of a strategic missile can be increased by 16 km if the mass of the third structure of the solid rocket engine is reduced by 1 kg [**3**].

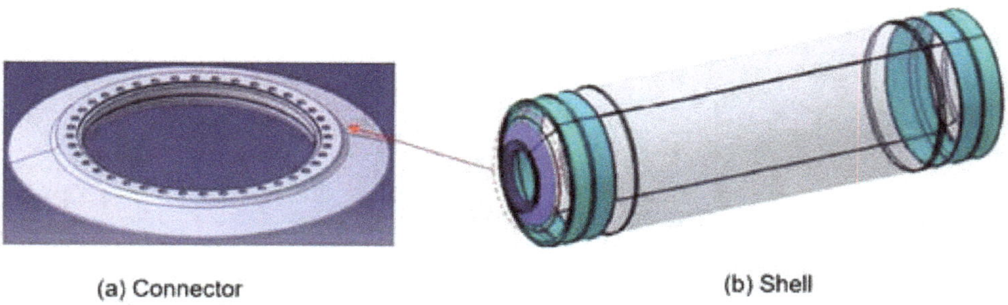

Figure 1. Schematic of SRMC connector and shell: (a) the connector to the shell body (b) the shell body of the rocket.

At present, carbon fiber-reinforced composites are widely used for strategic missiles and delivery systems, with several related papers published in this field. Ramanjaneyulu et al. [**4**] explored the SRMC using the finite element method, revealing that the hoop stress was gradually increased from the outer layer to the inner layer in all parts of the SRMC. Özaslan et al. [**5**] designed and analyzed a filament wound composite SRMC with finite element analysis and compared burst tests regarding the fiber direction strain distribution through the outer surface of the motor case to verify the analysis. Niharika et al. [**6**] used the simulation software ANSYS (R 18.0, ANSYS Inc., Canonsburg, PA, USA) to design a composite rocket motor casing. Hossam et al. [**7**] proposed that filament winding was the best technique for the production of composite pressure vessels (CPVs) in a short time, and different materials (including conventional and composite materials) were suitable for the design of SRMC structures. They also studied and summarized the optimum design of SRMC structures. Shaheen et al. [**8**] developed a 3D model of SRMC using CATIA V5R16 software (V5R16, Dassault Systems, Waltham, MA, USA) and conducted static structural analysis and linear buckling analysis for different stack-ups of a unidirectional carbon–epoxy composite and D6AC steel material rocket motor casing to specify the more efficient material. Prakash et al. [9] successfully designed and developed VEGA SRMC and discussed the effect of material mismatch on the static behaviour of the flex seal, which contributed imperatively to the development of composite rocket motor casings.

A large number of studies related to CERP have been carried out in other sectors. Juan [10] accurately modeled the winding layer of composite pressure vessels using the fiber winding pressure vessel plug-in WCM, as well as carried out a finite element simulation and blasting test verification on a type IV high-pressure hydrogen storage cylinder with the designed pressure of 70 MPa. Johansen et al. [11] designed a fiber winding analysis program and realized the winding analysis of any axisymmetric rotating body and its combination through an integrated CAD/CAE/CAM design method. Ambach [12] combined CFRP with steel and applied it to the manufacturing of an automobile roof, revealing that the mechanical properties of the material achieved good performance in terms of the crushing resistance of the automobile roof. Wang [13] explored the use of carbon fiber composite materials in biomedical science. Using barium titanate–hydroxyapatite (BT–HA) composite material as the matrix, Cf/BT–HA composite material was prepared to improve the artificial bone due to poor mechanical properties. Liang [14] studied the application of carbon fiber composite materials in bogies of rail transit vehicles, considering the properties of carbon fiber composite materials, such as high strength, high toughness, fatigue resistance, high temperature resistance, corrosion resistance, and light weight; he proposed a rectification plan for the use of carbon fiber composite material as a safety support in current vehicles. In terms of the spinning process, Kovarskii et al. [15] analyzed the structure of carbon fibers such as T800HB using EPR spectroscopy and X-ray diffraction, and they found that the microstructure of carbon fibers is directly related to their mechanical properties.

To date, many theoretical models related to SRMCs have been reported [16,17,18,19]. However, research on SRMC connectors is still insufficient. Generally, the case connector is the main force component, and the loading condition is complex. Meanwhile, the case connector is extremely sensitive to internal imperfections, necessitating methods to effectively improve its mechanical properties and dimensional accuracy. As is known, the SRMC connector operates in a high-temperature environment. Although the connector's external layer is protected by an insulating layer, the surface temperature can still reach up to a maximum of 250 °C. Furthermore, the dimensional stability of the connector is another basic item, and titanium alloy with excellent comprehensive properties has been broadly used in the connector. However, titanium alloy is expensive with high density, making it a single component with a large mass. The metal connector accounts for more than 15% of the total mass of the case. Furthermore, the process of manufacturing the metal connector is rather complex with a long cycle and high cost. Thus, it is urgent to develop a new material that can replace the metal material in the SRMC connector.

In this study, finite element analysis of the SRMC connector was performed. The lay-up and optimum structure designs of the connector were exported. The FEM simulation results were shown to be similar to experimental results. Thus, the performance of the optimum design was successfully improved.

2. EXPERIMENTAL ANALYSIS

The SRMC connector RS05A was made by using the mold pressing process with carbon fiber T300 fabric prepreg (Toray Inc., Lacq, France). The thickness and density of the single layer were 0.235 mm and 1.55 g/cm^3, respectively. Toray T700SC (12K, Toray Inc., Lacq, France)) carbon fiber was employed to produce a motor case with fiber winding technology. The fiber winding shell of the solid rocket motor was made of T700 fiber/epoxy resin composite material (Kosan Inc., Tokyo, Japan),, with a resin content of 32% and fiber content of 150 g, a viscosity of 300 mPa·s at room temperature, an opening period of 8–12 h, a curing temperature of 150 °C for 4 h, and a glass transition temperature of 170 °C. The bushing was embedded in the composite made of aluminum (AL7075-T6, Moju Inc., Shanghai, China) material, as shown in **Figure 2b**. The mechanical properties of carbon fiber, P700-1M resin, and AL are presented in **Table 1**.

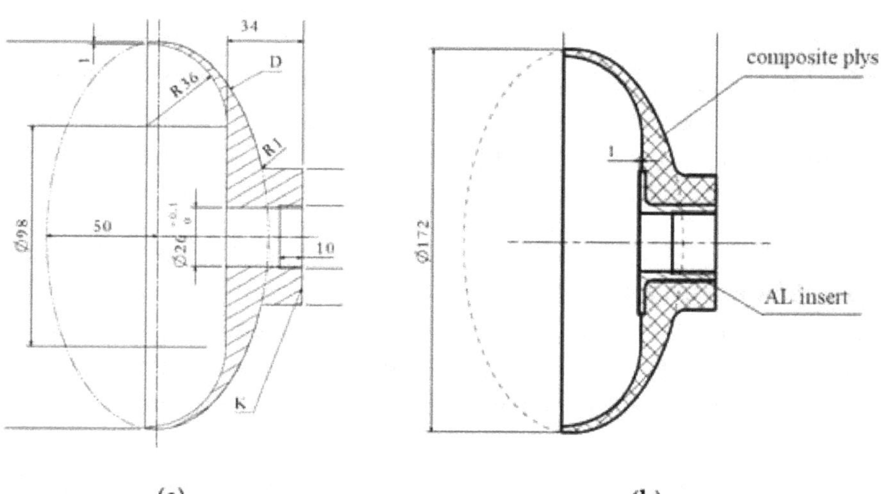

(a) (b)

Figure 2. Profile of the RS05A front connector: (a) primary front connector (b) after first-round optimization.

Table 1. Mechanical properties of the materials.

Property Items		Lamina Property	Resin	AL7075-T6
Tensile strength (MPa)	0°	665	70	570
	90°	552		
Tensile modulus (GPa)	0°	55.7	3.6	72
	90°	56.3		
Compression strength (MPa)	0°	500		
	90°	500		
Compression modulus (GPa)	0°	55.7		
	90°	56.3		

3. STRUCTURAL DESIGN

The primary connector structure of the RS05A front connector is illustrated in **Figure 1**a. After the first round of optimization using the finite element method, the length of the connector was adjusted to make the street longer, so as to fit more closely with the shell wall, as exhibited in **Figure 1**b. The primary front connector was 10 mm from the opening cut, while the optimized one was 20 mm from the opening cut.

Since the connector was fixed on the case using bolts, the primary RS05A metal connector was designed with M28 × 1.5 threaded hole in the middle. The bushing made of AL-7075-T6 was embedded in the composite connector to solve the problem of the composite being difficult to use as the metal connector.

According to the actual prepreg lay-up effect, the RS05A composite connector was designed to reduce the number of multilayer step structures during the production process, facilitating the insertion of the prepreg in the step structures. Meanwhile, the AL bushing embedded in the composite structure was adjusted into trapezoidal modes in order to ensure the flatness of the inner surface of the case and the thickness of the bushing root. In this way, an optimal design could be achieved.

4. RS05A LAY-UP MODE

Considering the operability of the actual production process, the final optimized lay-up is illustrated in **Figure 3**a. The optimized lamination of the composite was cured by a secondary co-curing process. The lamination was laid on the core die, and the lower half of the connector was cured by a molding process after lamination was completed. The secondary co-curing treatment was conducted when the top end face of the connector was laid up again. After curing, the connector was machined to the theoretical shape. Finally, the intermediate insert was embedded into the composite connector body using an adhesive.

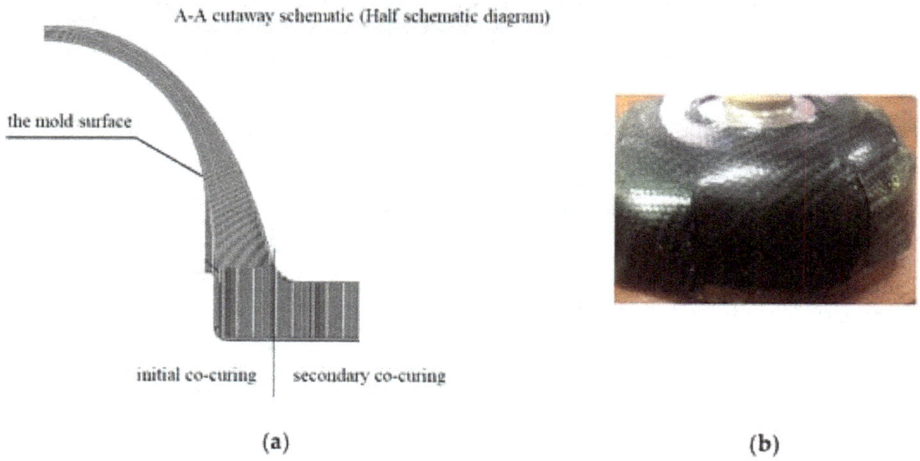

Figure 3. Lay-up diagram of the preferred scheme: (a) lay-up diagram (b) cutting opening mode.

The material used for the composite connector was carbon fiber T300 biaxial fabric, which was spread as an isotropic material in the form of a patchwork butt. The prepreg layering table of the connector structure is shown in **Table 2.** Given the large radian shape of the front connector, it was necessary to shear the pavement. The cutting opening mode is presented in **Figure 3**b.

Table 2. The layering table of the connector structure.

Serial Number	Lay Up	Thickness (mm)	Angle (°)	Layer Number
1	Twill weaves of carbon fiber T300	0.225	0/90	1001
2	Twill weaves of carbon fiber T300	0.225	45/-45	1002
3	Twill weaves of carbon fiber T300	0.225	0/90	1003
4	Twill weaves of carbon fiber T300	0.225	45/-45	1004
5	Twill weaves of carbon fiber T300	0.225	0/90	1005
6	Twill weaves of carbon fiber T300	0.225	45/-45	1006
...	...	...	...	...
163	Twill weaves of carbon fiber T300	0.225	0/90	1163
164	Twill weaves of carbon fiber T300	0.225	45/-45	1164
165	Prepreg of carbon fiber T300	0.145	0	2001
166	Prepreg of carbon fiber T300	0.145	45	2002
167	Prepreg of carbon fiber T300	0.145	-45	2003
168	Prepreg of carbon fiber T300	0.145	90	2004
169	Prepreg of carbon fiber T300	0.145	0	2005
170	Prepreg of carbon fiber T300	0.145	45	2006
171	Prepreg of carbon fiber T300	0.145	-45	2007

Serial Number	Lay Up	Thickness (mm)	Angle (°)	Layer Number
172	Prepreg of carbon fiber T300	0.145	90	2008
...	...	...	...	...
285	Prepreg of carbon fiber T300	0.145	0	2021
286	Prepreg of carbon fiber T300	0.145	45	2122
287	Prepreg of carbon fiber T300	0.145	-45	2123
288	Prepreg of carbon fiber T300	0.145	90	2124
289	Prepreg of carbon fiber T300	0.145	0	2125
Total thickness	**55.025 mm**			

5. FINITE ELEMENT MODEL

The shell layer was designed by grid theory [20], and the designed burst pressure was 15 MPa. The finite element computer software, ABAQUS (V6.13, Dassault Systems, Waltham, MA, USA), was employed for SRMC burst pressure simulation. The dimensions of the finite element model were the actual dimensions. The SRMC and connector were meshed with linear reduced integration solid elements (C3D8R), with a mesh size of approximately 10 mm. The model had a total of 23,355 cells and 29,020 nodes. Table 3 presents the front connector weight of different schemes. It can be seen that the front connector weight of the initial plan was 0.41 kg, while that of the optimized scheme was 0.408 kg; the corresponding weights of the AL insert were 0.056 kg and 0.0613 kg, respectively. The percentage weight loss of the optimized plan, final scheme, and experimental measurement was 31.4%, 31.0%, and 30.6%, respectively.

Table 3. Front connector weight of different schemes.

Location	AL Front Connector Scheme	The Initial Configuration	Optimized Scheme	Experimental Measurements
Front connector weight (kg)	0.68	0.41	0.408	0.411
AL insert weight (kg)	-	0.056	0.0613	0.0613
Total weight (kg)	0.68	0.466	0.469	0.472
Percentage weight loss (%)	-	31.4%	31.0%	30.6%

5.1. Lay-Up Information Table

The lay-up of the front connector of the RS05A composite was quasi-isotropic, the lay-up of the RS05A composite connector was divided into five directions (0°, 22.5°,

45°, 67.5°, and 90°), and the lay-up ratio was 1. In the actual lay-up, each layer was rotated by a certain angle to disperse the lap position and angle.

5.2. Loading and Constraints

Abaqus was used for linear loading calculation to achieve progressive failure analysis. In the finite element analysis, at each incremental step, the first equilibrium equation was solved, and the stress and strain of each layer of the element covered the stress and strain of the previous step. According to the damage mode, the stiffness could be reduced by changing the material parameters of the integral point. The equilibrium equation was re-established, and the next load increment step was substituted. If the structure relative stiffness value of the current load step (the ratio of the current stiffness to the initial stiffness) tended to zero and began to soften and enter the unloading state, the structure was considered to have lost the bearing capacity, necessitating the progressive failure analysis of the wound shell [**21**].

The RS05A Motor Case mainly bears internal pressure. Axial displacement constraints were applied to the middle part of the shell to avoid rigid body displacement in the finite element calculation, and cyclic symmetry conditions were applied to the sides of the shell and joint model. Uniformly distributed pressure was applied on the inner surface of the shell, increasing from 0 to 15 MPa. The boundary conditions for the SRMC and connector in ABAQUS are defined below. The X-axis translation of the RS05A motor case was constrained, in addition to the Y-direction translation of the upper and lower surface elements and the Z-direction translation of the front and rear surface elements. In other words, symmetric constraints were imposed on the motor case. The constraints are illustrated in **Figure 4**.

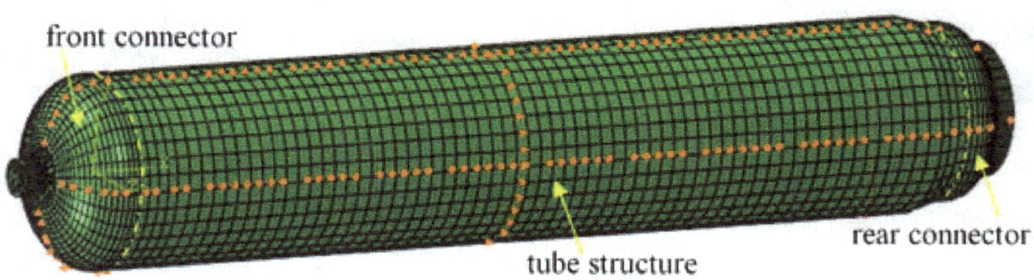

Figure 4. Schematic diagram of the RS05A motor case constraint.

The load condition was 15 MPa of internal blasting press.

Contact: Since the front connector and the insert are two parts, there may be relative friction between them. In this study, the contact constraint conditions were imposed on the bottom end face, the side of the front connector, and the bushing. The friction form was set using a friction coefficient of 1.5.

The SRMC failure criteria proposed by Hashin criteria [22] were applied to detect the failure modes in the fiber and matrix under both tension and compression failures, which involve four failure modes. The failure modes included in Hashin's criteria are expressed below.

Tensile fiber failure for $\sigma 11 \geq$ ٠:

$$\left(\frac{\sigma_{11}}{X_T}\right)^2 + \frac{\sigma_{12}^2 + \sigma_{13}^2}{S_{12}^2} \geq 1. \quad (1)$$

Compressive fiber failure for $\sigma 11 <$ ٠:

$$\left(\frac{\sigma_{11}}{X_c}\right)^2 \geq 1. \quad (2)$$

Tensile matrix failure for $\sigma 22 + \sigma 33 >$ ٠:

$$\frac{(\sigma_{22} + \sigma_{33})^2}{Y_T^2} + \frac{\sigma_{23}^2 - \sigma_{22}\sigma_{33}}{S_{23}^2} + \frac{\sigma_{12}^2 + \sigma_{13}^2}{S_{12}^2} \geq 1. \quad (3)$$

Compressive matrix failure for $\sigma 22 + \sigma 33 <$ ٠:

$$\left[\left(\frac{Y_C}{2S_{23}}\right)^2 - 1\right]\left(\frac{\sigma_{22} + \sigma_{33}}{Y_C}\right) + \frac{(\sigma_{22} + \sigma_{33})^2}{4S_{23}^2} + \frac{\sigma_{23}^2 - \sigma_{22}\sigma_{23}}{S_{23}^2} + \frac{\sigma_{12}^2 + \sigma_{13}^2}{S_{12}^2} \geq 1. \quad (4)$$

Interlaminar tensile failure for $\sigma 33 >$ ٠:

$$\left(\frac{\sigma_{33}}{Z_T}\right)^2 \geq 1. \quad (5)$$

$$\left(\frac{\sigma_{33}}{Z_C}\right)^2 \geq 1. \quad (6)$$

Here, the σij terms are components of the stress tensor, i and j are local coordinate axes parallel and transverse to the fibers in each ply, respectively, and the z-axis coincides with the through-thickness direction.

Statical analysis using FEM was performed for the RS05A Motor Case, where the connector received complicated stress under high internal pressure. The mechanical responses and damage morphology of the FE models were obtained.

6. RESULTS AND DISCUSSION

6.1. Analysis Results of the RS05A Front Connector

Pressure was applied on the shell; then, the shell was enlarged and deformation occurred in the middle of the front connector. This phenomenon was due to the existence of the pressure exerted internally. The deformation and maximum shear stress diagrams of the front connector under 15 MPa of blasting pressure are exhibited in **Figure 5.** It can be seen that the magnitude deformation of the front connector reached 5.376 mm. The maximum shear stress in the *XY*-direction was 2.57 MPa. **Table 4** presents the displacement and shear stress results of the design.

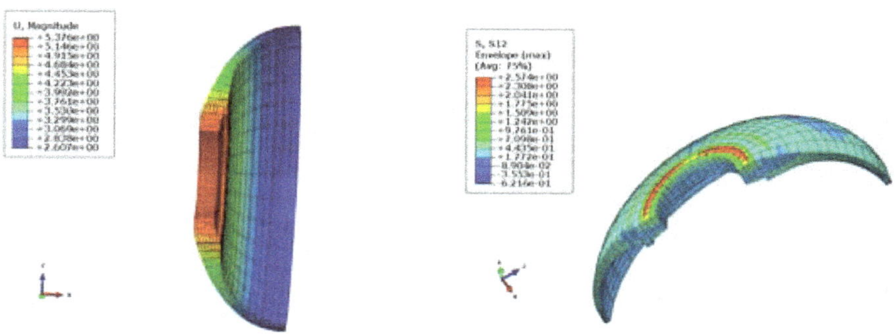

(a) (b)

Figure 5. Deformation and maximum shear stress diagrams of the front connector: (a) deformation diagram; (b) maximum shear stress in XY-direction (τxy = 257 × 10–2 MPa).

Table 4. Finite element calculation results.

		Final Preferred Solution		Experimental Measurements
	Allowable Value [23,24]	Calculated Value	Safety Factor	Experimental Value
Front connector deformation (mm)	-	1.92	-	1.98
Tensile stress in X-direction (MPa)	500	492.0	1.24	496.4
Compressive stress in X-direction (MPa)	-665	-158.9	1.02	-172.8
Tensile stress in Y-direction (MPa)	552	450.2	1.23	463.6
Compressive stress in Y-direction (MPa)	-500	-212.6	3.16	-235.8
Shear stress in XY-plane (MPa)	118	2.574	45.91	6.431

Von Mises stress of AL inserts (MPa)	505	332.4	1.52	362.3

The stress distribution of the front connector is presented in Figure 6 and Figure 7. As can be seen, the maximal tensile and compressive stress calculated using FEM was 492 MPa and -537 MPa, respectively. As shown in Table 4, the FEM results and experimental measurements were in agreement with the practical values.

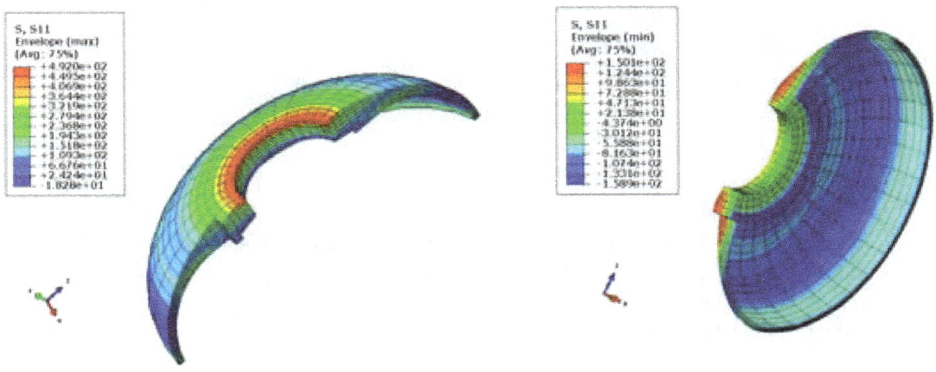

(a) (b)

Figure 6. Maximal tensile and compressive stress diagrams in the X-direction: (a) Tensile stress (σx = 4.92 × 102 MPa); (b) Compressive stress (σx = -1.589 × 102 MPa).

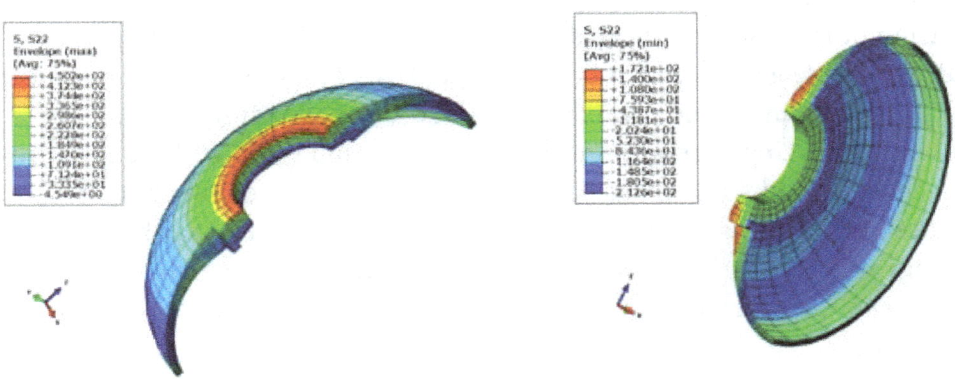

(a) (b)

Figure 7. Maximal tensile and compression stress diagram in the Y-direction: (a) Tensile stress (σy = 4.502 × 102 MPa); (b) Compression stress (σy = -2.126 × 102 MPa).

Figure 8 shows that the Von Mises stress of AL inserts was 332.4MPa. As shown in Table 4, the stress-strain values obtained from the simulations were all within the permissible limits obtained from the experiments.

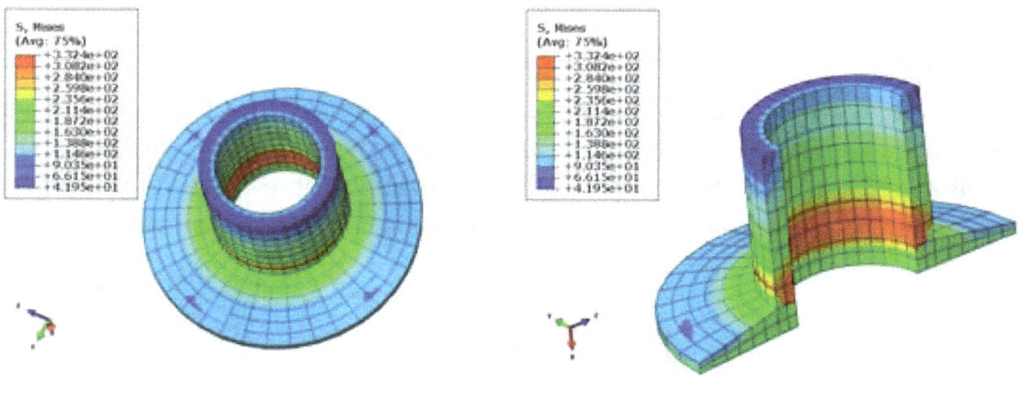

(a) (b)

Figure 8. Von Mises stress of AL inserts (σy = 332.4 MPa): (a) Von Mises stress (overall view); (b) Von Mises stress (partial view).

6.2. Experimental Results

A water pressure blasting experiment is designed for the shell to monitor the strain displacement change of the shell during blasting. Strain monitoring points were uniformly set on the shell and front connector, as shown in **Figure 9.** Deformation was relatively larger in the process of the booster, with resin shell cracking. Due to some damage of the strain gauge, the strain value could not be displayed. The complete results of the test points were generated, with each strain measuring point monitoring strain changes in both directions. The strain of point 8 at the small polar hole generated a sudden change. Two points were set for displacement change monitoring, which coincided with strain monitoring points 2 and 5. The results of extracting the two point shifts are shown in **Figure 10.** When the pressure reached 33 MPa, displacement occurred at both points. Combined with the strain displacement test results in the experiment, it can be seen that the actual burst pressure was 33 MPa. The hydrostatic test showed that the cylinder could meet the internal pressure of 15 MPa in working conditions and 33 MPa in blasting conditions.

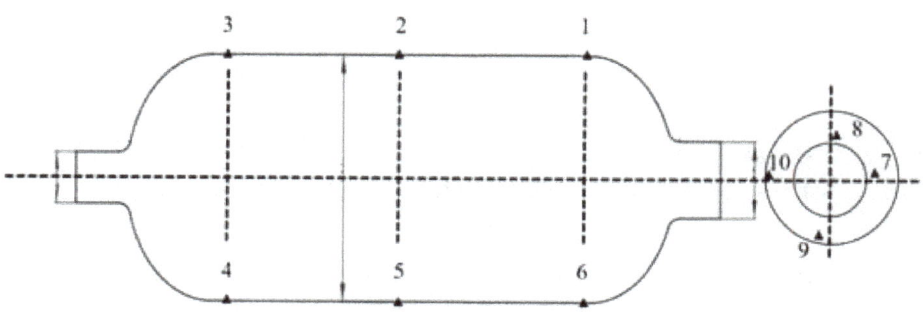

Figure 9. Stress measurement point.

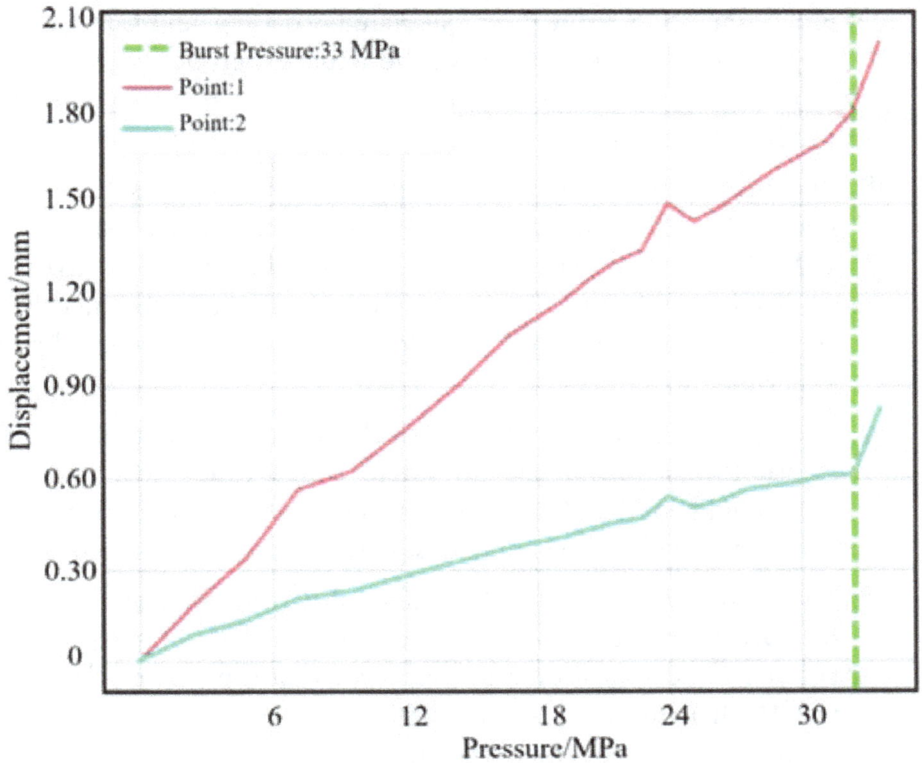

Figure 10. Tendency of displacement.

The calculation results of typical schemes are summarized in **Table 4**. As shown, the experimental measurements, such as the front connector deformation, tensile and compressive stress in the *XY*-direction, shear stress in *XY*-plane, and Von Mises stress of AL inserts, were preferred as the final solution.

The winding angle used in the calculation (18.5°) was the average winding angle, while the actual winding angle at the equator was about 26°, i.e., the winding angle from the middle part of the barrel to the equator of the back head changed from 18.5° to 26°, resulting in an increase in the actual torsional stiffness of the cylinder near the back head. Therefore, the measured circumferential strain value was small. In the actual working condition, the stress of the shell would be better than that of the proposed design, and no damage would occur with the calculated value. Both calculated stresses were within the range of allowable design values. The design scheme and calculation met the requirements, and the design proposal was reasonable.

7. CONCLUSIONS

- In this study, finite element analysis of the SRMC connector was performed. The lay-up and optimum structure designs of the connector were investigated. An experimental design was established, and the FEM simulation value was calculated. Loading and constrains were implemented in the FEM model. The actual experimental measurements were studied for a comparison. A blasting experiment was conducted to verify the simulation results.

- The maximum shear stress in the *XY*-direction was 2.57 MPa, the maximal tensile and compressive stress calculated using FEM was 492 MPa and -537 MPa, respectively, and the Von Mises stress of the AL insert was 332.4 MPa. The stress–strain values obtained from the simulations were all within the permissible limits obtained from the experiments.
- The results revealed that the calculated value of the final preferred solution was within the allowable range (**Table 4**), and at least 31% weight loss could be achieved (**Table 2**). This confirms that the performance of the optimum design was successfully improved.
- The accuracy of the modeling method was verified by analyzing the displacement and blasting pressure of the finite element simulation results. The comparison results showed that the FME result of blasting was 15 MPa, while the actual blasting was 33 MPa, suggesting that the simulated shell could meet the internal pressure in working conditions.

8. CONCISE SUMMARY

The connector is an essential component in the solid rocket motor case (SRMC), and its weight and performance can directly affect the blasting performance of SRMC. Considering the lightweight design of these structures, fiber-reinforced composite materials are used for the major components. In this study, the finite element analysis of the SRMC connector was performed. The lay-up design and structure optimum design of the connector were studied. Furthermore, the strain distribution on the composite body was compared with experimental measurements. The results demonstrate that the calculated value of the final preferred solution was within the allowable range, and at least 31% weight loss was achieved, suggesting that the performance of the optimum design was optimized. The comparison between the finite element calculation and the test results suggests that the design was within the allowable range and reasonable.

REFERENCES

1. Hou, X.; Qin, Y.; Ding, W.H. Load-bearing Capacity Analysis of Composite Case Structure of Solid Rocket Motor, Acta Materiae Compositae. *Acta Mater. Compos. Sin.* **2014**, *31*, 1343–1349. [**Google Scholar**]

2. Yang, M.; Chen, X.H.; Wu, C.B. Numerical Simulation for Burst Progressive Failure of Filament-Wound Composite Case in Solid Rocket Motor. *Mater. Mech. Eng.* **2012**, *36*, 92–96. **[Google Scholar] [CrossRef]**
3. Jiang, A.M.; Li, G.C.; Huang, W.D. Finite Element computation and parametric analysis on mechanical property of solid rocket motor bondline. *Chin. J. Explos. Propellants* **2012**, *35*, 54–59. **[Google Scholar]**
4. Ramanjaneyulu, V.; Murchy, V.B.; Mohan, R.C. Analysis of composite rocket motor case using finite element method. *Mater. Today Proc.* **2018**, *5*, 4920–4929. **[Google Scholar] [CrossRef]**
5. Özaslan, E.; Acar, B.; Yetgin, A. Design and validation of a filament wound composite rocket motor case. In Proceedings of the ASME 2018 Pressure Vessels and Piping Conference, Prague, Czech Republic, 20–15 July 2018; Volume 3A, pp. 7–1. **[Google Scholar]**
6. Niharika, B.; Varma, B.B. Design and analysis of composite rocket motor casing. *Mater. Sci. Eng.* **2018**, *455*, 1–8. **[Google Scholar] [CrossRef]**
7. Hossam, I.; Saleh, S.; Hamel, H. Review of challenges of the design of rocket motor case structures. In Proceedings of the 18th International Conference on Aerospace Sciences & Aviation Technology, Cairo, Egypt, 9–11 August 2021; pp. 1–13. **[Google Scholar]**
8. Shaheen, S.; Gupta, G.S. Design and analysis of carbon-epoxy composite rocket motor casing. *Int. J. Adv. Res. Innov. Ideas Educ.* **2015**, *1*, 859–871. **[Google Scholar]**
9. Prakash, D.; Murthy, V.B.; Mohan, R.C. Effect of material mismatch on static behavior of flex seal. *Mater. Today Proc.* **2017**, *4*, 2290–2297. **[Google Scholar] [CrossRef]**
10. Juan, P.B.R. 700 Bar Type IV High Pressure Hydrogen Storage Vessel Burst-Simulation and Experimental Validation. *Int. J. Hydrogen Energy* **2015**, *40*, 13183–13192. **[Google Scholar]**
11. Johansen, B.S.; Lystrup, A.; Jensen, M.T. CAD Path: A Complete Program for the CAD, CAE and CAM-winding of Advanced fibre composites. *J. Mater. Process. Technol.* **1998**, *77*, 194–200. **[Google Scholar] [CrossRef]**
12. Amach, M.F. Fibre Composite Strengthening of Thin Steel Passenger Vehicle Roof Structures. *Thin-Walled Struct.* **2014**, *74*, 1–11. **[Google Scholar]**
13. Wang, R.B.; Cheng, L.Q.; Lu, J.T.; Jin, R.Q.; He, M.Z.; Yan, S.H.; Jiang, D.M.; Deng, S.J.; Chu, X.C. Preparation and Properties of Carbon Fiber Reinforced BT-HA Composites. *Bull. Am. Ceram. Soc.* **2022**, *41*, 994–1001. **[Google Scholar]**

14. Liang, Y.; Chen, L.; Yang, J.Y.; Lyu, C.X.; Jia, H.L.; Wang, Y.S. Application of Carbon Fiber Composite Polymer on Urban Rail Transit Vehicle Bogy. *Urban Mass Transit.* **2020**, *23*, 5. **[Google Scholar]**
15. Kovarskii, A.L.; Kasparov, V.V.; Krivandin, A.V.; Shatalova, O.V.; Korokhin, R.A.; Kuperman, A.M. EPR spectroscopic and X-Ray diffraction studies of carbon fibers with different mechanical properties. *Russ. J. Phys. Chem.* **2017**, *11*, 233–241. **[Google Scholar] [CrossRef]**
16. Teja, P.S.; Sudhakar, B.; Dhass, A.D. Numerical and experimental analysis of hydroxyl-terminated poly-butadiene solid rocket motor by using ANSYS. *Mater. Today Proc.* **2020**, *33*, 308–314. **[Google Scholar] [CrossRef]**
17. Faber, J.; Schmidt-Eisenlohr, C.; Dickhut, T.; Ortmann, P. Concept study on optimized auxiliary material designs and application techniques for vacuum bagging of full-scale CFRP rocket boosters. In Proceedings of the 8th European Conference on Composite Materials, Athens, Greece, 25–28 June 2018; pp. 1–8. **[Google Scholar]**
18. Babu, P.M.; Krishna, G.B.; Prasad, B.S. Design & analysis of solid rocket motor casing for aerospace applications. *Int. J. Curr. Eng. Technol.* **2015**, *5*, 1947–1954. **[Google Scholar]**
19. Mohmmed, R.; Zhang, F.; Sun, B.Z.; Gu, B.H. Finite element analyses of low-velocity impact damage of foam sandwiched composites with different ply angles face sheets. *Mater. Des.* **2013**, *47*, 189–199. **[Google Scholar] [CrossRef]**
20. Chen, R.X. Netting analysis method for the filament-wound case design. *J. Solid Rocket. Technol.* **2003**, *26*, 30–32. **[Google Scholar]**
21. Gao, J.L.; Yu, J.S. Application of composite motor case in space carrier rocket. *Fiber Compos.* **2005**, *22*, 53–54. **[Google Scholar]**
22. Hashin, Z. Fatigue Failure Criteria for Unidirectional Fiber Composites. *J. Appl. Mech.* **1981**, *48*, 846–852. **[Google Scholar] [CrossRef]**
23. *GB 150-98*; Steel Presure Vessel. National Standardization Technical Committee on Pressure Vessels. Bejing China Standards Press: Beijing, China, 1998; pp. 26–51.
24. American Society of Mechanical Engineers. *Boiler and Pressure Vessel Code: Section V Division 1 and Division 2S*; Ju, D.Y., Translator; Chinese Mechanical Engineering Society, Pressure Vessel Society: Hefei, China, 1992; pp. 15–34. **[Google Scholar]**

CHAPTER-20

PREDICTION OF SIZE AND ELECTRICAL PERFORMANCE IN KNITTED STRAIN SENSORS

1. INTRODUCTION: AN OVERVIEW

In the burgeoning field of wearable technology, knitted strain sensors have emerged as pivotal components, offering a versatile means to capture and interpret physical deformation. This study delves into the intricate relationship between the size characteristics and electrical performance of knitted strain sensors. By exploring the predictive aspects of size and their influence on the electrical response, we aim to enhance the understanding of these sensors' behaviour under varying conditions. As wearables continue to evolve, optimizing the size of strain sensors becomes paramount for seamless integration into diverse applications, spanning from healthcare to sports and beyond.

With the foreseeable prosperity and integration of medical electrical devices, textile equipment, and human health monitoring equipment, wearable devices have attracted considerable attention from investigators due to their promising applications in human motion detection, soft robotics, electronic skins, and sensors [**1,2,3,4,5,6,7**]. Knitted strain sensors with the attributes of being lightweight, having good flexibility, and with a wide strain range have been emerging, promising a myriad of applications in the development of wearable sensing devices [**8,9,10,11,12,13**]. So far, plenty of knitted strain sensors are coated with conductive materials, such as graphene [**14,15,16**], polypyrrole [**17,18**], PEDOT: PSS [**19**], and nano-silver [**20,21**], showcasing the advantages of good stability, high sensitivity, and fast response. Various novel fabrication methods [**22,23,24**] have been proposed to knit the conductive yarn into the conductive fabric directly, forming an intelligent wearable device with sensing performance. However, these types of sensors are structurally unstable and have a small strain range, which definitely impedes their practical application [**25,26**].

In addition, yarn type, structure, and the interaction between loops are also important factors affecting the knitted sensor's performance. Production parameters, such as structural changes, spandex content, and the washing and ironing processes, play a fundamental role in determining the sensors' physical behaviour and sensing performance [**27**,**28**,**29**]. Most strain sensors based on knitted structures rely on resistance changes [**30**,**31**,**32**]. For instance, Liu et al. proposed a geometric model incorporated with a simplified resistive network, determining the resistance effect of conductive float stitches on knitted structures with different courses and wales [**33**]. To evaluate the overall degree of plane anisotropy of knitted fabrics, a new measurement procedure was established using the van der Pauw electrode configuration to solve the issue of measuring the edge resistance of fabrics [**34**]. Knitting elements with different proportions also have a considerable impact on the conductive fabric resistance [**35**]. Li et al. produced a resistance model of conductive knitted fabric under unidirectional stretching, superimposing length resistance and contact resistance to simulate the fabric resistance [**36**]. The as-developed resistance calculation systems reveal the discipline of resistance variation in conductive knitted fabrics, offering a theoretical reference in conductive fabric design.

The size prediction of knitted fabric is a considerable segment in the garment design and production process. For example, Liu et al. developed the size prediction model of warp-knitted Jacquard fabric to investigate the relationship between the yarn count, tensile density, and the size shrinkage rate, utilizing JavaScript and WebGL technologies to automatically generate clothing templates [**37**]. A structural model correlating the size of tubular knitted fabric with the loop geometric parameters was produced, deducing yarn-feeding parameters according to the elasticity and size requirements of the fabric [**38**]. Ulson et al. put forward a prediction system of circular-knitted cotton fabric, introducing an approach that saves time and money while improving knitted fabric quality for customers [**39**].

To date, the study on knitted strain sensors has mainly focused on the influence of elements of resistance, the resistance calculation model, and the sensor design and application, whereas there are few researches on size control and prediction. Intelligent garments monitoring human movements require that sensors be able to respond instantly and sensitively to changes in human joints and skin. In the actual development process, samples should be designed at least 2–3 times to determine the sensors' sizes at different parts of the human body, and the finishing process is also more complicated, resulting in a significant increase in cost and product losses. This makes the size control the focus and nodus in the design and fabrication of knitted strain sensors.

Herein, knitted strain sensors based on plain knit were successfully fabricated with nylon/spandex composite yarn and silver-plated nylon yarn using a flat knitting process. By investigating the influence of different knitting parameters on a sensor's size, the size prediction model was established. To assess its potential to serve as a strain sensor, electrical performance tests were carried out to probe its sensitivity, hysteresis deviation, working sense range, and repeatability under various stretching rates. Furthermore, the as-prepared strain sensor based on the size prediction model was applied to monitor different human movements in order to verify its accuracy, indicating its promising potential for use in intelligent wearable devices.

2. EXPERIMENTAL METHODS

2.1. Materials

In this experiment, silver-plated nylon yarn (222dtex/48F, the electrical resistance is 6.5 Ω/cm) was purchased from Qingdao Hengtong Weiye Special Fabric Technology Co., Ltd. (Qingdao, China). Nylon/spandex composite yarns (22/55dtex, 22/77dtex, and 44/77dtex) were purchased from Hubei Yutao Special Fiber Co., Ltd. (Chongyang, China). Elastic nylon filament (333dtex/24F) was obtained from Jiangsu Pingmei Yarn Industry Co., Ltd. (Hai'an, China).

2.2. Preparation of Knitted Strain Sensors

The plain knit structure has been proved to be a promising candidate as the fabric substrate due to its merits of compact structure, considerable flexibility, and relatively excellent electrical performance [40]. Figure 1 exhibits the pattern design and knitting process of the strain sensors. The machine parameters of a sensor's compression pattern transformed into an expanded pattern via automatic control instructions were set in SDS-ONE APEX3 pattern design system matching with a computerized flat knitting machine. Three yarn feeders were used in the knitting process, working from bottom to top.

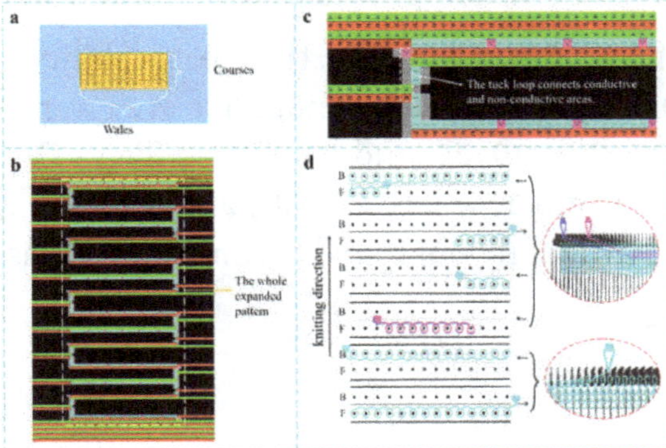

Figure 1. Pattern design and knitting process of the knitted strain sensor (a) compression pattern design (b) expanded pattern in the design system and (c) single pattern cycle (d) the knitting process and simulation of a single pattern.

The preparation process of the knitted strain sensor is illustrated in **Figure 2.** The elastic nylon filament was knitted to form the non-conductive area as a fabric substrate. Every loop in the conductive area consisted of two overlapping yarns, in which the silver-plated nylon yarn appeared at the technical surface of the strain sensors, with the nylon/spandex composite yarn at the back. The two areas were connected by a tuck loop.

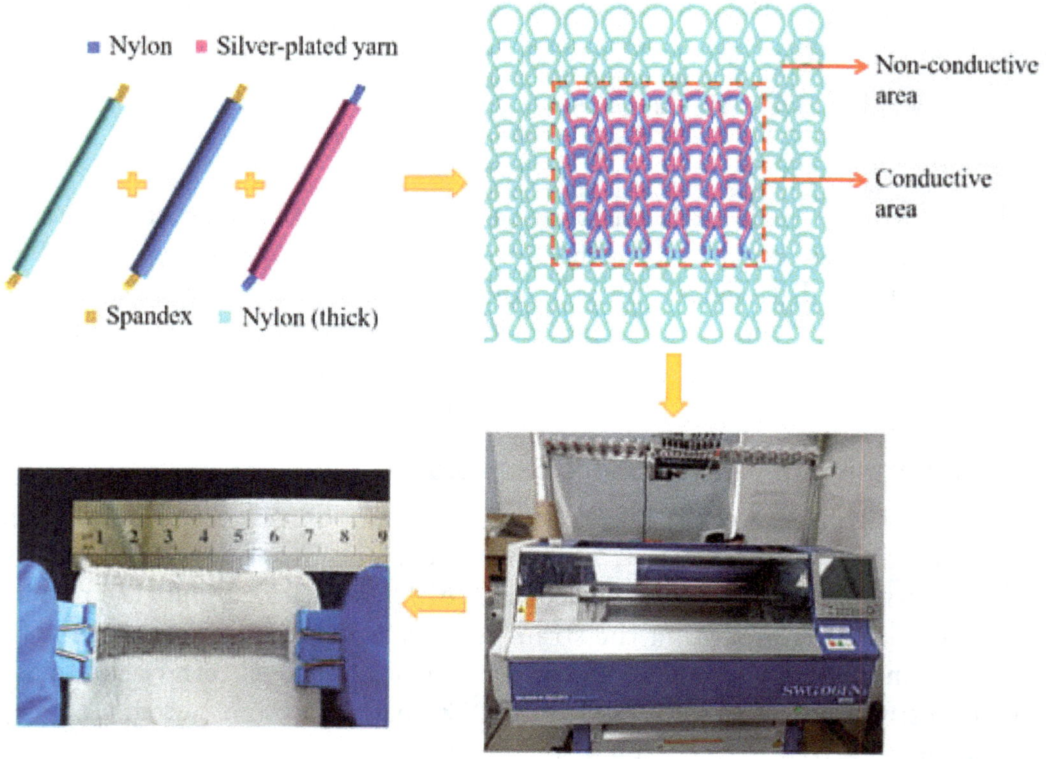

Figure 2. The fabrication process of the knitted strain sensor.

Considering the feasibility of knitting, the measurability, and the reduction in excess waste, the wales of the sensor's conductive area were set as 30, 40, and 50, and the courses were set as 10, 20, and 30, respectively. The specifications of all samples are shown in **Table 1**, where EN, SN, and NS are elastic nylon filament, silver-plated nylon yarn, and nylon/spandex composite yarns, respectively.

Table 1. Knitting parameters of the experimental samples.

Fabric No.	Wales	Courses	Yarn Composition			Spandex Content (%)
			Non-Conductive Area	Conductive Area		
				Plating Yarn	Ground Yarn	
F1	30	10	EN	SN	20/50 NS	28.5
F2					20/70 NS	22.2
F3					40/70 NS	36.3
F4		20			20/50 NS	28.5
F5					20/70 NS	22.2
F6					40/70 NS	36.3
F7		30			20/50 NS	28.5
F8					20/70 NS	22.2
F9					40/70 NS	36.3
F10	40	10			20/50 NS	28.5
F11					20/70 NS	22.2
F12					40/70 NS	36.3
F13		20			20/50 NS	28.5
F14					20/70 NS	22.2
F15					40/70 NS	36.3
F16		30			20/50 NS	28.5
F17					20/70 NS	22.2
F18					40/70 NS	36.3
F19	50	10			20/50 NS	28.5
F20					20/70 NS	22.2
F21					40/70 NS	36.3
F22		20			20/50 NS	28.5
F23					20/70 NS	22.2
F24					40/70 NS	36.3
F25		30			20/50 NS	28.5
F26					20/70 NS	22.2
F27					40/70 NS	36.3

2.3. Characterization and Measurements

2.3.1. Size-Change Test

Under the ironing condition of 150 °C and 0.4 MPa, the knitted strain sensors were ironed with a BOG energy-saving steam generator (Yancheng Xuanlang Machinery Equipment Co., Ltd., Yancheng, China). The size-change principle of the sensors was analyzed by SPSS regression analysis.

2.3.2. Electrical Performance Test

The electrical performance of a sensor largely depends on its strain. When the sensor is stretched, the contact conditions between the loops change, resulting in corresponding changes in its resistance [**41**]. In this paper, a sensor's resistance from 0-100% strain range was measured. The strain was implemented to the sensor by the electric stretching/compression test bench (Beijing Jipin Times Technology Co., Ltd., Beijing, China). The resistance was recorded using a CHI760 electrochemical analyzer (Shanghai Chenhua Instrument Co., Ltd., Shanghai, China), which presents the corresponding time-current (I-T) characteristic curve.

3. RESULTS AND DISCUSSIONS

3.1. Relationship between Knitting Factors and Sensor Size

The horizontal and vertical sizes of the conductive area in the sensors with different knitting factors were measured, including the gray and the finished samples. Size shrinkage of the samples was calculated by Formulas (1) and (2).

$$Sh=(Xh-Yh)/Xh\times 100\% \tag{1}$$

$$Sv=(Xv-Yv)/Xv\times 100\% \tag{2}$$

where Xh and Yh represent the horizontal size of the gray and the finished samples, respectively. Sh represents the horizontal size shrinkage. Xv and Yv represent the vertical size of the gray and the finished samples, respectively. Sv represents the vertical size shrinkage.

3.1.1. The Effect of Knitting Factors on Sensors' Horizontal Size

Figure 3 reveals the change rate of the sensors' horizontal size. When the wales and courses were constant, the horizontal change rate of the samples increased gradually with the increase of spandex content. There was inevitable residual stress in the nylon and spandex fibers, producing a shrinkage phenomenon when heated at a high temperature. A sensor with a high spandex content had a large shrinkage rate. It also increased with the augmentation of the wales and courses under the uniform yarn composition. This resulted mainly from the increase in the length of the needle and sinker loops, causing a higher shrinkage rate during the process.

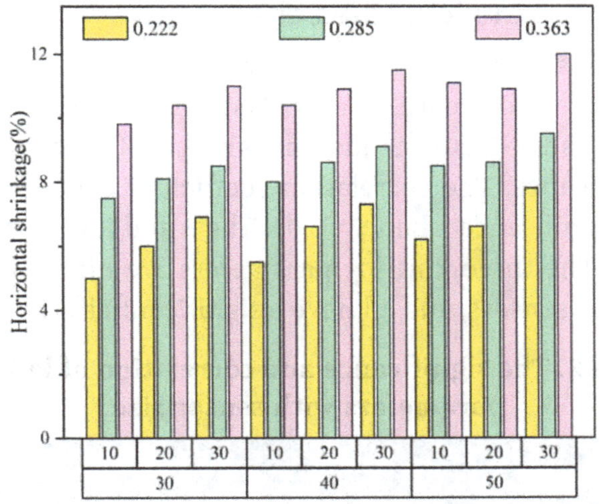

Figure 3. Change rate of sensors' horizontal size.

3.1.2. The Effect of Knitting Factors on Sensors' Vertical Size

The change rate of the sensors' vertical size, similar to that of the horizontal size, is shown in **Figure 4**. It increased with the addition of the spandex content, wales, and courses, which was mainly affected by the elastic shrinkage property of spandex and the number of leg yarn segments. The change rate of samples with vertical size exceeded that of those of horizontal size. The probable explanation is that the length of the needle and the sinker loop is shorter than the leg yarn segment in a loop with a compact arrangement. Therefore, the length change of the latter surpasses the former when subjected to heat contraction, leading to a higher vertical shrinkage rate.

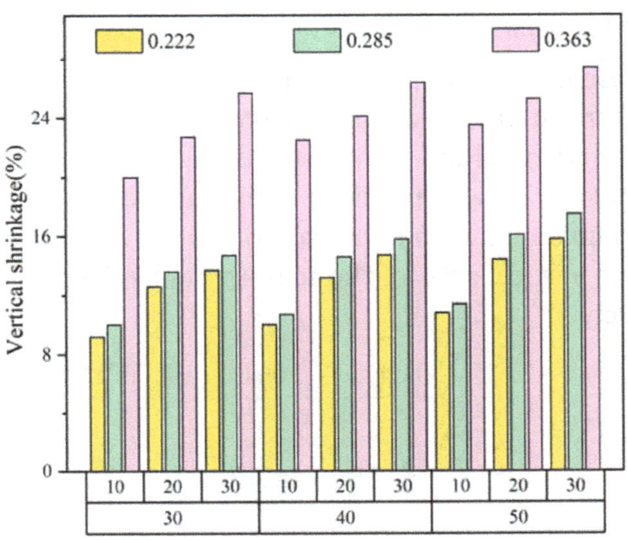

Figure 4. Change rate of sensors' vertical size.

3.1.3. Size Prediction Model of the Sensors

To probe the relationship between the spandex content, wales, courses, and sensor size, SPSS was utilized for regression analysis of the sorted data. S represents significance, which is the basis for judging whether R (correlation coefficient) is of statistical significance. The correlation coefficient between the two variables has no statistical significance when S > 0.05. The S and R values of the samples are shown in **Table 2**. The three knitting parameters are highly correlated with the size shrinkage of the sensors, and all have a significant effect.

Table 2. The significance and correlation of knitting parameters with sensor size.

Knitting Parameters	Factors	The Size Shrinkage (%)	
		Horizontal Direction	Vertical Direction
Spandex content	S	0.000	0.000
	R	0.987	0.936
Wales	S	0.000	0.038
	R	0.973	0.861
Courses	S	0.000	0.000
	R	0.978	0.877

In the light of the analysis by SPSS, the horizontal and vertical shrinkage rate of the sensor size was calculated using the following Formulas (3) and (4), respectively.

$$Sh=-3.617+0.044x+0.064y+31.567z, \tag{3}$$

$$Sv=-37.552+0.101x+0.215y+83.334z \tag{4}$$

where x represents the wales, y represents the courses, and z represents the spandex content. It can be observed from the equation that the dependent coefficients of the horizontal and vertical shrinkage rates are all positive, indicating that the size shrinkage in two directions of the samples increases with the three parameters. In combination with Formulas (1) and (2), the calculation methods of the horizontal and vertical sizes of the sensors can be described as Formulas (5) and (6), respectively.

$$Y_h = \frac{x}{P_x} - \frac{(-3.617+0.044x+0.064y+31.567z)\cdot P_x}{x}, \tag{5}$$

$$Y_v = \frac{y}{P_y} - \frac{(-37.552+0.101x+0.215y+83.334z)\cdot P_y}{y}, \tag{6}$$

In conclusion, the size prediction and control of the sensors can be realized by changing the number of wales, courses, and the spandex content, providing a certain guiding significance for future research.

3.2. Electrical Performance of the Knitted Strain Sensor

To illustrate the electrical properties of the knitted strain sensor, the sensing mechanism was systematically explored by simulating the relationship between resistance and deformation under the stretching–releasing process. As shown in **Figure** 5a, the knitting strain sensor connected to the CHI760 electrochemical analyzer through two collets (red wire and blue wire) was held at both ends of the ESM303 electric stretching/compression test bench along the horizontal direction. The sensor was stretched 0–100% at the stretching rate of 100 mm min^{-1}. Furthermore, a computer was affiliated to the CHI760 electrochemical analyzer through a multifunctional communication cable (black wire) to output the current change with time during the stretching–releasing process in real time. In the data processing software matched with the electrochemical analyzer, the applied AC voltage was set to 0.1 V. Then, the current data was processed by Excel software according to the formula: R = V/I, and the corresponding data of resistance change with time was obtained.

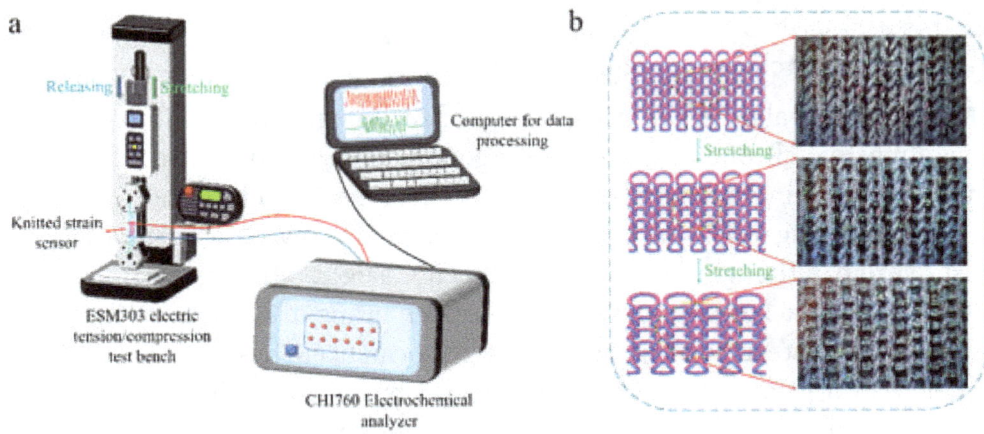

Figure 5. Research on sensing mechanism of the knitted strain sensor, (a) illustration of the measuring device, and (b) loop change of the strain sensor during the horizontal stretching process.

The loop structure of the knitted strain sensor can be regarded as a complex series and parallel electrical network involving the resistance of each yarn segment and contact resistance [**42**]. **Figure 5b** expounds on the loop changes of the sensor during the cycle test. The contact resistance change played a governing function in the horizontal stretching of the fabric, whereas the resistance change caused by loop transfer had little effect on the sensor's resistance. In the initial stage of stretching, the leg yarn segments transferred to the needle and sinker loops

[43], and the number of contact points changed rapidly, accounting for the higher sensitivity of the sensor. With the further increase in external stretching, the transfer of the yarn segment was no longer evident, resulting in a smaller change in the number of contact points. Herein, lower sensitivity was observed for higher stretching.

Sensitivity is one of the critical parameters in determining a sensor's performance. It usually represents the ratio of sensor output to input, elaborating the accuracy and effectiveness of the sensor [44]. GF is traditionally described as the gauge factor, which is defined as the following Formula (7).

$$\mathrm{GF} = \frac{(R - R_0)/R_0}{\varepsilon} = \frac{\Delta R/R_0}{\varepsilon}, \qquad (7)$$

where R0 is the initial resistance, R refers to the real-time resistance of the sensor in the stretching–releasing process, and ε represents the strain applied to the sensor. As depicted in Figure 6a, the variation of ΔR/R0 with strain in the horizontal stretching of the sensor can generally be divided into two phases. The ΔR/R0 of the sensor shows a rapid upward trend in phase I when the strain <25% (GF = 1.1452), whereas it slows down and gradually becomes gentle in phase-II in the strain range 25–100% (GF = 0.2673). This is consistent with the fabric-sensing mechanism described in Figure 5b. The more significant the change of contact points during stretching, the higher the sensitivity.

Figure 6b reveals the variation curves of ΔR/R0 and strain with time, which change almost synchronously, indicating a good response of the knitted strain sensor to the applied strain. The signal is from F1 for the following electrical performance test. The stretching–releasing curves of the sensor are revealed in Figure 6c. They almost overlap, although the two have a specific height difference. The maximum hysteresis of about 1.78% occurs when the strain is about 25%, indicating that the sensor has low hysteresis and good sensing performance. The elastic hysteresis and the energy absorption in the loops will affect the hysteresis of the sensor. Figure 6d reveals the dynamic response of the sensor under strain (0–100%). Evidently, the value of ΔR/R0 is practically identical under the same strain and increases gradually with the continuous augmentation of strain, demonstrating great repeatability of the sensor under different strains. Figure 6e declares the ΔR/R0 curve of the knitted strain sensor under four different stretching rates of 0.5, 1, 2.5, and 5 mm/s with a strain of 40%. The ΔR/R0 is highly consistent with the stretching curve and remains stable at each stretching rate. To illustrate the repeatability and durability of the sensor, 500 stretching–releasing cycles at 100% strain are shown in Figure 6f. It can be observed that the knitted strain sensor shows good repeatability and stability in electrical performance tests.

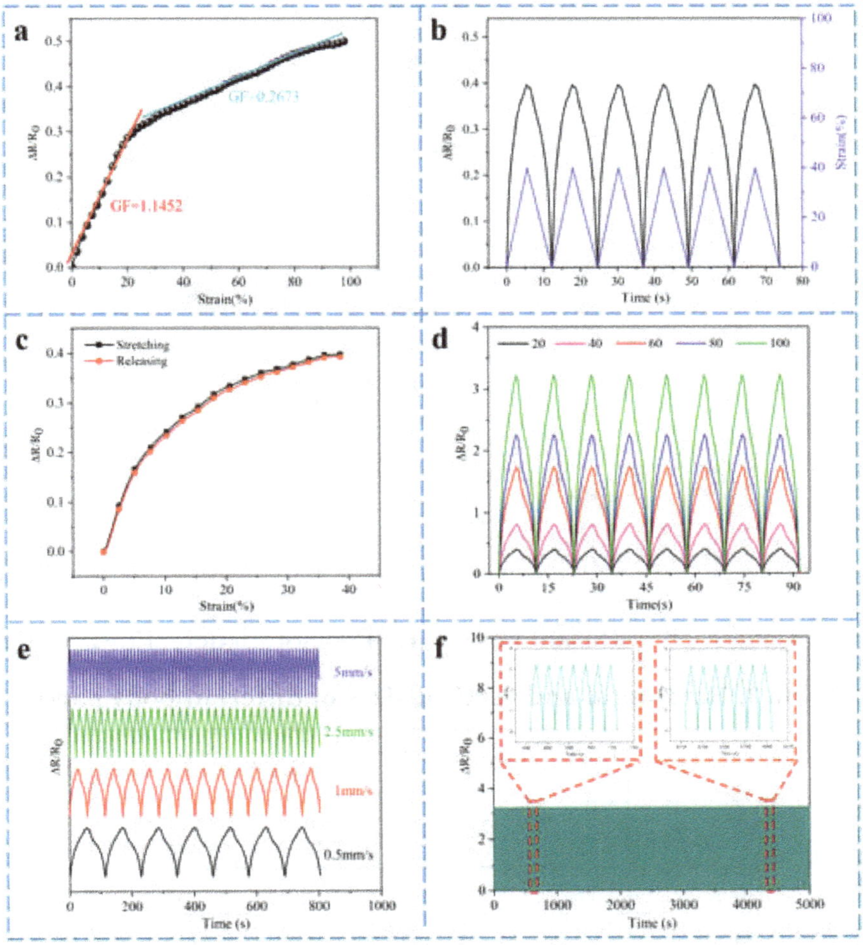

Figure 6. Sensing performance of the knitted strain sensor, (a) the ΔR/R0 curve of the sensor with strain, (b) the curves of ΔR/R0 and strain varying with time, (c) single stretching–releasing curve of the sensor, (d) the curves of ΔR/R0 under different strains (0–100%), (e) ΔR/R0 of strain sensor under different stretching rates of 40% strain, (f) the stretching–releasing tests under a cyclic strain of 100%.

3.3. Application of the Size Prediction Model

As illustrated in **Figure 7**, the as-designed knitted sensors under the size prediction model can be applied as a wearable device to four human body parts (wrist ①, elbow ②, throat ③, and forefinger ④) of a female volunteer to detect human movements. The horizontal and vertical sizes of different body parts are displayed in **Table 3**. Taking into account the effect of size changes and electrical properties, the parameters of the sensors' conductive area based on nylon/spandex composite yarn with better elasticity (44/77dtex, z = 0.36) in different body parts are revealed in Table 4. The size deviation rates of the four knitted strain sensors are within the acceptable range of 5%, verifying the correctness of the size prediction model.

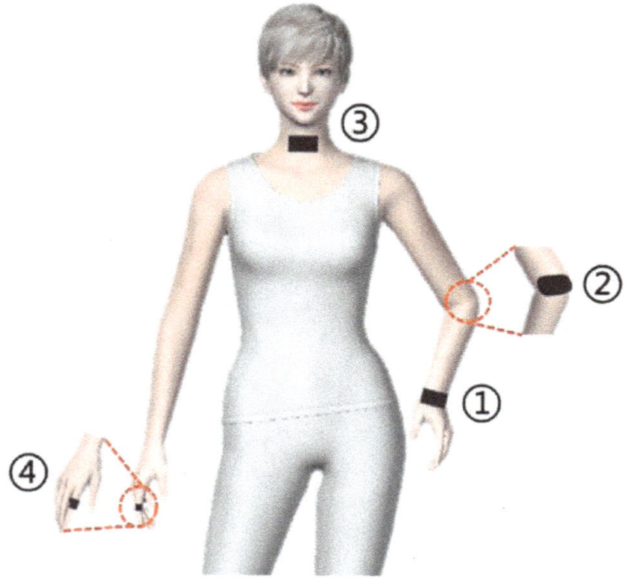

Figure 7. Different body parts for motion detection.

Table 3. Horizontal and vertical sizes of different body parts.

Body Part	Horizontal Size/cm	Vertical Size/cm
①	4.97	2.13
②	6.52	4.25
③	3.16	1.92
④	1.85	1.98

Table 4. The predicted and finished sizes of the sensors in different body parts.

Body Part	x	y	z	Predicted Size/cm		Finished Sizes/cm		Deviation Rate/%	
				Horizontal Direction	Vertical Direction	Horizontal Direction	Vertical Direction	Horizontal Direction	Vertical Direction
①	59	40	0.36	4.78	1.97	4.85	2.05	1.44	3.90
②	72	66	0.36	6.33	4.07	6.44	4.16	1.71	2.16
③	46	32	0.36	2.97	1.74	3.07	1.81	3.26	3.87
④	38	28	0.36	1.72	1.80	1.79	1.87	3.91	3.74

The joint bending displayed in Figure 8a,b indicates that the sensor with a wide workable strain range can detect human movements. The ΔR/R0 of the sensor is consistent with the joint motions. In addition, the strain sensor can detect subtle movements. Figure 8c indicates that the sensor was attached to the throat with the continuous swallowing of food, and the ΔR/R0 of the sensor had a relatively consistent response. Pronunciation could also be discriminated by repeatedly reading the words 'Jiangnan', 'K', 'T', and 'C', as demonstrated in Figure 8d,e. To further probe the strain sensor's capabilities in series motion detection, it was adhered to the forefinger to collect signals. As revealed in Figure 8f, the finger bent from small to large degrees. The resistance changed synchronously with the finger deformation when it bent, indicating a good discernible ability for subtle motion. All these indicate that the as-designed knitted strain sensor exhibits a potential application prospect in smart wearable devices in the near future.

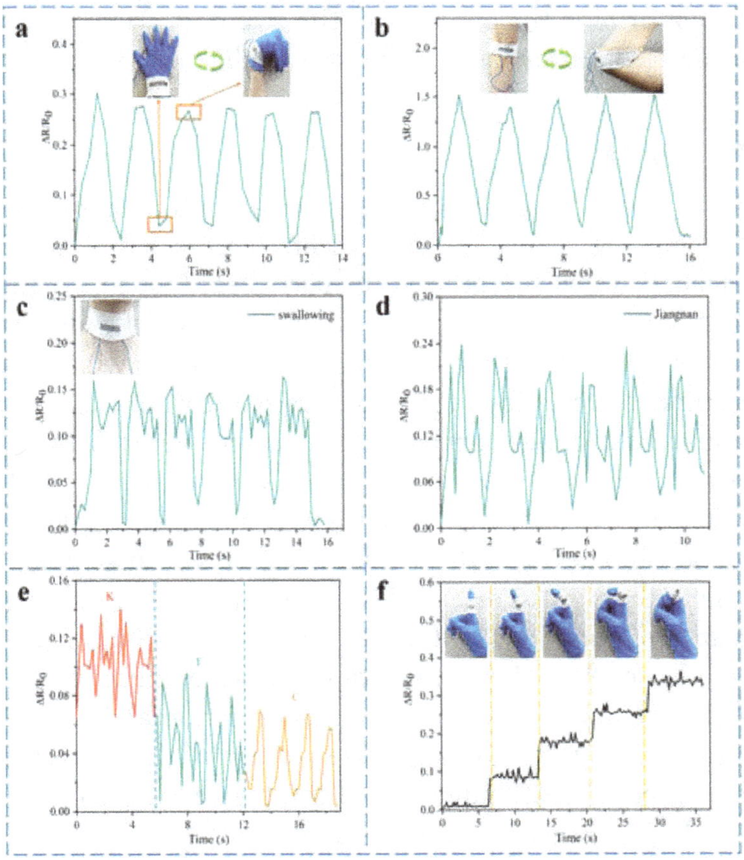

Figure 8. Application of the knitted strain sensors under the size prediction model in detecting different human movements, (a) wrist bending, (b) elbow bending, (c) swallowing of food, (d) speaking the word 'Jiangnan', (e) pronouncing words 'K', 'T', and 'C', (f) responsive curve of the sensor on the finger under diverse bending degrees.

4. CONCLUSIONS

In short, a knitted strain sensor of silver-plated nylon yarn and nylon/spandex composite yarn was successfully fabricated based on the knitting process. The numbers of wales, courses, and spandex content significantly affect the size of the knitted strain sensor. The change rate of samples with vertical size surpasses that of the horizontal size. The horizontal and vertical size prediction model of the knitted strain sensor was established. The sensor has a relatively good sensitivity of 1.1452 (strain ≤ 25%) with a rather large workable strain range (0–100%), good hysteresis, durability and stability over 500 cycles, ability to distinguish various tensile rates and strains, and good synchronization between output resistance and strains. The knitted strain sensors with different conductive areas based on the established size prediction model were designed and applied to human skin or clothing to monitor subtle and large-scale movements of the human body, such as pronunciation and joint bending, confirming the accuracy of the size prediction model. The results demonstrate that the sensor has a promising application prospect in intelligent wearable garments.

5. CONCISE SUMMARY

Benefitting from the multifunctional properties of knitted fabrics with elasticity, flexibility, and high resilience, knitted strain sensors based on structure and strain performance are widely utilized in sports health due to their adaptability to human movements. However, the fabrication process of common strain sensors mainly relies on experienced technicians to determine the best sensor size through repeated experiments, resulting in significant size errors and a long development cycle. Herein, knitted strain sensors based on plain knit were fabricated with nylon/spandex composite yarn and silver-plated nylon yarn using a flat knitting process. A size prediction model of knitted strain sensors was established by exploring the linear relationship between the conductive area size of samples and knitting parameters via SPSS regression analysis. Combined with stable structures and high performance of good sensitivity, stability, and durability, the knitted strain sensors based on size prediction models can be worn on human skin or garments to monitor different movements, such as pronunciation and joint bending. This research indicated that the reasonable size control of the knitted strain sensor could realize its precise positioning in intelligent garments, exhibiting promising potential in intelligent wearable electronics.

REFERENCES

1. Khoshmanesh, F.; Thurgood, P.; Pirogova, E.; Nahavandi, S.; Baratchi, S. Wearable sensors: At the frontier of personalised health monitoring, smart prosthetics and assistive technologies. Biosens. Bioelectron. 2021, 176, 112946. [Google Scholar] [CrossRef] [PubMed]

2. Jiang, S.W.; Yu, J.T.; Xiao, Y.; Zhu, Y.Y.; Zhang, W.L. Ultrawide sensing range and highly sensitive flexible pressure sensor based on a percolative thin film with a knoll-like microstructured surface. ACS Appl. Mater. Interfaces 2019, 11, 20500–20508. [Google Scholar] [CrossRef] [PubMed]

3. Tao, L.Q.; Zhang, K.N.; Tian, H.; Liu, Y.; Wang, D.Y.; Chen, Y.Q.; Yang, Y.; Ren, T.L. Graphene-paper pressure sensor for detecting human motions. ACS Nano 2017, 11, 8790–8795. [Google Scholar] [CrossRef] [PubMed]

4. Chen, X.Y.; Liu, H.; Zheng, Y.J.; Zhai, Y.; Liu, X.H.; Liu, C.T.; Mi, L.W.; Guo, Z.H.; Shen, C.Y. Highly compressible and robust polyimide/carbon nanotube composite aerogel for high-performance wearable pressure sensor. ACS Appl. Mater. Interfaces 2019, 11, 42594–42606. [Google Scholar] [CrossRef]

5. Park, H.; Kim, J.W.; Hong, S.Y.; Lee, G.; Lee, H.; Song, C.; Keum, K.; Jeong, Y.R.; Jin, S.W.; Kim, D.S.; et al. Dynamically Stretchable Supercapacitor for Powering an Integrated Biosensor in All-in-One Textile System. ACS Nano 2019, 13, 10469–10480. [Google Scholar] [CrossRef]

6. Zhang, F.J.; Zang, Y.P.; Huang, D.Z.; Di, C.A.; Zhu, D.B. Flexible and self-powered temperature-pressure dual-parameter sensors using microstructure-frame-supported organic thermoelectric materials. Nat. Commun. 2015, 6, 8356. [Google Scholar] [CrossRef]

7. Chen, H.T.; Song, Y.; Guo, H.; Miao, L.M.; Chen, X.X.; Su, Z.M.; Zhang, H.X. Hybrid Porous Micro Structured Finger Skin Inspired Self-Powered Electronic Skin System for Pressure Sensing and Sliding Detection. Nano Energy 2018, 51, 496–503. [Google Scholar] [CrossRef]

8. Li, Y.T.; Miao, X.H.; Niu, L.; Jiang, G.M.; Ma, P.B. Human motion recognition of knitted flexible sensor in walking cycle. Sensors 2019, 20, 35. [Google Scholar] [CrossRef][Green Version]

9. Li, Y.T.; Miao, X.H.; Chen, J.Y.; Jiang, G.M.; Liu, Q. Sensing performance of knitted strain sensor on two-dimensional and three-dimensional surfaces. Mater. Design 2021, 197, 109273. [Google Scholar] [CrossRef]

10. Wang, C.Y.; Zhang, M.C.; Xia, K.L.; Gong, X.Q.; Wang, H.M.; Yin, Z.; Guan, B.L.; Zhang, Y.Y. Intrinsically stretchable and conductive textile by a scalable process for elastic wearable electronics. ACS Appl. Mater. Interfaces 2017, 9, 13331–13338. [Google Scholar] [CrossRef]

11. Ali, A.; Baheti, V.; Militky, J.; Khan, Z. Utility of silver-coated fabrics as electrodes in electrotherapy application. J. Appl. Polym. Sci. 2018, 135, 46357. [Google Scholar] [CrossRef]

12. Li, Y.D.; Li, Y.N.; Su, M.; Li, W.B.; Li, Y.F.; Li, H.Z.; Qian, X.; Zhang, X.Y.; Li, F.Y.; Song, Y.L. Electronic textile by dyeing method for multiresolution physical

kineses monitoring. Adv. Electron. Mater. 2017, 3, 1700253. [Google Scholar] [CrossRef]

13. Han, X.X.; Miao, X.H.; Chen, X.; Jiang, G.M.; Niu, L. Research on finger movement sensing performance of conductive gloves. J. Eng. Fiber. Fabr. 2019, 14, 1–7. [Google Scholar] [CrossRef][Green Version]
14. Lee, H.; Glasper, M.J.; Li, X.D.; Nychka, J.A.; Batcheller, J.; Chung, H.J.; Chen, Y. Preparation of fabric strain sensor based on graphene for human motion monitoring. J. Mater. Sci. 2018, 53, 9026–9033. [Google Scholar] [CrossRef]
15. Miankafshe, M.A.; Bashir, T.; Persson, N.K. Electrostatic grafting of graphene onto polyamide 6, 6 yarns for use as conductive elements in smart textile applications. New J. Chem. 2020, 44, 7591–7601. [Google Scholar] [CrossRef]
16. Sun, L.F.; Wang, F.; Jiang, J.J.; Liu, H.C.; Du, B.L.; Li, M.Z.; Liu, Y.X.; Li, M.H. A wearable fabric strain sensor assemblied by graphene with dual sensing performance approach to practice application assisted by wireless Bluetooth. Cellulose 2020, 27, 8923–8935. [Google Scholar] [CrossRef]
17. Chen, X.D.; Li, B.T.; Qiao, Y.; Lu, Z.S. Preparing polypyrrole-coated stretchable textile via low-temperature interfacial polymerization for highly sensitive strain sensor. Micromachines 2019, 10, 788. [Google Scholar] [CrossRef][Green Version]
18. Hao, D.D.; Xu, B.; Cai, Z.S. Polypyrrole coated knitted fabric for robust wearable sensor and heater. J. Mater. Sci-Mater. El. 2018, 29, 9218–9226. [Google Scholar] [CrossRef]
19. Åkerfeldt, M.; Lund, A.; Walkenström, P. Textile sensing glove with piezoelectric PVDF fibers and printed electrodes of PEDOT: PSS. Text. Res. J. 2015, 85, 1789–1799. [Google Scholar] [CrossRef]
20. Amjadi, M.; Pichitpajongkit, A.; Lee, S.; Ryu, S.; Park, I. Highly stretchable and sensitive strain sensor based on silver nanowire—elastomer nanocomposite. ACS Nano 2014, 8, 5154–5163. [Google Scholar] [CrossRef]
21. Gurarslan, A.; Özdemir, B.; Bayat, İ.H.; Yelten, M.B.; Kurt, G.K. Silver-nanowire coated knitted wool fabrics for wearable electronic applications. J. Eng. Fiber. Fabr. 2019, 14, 1–8. [Google Scholar] [CrossRef][Green Version]
22. Wang, J.F.; Soltanian, S.; Servati, P.; Ko, F.; Weng, M. A knitted wearable flexible sensor for monitoring breathing condition. J. Eng. Fiber. Fabr. 2020, 15, 1–13. [Google Scholar] [CrossRef]
23. Dawit, H.W.; Zhang, Q.; Li, Y.M.; Islam, S.R.; Mao, J.F.; Wang, L. Design of Electro-Thermal Glove with Sensor Function for Raynaud's Phenomenon Patients. Materials 2021, 14, 377. [Google Scholar] [CrossRef] [PubMed]

24. Son, Y.; Lee, S.; Choi, Y.; Han, S.; Won, H.; Sung, T.H.; Choi, Y.; Bae, J. Design framework for a seamless smart glove using a digital knitting system. Fash. Text. 2021, 8, 6. [Google Scholar]

25. Ma, J.H.; Wang, P.; Chen, H.Y.; Bao, S.J.; Chen, W.; Lu, H.B. Highly Sensitive and Large-Range Strain Sensor with a Self-Compensated Two-Order Structure for Human Motion Detection. ACS Appl. Mater. Interfaces 2019, 11, 8527–8536. [Google Scholar] [CrossRef]

26. Liu, H.; Li, Q.M.; Zhang, S.D.; Yin, R.; Liu, X.H.; He, Y.X.; Dai, K.; Shan, C.X.; Guo, J.; Liu, C.T.; et al. Electrically conductive polymer composites for smart flexible strain sensors: A critical review. J. Mater. Chem. C 2018, 6, 12121–12141. [Google Scholar] [CrossRef]

27. Atalay, O.; Kennon, W.R. Knitted strain sensors: Impact of design parameters on sensing properties. Sensors 2014, 14, 4712–4730. [Google Scholar] [CrossRef]

28. Raji, R.K.; Miao, X.H.; Zhang, S.; Li, Y.T.; Wan, A.L. Influence of rib structure and elastic yarn type variations on textile piezoresistive strain sensor characteristics. Fibres Text. East. Eur. 2018, 26, 24–31. [Google Scholar] [CrossRef]

29. Han, X.X.; Miao, X.H.; Chen, X.; Niu, L.; Wan, A.L. Effect of Elasticity on Electrical Properties of Weft-Knitted Conductive Fabrics. Fibres Text. East. Eur. 2021, 29, 47–52. [Google Scholar] [CrossRef]

30. Frydrysiak, M.; Zięba, J. Textronic sensor for monitoring respiratory rhythm. Fibres Text. East. Eur. 2012, 20, 74–78. [Google Scholar]

31. Wang, J.F.; Long, H.U. Research on wearable sensors based on knitted fabrics with silver plating fiber. Adv. Mat. Res. 2011, 331, 36–39. [Google Scholar] [CrossRef]

32. Zhang, Y.J.; Long, H.R. Resistive network model of the weft-knitted strain sensor with the plating stitch-Part 1: Resistive network model under static relaxation. J. Eng. Fiber. Fabr. 2020, 15, 1–16. [Google Scholar] [CrossRef]

33. Liu, S.; Yang, C.X.; Zhao, Y.F.; Tao, X.M.; Tong, J.H.; Li, L. The impact of float stitches on the resistance of conductive knitted structures. Text. Res. J. 2016, 86, 1455–1473. [Google Scholar] [CrossRef]

34. Tokarska, M.; Orpel, M. Study of anisotropic electrical resistance of knitted fabrics. Text. Res. J. 2019, 89, 1073–1083. [Google Scholar] [CrossRef]

35. Liu, S.; Liu, Y.P.; Li, L. The impact of different proportions of knitting elements on the resistive properties of conductive fabrics. Text. Res. J. 2019, 89, 881–890. [Google Scholar] [CrossRef]

36. Li, L.; Liu, S.; Ding, F.; Hua, T.; Au, W.M.; Wong, K.S. Electromechanical analysis of length-related resistance and contact resistance of conductive knitted fabrics. Text. Res. J. 2012, 82, 2062–2070. [Google Scholar] [CrossRef]

37. Liu, H.S.; Jiang, G.M.; Dong, Z.J.; Xia, F.L.; Cong, H.L. The size prediction and auto-generation of garment template. Int. J. Cloth. Sci. Technol. 2020, 33, 74–92. [Google Scholar] [CrossRef]

38. Yuan, Y.H.; Zhong, J.; Ru, X.; Liu, B. Correlation of yarn feeding to the dimensional and elastic parameters of tubular knitted fabric. Text. Res. J. 2022, 92, 446–455. [Google Scholar] [CrossRef]

39. Ulson, A.A.; Cabral, L.F.; Souza, S.M. Prediction of dimensional changes in circular knitted cotton fabrics. Text. Res. J. 2010, 80, 236–252. [Google Scholar] [CrossRef]

40. Wicaksono, I.; Tucker, C.I.; Sun, T.; Guerrero, C.A.; Liu, C.; Woo, W.M.; Pence, E.J.; Dagdeviren, C. A tailored, electronic textile conformable suit for large-scale spatiotemporal physiological sensing in vivo. npj Flex. Electron. 2020, 4, 1–13. [Google Scholar] [CrossRef][Green Version]

41. Zhao, B.Y.; Cong, H.L.; Dong, Z.J. Highly stretchable and sensitive strain sensor based on Ti3C2-coated electrospinning TPU film for human motion detection. Smart Mater. Struct. 2021, 30, 095003. [Google Scholar] [CrossRef]

42. Isaia, C.; McNally, D.S.; McMaster, S.A.; Branson, D.T. Effect of mechanical preconditioning on the electrical properties of knitted conductive textiles during cyclic loading. Text. Res. J. 2019, 89, 445–460. [Google Scholar] [CrossRef]

43. Šafárová, V.; Malachová, K.; Militký, J. Electromechanical analysis of textile structures designed for wearable sensors. In Proceedings of the 16th International Conference on Mechatronics-Mechatronika, Brno, Czech Republic, 3–5 December 2014; pp. 416–422. [Google Scholar]

44. Li, X.P.; Li, Y.; Li, X.F.; Song, D.K.; Min, P.; Hu, C.; Zhang, H.B.; Koratkar, N.; Yu, Z.Z. Highly sensitive, reliable and flexible piezoresistive pressure sensors featuring polyurethane sponge coated with MXene sheets. J. Colloid Interf. Sci. 2019, 542, 54–62. [Google Scholar] [CrossRef] [PubMed]

CHAPTER-21

INVESTIGATION INTO THE TENSILE CHARACTERISTICS OF CARBON FIBER COMPOSITE LAMINATES

1. INTRODUCTION: AN OVERVIEW

The pursuit of lightweight and high-strength materials has become a focal point in contemporary materials science and engineering. Among these, carbon fiber composite laminates stand out as a promising candidate due to their exceptional tensile properties. This research embarks on a comprehensive exploration of the tensile characteristics inherent in carbon fiber composite laminates. By scrutinizing their mechanical behaviour under tension, this study aims to unravel the intricate details that govern the tensile properties of these advanced materials. The insights gained from this research hold significance not only for material scientists and engineers but also for industries seeking robust and lightweight solutions for various applications, ranging from aerospace to automotive and beyond.

Carbon fiber reinforced polymer (CERP), with its high specific strength, tailored modulus, and sound stealth absorbing performance, has replaced some traditional metal materials and material structures. As an indispensable assembly of the modern aviation industry [1], carbon fiber and its composite materials represent significant advantages with regard to fatigue resistance, tensile resistance, vibration damping, volatile temperature resistance, and corrosion resistance. However, the composite structures also have some defects and are easily damaged under long-term loading, machining, and corrosion due to factors such as production process, design scheme, and retention environment. In order to effectively prevent secondary accidents, including internal and external structural damage, it is necessary to carry out real-time quality and safety inspections when they are put into use to ensure mechanical integrity and production safety.

The early research work on the dynamic failure behaviour of fiber reinforced composite laminates is mainly based on the contact tensile loading formed by the drop weight impact and the penetration of the projectile. The research [**2**] results show that the main failure modes of fiber reinforced composite laminates include matrix and fiber fracture, spalling, and so on. Schiffer et al. [**3**] used an underwater tensile loading simulation device to explore the response of laminates under high-strength underwater tensile loads and established a theoretical analysis model for the dynamic response of composite laminates. Yang et al. [**4**] leveraged three-dimensional DIC to mine the transverse dynamic response process of carbon fiber woven laminates under projectile penetration.

Rajput et al. [5] studied the impact of laminate thickness on the impact response and damage mechanism through experiments and numerical analysis, and the results showed that the depth of pits had a bilinear response to the strength exerted. Karalis, G [6] studied the dynamic behaviour and failure mechanism of laminates under underwater tensile loads. Lin et al. [7] conducted a comparative analysis of the dynamic failure behaviour of flat and curved panel structures of woven basalt/epoxy laminates under blast loading, emphasizing the important influence of structural curvature on the dynamic response. Wei et al. [8] carried out experiments and numerical simulations on the high-speed penetration of laminates and concluded that, in addition to fiber failure, laminates under penetration loads also include spalling and matrix cracking. Penetration velocity, penetration angle, fiber layup, and laminate structure form generate rich research results on the dynamic behaviour and failure of laminate structures under tensile loads [9]. Based on the Hashin failure criterion, Wang Danyong [10] and Su Rui [11] established a progressive damage analysis model for composite materials with similar ideas and also provided a life calculation result that was close to the experiment. Saeedifar et al. and Yang [12,13] established the finite element model of interlaminar fracture toughness test of fiber reinforced composites based on integral and VCCT technology, respectively, to study the delamination processes, such as crack generation and crack propagation, and the energy release rate during interlaminar failure. Three-point bending tests with complex failure forms are rarely studied, and the simulation is not accurate. Camanho [14] proposed that the initial damage inside the composite structure was mainly caused by the separation between fiber layers. Combining with the maximum stress theory, and based on the three-dimensional progressive failure constitutive model, Pearce [15] explored the stiffness degradation characteristics of composite connection structures under out-of-plane loads.

Pravalika and Kashi Ishikawa [16] studied the strength of composite joints through experiments and divided the failure types of carbon fiber composite into four types, namely fiber failure, matrix failure, interlaminar delamination, and normal shear failure, using ultrasonic scanning imaging system and observing the internal damage of the composite material. Based on the observation results, the damage process of the specimen was divided into four stages, namely the appearance of damage, the extension of local cracks, the expansion of damage, and the appearance of structural cracks. Ostapiuk and Orifici [17,18] studied the effect of the ratio of aperture to plate thickness on the connection performance. The results show that when the ratio of aperture to plate thickness is equal to 1, the bearing capacity of the composite plate is the strongest. MA Mc Patel [19] and Whitworth [20] analyzed the stress of single-nail and multi-nail connections by means of tests. The test results show that as the number of nail holes in the laminate increases, the strength of the laminate decreases. Ang [21] combined the failure criterion and the maximum stress criterion proposed by Hashin to study the mechanical connection characteristics of composite materials through experimental research and numerical simulation and predicted the strength and failure process. Zhou [22] and Zhang [23] pointed out that the pull-off failure mode is the same as the internal failure characteristics caused by low-speed stretching, mainly manifested as matrix failure and delamination; the damage extends from the edge of the hole to the surrounding area, and the damage area is distributed in a network along the thickness direction.

It can be derived from the above statement that the process of pull-off failure is very complex, and there are many factors affecting the pull-off strength of laminates, such as temperature and humidity conditions, ply ratio, material geometric parameters and constraints, etc. Due to many other reasons, there are relatively few studies on the pull-off characteristics of composite materials, and there are still many problems that need to be further explored.

This work aims at the dynamic response and failure mode of the carbon fiber composite laminates under the tensile load generated in the thread pull-off. Along with the ultrasonic characteristic scanning imaging system, the damaged specimens are tested. The pull-off test of composite laminates is carried out to obtain the pull-off strength; the load–displacement curve of the entire pull-off behaviour is recorded; the layer-based damage are studied; the failure processes and the mechanical response of the laminate are analyzed. The numerical simulation results are to be further explored, focusing on the main failure modes of the laminates, i.e., the initiation and expansion of damage in the laminate layers, and their effects on the pull-off strength of the composite laminates.

2. EXPERIMENTS

2.1. Experimental Materials

The carbon fiber composite laminate used in this study is the T700 carbon fiber composite single-layer board. The material properties of the T700 carbon fiber composite single-layer panels are as follows: longitudinal stiffness $E1$=100 GPa transverse stiffness $E2$=80 GPa, Poisson's ratio $\nu12$=0.21, shear modulus $G12$=4 GPa, longitudinal tensile strength XT=2100 MPa, longitudinal compressive strength XC=700 MPa, transverse tensile strength YT=42 MPa, transverse compressive strength YC=160 MPa, interlaminar shear strength S=104 MPa, density ρ=1500 kg/m^3.

The thread pull-off (rear joint) used in this study adopts high-temperature T700 spreading cloth to form the RS03A composite material rear joint according to the requirements specified in the task book. The rear joint of the RS03A composite material is selected for the thread pull-off test. The composite material layer (matrix) and the bushing are composed of TC4 material and equipped with a boss at the bottom to bear the axial load of the bushing. The measured axial tensile properties and stress–strain curves are shown in **Figure 1** and **Table 1**.

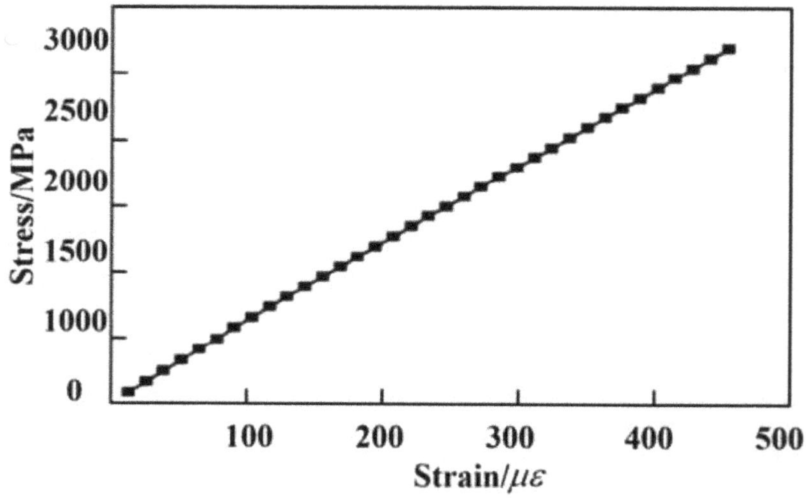

Figure 1. Thread axial tensile stress–strain curve.

Table 1. Axial tensile properties of threads.

Sample	Module/GPa	Tensile/MPa
1#	165.6	1615
2#	178.8	1497
3#	166.8	1544
4#	170.9	1604
Avg. value	170.5	1565
Dispersion coefficient	3.50	3.52

Note: "#"stands for nothing here. It is a symbol marked on the sample part to tell one digit apart from another.

Figure 1 illustrates that under the action of the axial tensile load, the stress and strain of the threaded specimen result in a smooth linear relationship. No fracture signs occur in the figure. According to the measured stress–strain curve, the axial tensile fracture strain is about 0.9%. It can be seen from Table 1 that the average axial tensile fracture strength of the thread is 1565 MPa, and the average modulus is 170.5 GPa.

2.2. Experiment Equipment

The test piece and the adapter are connected and tightened by bolts; the adapter is connected with the slider, and then, the adapter is inserted into the indenter. The above-mentioned overall structures are placed on the inclined blank component fixed on the load-bearing ground rail, and the actuator is connected with the slider through a multi-strand wire rope, as shown in **Figure 2**. The loading control equipment for the pull-off and bending test of the laminate test piece adopt a multi-channel coordinated loading control system. The error of the coordinated loading control system is less than 1%, which met the requirements of the task book for loading accuracy.

Figure 2. Overall condition after pull-off test installation.

2.3. Experimental Installment

During the experiment, the INSTRON1346 electro-hydraulic servo-controlled material testing machine was used to carry out tensile experiments and laminate compression experiments, as shown in **Figure 2** and **Figure 3.** The data acquired in real time during the loading process of the test system include load, displacement, and transverse deflection. Two billets are installed at the appropriate position below the actuator with anchor bolts. The combination of the pull-head adapter and the test part is placed in the pressure head and placed in the center.

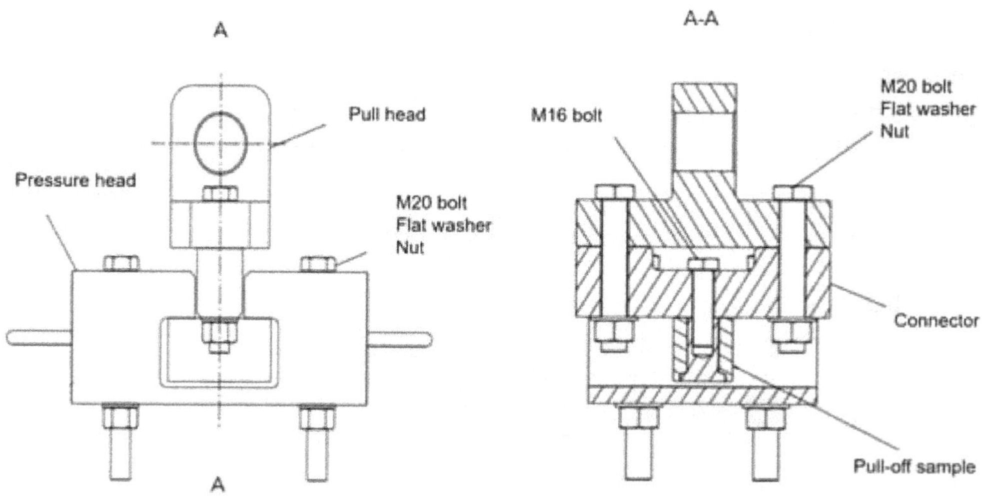

Figure 3. Schematic diagram of the installation of the thread pull-off test piece.

3. PROCESS

3.1. Test Process

Before starting the test, a static load spectrum is created. The loading system and displacement (test parts are required to be close to the fixture) are set to zero. Different load intensities are exerted to monitor the displacement. During the tensile process, the initial failure load, maximum load, and corresponding displacement are recorded with displacement control until the specimen is fractured.

3.2. Test Status

The ultimate tensile strength is the maximum tensile stress that carbon fiber composites can withstand before reaching failure under the tensile test load.

Tensile strength calculation formula

$$\sigma_t = \frac{P_{\max}}{bh} \qquad (1)$$

Tensile modulus calculation formula

$$E_1 = \frac{\Delta Pl}{bh\Delta l} \quad (2)$$

In the formula, the ultimate tensile strength is σt, the maximum load $Pmax$, the specimen width b, the specimen thickness hh, the tensile modulus Et. The maximum of the displacement load of the tensile test is shown in **Figure 4.** The test data of each group of test pieces are calculated to obtain an average tensile strength of the laminate of 783.23 MPa, an average limit load of 41.97 KN, and an average Poisson's ratio of 0.317, as shown in **Table 2.** The tensile strength, tensile elastic modulus, and Poisson's ratio data obtained from the sample are similar, which shows that the heterogeneity of the carbon fiber composite sample is within the experimental error range. The obtained test results are well performed.

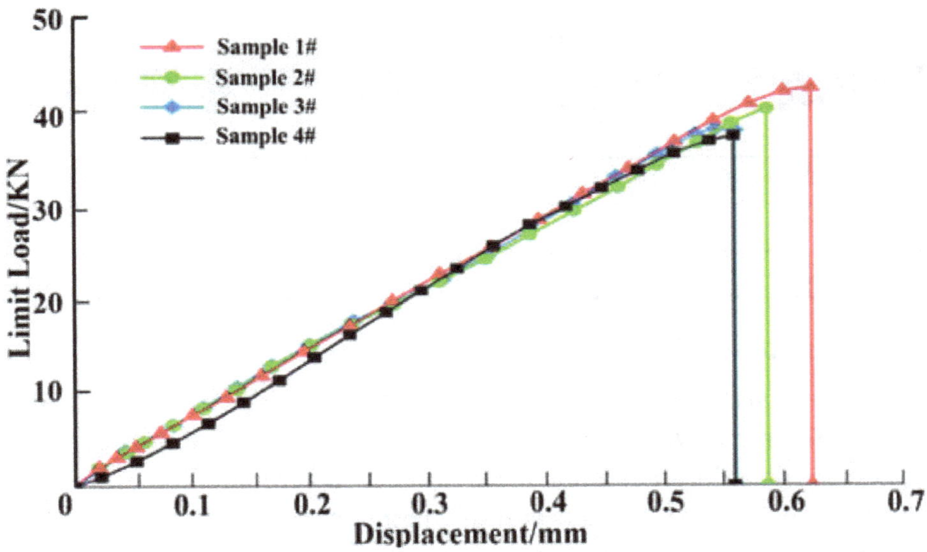

Figure 4. Limit load–displacement curve.

Table 2. Laminate tensile test results data.

Sample	Compressive Strength (MPa)	Limit Load (KN)
1#	830.43	43.01
2#	884.90	43.59
3#	743.46	41.21
4#	690.16	40.08
Average	787.23	41.97

Note: "#"stands for nothing here. It is a symbol marked on the sample part to tell one digit apart from another.

According to the literature [24,25], the bending strength can be expressed as

$$\sigma f=3PL/(2bh^2) \quad (3)$$

In the formula: Off is the bending strength, MPa; *P* is the maximum load value when the sample fails, N; *L* is the span, mm; *hh* is the thickness of the sample, mm.

The flexural modulus of elasticity is

$$Ef=\Delta PL3/(4bh3\Delta f) \quad (4)$$

Shown in the Equation (4) is the calculation of the flexural modulus of elasticity. *Ef* is the flexural modulus of elasticity, MPa; Δ*P* is the load increment of the initial straight-line segment on the load-deflection curve, N; Δf is the corresponding deflection increment, mm, at the midpoint of the sample span. The test results are shown in Table 3.

Table 3. Test results.

Sample	Bending Strength/MPa	Flexural Modulus of Elasticity/GPa
1#	1068	162
2#	1053	137
3#	995	155
4#	989	131

3.3. Finite Element Model

Three-dimensional finite model is constructed to analyze the laminate, as shown in **Figure 5**. The element type adopts the eight-node reduction integral solid element C3D8R, and the orientation of lamination is realized through material orientation. When the load is applied, the left end of the model is fixed, and the right end is applied with axial displacement load. Different strain rate conditions are realized by adjusting the step length of the analysis. The time domain is set as 0.05 to simulate the quasi-static loading mode of force in the pull test.

Figure 5. Finite element model of composite laminate assembly.

3.4. Damage Monitoring

As the composite laminate is damaged by pulling, the failure mode of the hole edge is fuzzy, and the vidual defects are not very distinct from each other. Therefore, it needs to be observed by non-destructive testing equipment. In this test, the ultrasonic characteristic scanning imaging system (UTF-SCAN-1 Water immersion C-scan detection system) performs non-destructive testing on the damaged specimens. The location, size, and damage plan of the damaged area of the laminate can be obtained by ultrasonic C-scanning. The C-scan non-destructive testing test is carried out on the test pieces of different thicknesses after the pull-off test, and the damage of each layer of the test pieces with different thicknesses, area, and depth is obtained. As shown in **Figure 6**a–c above, the maximum failure surface of the specimen with a thickness of 1.25 mm is located in the lower layer of the laminate, that is, the fourth layer (90-degree direction), and the specimen with a thickness of 3.25 mm has the largest failure surface, while the failure surface is located in the middle layer of the laminate, that is, the ninth layer (90-degree direction). The maximum failure surface of the specimen with a thickness of 5.10 mm is located in the upper layer of the laminate, that is, the seventh layer (0-degree direction). It can be seen that with the increase in the thickness of the specimen, the position of the maximum failure surface of the specimen moves along the thickness direction from the upper layer position (including the straight hole surface) of the laminate to the lower layer position (including the countersunk hole surface).

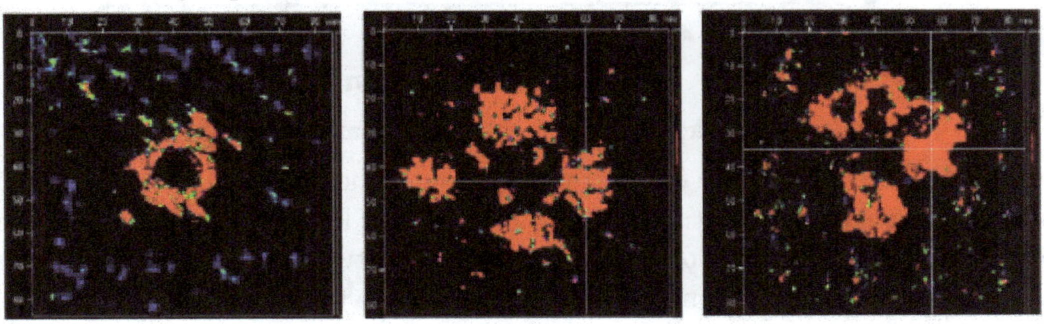

(a) (b) (c)

Figure 6. The maximum failure surface and depth of the specimen with different thickness: (a) the fourth layer, 1.25 mm, (b) the ninth layer, 3.25 mm, (c) the seventh layer, 5.10 mm.

4. RESULTS AND DISCUSSION

4.1. Analysis of Tensile Properties of Carbon Fiber Composites

After, respectively, exploring the axial tensile properties of composite materials and threaded composite materials, the rear joint of RS03A composite material is selected for thread tensile testing. The load–displacement curve during the tensile process is shown in **Figure 7**. As can be seen from **Figure 7**, when the load is applied to a certain level (point A in **Figure 7**), it will suddenly drop, and the composite material will emit a crisp sound during the experiment; the load at this time is defined as the initial failure load [**26**]. As the displacement increases, the load continues to increment, and the sound continues to appear during the period until the last loud sound. The cylinder composite material breaks as a whole, and the maximum load at this time is the breaking load. The initial failure load and fracture load are used to calculate the stresses, which are recorded as the initial failure stress ($\sigma 1$) and the final failure stress ($\sigma 2$).

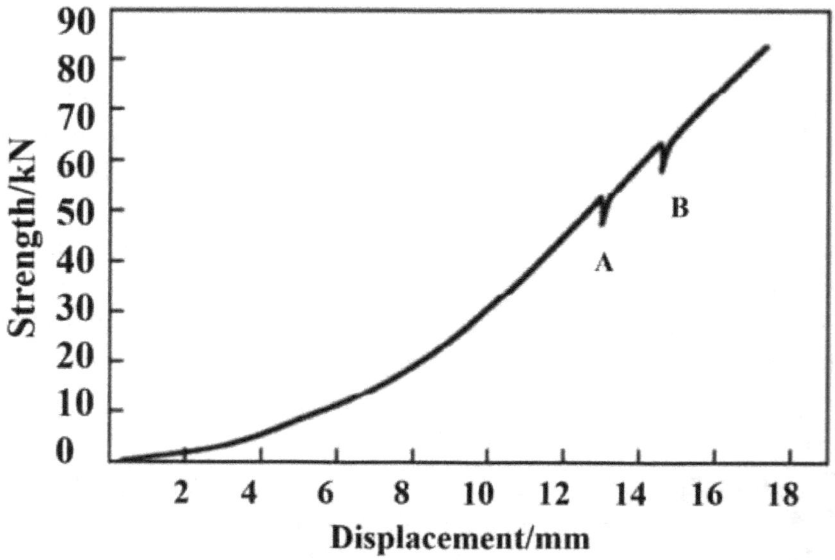

Figure 7. Load–displacement curve of axial tension.

At the beginning of the experiment, the appearance of the laminate basically does not change, and the stress–strain curve shows a linear rise. When the strain reaches 4500 $\mu\varepsilon$, the edge of the hole in contact between the laminate and the adapter begins to have a small fiber uplift, and there is a clear and crisp squeak, indicating that the laminate begins to damage. As the displacement continues to increase, the area of the uplifted fiber increases slowly, and the damage range expands from the center of the circular hole to the edge of the laminate, and some fibers are pulled off at the edge of the hole. The load drops slightly, the curve of the increase in stress continues to rise slowly. As the load continues to drop, it is accompanied by a continuous brittle sound. Finally, the "explosion" is carried out in the "middle section" of the failure mode until the laminated plate is completely destroyed. Fiber fracture, delamination, fiber pull, and debonding are found at all locations, indicating that the matrix and interface are severely damaged during loading.

The stress–strain curve of axial tension is shown in **Figure 8**. It can be seen that the axial tensile modulus of the composite samples is relatively stable, with an average value of about 90 GPa. The average initial failure stress is about 470 MPa, and the final failure stress average is about 800 MPa. Shown in **Figure 9** is the photo of the composite material after fracture. It can be seen from the figure that when the composite material is finally damaged, the fiber is fractured. At the same time, it can be seen that the fiber layer also cracks many times during the fracture. The final breaking load of the composite material is caused by the fracture of the helical fiber as the main force carrier. When the first load occurs, the hoop layer composite material has already cracked. At this time, its initial failure stress is about 470 MPa, and the axial tensile fracture strain is about 0.5%, manifesting that the decrease in the first load force in value is due to the failure of the composite laminate reaching the breaking strain. In the actual working conditions of the composite material used in this experiment, if cracking occurs, it signifies a functional failure. Therefore, the initial failure stress when the load decreases for the first time is used as the criterion for judging whether the composite material fails.

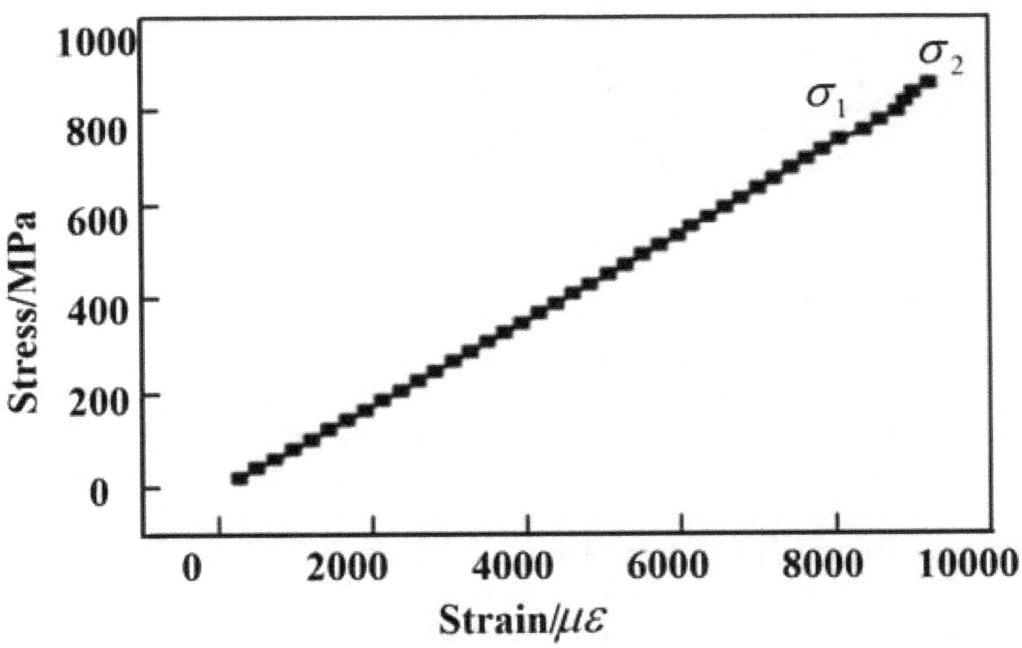

Figure 8. Stress–strain curve of axial tension.

The picture after fracture of composite material is shown in **Figure 9**. It can be seen from the picture that fiber fracture occurs when the composite material is finally destroyed. At the same time, it can be seen that the fiber layer of the fracture also occurs several times. The final fracture load of the composite is large, which is caused by the fracture of the helical fiber. The test curve shows that the tensile failure of composite laminates is divided into two stages, and it is a nonlinear and progressive failure process.

Figure 9. Fracture photo.

4.2. Layer-Based Damage

Figure 10a,b shows the interfacial delamination damage at 5–100% loading percentage, respectively. The matrix cracking first appears on the back of the laminate, the damage occurs in the middle area of the laminate, and the damage area of the unit layer farther from the tensile side is larger than that near the tensile side, which can be explained by the deformation and failure principle of the laminate, that is, matrix tensile damage starts from the back and extends to the upper layer [27]. With the increase in tensile energy, the cracked area of the matrix gradually expands. The damage profile of each layer is roughly an irregular ellipse and expands along the fiber direction. This is because the stress is transmitted faster in the fiber direction, so the damage profile in the fiber direction is larger, which is consistent with the experimental results.

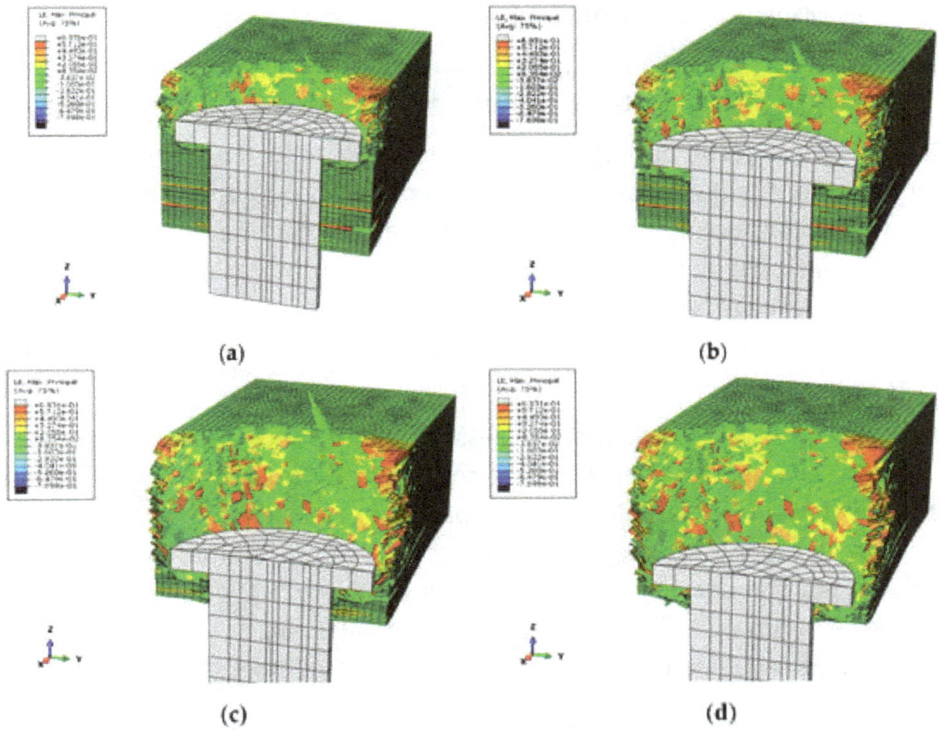

Figure 10. Matrix tensile damage distribution under load of different percentage: (a) load intensity 5 × 10-2 to 25 × 10-2, (b) load intensity 25 × 10-2 to 50 × 10-2, (c) load intensity 50 × 10-2 to 75 × 10-2, (d) load intensity 75 × 10-2 to 100 × 10-2.

The figures indicate that delamination occurs at each interface with varying degrees of damage. When the damage variable is equal to 1, complete delamination is indicated. The main axis of the delamination area (45°/-45°) is along the -45° direction, and the main axis of the delamination area (-45°/45°) is along the 45° direction, that is, the main direction of the delamination damage is along the laying of the fibers close to the direction of the lower layer. It can also be seen that, regardless of the load percentage, the interface near the back of the tensile point is the first to experience delamination damage with the largest damage area. However, the damage variables of the other layers are between 0 and 1, which only achieves a partial delamination effect. With the same energy, the closer the sublayer to the tensile side, the smaller the degree of delamination damage. This is because when the laminate is stretched, the sublayers farther from the stretched side are subjected to greater tensile stress than the sublayers adjacent to the stretched side, so the damage propagates from the bottom layer to the top layer. Additionally, with the increase in energy, except for the bottom layer, the delamination damage area of other layers also becomes larger.

4.3. Based on Mechanical Response

Under the pressure of out-of-plane load, the mechanical response of the pull-out failure process of the connected structures demonstrates the following four stages.

The bolt stays in the pre-tightening stage, and the load is transferred through the contact pair of the model. The materials of each component are in the elastic stage, and the connecting structure experiences no macro damage. In the model, the damage variable value of the cohesive force unit increases from 0, and there is microcrack damage between material interfaces. At this stage, during the initial loading process, the load and displacement increase rapidly until the maximum load is reached.

During the plastic stage, the stiffness of the connecting structure decreases [**28**]. Weak plastic deformation occurs at both ends of the screw hole, the increasing rate of load slows down, and the contact pair between the bolt and the screw hole produces weak slip. At this stage, the displacement of the laminated plate continues to increase, but the bearing capacity fluctuates up and down, indicating that the matrix of the laminated plate had been damaged and can no longer bear the load, and internal damage had occurred, leading to the decline of stiffness.

The carbon fiber composite has anisotropic characteristics, and the pulling load is in the direction of the laminate method, with bearing performance [29]. When the load displacement is increased to about 0.58 mm, the damage variable value in the model reaches the failure limit, and the fiber layer is separated in the area near the screw hole, resulting in tension, compression, and shear failure of fiber and matrix material. When the stress of the cohesive element reaches the fracture toughness value of the interface, the adapter disconnects from the fiber layer. The internal micro-clearance of laminates and partial contact failure at the screw hole lead to the deterioration of the stiffness of the connecting structure. Figure 11 shows the fiber damage status after tensile failure.

Conical uplift occurs near the hole of the laminates. Due to the extrusion of the bolt contact surface, the micro-gap inside the composite is compressed, the stiffness of the connection structure is appropriately increased, and the slope of the response curve is slightly increased. When the load reaches the ultimate stress value, the metal in the ring region of the head produces crushing failure [**30**]. The ultimate load is the tensile strength of the connecting structure.

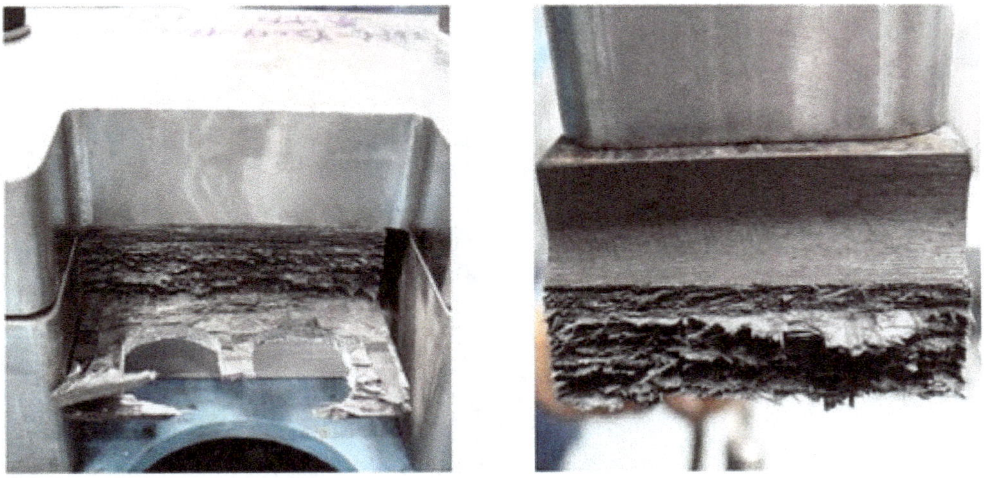

(a) (b)

Figure 11. Fiber damage status (a,b).

4.4. Failure Mode Based

Due to the continuous loading of the equipment, the laminate will slowly deform and fail as the loading strength of the jack increases. The central loading area of the laminate beam is subject to the maximum stress due to bending/tensioning [**31,32**]. Being exposed to the action of incident compressive stress waves and bending waves, the threads will form lateral deformations and localized wrinkles at the edges in contact with the laminate beams. The compressive stress wave is reflected by the backside of the laminate to form a tensile wave. When the tensile wave intensity is large enough, the laminate will undergo spalling between the fiber and the matrix. Under sufficient loading strength, with the increase in transverse deformation and axial tensile, the fracture failure of the matrix and fibers occurs in the laminate.

Figure 12 compares the deformation profile with time and strength at 50% and 100%, showing the obvious localization of deformation. When the loading strength is between 5% and 50%, and the velocity is 1.5 mm/min, the two sides of the laminate always slip along the radial direction without lateral deformation. However, when the velocity is set at 1 mm/min, the transverse defection will move sharply. When the loading percentage is 40–50%, the deformation gradually diminishes. With the increment of structural deformation, the target plate finally leaves the fixture and continues to move with a certain kinetic energy. When the loading percentage is greater than 50%, the laminate mainly undergoes elastic deformation, and no obvious failure occurs on the surface of the laminate. When the loading percentage reaches 60%, the laminate fails with the increase in tensile strength. The laminate is completely broken when the load reaches 90%.

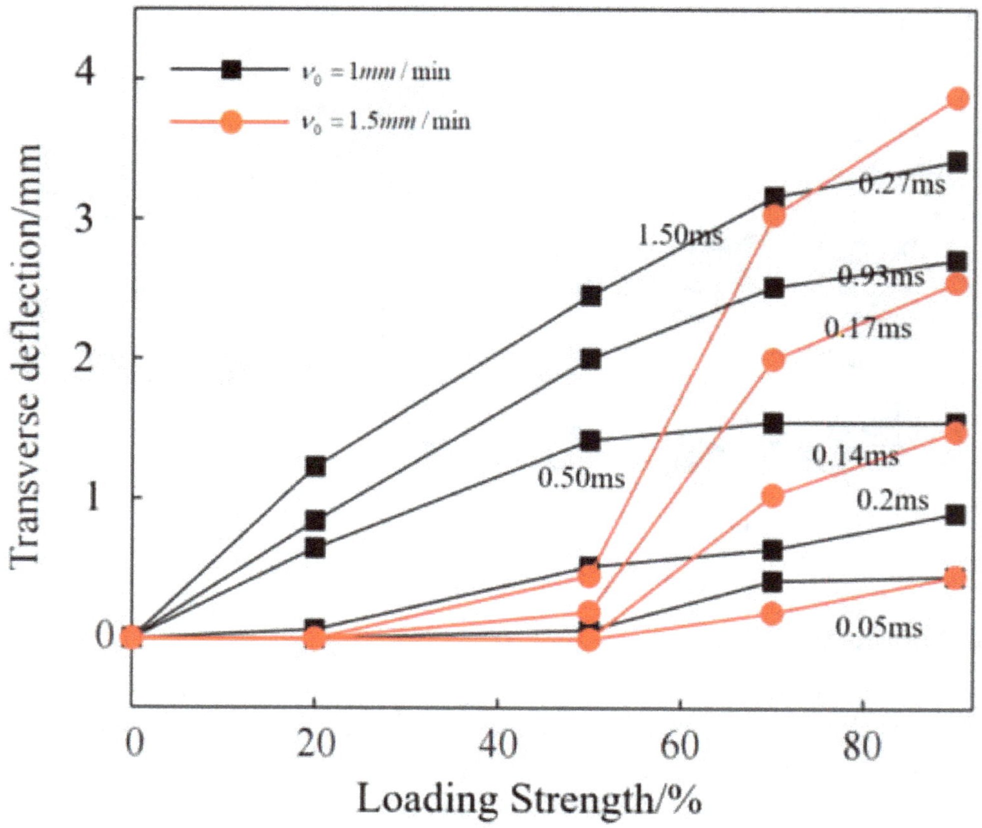

Figure 12. Deformation profiles of laminates at different loading strengths and velocities.

5. CONCLUSIONS

An experimental study on tensile properties of carbon fiber composite laminates was carried out. The damage condition of the laminates' layer, the mechanical response, and failure mode were discussed.

- It can be concluded from the load–displacement curve of the pull-off that the pull-off failure of the composite laminate is nonlinear and conforms to the principle of progressive damage. With the pressure of out-of-plane load, the mechanical response of the screw structure presents a nonlinear trend and exhibits the process in four stages. The characteristics of pull-out failure are similar to impact failure. The conical uplift of the connecting hole area and the separation between the layers of the fiber–metal interface are the main factors leading to pull-out failure.
- As the tensile load increases, the damage area of various damage types

also becomes larger. When the tensile energy increases to a certain degree, the load will drop twice, and the damage will occur twice. In this paper, for the first time, when the load increased from the initial value to 55 KN, the contact edge between the laminate and the pull rod is raised, resulting in large area damage. At the same time, the load decreases, which is the initial failure. The second time is when the load reaches 65 KN, the fiber at the hole edge of the laminates is pulled off, and the damaged area expands from the hole edge to the surrounding area. The failure part is mainly concentrated near the edge of the hole on the surface of the laminate (including the straight hole surface), and the fiber at the failure witnesses a whole piece of bulge.

- With the increase in pull-off strength, the failure of carbon fiber composite panels is mainly divided into micro-deformation under low-load pull-off, half-fold fracture under medium-load strength impact, and complete fracture under high-strength load tensile, and exhibits structural failure modes. The failure mode of the composite laminate is a non-fracture failure of local nature, with high safety.
- The deformation velocity of the midpoint increases with the increase in the tensile rate. When the rate and load increase simultaneously, localization failure occurs, and the critical maximum deformation of laminates decreases with the increase in the rate. Both the loading strength and tensile rate can be used as factors to assess the tensile properties of carbon fiber composites.

6. CONCISE SUMMARY

In order to study the thread tensile performance of carbon fiber composite laminates, the connection between the test piece, connecting bolts, bushings, and the composite matrix, was leveraged for loading, and combined with an ultra-sound scanning imaging system, experiments were carried out on the dynamic response to record the failure behaviour of the laminate structure of equal thickness. The effects of different pull-off loading strengths on the dynamic failure process, deformation profile, midpoint deformation, failure mode, and energy dissipation ratio of the thread were studied. The results show that (1) with the increase in pull-off strength, the response speed of mid-point deformation increases, the thread deformation mode changes from overall deformation to partial deformation, and the localized effect increases, accompanied by severe matrix and fiber fracture failure; (2) the thread energy dissipation ratio ascends with increasing pull-off strength and exhibits three distinct stages, i.e., elastic deformation, central fracture, and complete failure, which are directly related to the structural failure mode; (3) the failure load increases with the increment of the thickness of the laminate, and the maximum failure surface of the specimen will move from the upper layer of the laminate to the lower layer along the thickness direction; (4) the deformation velocity of the midpoint augments with the increase in the tensile rate, which can be included as a factor to assess the tensile properties of carbon fiber composites.

REFERENCES

1. Choi, H.S.; Ahn, K.J.; Nam, J.D.; Chun, H.J. Hygroscopic aspects of epoxy/carbon fiber composite laminates in aircraft environments. *Compos. Part A Appl. Sci. Manuf.* **2001**, *32*, 709–720. **[Google Scholar]** **[CrossRef]**
2. Liu, Q.; Guo, B.; Liu, W. Mechanical response analysis of carbon fiber composite laminates under low velocity impact. *Sci. Technol. Eng.* **2019**, *19*, 97–102. **[Google Scholar]**
3. Schiffer, A.; Tagarielli, V.L. The dynamic response of composite plates to underwater blast: Theoretical and numerical modelling. *Int. J. Impact Eng.* **2014**, *70*, 1–13. **[Google Scholar]** **[CrossRef][Green Version]**
4. Yang, Y.Q.; Zhang, L.; Guo, L.C.; Zhang, W.; Zhao, J.; Xie, W. Dynamic response and research of 3D braided carbon fiber reinforced plastics subjected to ballistic impact loading. *Compos. Struct.* **2018**, *206*, 578–587. **[Google Scholar]** **[CrossRef]**
5. Rajput, M.S.; Burman, M.; Forsberg, F.; Hallström, S. Experimental and numerical study of the response to various impact ener-gy levels for composite sandwich plates with different face thicknesses. *J. Sandw. Struct. Mater.* **2019**, *21*, 1654–1682. **[Google Scholar]** **[CrossRef]**
6. Karalis, G.; Tzounis, L.; Tsirka, K.; Mytafides, C.K.; Liebscher, M.; Paipetis, A.S. Carbon fiber/epoxy composite laminates as through-thickness thermoelectric generators. *Compos. Sci. Technol.* **2022**, *220*, 109291. **[Google Scholar]** **[CrossRef]**
7. Lin, T.; Xu, J.; Ji, M.; Chen, M. Drilling performance of uncoated brad spur tools for high-strength carbon fiber-reinforced polymer laminates. *Proc. Inst. Mech. Eng. Part L J. Mater. Des. Appl.* **2021**, *5*, 123–125. **[Google Scholar]** **[CrossRef]**
8. Wei, Z.; Fernandes, H.C.; Herrmann, H.G.; Tarpani, J.R.; Osman, A. A Deep Learning Method for the Impact Damage Segmentation of Curve-Shaped CFRP Specimens Inspected by Infrared Thermography. *Sensors* **2021**, *21*, 395. **[Google Scholar]** **[CrossRef]** **[PubMed]**
9. Patel, H.V.; Patel, S.M.; Dave, H.K. Influence of Fiber Orientation and Number of Layer on Tensile and Flexural Strength of Carbon Fiber-Reinforced Composites Fabricated by VARTM Process. In *Advances in Manufacturing Processes*; Springer: Singapore, 2021. **[Google Scholar]**

10. Wang, D.Y. Research on Prediction of Damage Failure and Fatigue Life for Composite Bolted Joints. Ph.D. Thesis, Nanjing University of Aeronautics & Astronautics, Nanjing, China, 2006. [**Google Scholar**]

11. Su, R. Study on the Life Prediction of Composite-to-Titanium Bolted Joints. Master's Thesis, Shanghai Jiaotong University, Shanghai, China, 2013. [**Google Scholar**]

12. Saeedifar, M.; Fotouhi, M.; Najafabadi, M. Interlaminar fracture toughness evaluation in glass/epoxy composites using acoustic emission and finite element methods. *J. Mater. Eng. Perform.* **2015**, *24*, 373–384. [**Google Scholar**] [**CrossRef**]

13. Yang, H. Simulation of the damage progression of composite panels under three-point bending load. *Gas Turbine Exp. Res.* **2006**, *19*, 38–42. [**Google Scholar**]

14. Catalanoti, G.; Camanho, P.P. A semi-analytical method to pre-dict net-tension failure of mechanically fastened joints in com-posite laminates. *Compos. Sci. Technol.* **2013**, *76*, 69–76. [**Google Scholar**] [**CrossRef**][**Green Version**]

15. Pearce, G.M.K.; Johnson, A.F.; Hellier, A.K.; Thomson, R.S. A study of dy-namic pull-through failure of composite bolted joints using the stacked-shell finite element approach. *Compos. Struct.* **2014**, *118*, 86–93. [**Google Scholar**] [**CrossRef**][**Green Version**]

16. Pravalika, P.; Chamakura, S.; Govardhan, D. Optimization of Lay-Up Sequence of Composite Laminates under Tensile and Compressive Loading. *Int. J. Eng. Res. Technol.* **2021**, *10*, 666–675. [**Google Scholar**]

17. Fontaine, D.; Leblanc, J.; Shukla, A. Blast response of carbon-fiber/epoxy laminates subjected to long-term seawater exposure at sea floor depth pressures. *Compos. Part B Eng.* **2021**, *215*, 108647. [**Google Scholar**] [**CrossRef**]

18. Ostapiuk, M.; Loureiro, M.; Bienia, J.; Marques, A.C. Interlaminar shear strength study of Mg and carbon fiber-based hybrid laminates with self-healing microcapsules. *Compos. Struct.* **2021**, *255*, 113042. [**Google Scholar**] [**CrossRef**]

19. Orifici, A.C.; Krueger, R. Benchmark assessment of automated delamination propagation capabilities in finite element codes for static loading. *Finite Elem. Anal. Des.* **2012**, *54*, 28–36. [**Google Scholar**] [**CrossRef**][**Green Version**]

20. McCarthy, M.A.; McCarthy, C.T.; Lawlor, V.P.; Stanley, W.F. Three-dimensional finite element analysis of single-bolt, single-lap composite bolted joints: Part I: Model development and validation. *Compos. Struct.* **2005**, *71*, 140–158. [**Google Scholar**] [**CrossRef**]

21. Whitworth, H.A.; Othieno, M.; Barton, O. Failure analysis of composite pin loaded joints. *Compos. Struct.* **2003**, *59*, 261–266. **[Google Scholar] [CrossRef]**

22. Ang, B.C. Fabrication and Thermo-Electro and Mechanical Properties Evaluation of Helical Multiwall Carbon Nanotube-Carbon Fiber/Epoxy Composite Laminates. *Polymers* **2021**, *13*, 1437. **[Google Scholar]**

23. Zhou, J.; Wang, S. A progressive damage model of composite laminates under low-velocity impact. *J. Northwestern Polytech. Univ.* **2021**, *39*, 37–45. **[Google Scholar] [CrossRef]**

24. Zhang, H.; Hou, B.; He, Y.T.; Feng, Y.; Tan, X.F. Tensil Property of Aeronautical Composite-Metal Joint Structure and Its Progressive Damage. *Mater. Mech. Eng.* **2017**, *41*, 87–91. **[Google Scholar]**

25. Wang, S.X.; Wu, L.Z.; Li, M. Low-velocity impact and residual tensile strength analysis to carbon fiber composite laminates—ScienceDirect. *Mater. Des.* **2010**, *31*, 118–125. **[Google Scholar] [CrossRef]**

26. Hsueh, C.H. Interfacial Debonding and Fiber Pull-Out Stresses of Fiber-Reinforced Composites. *Mat. Sci. Eng. A—Struct.* **1990**, *123*, 1–11. **[Google Scholar] [CrossRef]**

27. Zhang, D.; Wang, H.; Shen, W.; Rao, C.; Xie, Y. Research on the axial direction mechanical property of carbon fiber composite cylinder with multilayer structure. *New Chem. Mater.* **2021**, *5*, 152. **[Google Scholar]**

28. Filik, K.; Karnas, G.; Masowski, G.; Oleksy, M.; Oliwa, R.; Bulanda, K. Testing of Conductive Carbon Fiber Reinforced Polymer Composites Using Current Impulses Simulating Lightning Effects. *Energies* **2021**, *14*, 7899. **[Google Scholar] [CrossRef]**

29. Kim, S.H.; Park, Y.G.; Kim, S.S. Double-layered microwave absorbers composed of ferrite and carbon fiber composite laminates. *Phys. Status Solidi* **2007**, *4*, 4602–4605. **[Google Scholar] [CrossRef]**

30. Liu, P. Localized damage models and implicit finite element analysis of notched carbon fiber/epoxy composite laminates under tension. *Damage Modeling Compos. Struct.* **2021**, *1*, 69–107. **[Google Scholar]**

31. Mohammadi, R.; Najafabadi, M.A.; Saghafi, H.; Saeedifar, M.; Zarouchas, D. A quantitative assessment of the damage mechanisms of CFRP laminates interleaved by PA66 electrospun nanofibers using acoustic emission. *Compos. Struct.* **2021**, *258*, 113395. **[Google Scholar] [CrossRef]**

32. Shan, S.H.; Karuppanan, S.; Megat-Yusoff, P.S.M.; Sajid, Z. Impact resistance and damage tolerance of fiber reinforced composites: A review. *Compos. Struct.* **2019**, *217*, 100–121. **[Google Scholar]**

INDEX

T

www.ingramcontent.com/pod-product-compliance
Lightning Source LLC
LaVergne TN
LVHW062213200926
848484LV00006B/1382
9781787152724